Advances in Computational Algorithms and Data Analysis

Lecture Notes in Electrical Engineering

Volume 14

For other titles published in this series, go to
http://www.springer.com/7818

Sio-Iong Ao • Burghard Rieger • Su-Shing Chen
Editors

Advances in Computational Algorithms and Data Analysis

 Springer

Editors

Sio-Iong Ao
International Association of Engineers
Unit 1, 1/F, 37-39 Hung To Road
Hong Kong
Hong Kong/PR China

Su-Shing Chen
Department of Computer & Information
Science & Engineering (CISE)
University of Florida
PO Box 116120
Gainesville FL 32611-6120
E450, CSE Building
USA

Burghard Rieger
Universität Trier
FB II Linguistische
Datenverarbeitung
Computerlinguistik
Universitätsring 15
54286 Trier
Germany

ISBN: 978-90-481-8027-1   e-ISBN: 978-1-4020-8919-0

Printed on acid-free paper

9 8 7 6 5 4 3 2 1

springer.com

Contents

Chapter 1
Scaling Exponent for the Healthy and Diseased Heartbeat
Quantification of the Heartbeat Interval Fluctuations

Toru Yazawa and Katsunori Tanaka[*]

Abstract "Alternans" is an arrhythmia exhibiting alternating amplitude or alternating interval from heartbeat to heartbeat, which was first described in 1872 by Traube. Recently alternans was finally recognized as the harbinger of a cardiac disease because physicians noticed that an ischemic heart exhibits alternans. To quantify irregularity of the heartbeat including alternans, we used the detrended fluctuation analysis (DFA). We revealed that in both, animal models and humans, the alternans rhythm lowers the scaling exponent. This correspondence describes that the scaling exponent calculated by the DFA reflects a risk for the "failing" heart.

Keywords Alternans · Animal models · Crustaceans · DFA · Heartbeat

1.1 Introduction

My persimmon tree bears rich fruits every other year. A climatologist report that global atmospheric oxygen has bistability [1]. Period-2 is such an intriguing rhythm in nature. The cardiac "Alternans" is another period-2. In cardiac period-2, the heartbeat is alternating the amplitude/interval from beat to beat on the electrocardiogram (EKG). Alternans has remained an electrocardiographic curiosity for more than three quarters of a century [2, 3]. Recently, alternans is recognized as a marker for patients at an increased risk of sudden cardiac death [2–7]. In our physiological experiments on the hearts in the 1980s, we have noticed that alternans is frequently observable with the "isolated" hearts of crustaceans (Note: At this isolated

T. Yazawa (✉) and K. Tanaka
Department of Biological Science, Tokyo Metropolitan University, Tokyo, Japan.
Bio-Physical Cardiology Reseach Group
e-mail: yazawa-tohru@c.metrou.ac.jp; yazawatorujp@yahoo.co.jp

[*]The contact author mailing address:1705-6-301 Sugikubo, Ebina, 243-0414 Japan, telephone and fax number:+81-462392350. Present address, 228-2 Dai, Kumagaya, Saitama, 360–0804 Japan.

S.-I. Ao et al. (eds.), *Advances in Computational Algorithms and Data Analysis*,
Lecture Notes in Electrical Engineering 14,
© Springer Science+Business Media B.V. 2009

1

condition the heart sooner or later dies in the experimental dish). We soon realized that alternans is a sign of future cardiac cessation. Presently, some authors believe that it is the harbinger for sudden death [2, 6]. So we came back to the crustaceans, because we knew the crustaceans are outstanding models. However, details of alternans have not been studied in crustaceans. So we considered that we could study this intriguing rhythm by the detrended fluctuation analysis (DFA), since we have already demonstrated that the DFA can distinguish a normal heart (intact heart) from an unhealthy heart (isolated heart) in animal models [8]. We finally revealed that alternans and lowered scaling exponent occurred concurrently. In this report, we demonstrate that the DFA is an advantageous tool in analyzing, diagnosing and managing the dysfunction of the heart.

1.2 Procedure

1.2.1 DFA Methods: Background

The DFA is an analytical method in physics, based on the concept of "scaling" [9, 10]. The DFA was applied to understand a "critical phenomenon" [9, 11, 12]. Systems near critical points exhibit self-similar properties. Systems that exhibit self-similar properties are believed to be invariant under a transformation of scale. Finally the DFA was expected to apply to any biological system, which has the property of scaling.

Stanley and colleagues have considered that the heartbeat fluctuation is a phenomenon, which has the property of scaling. They first applied the scaling-concept to a biological data in the late 1980s to early 1990s [11, 12]. They emphasized on its potential utility in life science [11]. However, although the nonlinear method is increasingly advancing, a biomedical computation on the heart seems not to have matured technologically. Indeed we still ask us: Can we decode the fluctuations in cardiac rhythms to better diagnose a human disease?

1.2.2 DFA Methods

We made our own programs for measuring the beat-to-beat intervals and for calculating the approximate scaling exponent of the interval time series. These DFA-computation methods have already been explained elsewhere [13]. We describe it here briefly on the most basic level.

Firstly, we obtain the heartbeat data digitized at 1 KHz. About 3,000 beats are necessary for a reliable calculation of an approximate scaling exponent. Usually a continuous record for about 50 min at a single testing is required. We use an EKG or finger pressure pulses.

Secondly, our own program captured pulse peaks. Using this program, we identified the true-heartbeat-peaks in terms of the time of peaks appearances. In this procedure we were able to reject unavoidable noise-false peaks. As a result we obtained a sequence of peak time $\{Pi\}$. This $\{Pi\}$ involved all true-heartbeat-peaks from the first peak to the ending peaks, usually ca. 2,000 peaks.

Practically, by eye-observation on the PC screen, all real peaks were identified and all noise peaks (by movement of the subject) were removed. Experiences on neurobiology and cardiac physiology are necessary when determining whether a spike-pulse is a cardiac signal or a noise signal. Finally our eyes confirmed, read, these entire beats.

Thirdly, using our own program, intervals of the heartbeat $\{I_i\}$, such as the R-R intervals of an EKG, were calculated, which is defined as:

$$\{P_{i+1} - P_i\} = \{I_i\} \tag{1.1}$$

Fourth, the mean-interval was calculated, which is defined as:

$$\langle I \rangle = \frac{1}{N} \sum_{k=1}^{N} I_k \tag{1.2}$$

where N is the total number of interval data.

Fifth, the fluctuation value was calculated by removing a mean value from each interval data, which is defined as:

$$Y_i = I_i - \langle I \rangle \tag{1.3}$$

Sixth, a set of beat-interval-fluctuation data $\{B_i\}$ upon which we do the DFA, was obtained by adding each value derived from the Eq. (1.3), which is defined as:

$$B_i = \sum_{k=0}^{i} Y_k \tag{1.4}$$

Here the maximum number of i is the total number of the data point, i.e., the total number of peaks in a recording. In the next step, we determine a box size, τ (*TAU*), which represents the numbers of beats in a box and which can range from 1 to a maximum. Maximum *TAU* can be the total number of heartbeats to be studied. For the reliable calculation of the approximate scaling exponent, the number of total heartbeats is hopefully greater than 3,000. If *TAU* is 300, for example, we can obtain ten boxes.

Seventh, we calculate the variance, successively, as expressed by Eqs. (1.5) and (1.6).

$$B_i' = B_i - f_{n,\tau}(i) \tag{1.5}$$

Here, n is the number of boxes at a box size τ (*TAU*). In each box, we performed the linear least-square fit with a polynomial function of fourth order. Then we made a detrended time series $\{B_i'\}$ as Eq. (1.5).

$$F^2(\tau) = \langle (B_{i+\tau}' - B_i')^2 \rangle \tag{1.6}$$

Here we adopted a method for "standard deviation analysis." This method is probably the most natural method of variance detection (see ref [14]). Mathematically, this is a known method for studying "random walk" time-series.

Meantime, Peng et al. [12] analyzed the heartbeat as we did. But they used a different idea than ours. They considered that the behavior of the heartbeat fluctuation is a phenomena belonging to the "critical" phenomena. (A "critical" phenomenon is involved in a "Random-walk" phenomenon.) Our consideration is that the behavior of a heartbeat fluctuation is a phenomenon, involved in a "random walk" type phenomenon. Peng et al. [12] used a much a stricter concept than ours. There was no mathematical proof, whether Peng's or ours is feasible for the heartbeat analysis, since the reality of this complex system is still under study. Probably the important point will be if the heartbeat fluctuation is a "critical" phenomenon or not. What we wondered is that the physiological *homeostasis* is a "critical" phenomena or not. As for the biomedical computing technology, we prefer a tangible method to uncover "*something* is wrong with the heart" instead of "*what* is the causality of a failing heart." So far, no one can deny that the heartbeat fluctuation belongs to the "critical" phenomena, but also there is no proof for that. We chose technically "Random walk." The idea of "random walk" is applicable to biology; a broad area from a biochemical reaction pathway to animal's foraging strategies [15]. "Random walk time series" was made by the Eq. (1.4). The present DFA study is a typical example for the "random walk" analysis on the heartbeats.

Eighth, finally, we plotted a variance against the box size. Then the scaling exponent is calculated, by looking at the slope (see Fig. 1.8 for an example).

Most of computations mentioned above are automated although we still need, in part, keyboard manipulations on PC. Our automatic program is helpful and reliable to distinguish a normal state of heart (scaling exponent exhibits near 1.0) from a sick state of heart (high or low exponent). In this report, we mention the three categories in differentiation: Normal, high, and low.

1.2.3 EKG and Finger Pulse

From human subjects we mostly used the finger pulse recording with a Piezo-crystal mechanic-electric sensor, connected to a Power Lab System (AD Instruments, Australia). The EKG recordings from crustacean model animals were done by implanted permanent metal electrodes, which are connected to the Power Lab System. By this recording, animals walked around in the container.

It is most important to us, that animal models are healthy before an investigation. We captured all specimens from a natural habitat by ourselves. We observed with our very own eyes that all animals, which were used, were naturally healthy, before starting any experiments.

During the decades of biological research on crustaceans, many famous invertebrate scientists (i.e. some distinguished cardiac researchers: J. S. Alexandrovicz in 1930s, C. A. G. Wiersma in 1940s, D. M. Maynard in 1950s, I. M. Cooke in 1970s, and McMahon and Wilkens in 1980s) have studied the hearts and its surrounding organs, investigating every accessible detail of its physiology and morphology, from nerves to muscles, from transmitters to hormones and from circulatory/respiratory control, to relevance and to behavioral integration. Thus it was that the crustacean cardiovascular systems have been well documented, as reviewed by Cooke [16] and Richter [17]. We, ourselves, have also been studying the heart of this animal [18].

1.3 Results

It is known that the human heart rate goes up to over 200 beats per min. when an life comes to an end (Dr. Umeda, Tokyo Univ. Med.; Dr. Shimoda, Tokyo Women Med. Univ., personal communication). Literally a similar observation has been reported during this period accompanied by a brain death (see Fig. 1.1 of ref [19]).

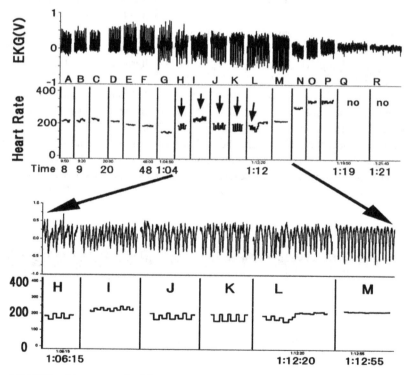

Fig. 1.1 EKG and heart rate of a dying crustacean, isopods, *Ligia exotica*. Two metal electrodes, 200 μm in diameter, were placed on the heart by penetrating them through the dorsal carapace. A sticky tape on a cardboard immobilized the animal. Records are shown intermittently for about 1 h and 20 min. From H to M, the EKG and heart rates are enlarged. Small five arrows indicate alternans, which is observable at H–L. From Q to R, no EKG signals were observed. Only small sized signals with sporadically appearance were seen

In an animal model we observed an increase of heart rates during the dying period (Fig. 1.1). Here, a healthy heart rate was about 200 (see A, B, and C). The heart rate is normally high in small animals. The body size of this animal was 3 cm in length. At terminal condition (see N, O, and P in Fig. 1.1), the heart rate attained over 300 beats per min. This model experiment indeed demonstrated a strong resemblance of a cardiac control mechanism, between lower animals and humans.

The aforementioned examples give evidence that: "Animals are models of humans." The similarity (a general idea that animals evolved from a common ancestor) was first noticed by a German scientist, who noted: "Biogenetische Grundregel" ("Recapitulation theory" by Ernst Heinrich Philipp August Haeckel). There is another reason to use lower animals. Ethics is of course a big requisition. But we know Gehring's discovery of a gene, named *homeobox*: To our surprise at that time, an identical gene named "pax-6" was found to work in both, in fly's eyes and mouse's eyes at a embryonic stage for developing an optical sensory organ [20, 21]. In 2007, further strong evidence has been presented with a new data of the origin of the central nervous system: The role of genes, which patterns the nervous system in embryos of chordates (like humans) and annelids (a lower animal) are surprisingly similar and the mechanism is inherited almost unchanged from lower animals to higher animals through a long period of geology [22, 23].

In this isopod specimen (Fig. 1.1), interestingly, a significant alternans is seen when they are dying (H, I, J, and K, Fig. 1.1). Its alternans lasted for not long, therefore we did not perform the DFA on this data.

EKG from a dying crab also exhibited alternans. Alternans appeared intermittently but very densely (Figs. 1.2a, 1.3). The alternans was again followed by a period of high-rate heartbeats before it died (Fig. 1.2b).

Finally we succeeded to calculate the scaling exponent of "alternans." The DFA revealed that the alternans exhibits a low approximate scaling exponent (Fig. 1.3).

The crab had a normal/healthy-scaling exponent (Fig. 1.4, which is 11 days before Fig. 1.3) when an EKG recording was first done, right after its collection in the

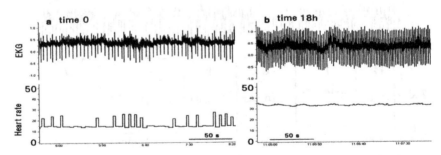

Fig. 1.2 EKG from a dying Mitten crab, *Eriocheir japonicus*. **a** A recording started at time zero. An irregular rate and alternans can be seen. The base line heart rate is about 15 beat per min. **b** 18 h after **a**, no alternans is seen. The heart rate increased to about 35 beats per min. This crab died 8.5 h after the recording **b**

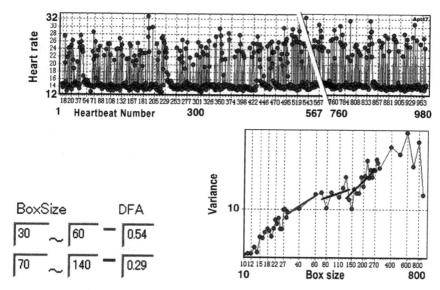

Fig. 1.3 Mitten crab DFA. The same crab shown in Fig. 1.2. About 980 beats for the DFA; the middle part was omitted. The approximate scaling exponent for alternans was low. Short-term box-size "30 beat–60 beat" and "70 beat–140 beat" were calculated, 0.54 and 0.29, respectively

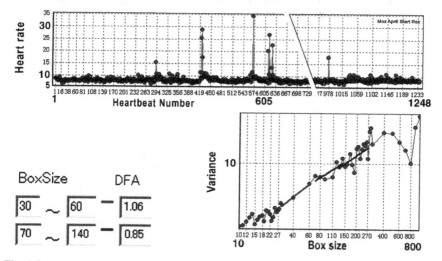

Fig. 1.4 Mitten crab DFA. The same crab shown in Figs. 1.2 and 1.3. But the recording was immediately after the specimen was captured. The approximate scaling exponent was not low but about 1.0 (cf., Fig. 1.3). The crab's heart seems to be normal on the first day of the experiment

South Pacific, on Bonin Island. Therefore, alternans and low exponents would be a sign of illness.

In models, we found that the isolated heart, which can repeat contractions for hours in a dish, often exhibits alternans (Fig. 1.5). The DFA again revealed that the

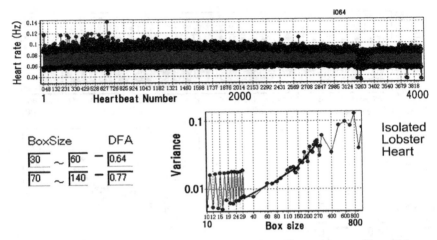

Fig. 1.5 Isolated heart of a spiny lobster, *Panulirus japonicus*. Alternans appeared here all the way down from the first beat to 4,000th beat. The scaling exponent was found to be low

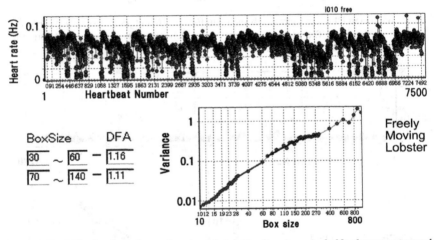

Fig. 1.6 The DFA of an intact heart of a spiny lobster, *Panulirus japoniculs*. No alternans appeared. The heart rate (shown in Hz) frequently dropped down, so-called bradycardia. This is well known in normal crabs and lobsters, first reported by Wilkens et al. [28]. The present DFA revealed that the scaling exponent is normal (nearly equal to 1.0) when the lobster is healthy and freely moving in the tank

scaling exponent of alternans is low (Fig. 1.5). We therefore tested another three-isolated hearts of these lobster species, all of which exhibited alternans (data not shown), and the scaling exponent of the alternans' heart was low. A healthy lobster before a dissection, however, exhibits a normal scaling exponent (Fig. 1.6).

After model experiments, we studied the human heartbeat. The finger pulse of a volunteer was tested (Figs. 1.7 and 1.8). Similar to the models, human alternans exhibited a low exponent (Fig. 1.8). This subject, a 65 years old female, is physically

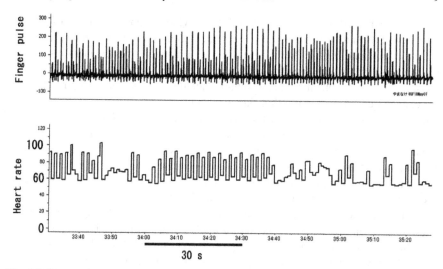

Fig. 1.7 Human alternans. A woman, who volunteered, age 65. Upper trace, recording of finger pulses. Lower trace, heart rate. Both amplitudes alternans and intervals alternans can be seen

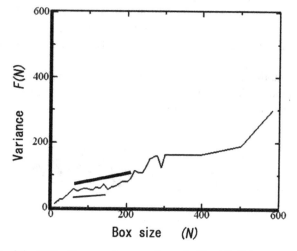

Fig. 1.8 Result of the DFA of human alternans shown in Fig. 1.7. Thick and thin straight lines represent a slope obtained from a different box-size-length. The slope determines the scaling exponent. The 45° slope (not shown here) gives the scaling exponent 1.0 which represents that the heart is totally healthy. The two lines are obviously less steep than the 45° slope. This argument draws a conclusion: The alternans significantly lowers the scaling exponent

weak and she cannot walk a long distance. However, she talked with an energetic attitude. She was at first nervous because of us (we did not realize it and even she herself didn't notice it), but finally she got accustomed to our finger pulse testing task and then she was relaxed. Hours later, we were surprised to note that her alternans decreased in numbers. The heart reflects the mind. We observed that alternans is coupled with the psychological condition. This is evidence that an impulse discharge frequency of the autonomic nervous system changes the condition of the heart.

It is known that healthy human hearts exhibit a scaling exponent of 1.0 [11]. Our analysis revealed the same results (data not shown, from age 9 to 82, about 30 subjects). So, the exponent 1.0 is the sign of "healthy heart in healthy body." In the future, health checkup may use this diagnostic idea.

It is not known what exponents the sick human hearts show. As aforementioned, a human alternans heart exhibits low scaling exponent (Figs. 1.7 and 1.8). It still remains to be investigated, because there may be other types of a diseased heart. In crab hearts we have already found out that a sudden-death heart exhibited a high exponent and a dying heart exhibited a gradual decay of exponent [24]. Here we tested our DFA computation on diseased human hearts.

A subject (male age 60), whose heartbeats are shown in Fig. 1.9, once visited to us and told us that he had a problem with his heart. He did not give us any other information about his heart, but challenged us and ordered: "Check my heartbeat and tell me what is wrong with my heart." The result of analysis is shown in Fig. 1.9. I replied to him: "According to our experience, I am sorry to tell you, that I diagnosed your heart and is indeed has a problem with its heartbeats. I wonder if there is a partly injured myocardium in your case, such as the ischemic heart." After listening to my opinion he explained that his heart has an implanted defibrillator, because of the damage of an apex of the heart in its left ventricle. He continued: "Please call me if your analysis machine will be commercially available in the future. I will buy it."

Figure 1.10 is another example analysis of a human diseased heart. A professor of Physiology at an Indonesian National University came to meet us. He told us that he had a bypass surgery of his heart, before we started to analyze his heartbeat. Finally

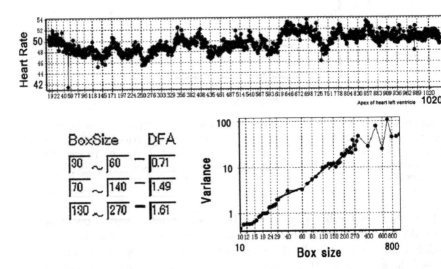

Fig. 1.9 Ischemic heart disease. The scaling exponent is high, see Box size 70–270

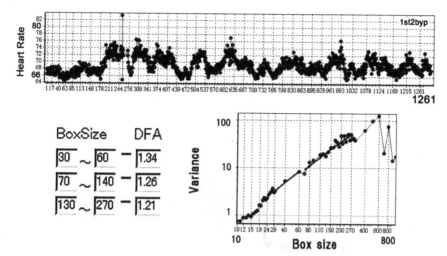

Fig. 1.10 A bypass surgery subject. The scaling exponent is high, see Box size 30–270

we found out that he had a high exponent, way over 1.0 (not shown). The subject shown in Fig. 1.10 is a case study: He (male age 65) has had a bypass operation 10 months before this measurement. One of the coronary artery had received a *stent*. Two other coronary arteries received bypass operations. This heart again exhibited a high exponent as shown in Fig. 1.10.

1.4 Discussion

A real time observation of both, EKG-signal and biological condition of a specimen, is important to deduce the physiological meaning from the scaling exponent. We performed all experimentations with our own eyes in front of us, together with the on-line EKG or on-line pulse recordings. We have engaged in both, recordings and analysis, by the same scientists. We thus were able to interpret the relationship between subjects' behavior, EKG waveform and DFA exponent. This is one important point of advantage in this study.

Other important points of our present studies are, that we made our own PC program, which assisted the accuracy of the peak-identification of heartbeats and then the calculation of the scaling exponent. Our DFA program shortened the period length of time-consuming analysis. We are currently developing a much simpler program for a practical use. It is a series of automated calculations, intended to be used by non-physicists who have no physical training. The program freed us from complicated PC tasks before obtaining the result of a calculation of the scaling exponent. As a result, we will be able to handle many data, sampled from various subjects.

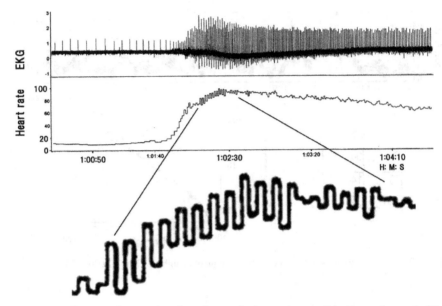

Fig. 1.11 EKG and heart rate recordings from a freely moving crayfish, *Procambarus clarkii*. Alternans can be seen at a rising phase of a heart rate tachogram. This occurred spontaneously when the animal was in the shelter. An arousal from sleep might happen. Generally, alternans occurs at a top speed of a cardiac acceleration

It is said that alternans is the harbinger of a sudden death of humans. That was true in dying models. However, alternans was also detectable in non-dying occasions in models, for example, during emotional changes (Fig. 1.11) [25]. This is evidence for that stressful psychological circumstances invoke autonomic acceleratory commands, leading to trigger alternans in the heart. In the early to mid-1990s a series of clinical trials demonstrated that an adrenergic blockade (that is a pharmacological blockade of the sympathetic acceleratory effects on the heart) was actually cardio-protective, particularly in post-myocardial infarction (MI) patients.

Physiological rhythms are considered to be generated by nonlinear dynamical systems [26]. There is evidence that physiological signals under healthy condition have a fractal temporal structure [27]. Free-running physiological systems are often characterized by 1/f-like scaling of the power spectra, $S(f)$, where f is the frequency [15]. This type of spectral density is often called the 1/f spectrum and such fluctuations are called 1/f fluctuations. The 1/f fluctuations were well documented in the heartbeat of normal persons [15]. After all, we noticed that a disease often leads to alterations from a normal to a pathological rhythm, and then we distinguished normal conditions from pathological conditions with our DFA program, by computing the scaling exponent for the healthy and diseased heart [10].

1.5 Concluding Remarks

Alternans lowers the approximate scaling exponent. An ischemic heart pushes the scaling exponent up, way over 1.0. In our present study, the DFA identified a typical diseased heart, the ischemic heart, which has partially damaged myocardium. This might be good news in the future for persons, who might potentially have a sudden death, to prevent it! Indeed cardiac failure has a principal underlying aetiology of ischemic damage from a vascular insufficiency (that is a decreased oxygen supply, particularly from coronary arteries). The life science may have profited from the ability of the DAF. The DFA provides analytical strategies, if models and human beings live on all functions under the same set of physical laws. This is still a testable hypothesis.

Acknowledgements This work was supported by a Grant-In-Aid for Scientific Research, No. 1248217 (TY) and No. 14540633 (TY). We thank G. W. Channell for her English revision. We are grateful to Professors T. Katsuyama at the Physics Department of the Numazu National College of Technology and I. Shimada at the Physics Department of Nihon University for their continuous and critical discussions on the DFA. We are also grateful to Mr. A. Kato and Mr. T. Nagaoka for their technical assistance and support throughout these experiments. Part of this bio-computation method has been submitted to a patent No.2007–6915 by the Tokyo Metropolitan University. We are very grateful to all volunteers, for allowing us to test their heartbeats by finger pulse tests.

References

1. C. Goldblatt, T. M. Lenton, and A. J. Watson, Bistability of atmospheric oxygen and the great oxidation. *Nature* **443**, 683–686 (2006).
2. D. S. Rosenbaum, L. E. Jackson, J. M. Smith, H. Garan, J. N. Ruskin, and R. J. Cohen, Electrical alternans and vulnerability to ventricular arrhythmias. *New Eng. J. Med.* **330**, 235–241 (1994).
3. B. Surawicz and C. Fish, Cardiac alternans: Diverse mechanisms and clinical manifestations. *J. Am. Coll. Cardiol.* **20**, 483–499 (1992).
4. M. R. Gold, D. M. Bloomfield, K. P. Anderson, N. E. El-Sherif, D. J. Wilber, W. J. Groh, N. A. M. Estes, III, E. S. Kaufman, M. L. Greenberg, and D. S. Rosenbaum, A comparison of T-wave alternans, signal averaged electrocardiography and programmed ventricular stimulation for arrhythmia risk stratification. *J. Am. Coll. Cardiol.* **36**(7), 2247–2253 (2000).
5. A. A. Armoundas, D. S. Rosenbaum, J. N. Ruskin, H. Garan, and R. J. Cohen, Prognostic significance of electrical alternans versus signal averaged electrocardiography in predicting the outcome of electrophysiological testing and arrhythmia-free survival. *Heart* **80**, 251–256 (1998).
6. B. Pieske and K. Kockskamper, Alternans goes subcellular: A "disease" of the ryanodine receptor? *Circ. Res.* **91**, 553–555 (2002).
7. K. Hall, D. J. Christini, M. Tremblay, J. J. Collins, L. Glass, and J. Billette, Dynamic control of cardiac alternans. *Phys. Rev. Lett.* **78**, 4518–4521 (1997).
8. T. Yazawa, K. Kiyono, K. Tanaka, and T. Katsuyama, Neurodynamical control systems of the heart of Japanese spiny lobster, *Panulirus japonicus*, *Izvestiya VUZ. Appl. Nonlin. Dynam.* **12**(1–2), 114–121 (2004).
9. H. E. Stanley, Phase transitions: Power laws and universality. *Nature* **378**, 554 (1995).

10. P. Ch. Ivanov, A. L. Goldberger, and H. E. Stanley, Fractal and multifractal approaches in physiology, *The Science of Disasters: Climate Disruptions, Heart Attacks, and Market*, A. Bunde et al. (Eds.) (Springer, Berlin, 2002).

11. A. L. Goldberger, L. A. N. Amaral, J. M. Hausdorff, P. C. Ivanov, and C.-K. Peng, Fractal dynamics in physiology: Alterations with disease and aging. *PNAS* **99**(Suppl. 1), 2466–2472 (2002).

12. C.-K. Peng, S. Havlin, H. E. Stanley, and A. L. Goldberger, Quantification of scaling exponents and crossover phenomena in nonstationary heartbeat time series. *Chaos* **5**, 82–87 (1995).

13. T. Katsuyama, T. Yazawa, K. Kiyono, K. Tanaka, and M. Otokawa, Scaling analysis of heart-interval fluctuation in the in-situ and in-vivo heart of spiny lobster, *Panulirus japonicus*. *Bull. Univ. Housei Tama* **18**, 97–108 (2003) (in Japanese).

14. N. Scafetta and P. Grigolini, Scaling detection in time series: Diffusion entropy analysis. *Phys. Rev. E* **66**, 1–10 (2002).

15. M. F. Shlesinger, Mathematical physics: First encounters. *Nature* **450**, 40–41 (2007).

16. I. M. Cooke, Reliable, responsive pacemaking and pattern generation with minimal cell numbers: The crustacean cardiac ganglion. *Biol. Bull.* **202**, 108–136 (2002).

17. Von K. Richter, Structure and function of invertebrate hearts. *Zool. Jb. Physiol. Bd.* **77S**, 477–668 (1973) (in German).

18. T. Yazawa, T. Katsuyama, A. Katou, H. Kaizaki, M. Yasumatsu, T. Ishiwata, H. Hasegawa, and M. Otokawa, Fourier spectral analysis and micro-bore column HPLC analysis of neuronal and hormonal regulation of crustacean heart. *Bull Housei Univ. Tama* **16**, 29–40 (2001) (in Japanese).

19. F. Conci, M. Di Rienzo, and P. Castiglioni, Blood pressure and heart rate variability and baroreflex sensitivity before and after brain death. *J. Neurol. Neurosurg. Psychiat.* **71**, 621–631 (2001).

20. S. B. Carroll, *Endless Forms Most Beautiful* (W.W. Norton, New York, 2005).

21. W. J. Gehring, *Master Control Genes in Development and Evolution: The Homeobox Story* (Yale University Press, New Haven, CT, 1998).

22. M. J. Telford, A single origin of the central nervous system? *Cell* **129**, 237–239 (2007).

23. A. Denes, G. Jékely, P. Steinmetz, F. Raible, H. Snyman, B. Prud'homme, D. Ferrier, G. Balavoine, and D. Arendt, Molecular architecture of annelid nerve cord supports common origin of nervous system centralization in bilateria. *Cell* **129**(2), 277–288 (2007).

24. T. Yazawa, K. Tanaka, and T. Katsuyama, Neurodynamical control of the heart of healthy and dying crustacean animals, The 3rd International Conference on Computing, Communications and Control Technologies, CCCT2005, July 24–27, Austin, TX, Proceedings, Vol. 1. International Institute of Informatics and Systemics, pp. 367–372.

25. T. Yazawa, K. Tanaka, A. Kato, T. Nagaoka, and T. Katsuyama, Alternans lowers the scaling exponent of heartbeat fluctuation dynamics in animal models and humans, The World Congress on Engineering and Computer Science, WCECS2007, San Francisco, CA, Proceedings, pp. 1–6.

26. L. Glass, Synchronization and rhythmic processes in physiology. *Nature* **410**, 277–284 (2001).

27. P. C. Ivanov, L. A. N. Amaral, A. L. Goldberger, S. Havlin, M. G. Rosenblum, Z. R. Struzik, and H. E. Stanley, Multifractality in human heartbeat dynamics. *Nature* **399**, 461–465 (1999).

28. J. L. Wilkens, L. A. Wilkens, and B. R. McMahon, Central control of cardiac and scaphognathite pacemakers in the crab, Cancer magister. *J. Comp. Physiol. A* **90**, 89–104 (1974).

Chapter 2
CLUSTAG & WCLUSTAG: Hierarchical Clustering Algorithms for Efficient Tag-SNP Selection

Sio-Iong Ao

Abstract More than 6 million single nucleotide polymorphisms (SNPs) in the human genome have been genotyped by the HapMap project. Although only a proportion of these SNPs are functional, all can be considered as candidate markers for indirect association studies to detect disease-related genetic variants. The complete screening of a gene or a chromosomal region is nevertheless an expensive undertaking for association studies. A key strategy for improving the efficiency of association studies is to select a subset of informative SNPs, called tag SNPs, for analysis. In the chapter, hierarchical clustering algorithms have been proposed for efficient tag SNP selection.

Keywords Single nucleotide polymorphisms · Tag-SNP · Clustering · HapMap

2.1 Introduction

In the genome, a specific position is called a locus [1]. A genetic polymorphism refers to the existence of different DNA sequences at the same locus among a population. These different sequences are called alleles. In each base of the sequence, there can be any one of the four different chemical entities, which are adenine (A), cytosine (C), guanine (G) and thymine (T). Inside these genomic sequences, there contain the information about our physical traits, our resistance power to diseases and our responses to outside chemicals. The differences in sequences can be grouped into large-scale chromosome abnormalities and small-scale mutations. The abnormalities include the loss or gain of chromosomes, and the breaking down and rejoining of chromatids.

S.-I. Ao
Oxford University Computing Laboratory, University of Oxford, Wolfson Building,
Parks Road, Oxford OX1 3QD, UK
e-mail: siao@graduate.hku.hk

S.-I. Ao et al. (eds.), *Advances in Computational Algorithms and Data Analysis*,
Lecture Notes in Electrical Engineering 14,
© Springer Science+Business Media B.V. 2009

The single nucleotide polymorphisms (SNP) is a common type of this small-scale mutation, and is estimated to occur once every 100–300 base pairs (bp) and the total number of SNPs identified reached more than 1.4 million. A candidate SNP is a SNP that has a potential for functional effect. It includes SNPs in regulatory regions or functional regions, and even in some non-synonymous regions. Different kinds of the SNP variations can provide different useful information about the diseases in different ways:

1. Functional variation refers to the situation when the SNP is with a nonsynonymous substitution in a coding region.
2. Regulatory variation happens when the SNP is in a non-coding region, but it can influence the properties of gene expressions [2].
3. Associations of the SNP with the disease become useful when there are some SNPs close enough to the mutations that cause the diseases. These SNPs can then be utilized in the association studies with the diseases [3].
4. Construction of the haplotype maps becomes possible with the collection of the information of the SNPs. The map is helpful for selecting SNPs that can be informative for explaining the differences in different ethnic groups and populations.

2.1.1 Methods for Selecting Tag SNPs

As described in the previous discussion, there exist redundant information in the whole set of SNPs and it is expensive to genotype this whole set. Different approaches have been developed to reduce the set of SNPs that are to be genotyped. These selected subsets are called haplotype tagging SNP (htSNPs) or tag SNPs. The approaches can be divided into two main categories [4]: (1) The block based tagging, and (2) The entropy based tagging (or called non-block based tagging).

With the block based tagging, we need to define the haplotype block first. Inside each haplotype block, the SNPs are in strong LD with each other, while, for SNPs of different blocks, they are of low LD. The disadvantage of this type of tagging SNPs is that the definition of the haplotype block is not unique and sometimes ambiguous, as we will see later. Also, it is true that the coverage of the haplotype block is not enough in some genomic region.

Because of the problems associated with the haplotype blocks, alternative methods have been developed and they can collectively called entropy based tagging. The term entropy is used loosely as the measure for assessing the amount of information that can be captured or represented by these tag SNPs. In this approach, it is not necessary to define the haplotypes and then to define the haplotype blocks. Instead, the goal of this approach is to select a subset of SNPs (the tag SNPs) that can capture the most information across the genomic region. Different multivariate statistical techniques have been applied to achieve this task. Byng et al. [5] proposed the use of single and complete linkage hierarchical cluster analysis to select tag SNPs. Hierarchical clustering starts with a square matrix of pair-wise distances between the objects to be clustered. For the problem of tag SNP selection, the objects to be clustered are the SNPs, and an appropriate measure of distance is $1 - R^2$, where R^2 is the squared correlation between two SNPs. The rationale is this: the required sample size for a tag SNP to detect an indirect association with a disease is inversely proportional to the R^2 between the tag SNP and the causal SNP.

2.1.2 Motivations for Developing Non-block based Tagging Methods

Meng et al. [6] have noticed that all the block-detecting methods can result in different block boundaries. In fact, the existence of the block is still conflicting [7]. As a result, Meng et al. have developed a method that use the spectral decomposition to decompose the matrix of pairwise LD between markers. The selection of the markers is based on their contributions to the total genetic variation. Meng et al. have used the sliding window approach for dealing with large genomic regions.

In the experimental data result analysis, Meng et al. have found that, for the chromosome 12 dataset, when selecting 415 markers (63.9%) out of a total 649, the spectral decomposition method can explain 90% of the variation. For the chromosome 22 dataset that is used for association study with the CYP2D6 poor-metabolizer phenotype, 20 out of 27 markers are selected by the method and they are shown to retain most of the information content of the full data well.

Meng et al. have also pointed out the differences between the method and that based on the haplotype. It can be generalized to the comparison of methods based on two-locus LD (i.e., pairwise correlation between single-markers) and methods based on haplotypes. Haplotypes can provide more information than the pairwise LD measures, if the LD measure involving with more than two markers is making a significant contribution to the overall LD measure. If not, then, haplotype frequencies are just linear combinations of pairwise LD frequencies. It has been found that, in the experimental study of chromosome 12 and 22, the LD based on the three-locus LD decays more quickly than the two-locus LD, and that the extent of three-locus LD is relatively small. Thus, it can justify the approach with two-locus LD of single-markers.

Meng et al. have also found that the two-locus approach of the spectral decomposition is of similar performance to that based the haplotypes. Nevertheless, the two-locus approach has the advantage that techniques, like sliding windows, can be applied for easing the computational burden, as this approach only requires the pairwise LD. The haplotype information is not required here, in contrast with the haplotype-based method, which require the estimation of the haplotype frequencies with the numerical methods like EM algorithms. The computational time for such algorithms will increase dramatically when the number of markers increases.

2.2 CLUSTAG: How Tag SNPs are Selected Efficiently

2.2.1 Agglomerative Clustering

The clustering algorithms of the CLUSTAG [8, 9] are of agglomerative cluster-ing, where the two clusters with the smallest inter-cluster distance are successively merged until all the objects have been merged into a single cluster. Different forms of agglomerative clustering differ in the definition of the distance between two clus-ters, each of which may contain more than one object. In single-linkage or nearest-neighbour clustering, the distance between two clusters is the distance between the nearest pair of objects, one from each cluster. In complete linkage or farthest neighbour clustering, the distance between two clusters is the distance between the farthest pair of objects, one from each cluster. The clustering process can be rep-resented by a dendrogram. The dendrogram can show how the individual objects are successively merged at greater distances into larger and fewer clusters. All dis-tinct clusters that have been generated at or below a certain user-defined distance are considered (see Fig. 2.1). For the example of complete linkage clustering, the dis-tances between rs2103317, rs2354377 and rs1534612 are less than the user-defined distance. So are the distances between rs7593150 and rs7579426.

2.2.2 Clustering Algorithm with Minimax for Measuring Distances between Clusters, and Graph Algorithm

A desirable property for a clustering algorithm, in the context of tag-SNP selection, would be that a cluster must contain at least one SNP (the tag SNP) that is no more than the merging distance from all the other SNPs from the same cluster. If this is the case, then by setting a cutoff merging distance of C, one can ensure that no SNP is further than C away from the tag SNP in its cluster. As said, neither of the

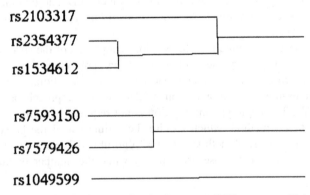

Fig. 2.1 Sample illustrative dendrogram showing how seven SNPs are merged into three clusters at or below the cutoff merging distance

methods proposed by Byng et al. [5] is ideal, since the single-linkage method does not guarantee the existence of a tag SNP with distance less than C from all SNPs in the same cluster, while complete-linkage is too conservative in that all SNPs have distance under C from all other SNPs in the same cluster.

In order to achieve the desired property described above, we propose a new definition of the distance between two clusters, as follows:

1. For each SNP belonging to either cluster, find the maximum distance between it and all the other SNPs in the two clusters.
2. The smallest of these maximum distances is defined as the distance between the two clusters.
3. The corresponding SNP is defined as the tag SNP of the newly merged cluster.

We call this method minimax clustering, which is an agglomerative method. There is a parallel in topology in which the distance between two compact sets can be measured by a sup-inf metric known as Hausdorff distance [10].

For comparison we have also implemented an algorithm based on the NP-complete minimum dominating set of the set-cover problem in the graph theory, similar to the greedy algorithm developed by Carlson et al. [11]. The set of SNPs are the nodes of a graph, which are connected by edges where their corresponding SNPs have $R^2 > C$. The objective is to find a subset of nodes such that that all nodes are connected directly to at least one SNP of that subset. The details of this heuristic algorithm can be found in Reuven and Zehavit [12] and Johnson [13]. The one by Johnson is on the studies of the error bound of the algorithm. Briefly, at the beginning of the method, all the SNPs belong to the untagged set. The algorithm picks the node with the largest number of nodes that are connected directly to it (without passing through any other nodes) from the untagged set. Then the SNPs inside the selected subset are deleted from the untagged set, and the next largest connected subset is chosen from the untagged set. The algorithm terminates when the untagged set becomes empty.

2.3 Experimental Results of CLUSTAG

We have implemented the complete linkage, minimax linkage and set cover algorithms in the program CLUSTAG. The program takes a file of R^2 values produced, for example, by HAPLOVIEW [14], and outputs a text file containing one row per SNP and the following columns (Fig. 2.2): (i) SNP name, (ii) cluster number, (iii) chromosomal position, (iv) minor allele frequency, (v) maximal distance $(1 - R^2)$ from other SNPs in the same cluster, and (vi) average distance $(1 - R^2)$ from other SNPs in the cluster. Both (v) and (vi) are useful for providing alternative SNPs that can serve as the tag SNP of the cluster, allowing some flexibility in the construction of multiplex SNP assays. A visual display (in html format) provides a representation of the SNPs in their chromosomal locations, color-labeled to indicate cluster membership (Fig. 2.2). The tag SNP is highlighted and hyperlinked to a text box containing columns (i)–(vi) on the cluster.

	A	B	C	D	E	F	G	H	I	J
1	SNP NAME	Cluster	Tagged	SNP Positions	MAF	Max(1-R^2)	Order by Max(1-R^2)	Average(1-R^2)	Order by Average(1-R^2)	
2	rs2305634	0	0	47004130	0.36	0.19	20	0.091666676	19	
3	rs1079276	0	0	47006331	0.45	0.14999998	9	0.076666676	16	
4	rs7646799	0	0	47113159	0.45	0.14999998	9	0.077916674	18	
5	rs11130115	0	0	47142253	0.45	0.14999998	9	0.076666676	16	
6	rs2305638	0	0	47007434	0.46	0.18	17	0.09625002	21	
7	rs7610636	0	0	47025028	0.47	0.19	20	0.102500014	23	
8	rs4078466	0	0	47087135	0.46	0.18	17	0.09583335	20	
9	rs6785790	0	0	47107524	0.46	0.19	20	0.097916685	22	
10	rs4315703	0	0	47283373	0.42	0.13999999	3	0.04833333	4	
11	rs295441	0	0	47293887	0.43	0.13	1	0.047916662	2	
12	rs4683327	0	0	47319854	0.42	0.13999999	3	0.04833333	4	
13	rs2159400	0	0	47336482	0.43	0.13	1	0.047916662	2	
14	rs295458	0	0	47346177	0.42	0.13999999	3	0.04833333	4	
15	rs807931	0	0	47349909	0.4	0.14999998	9	0.05166666	9	
16	rs7613282	0	0	47350001	0.42	0.13999999	3	0.04833333	4	
17	rs11130127	0	0	47359385	0.39	0.13999999	3	0.049999993	8	
18	rs922957	0	1	47382744	0.42	0.13999999	3	0.040000003	1	
19	rs2062278	0	0	47377353	0.42	0.17000002	13	0.060416657	10	
20	rs10865946	0	0	47385652	0.41	0.17000002	13	0.064583324	13	
21	rs11712445	0	0	47395144	0.41	0.19	20	0.070833325	15	
22	rs4858811	0	0	47441116	0.42	0.17000002	13	0.060416657	10	
23	rs11130128	0	0	47447224	0.42	0.17000002	13	0.060416657	10	
24	rs2101247	0	0	47454895	0.4	0.18	17	0.06624999	14	
25	rs6442055	1	0	47071314	0.43	0.13999999	1	0.051999997	2	
26	rs6766230	1	1	47156801	0.42	0.13999999	1	0.043999992	1	

Fig. 2.2 Text output of the CLUSTAG

We have compared the performance of the three implemented algorithms, using SNP data from the ENCODE regions of the HapMap project, according to three criteria:

1. Compression, the ratio of clusters to SNPs
2. Compactness, the average distance between a SNP and the tag SNP of its cluster $(1 - R^2)$ and
3. Run time

Our results show that the compression ratio is roughly equivalent for the set cover and minimax clustering algorithms but substantially higher for the complete linkage (Table 2.1). The minimax algorithm produces more compact clusters than the set cover algorithm (Table 2.2), but takes approximately twice as long to run. The run times of all three algorithms are expected to increase in proportion to the square of the number of SNPs.

The complexity of the clustering methods are of order $O(n^2)$. With the run time information in our table of several hundred SNPs and this complexity information, the users can estimate roughly the expected run time for their samples before the program's execution. The run time will not be an issue for data of several hundred to a hundred thousand SNPs. But, it will be a constraint when we are studying the whole genome at one time, when the size may be of several million SNPs. This is an area of further work as the HAPMAP project is producing the whole genome haplotype information.

We have also the tested the different threshold values C for the chromosome 9 of the ENCODE data in the following two figures. The values of the threshold C

Table 2.1 Properties of three tag SNP selection algorithms, evaluated for ENCODE regions

Encode region (SNP no.)	Compression			Run time (s)		
	Complete	Minimax	Set cover	Complete	Minimax	Set cover
2A (519)	0.277	0.245	0.247	3.94	5.42	3.20
2B (595)	0.291	0.255	0.261	5.44	6.92	4.03
4 (665)	0.242	0.211	0.209	6.53	13.30	5.25
7A (417)	0.314	0.281	0.281	2.56	3.39	2.00
7B (463)	0.186	0.166	0.171	3.53	5.03	2.84
7C (433)	0.240	0.217	0.215	2.38	3.28	1.80
8A (364)	0.269	0.245	0.245	2.39	2.94	1.83
9 (258)	0.360	0.318	0.314	1.47	1.74	0.98
12 (454)	0.260	0.227	0.227	2.69	3.69	2.03
18 (350)	0.283	0.254	0.254	2.17	2.81	1.64

Table 2.2 Compactness of three tag SNP selection algorithms, evaluated for ENCODE regions

Encode region (SNP no.)	Compactness Complete	Minimax	Set cover
2A (519)	0.021	0.033	0.037
2B (595)	0.018	0.033	0.032
4 (665)	0.016	0.031	0.035
7A (417)	0.013	0.028	0.032
7B (463)	0.020	0.030	0.035
7C (433)	0.018	0.019	0.021
8A (364)	0.019	0.035	0.040
9 (258)	0.012	0.025	0.031
12 (454)	0.017	0.028	0.034
18 (350)	0.014	0.033	0.037

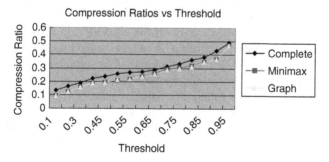

Fig. 2.3 Compression ratios vs. different threshold values

are 0.7, 0.75, 0.8, 0.85, 0.9 and 0.95, which cover the range of reasonable threshold values. The results show that the compression ratio and the compactness are quite stable over the range from 0.7 to 0.8 (Figs. 2.3 and 2.4).

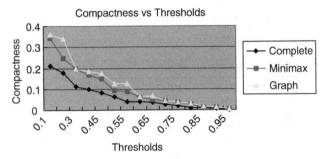

Fig. 2.4 Compactness vs. different threshold values

2.4 WCLUSTAG: Motivations for Combining Functional and sxLD Information in the Tag SNP Selection

In the association studies for complex diseases, there are mainly two approaches for selecting the candidate polymorphisms. In the functional approach, the candidate polymorphisms are selected if they are found to cause a change in the amino acid sequence or gene expressions. The second approach, the positional approach, is to systematically screen polymorphisms in a particular genome region by using the linkage disequilibrium information with the disease-related functional variants. The functional approach is direct approach, while the positional approach is indirect approach. The algorithms and programs that we have described in the above sections are basically constructed with the positional approach. The candidate tag SNPs are selected for genotyping by utilizing the redundancy between near-by SNPs through the LD information. The purpose is to improve the efficiency of the analysis with minimal loss of information while reducing the genotyping costs at the same time.

In order to further utilize the genomic information for improving the tag-SNP selection efficiency, it would be desirable if the tag-SNP selection algorithm can take account of the functional information, as well as the LD information. In the human genome, it is well known that different kinds of polymorphisms have different effects on the gene expressions and importance. The SNPs can attach more importance when their positions are within the coding, regulatory regions. Similarly, for SNPs in the non-coding regions, they are attached with less biological importance. Furthermore, it is also desirable for the tag-SNP selection algorithm to take care of practical laboratory considerations like the readiness of the SNPs for assaying and the existing genotyped results in the previous laboratory experiments.

2.4.1 Constructions of the Asymmetric Distance Matrix for Clustering

The WCLUSTAG [15] is developed in order to take care of the functional information and LD information, as well as the laboratory consideration. The development of the WCLUSTAG is based on the previous CLUSTAG, by adding the variable tagging threshold and other functions, and the web-based interface. As described above, the CLUSTAG is of agglomerative hierarchical clustering and starts with the constructing of a square matrix of pair-wise distance between the objects to be clustered. An appropriate distance measure for the LD tagging is $1 - R^2$, where the second term is the square of the correlation between the SNPs. The clusters with the least inter-cluster distance are successively merged with each other. A cutoff merging distance, denoted by C, is required for the terminating of the algorithm and for ensuring that, in each cluster, it contains no SNP further than C away from the corresponding tag SNP.

In CLUSTAG, this cutoff merging distance C is the same for all the SNPs. In WCLUSTAG, the program has been modified so that the tagging threshold C can be specified by the user for each SNP, and can be variable among the SNPs. Then, the factors like the positional and functional information, as well as the practical laboratory information, can be utilized for the assigning of tagging threshold values for SNPs individually. For example, for SNPs in the coding or regulatory regions, a high value of C (e.g. 0.8) can be assigned to these SNPs. On the other hand, for the other SNPs in the genome regions (like the non-coding regions), a low value of C (like 0.4) can be given. With this modification, unlike the CLUSTAG, the square matrix of pair-wise distances between the objects becomes asymmetric for WCLUSTAG. For example, let a coding SNP have a C of 0.8, and another non-coding SNP of C value 0.4, and let the R^2 between these two SNPs be 0.5. It can be observed that the first SNP can serve as the tag SNP for the second. On the hand, the second SNP is not able to tag the first one. Thus, the WCLUSTAG has been built with the capability for handling of asymmetric distance matrix, such that the distance from object h to object k is not required to be the same as the distance from object k to object h.

With these considerations, the WCLUSTAG has been modified from CLUSTAG and works as followed:

Firstly, a user-define value C is assigned for each SNP;

Secondly, let C_k be the value of C for SNP k, and, let the distance from SNP h to SNP k be $C_k - R_{hk}^2$. Then, for $C_k - R_{hk}^2 < 0$, SNP h can serve as a tag SNP for SNP k.

Thirdly, the minimax clustering method is applied with this new asymmetric distance matrix, and the cut-off merging distance is zero.

Then, cluster is formed for the case that there is a tag SNP that has a distance of zero or less with its cluster members respectively. The set-cover algorithm has undergone similar modifications in WCLUSTAG.

2.4.2 Handling of the Additional Genomic Information

As discussed above, it is desirable that the tag-selection algorithm can initially select all SNPs that have already been genotyped, and then remove these SNPs and the SNPs tagged by these SNPs from the next genotyping experiment. The algorithm will provide the laboratory users with more flexibility if the algorithm can exclude those SNPs that have problems with assay design etc. In order to achieve these properties, the algorithm has been subjected to the below further modifications, which can be done by changing the values of certain elements in the matrix similarities $[R_{hk}^2]$.

For the case that the SNP t has already been genotyped, all the elements of column t in the matrix are set to zeros, except for the diagonal element of the column t which remains one. This setting can ensure that the SNP t can not be tagged by any other SNPs, and, therefore, it will be included as one of the tag SNPs in the clustering and graph algorithms. For the case that the SNP t has problem with assay design, all the elements of the row t in the matrix are set to zero. Therefore, the SNP t can never serve as one of the tag SNPs in the algorithms. There is one problem associated with these settings. With these settings, it does not ensure that all the problematic SNPs for assay design can be tagged in the algorithms. This is because some non-assayable SNPs can only be tagged by certain SNPs, while these SNPs may not be selected as the tag SNPs with the algorithms. This problem can be solved with the following further modification, which forces the selection of certain SNPs for tagging these non-assayable SNPs.

1. Firstly, for non-assayable SNPs that can not be tagged by any assayable SNP, as there do not exist any assayable tag SNP for them, they are listed and excluded from further processing. Then, the remaining non-assayable SNPs are subjected to following procedure to ensure that there will exist at least one tag SNP for each of them.
2. The set of already-genotyped SNPs (if existed) are checked if the SNPs there can tag the non-assayable SNPs. The SNPs of the non-assayable SNPs that can not be tagged by these already-genotyped SNPs are called the set of untagged non-assayable SNPs.
3. Each assayable SNPs (but not those already genotyped) is checked against the untagged non-assayable SNPs for the number of untagged non-assayable SNPs that each assayable SNP can tag. The one with the largest number is assigned as a SNP for forced selection, and the non-assayable SNPs that can be tagged by this SNP are removed from the set of untagged non-assayable SNPs.

For cases that there still exist untagged non-assayable SNPs, the above step (2) is repeated until there exist no untagged non-assayable SNP.

The SNPs selected in the above steps (2) and (3) are treated in the same way as the SNPs that have been already genotyped, and are subjected to the same procedure for forced selection.

2.5 WCLUSTAG Experimental Genomic Results

To illustrate the performance of the new algorithms, the CEPH sample genotype data from the International Haplotype Map Project was tested with the algorithms.

Table 2.3 Properties of the tag SNP selection algorithms, weighted with 0.8 for gene regions and 0.4 for other regions

Encode region (SNP no.)	Compression (uniform)			Compression (weighted)	
	Complete	Minimax	Set cover	Minimax	Set cover
2A (519)	0.277	0.245	0.247	0.104	0.104
2B (595)	0.291	0.255	0.261	0.197	0.198
4 (665)	0.242	0.211	0.209	0.089	0.089
7A (417)	0.314	0.281	0.281	0.149	0.139
7B (463)	0.186	0.166	0.171	0.114	0.114
7C (433)	0.240	0.217	0.215	0.189	0.185
8A (364)	0.269	0.245	0.245	0.190	0.190
9 (258)	0.360	0.318	0.314	0.221	0.225
12 (454)	0.260	0.227	0.227	0.167	0.163
18 (350)	0.283	0.254	0.254	0.186	0.189
Average	0.267	2.237	0.238	0.154	0.153
Additional saving	–	–	–	35.2%	35.9%

Table 2.4 Effect of weighting scheme (intragenic versus other SNPs) on the comparison ratios for tag-SNP selection algorithms in the Chromosome 9 Encode data

Weighted ratio	Compression Minimax	Set cover
0.8: 0.4	0.221	0.225
0.8: 0.3	0.198	0.198
0.8: 0.5	0.240	0.244
0.7: 0.4	0.217	0.221
0.9: 0.4	0.240	0.244
0.8: 0.8	0.318	0.314

The ENCODE regions were selected because genotyping were undertaken for all known SNPs in these regions. Intragenic regions were identified from the start and end points of the coding sequences for the 33 K Ensembl genes in NCBI build 34. Intragenic SNPs are given a C weighting of 0.8, and other SNPs 0.4. The compression ratios (the number of tag-SNPs over the total number of SNPs) of the various ENCODE regions are compared with the original procedure which used a uniform C value of 0.8. Our results show that there can be a further 35.2% saving with our weighted minimax algorithm, and 35.9% with the set cover method (Table 2.3). We also explored the impact of using different weighting schemes. Some additional saving can be obtaining by lowering the weights for either intragenic or other SNPs, although the compression ratios remain in the region of 0.2 (Table 2.4). The average ratio of the SNPs in the intragenic regions to the overall SNPs is 32.3% in the dataset (Table 2.5).

Table 2.5 The number of SNPs in the intragenic regions and the other regions. The average ratio of the SNPs in the intragenic regions to the overall SNPs is 32.3%

SNPs no.	SNPs in intragenic regions	SNPs in other regions
chr2A	0	519
chr2B	273	322
chr4	0	665
chr7A	21	396
chr7B	159	304
chr7C	299	134
chr8A	203	161
chr9	66	192
chr12	180	274
chr18	167	183

2.6 Result Discussions

With the necessary modifications, the WCLUSTAG can enable the users to select tag SNPs, complying the advantage of the functional approach and positional approach of the association studies. The choice of the threshold values can be made according to the budget for the disease data. Currently, the users can use the downloadable program version, which may be convenient for running scripts for multiple data sets. Or, the users can assess our web interface for importing their own genotype data. The web interface also has the capability of downloading the HapMap data directly from its mirror database for further computation.

There are factors that can affect the overall effectiveness of the tagging strategy. They include the functional information like the comprehensiveness of SNP maps, the quality of functional annotation of the genome, and the linkage disequilibrium information between the polymorphisms and the complex human diseases, and the underlying genetic architecture of the complex diseases. Many of these have not been fully understood by researchers and remain to be explored in the future studies.

References

1. Sham, P., "Statistics in human genetics". Arnold, UK, 1998.
2. Cowles, C., Joel, N., Altshuler, D., and Lander, E., "Detection of regulatory variation in mouse genes". Nat. Genet. 32, 432–437, 2002.
3. Sherry, S., Ward, M., and Sirotkin, K., "Use of molecular variation in the NCBI dbSNP database". Hum. Mutat. 15, 68–75, 2000.
4. CIGMR 2005, "Tagging SNPs". Web Address: http://slack.ser.man.ac.uk/theory/tagging.html. Modified date: March 22 2005.
5. Byng, M. et al., "SNP subset selection for genetic association studies". Ann. Hum. Genet. 67, 543–556, 2003.
6. Meng, Z. et al., "Selection of genetic markers for association analysis, using linkage disequilibrium and haplotypes". Am. J. Hum. Genet. 73, 115–130, 2003.
7. Couzin, J., "New mapping projects splits the community". Science 296, 1391–1393, 2002.

8. Ao, S. I., Yip, K., Ng, M. et al., "CLUSTAG: Hierarchical clustering and graph methods for selecting tag SNPs". Bioinformatics 21(8), 1735–1736, 2005.
9. Ao, S. I., "Data Mining Algorithms for Genomic Analysis". Ph.D. thesis, The University of Hong Kong, Hong Kong, May 2007.
10. Wucklidge, W., "Efficient visual recognition using the Hausdorff distance". Springer, 1996.
11. Carlson, C. et al., "Selecting a maximally informative set of single-nucleotide polymorphisms for association analyses using linkage disequilibrium". Am. J. Hum. Genet. 74, 106–120, 2004.
12. Reuven, Y. and Zehavit, K., "Approximating the dense set-cover problem". J. Comput. Syst. Sci. 69, 547–561, 2004.
13. Johnson, D., "Approximation algorithms for combinatorial problems". Ann. ACM Symp. Theor. Comput. 38–49, 1973.
14. Barrett, J. et al., "Haploview: Analysis and visualization of LD and haplotype maps". Bioinformatics 21(2), 263–265, 2005.
15. Sham, P., Ao, S. I. et al., "Combining functional and linkage disequilibrium information in the selection of tag SNPs". Bioinformatics 23(1), 129–131, 2007.

Chapter 3
The Effects of Gene Recruitment on the Evolvability and Robustness of Pattern-Forming Gene Networks

Alexander V. Spirov and David M. Holloway

Abstract Gene recruitment or co-option is defined as the placement of a new gene under a foreign regulatory system. Such re-arrangement of pre-existing regulatory networks can lead to an increase in genomic complexity. This reorganization is recognized as a major driving force in evolution. We simulated the evolution of gene networks by means of the Genetic Algorithms (GA) technique. We used standard GA methods of point mutation and multi-point crossover, as well as our own operators for introducing or withdrawing new genes on the network. The starting point for our computer evolutionary experiments was a 4-gene dynamic model representing the real genetic network controlling segmentation in the fruit fly *Drosophila*. Model output was fit to experimentally observed gene expression patterns in the early fly embryo. We compared this to output for networks with more and less genes, and with variation in maternal regulatory input. We found that the mutation operator, together with the gene introduction procedure, was sufficient for recruiting new genes into pre-existing networks. Reinforcement of the evolutionary search by crossover operators facilitates this recruitment, but is not necessary. Gene recruitment causes outgrowth of an evolving network, resulting in redundancy, in the sense that the number of genes goes up, as well as the regulatory interactions on the original genes. The recruited genes can have uniform or patterned expressions, many of which recapitulate gene patterns seen in flies, including genes which are not explicitly put in our model. Recruitment of new genes can affect the evolvability of networks (in general, their ability to produce the variation to facilitate adaptive evolution). We see

A.V. Spirov (✉)
Applied Mathematics and Statistics, and Center for Developmental Genetics,
State University of New York, CMM Bldg, Rm481, South Loop, SUNY at Stony Brook,
Stony Brook, NY 11794-5140, USA
e-mail: Alexander.Spirov@sunysb.edu

D.M. Holloway
Mathematics Department, British Columbia Institute of Technology, Burnaby, B.C.,
Canada, and with the Biology Department, University of Victoria, B.C., Canada
e-mail: David_Holloway@bcit.ca

S.-I. Ao et al. (eds.), *Advances in Computational Algorithms and Data Analysis*,
Lecture Notes in Electrical Engineering 14,
© Springer Science+Business Media B.V. 2009

this in particular with a 2-gene subnetwork. To study robustness, we have subjected the networks to experimental levels of variability in maternal regulatory patterns. The majority of networks are not robust to these perturbations. However, a significant subset of the networks do display very high robustness. Within these networks, we find a variety of outcomes, with independent control of different gene expression boundaries. Increase in the number and connectivity of genes (redundancy) does not appear to correlate with robustness. Indeed, removal of recruited genes tends to give a worse fit to data than the original network; new genes are not freely disposable once they acquire functions in the network.

Keywords Complexification of gene networks · Gene co-option · Gene recruitment · Pattern formation · Modeling of biological evolution by Genetic Algorithms · Redundancy and robustness of gene networks

3.1 Introduction

Early in metazoan evolution, gene networks specifying developmental events in embryos may have consisted of no more than 2 or 3 interacting genes. Over time, these were augmented by incorporating new genes and integrating originally distinct pathways [1]. While it may initially be thought that new functions require novel genes, whole genome sequencing has shown that apparent increases in developmental complexity do not correlate with increasing numbers of genes [2]: the number of genes in the human genome is somewhat higher than in fruit flies and nematodes, but lower than in pufferfish and cress and rice plants. Therefore, evolution of developmental pathways may most commonly proceed by recruitment of preexisting external genes into preexisting networks, to create novel functions and novel developmental pathways [3]; developmental evolution may act primarily on genetic regulation [4, 5].

Specifically, gene recruitment may occur through mutational changes in the regulatory sequences of a gene in an established pathway, enabling a new transcriptional regulator (or regulators) to bind. This regulator may be from a newly evolved gene (say via duplication and subsequent change), in which case it simply adds to the existing pathway, or it may have already been part of a pre-existing pathway, in which case the two pathways become integrated. In either case, the developmental function of the pathway may be significantly altered. Similarly significant alterations can arise by inserting regulatory sequences for an existing gene at new loci, transferring transcriptional control of the original gene to other members of the genome [1, 6].

In insects, two distinct modes of segmenting the body have evolved. In primitive insects, such as the grasshopper, the short germ band mode lays out body segments sequentially. Many more highly derived insects, such as flies, use the long germ band mode to establish all body segments simultaneously. This simultaneous mechanism must act quickly during development; it has been proposed that it evolved by co-option of new genes to the short germ band mechanism, in order to maintain accurate regulation of patterned gene transcription over the whole embryo in a condensed time frame 1. The invertebrate segmentation network is one of the best-studied gene ensembles, in which the amount of diverse experimental data provides a unique opportunity for studying known and hypothetical scenarios of its evolution in detail. In particular, the level of detail for the segmentation gene network for the fruit fly (*Drosophila melanogaster*) has made it for many years the most popular object for computer simulations of its function and evolution [7–12].

In this publication, we investigate the interrelations between redundancy (addition of extra genes to a network), evolvability (ability of a network to change), and robustness (ability of a network to remain fit in a variable environment). We use an *in silico* approach to simulate evolution of a dynamic model of the gap gene network, central to fly segmentation (specifically). This model (adapted from [9,13]) is a system of differential equations describing the regulatory interactions of 4 gap genes (*giant, gt; hunchback, hb; Krüppel, Kr; knirps, kni*), under the control of gradients of maternal proteins (Bicoid, Bcd, in our basic model; plus maternally-derived Hb (Hb_{mat}), Caudal (Cad), and Tailless (Tll) in our extended model). Figure 3.1A shows the integrated (averaged) spatial patterns of the gap genes along the antero-posterior (A-P; head to tail) axis of the fly embryo in early nuclear cleavage cycle 14A (*even-skipped, eve*, is a pair-rule gene, regulated by the maternal and gap genes). Figure 3.1B shows the gap patterns slightly later in development, at mid cleavage cycle 14A. Figure 3.1C shows the patterns of the maternal input factors. Model parameters for gene interaction strengths are varied and solutions selected by a Genetic Algorithms method (details below) based on how well they fit the gap gene data. This produces networks describing particular interactions (and quantitative strengths) between the component genes (e.g., Fig. 3.1D). In this way, we can use a model of our current understanding of fly segmentation to study the evolutionary dynamics of how the segmentation network may have arisen, and how this might reflect on its current characteristics.

In particular, we are interested in what genetic mechanisms are necessary for recruiting (co-opting) new genes to small networks, what characteristics these recruits have (e.g., spatial patterns, regulatory interactions), and how they might change the behavior of the network. There is currently much discussion in evolutionary biology on these topics, and it is expected that the outgrowth of preexisting networks through gene recruitment should cause structural (genes duplicating existing ones) and functional (development of compensatory pathways) redundancy of the networks [14]. Cases of such redundancy have been found in many genetic ensembles in many organisms [14]. One of the common conclusions from these cases is that the redundancy could affect such key species characteristics as evolvability or robustness to perturbations and variability during development.

The segmentation network lays down the spatial order of the developing embryo, so the fitness of any network depends on how reliably it establishes spatial position. In our computations, we establish this type of fitness by scoring model solutions on how well they reproduce experimental pattern. By doing hundreds of simulations, we generate a large sample of networks for studying the mechanisms of gene recruitment and how these relate to evolvability and robustness (in particular for making reproducible output in the face of biological levels of variability in the upstream maternal control gradients).

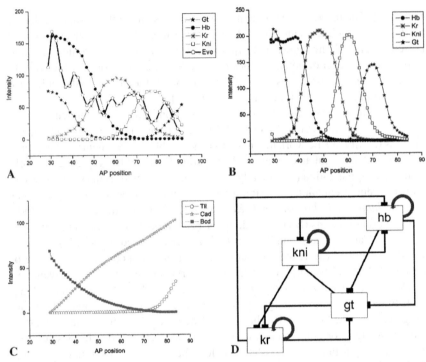

Fig. 3.1 Biological data used to fit ODE model by GA. **A**. Integrated gene expression profiles for early cleavage cycle 14A. Vertical axis represents relative protein concentrations (proportional to intensity), horizontal axis represents position along the anteroposterior (A-P) embryo axis (where 0% is the anterior pole). Data from the FlyEx database [26]. **B**. Integrated gene expression profiles for mid cleavage cycle 14A. **C**. Integrated gene expression profiles for the external, maternal inputs used in this paper, from the very beginning of cleavage cycle 14A. **D**. Overview of the gap gene network (After [14]). The gap genes are represented as boxes. Repressive interactions are represented by T-bar connectors. Looped arrows mean self-regulation

We investigate the mechanisms of gene recruitment through the Genetic Algorithms (GA) technique. Run on our fly segmentation model, it is a simulation of how this network may have evolved in nature. We use standard GA operators (mutation and crossover), as well as our own operators for introducing and removing new genes on the networks.

In computing evolutionary searches, we have found that the standard operator for point mutations, in combination with the gene introduction operator, is enough to support recruitment of new genes to pre-existing networks. This is in contrast to a mainstream view in evolutionary biology, that the main mechanism facilitating recruitment is the sophisticated shuffling of genetic material, such as unequal crossover (recombination), or the activity of transposons [15]. A computational approach allows us to systematically compare recruitment by these different mechanisms, specifically point mutation versus one- and multi-point crossover.

Our results indicate how complexification or "outgrowth" of gene networks can proceed, by recruiting new genes to make new connections between old and new members of the network. We have characterized the structure of the evolved networks, as well as the possible influence of gene recruitment on evolvability. In particular, we found that for a 2-gene subnetwork evolvability is clearly raised via co-option of new genes. Evolvability was not so clearly raised when starting with the 4-gene networks.

We also studied the effects of gene recruitment on the robustness of the computer networks to variability in external control parameters. Specifically, we simulated variability in maternal morphogenetic factors which are upstream (in terms of regulatory control) of the simulated networks. The gap gene network has been shown to be quite robust to this sort of variability, spurring a great deal of interest in the biology community on how embryos might filter maternal or environmental variability or noise [16–21]. We tested robustness in our basic 4-gene network (Bcd control only, which has been most extensively studied experimentally) and our extended 4-gene network (additional control by $Hb_{maternal}$, Cad and Tll). By simulating variability in each of the upstream factors we can see which computed solutions have experimentally observed levels of robustness, or better, and whether networks evolve with more robustness to particular factors. Computation allows us to understand the experimentally well-characterized factors, such as Bcd, and extend results to the other, less well-characterized maternal factors.

Analyzing several hundred high-scoring solutions of the three variants of our model, we found very diverse ways for the networks to solve the pattern fit, and these had quite different levels of robustness to variability in maternal factors. We did not, however, find a clear correlation between the types of new connections in the evolved networks and robustness. Many of the recruited genes are, however, spatially patterned like known *Drosophila* genes. These patterns can either be like those for members of our 4-gene model, or our evolutionary searches also recruit genes with patterns like real segmentation genes that aren't one of the original model genes.

3.2 Methods and Approaches

3.2.1 The Segmentation Gene Network and its Modeling

Four gap genes, *Kr, gt, kni* and *hb*, are the core elements in our segmentation model. In *Drosophila*, these are transcriptionally activated by the maternal Bcd protein gradient in a concentration dependent manner, a classic example of a morphogen as characterized by Wolpert [22]. Three other gradients, Hb_{mat}, Cad, and Tll, help determine the positions of the gap genes. The combination of this upstream specification and gap-gap cross-regulation results in sharp and precise gap patterns.

We model these genes (and proteins) and their interactions using the gene circuit framework [23, 24], to produce A-P concentration patterns (fitting data such as in Fig. 3.1A–B). The model is computed for a one-dimensional row of nuclei, between 30% and 94% A-P position (where 0% is the anterior pole) during nuclear cleavage cycles 13 and 14A. The gap gene proteins (Kr, Gt, Kni and Hb) are variables in the model, with the rates of change of their concentrations dv_i^a/dt (for each gene product a in each nucleus i) defining a system of *number of proteins* times *number of nuclei* ODEs (Ordinary Differential Equations) given byx

$$dv_i^a/dt = R_a g(u^a) + D^a \left[(v_{i-1}^a - v_i^a) + (v_{i+1}^a - v_i^a) \right] - \lambda_a v_i^a. \tag{3.1}$$

The main terms on the right hand side of (3.1) represent protein synthesis (R_a), diffusion (D^a) and decay $(\lambda_a).g(u^a)$ is a sigmoid regulation-expression function. For values u^a below -1.5 and above $1.5 g(u^a)$ rapidly approaches zero. u^a is given by $u^a = \sum_b T^{ab} v_i^b + m^a v_i^{Bcd} + h^a$. Parameters T^{ab} constitute a genetic interconnectivity matrix, representing activation of gene a by the product of gene b (with concentration v_i^b) if positive, repression if negative, and no interaction if close to zero. v_i^{Bcd} represents the concentration of Bcd in nucleus i, which is constant in time. m^a describes the regulatory input of Bcd to each gene. Bcd is a general activator for all 4 gap genes considered. h^a represents regulatory input from ubiquitous factors. Our extended model includes Hb_{mat}, Cad, and Tll in a similar manner to Bcd, as time-independent parameters.

The full extended model involves heavy computation, which can greatly delay evolutionary searches and the generation of large samples of networks. For this reason, we have focused on reduced networks to study robustness and evolvability, either the basic 4-gene model (under Bcd control only), or a 2-gene subnetwork of Hb and Kr, with Bcd, Cad and Tll maternal control. This subnetwork can serve as a core for the whole gap network, and allow us to compare 4-gene networks evolved from this core to experimental results and results from the original (non-evolved) 4-gene models.

After networks are created through the Genetic Algorithm selection, we can analyze the robustness of each solution to maternal variability (in Bcd, Hb_{mat}, Cad and Tll). For this, we take any particular parameter set (network) and rerun the solution many times with different Bcd, Hb_{mat}, Cad and Tll gradients. The gradient variability is biological: the different gradients are data obtained from individual embryos. We used 89 individual Bcd gradients, 38 Cad gradients, 35 Hb_{mat}, and 27 Tll ones.

3.2.2 Experimental Data for Fitting

The data we used to fit our models is the result of a large-scale project we are engaged in, aimed at collecting, processing and analyzing the expression of the *Drosophila* segmentation genes [17, 20, 25]. Most of this dataset is now available publicly [26]. In this paper, we use expression data from early and mid cleavage cycle 14 (prior to full cellularization). This period of development is the stage during which segmentation patterns become mature, and also progressively more complicated, due to activation of more and more genes that interact with the 4 gap genes in our model (this is called the mid-blastula transition). It is unknown precisely how many newly activated genes begin to interact with our 4 gap genes, nor do we know the precise activation times of these new genes. Therefore, new genes which are recruited in our simulations may shed light on the spatial patterns and regulatory features of real genes activated during the mid-blastula transition.

We have found that our models have faster and better fits to early patterns than to later, more mature ones. We believe this reflects that early gap patterns are chiefly under the control of the genes and maternal factors explicitly included in the model, while later, more complicated patterns begin to reflect interactions with other, newly activated genes recruited to the basic network.

We have also found, with quantitative data analysis [20, 25] that segmentation patterns become more precise and robust from early to mid cycle 14. Hence, it is instructive to fit our models separately to early patterns and to mid cycle 14 patterns, to see if the robustness of the solutions reflects the trend in the data.

3.2.3 GA to Simulate Evolution of Gene Networks

The set of ODEs (3.1) was solved numerically by Euler's method [27]. We minimized the following cost function E by adjusting parameters T^{ab} in Eq. (3.1):

$$E = \sum \left(v_i^a(t)_{model} - v_i^a(t)_{data} \right)^2. \tag{3.2}$$

For the remaining parameters, m^a and h^a were found in preliminary runs and then used as fixed parameters; R_a, D^a, λ_a were determined similarly for the core 4- and 5-gene networks, but were found by GA in the reduced 2-gene networks; for the extended 4-gene model, R_a was found by GA and the rest were fixed.

Our approach followed the general scheme of population dynamics, by using repeated cycles of mutation, selection and reproduction. This is common to both GA [28] and general simulations of biological evolution.

Following the standard GA approach, the program generates a population of floating-point chromosomes, one chromosome for each gene a. The value of a given floating-point array a (chromosome a) at index b corresponds to a T^{ab} value (see Eq. (3.1)). The task of the evolutionary search is to optimize the T^{ab} to fit to the experimental patterns (e.g. Fig. 3.1A, B).

The initial chromosome values are generated at random. The program then calculates the v_i by Eq. (3.1) and scores each chromosome set (T matrix) by the cost function E (Eq. (3.2)). An average score is then calculated for all the chromosome sets run. Chromosome sets with worse-than-average scores are replaced by randomly-chosen chromosome sets with better-than-average scores. A proportion (from 5–25%, depending on computation) of the chromosomes are then selected to reproduce, undergoing the standard operations of mutation and crossover (defined below; 1/10 of these operations are crossover), giving changes to one or more of the T^{ab} values. The complete cycle of ODE solution, scoring, replacement of below-average chromosome sets, and mutation and crossover is repeated until the E score converges below a set threshold, typically 4,000–5,000 generations.

In GA, mutation is a genetic operator used to maintain genetic diversity from one generation of a population of chromosomes to the next, analogous to biological mutation. Point mutation in GA involves a probability that a T^{ab} value on a chromosome will be changed from its original state (compared to changing a nucleotide in biological point mutation).

GA crossover is a genetic operator used to vary chromosomes from one generation to the next, by swapping strings of values between chromosomes, analogous to crossover in biological reproduction. In one-point crossover, a point on a parent chromosome is selected. All data beyond that point is swapped between two parent chromosomes. Two-point crossover calls for two points to be selected on the parent strings. Everything between the two points is then swapped between the parent strings. Multi-point crossover is defined by analogy with the two-point case.

The model is implemented in Delphi (Windows) and GNU Pascal (Linux) and available from the authors upon request. Each run of the algorithm requires about 3 h CPU time on a Dell workstation (Intel Xeon CPU 2.80 GHz).

3.2.3.1 Introduction and Withdrawal of New Genes

In biology, one can imagine at least two scenarios for how new genes could become available for recruitment into a network [3]. First, a new gene could appear in the genome by the process of gene duplication. Second, a given gene from another network could become available for recruitment. In our model we do not distinguish these two cases, but introduce a Gene Introduction operator which adds a new gene to the network (at a rate of 5–10% per generation, depending on computation). Specifically, this adds a new row and column to the T^{ab} matrix, which can be then be operated on by mutation and crossover. To study the importance of this one-way process forcing networks to recruit new genes, we introduced a Gene Withdrawal operator which removes a row and column from the T^{ab} matrix (at a rate of 2–10% per generation, depending on computation). Gene Withdrawal does not operate if the network is minimal (N = 4 genes).

3.3 Results and Discussion

Recruitment of new genes into the preexisting network is typical for our model. We have found that even with point mutation alone, the network will recruit small numbers (from one to four) of new genes by the time it converges below the threshold E score. If mutation is reinforced with crossover, the number of recruits increases slightly (but statistically significantly). Increasing the rate of crossover leads to continual recruitment up to convergence, with some dozens of genes in the final networks.

We also find that cooption of a new gene can facilitate the evolutionary search, i.e. increase evolvability, in the sense of giving faster and better fits to the data. This was most apparent for a subnetwork 2-gene model.

Recruited genes can be uniform or spatially patterned. These patterns can either recapitulate patterns of existing network genes, or introduce patterns novel to the model network, but like patterns seen in the full biological network.

When we test the evolved networks against variability in maternal control factors, we find a significant minority display high robustness. These network solutions are varied, and show that robustness can be local to particular gene pattern boundaries. We did not find, however, that gene recruitment was associated with the robustness of the networks.

We expand on these findings in detail in the subsections below.

3.3.1 Point Mutations are Enough to Recruit New Genes

In our first series of runs, we studied recruitment events in detail and checked if crossover can raise the efficacy of recruitment. Several sets of runs under different conditions (point mutations only; point mutations + multi-point crossover; etc.) were performed, with each set including $\sim$200 runs. For runs with both mutation and crossover, the mutation rate was adjusted so that total change per generation stayed comparable to runs with mutation only (e.g. if crossover, with rate 2% per generation was added to mutation, which had run at a 20% rate, the mutation rate would be adjusted to 18%). Runs with E (Eq. (3.2)) scores below a threshold level were picked as winners. The threshold was established by visual inspection of the quality of fits to the expression patterns, and resulted in about half of the runs being winners. These winners were analyzed further to see what qualities they had.

We found new genes recruited to the network formed two distinct types of pattern. In the first type, recruits formed flat or nearly flat patterns (uniform distributions); they were incorporated into the network as ubiquitous activators or inhibitors. In the second type, recruits produced monotonic gradients, or even more sophisticated spatial patterns, influencing the patterns of the obligatory, minimal 4 genes of the network (*gt, hb, Kr & kni*). Figure 3.2 shows a representative example of such a network. The obligatory 4 genes all fit well to the experimental data in Fig. 3.1. All good-score networks studied (112 point mutation only + 94 also with crossover) included at least one new recruit acting upstream of the obligatory genes (i.e., the obligatory genes were regulatory targets of the recruits). Nearly all networks studied included at least one (but usually more) upstream recruit that formed an AP gradient, such as Bcd. But most networks also included one or more upstream recruits that formed an opposing, postero-anterior gradient (Fig. 3.2B, patterns A, B). This is especially interesting because the minimal 4-gene ensemble we fitted in these

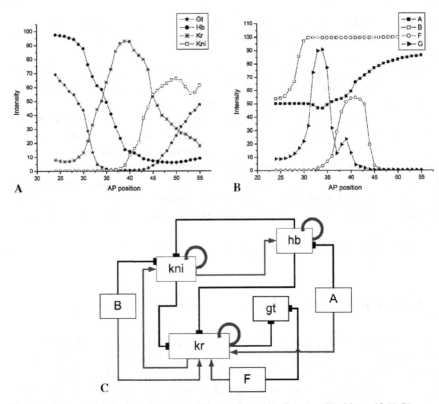

Fig. 3.2 An example of a redundant gene network selected by Genetic Algorithms: 12 (A-L) genes have been recruited to the original 4 model genes. **A**. Representative patterns for the 4 obligatory genes. **B**. Patterns for some (A,B,F,G) of the genes recruited upstream of the 4 obligatory genes (in **A**). **C**. Overview of the gene network in A-B, showing some of the interactions of the recruited genes. The genes are represented as boxes. Repressive interactions are represented by T-bar connectors. Looped arrows mean self-regulation. Cf. with Fig. 3.1C. In this simulation, 4,000 networks were generated for each generation; the point mutation rate was 18% per generation, plus 2% crossover rate; 20% of individuals with the best scores were marked for reproduction; and the rate for new gene recruitment was 5% per generation

runs did not posses such postero-anterior gradients. Hence, recruitment produced a kind of compensation for this lack of essential external output: in real fly embryos postero-anterior gradients of proteins such as *caudal* and *nanos* are essential for early segmentation.

In some cases, upstream recruits formed not simple monotonic gradients, but more sophisticated patterns with sub-domains (Fig. 3.2B, patterns F, G). These patterns are reminiscent of the mature patterns of *Drosophila* gap genes and demonstrate how recruitment could supply new gap genes for an evolving segmentation network (as in the transition from short to long germ band mechanisms).

We found that the point mutation operator is enough to recruit at least 1 new gene to the network; i.e., not one of the evolved high-score networks had just the

Table 3.1 Outgrowth of networks by evolutionary search, with point mutations only and point mutations plus crossover[a]

	N runs	Mean score	Recruits, in toto	Recruits upstream of 4 obligatory genes	Upstream recruits expressed ~ubiquitously	Upstream recruits forming patterns
Point mutations	112	188.00 ± 69.61	2.99 ± 0.93	1.98 ± 0.75	0.33 ± 0.49	1.65 ± 0.57
Point mutations + crossover	94	171.18 ± 63.70	2.95 ± 0.60	2.22 ± 0.92	0.20 ± 0.43	2.02 ± 0.98

[a] Results are mean $\pm$ standard deviation.

obligatory 4 genes (Table 3.1). The mean number of recruits was around 3, while the average number of recruits upstream of (controlling) the obligatory 4 genes was about two. As shown in Table 3.1, it appears that crossover selects networks with a slightly better score, gives a higher average number of upstream recruits and a lower number of these recruits have uniform distributions (all differences statistically significant). These upstream recruits more often form gradients or more complicated patterns (highly statistically significant). So, while we find crossover facilitates recruitment, point mutation is certainly sufficient for this, in contrast to a mainstream view in evolutionary biology, that complex recombination of genetic material is required for recruitment [15].

3.3.2 Addition and Subtraction of New Genes

A simple explanation for why the number of recruits rises during evolution of a network is that addition of new potential recruits to the system creates an implicit pressure facilitating that recruitment. More specifically, a new recruit becomes incorporated into a network as its T matrix values begin to deviate from zero. Holding the T values at their initial zero state would involve a cost to the existing network, hence the presence of a new recruit causes pressure to evolve its T values and become incorporated into the network. With this tendency towards incorporation, the mean number of recruits should depend on the introduction rate. To test this, we introduced the Gene Withdrawal operator into our computations, as a way to control the net introduction rate. In conditions where addition is higher than subtraction, mutation and crossover operators still ensure recruitment. However, if the subtraction rate is equal to or greater than the addition rate, then recruitment is reduced compared to the Table 3.1 results; due to the random nature of the mutation, some networks can still gain recruits under these conditions. Hence, by using the Gene Introduction and Withdrawal operators to control net addition, we can show that addition of a new recruit creates implicit pressure for incorporation, facilitating recruitment.

3.3.3 Network Redundancy and Evolvability

The minimal, obligatory 4-gene network fits experimental pattern with good quality. Introduction of new recruits to this network does not generally raise the quality of the fits. In this sense, the new interactions with the recruited gene can be considered redundant. However, this is not what is frequently called structural redundancy, in which repeated elements (genes) can substitute for lost elements, providing a type of robustness in networks. We find that withdrawal of a recruited gene from a good-scoring network (solution) makes its fit worse. Therefore, recruitment tends to alter the interactions of the original network; it is not advantageous to remove a gene once it has acquired functionality in the network. In this and the next section, we evaluate how recruitment affects a network's properties of evolvability and robustness. In terms of evolvability, we investigate whether recruitment of additional genes aids a network's capacity to evolve further. In particular, we can see if recruitment leads to faster (less generations) or better fits of the network to the data.

3.3.3.1 Evolvability of the 4-Gene Models

To begin to investigate what potential role these added interactions provide, we tested whether they might help a network recruit more new genes. In these runs, we constrained the model to keep 5 obligatory genes: *gt, Kr, kni, hb,* and one new recruit. We first fit the model to the usual gap gene data of Fig. 3.1, during which process new genes were recruited. Once a good fit was attained, the fit criteria were changed to require the model to fit an expression pattern for 5 genes, by including the pattern for the primary pair-rule gene *even-skipped*. The 5th pattern could be fit by any of the newly recruited genes. Our expectation was that higher redundancy of networks could facilitate the evolutionary search for the gene to fit this pattern. We performed runs with point mutations only, and with point mutations and crossover (Table 3.2). The parameters for these runs were exactly as for Section 3.3 (see caption for Fig. 3.2). To our surprise, we did not find any difference in efficacy between these runs and the previous runs of Table 3.1, as measured by the average number of

Table 3.2 Efficacy of evolutionary search with redundant networks

	N runs	Mean score (averaged)	Recruits, in toto (averaged)	Recruits upstream of *even-skipped* (*eve*) (averaged)	Recruits upstream *eve*, expressed $\sim$ ubiquitously (averaged)	Recruits upstream *eve*, forming gradients (averaged)
Point mutations	98	235.23 ± 65.68	3.06 ± 0.96	0.50 ± 0.56	0.40 ± 0.49	0.10 ± 0.39
Point mutations + crossover	73	224.35 ± 76.45	2.92 ± 0.57	0.67 ± 0.67	0.23 ± 0.43	0.44 ± 0.52

recruits. We did find, however, that the average number and character of the recruits upstream of the *even-skipped* gene were significantly different: upstream recruits are far fewer in Table 3.2, and recruits form far fewer patterns. As in Table 3.1, crossover still tends to favor patterned recruits, compared with mutation alone.

Hence, gene networks with four obligatory genes do not show evident correlation of evolvability and the extent of redundancy.

3.3.3.2 Evolvability of the 2-Gene Model

To further investigate how the number of genes in the network might affect evolvability, we did a similar study, but starting from a 2-gene network, with *hb* and *Kr* only. As a control, we ran 500 simulations with this simple network, and computed an average score for how well Hb and Kr fit the biological patterns (both the mid cycle 14A patterns and the early cycle 14A ones, see Methods). Then, we did a series of 579 test simulations, in which 2 new genes were added at the onset of the evolutionary computation. Further addition/withdrawal operators were not used during the course of the computations. Again, the test networks were only required to fit the Hb and Kr patterns, but we wanted to see whether the two introduced genes would be incorporated into the network in such a way as to affect these pattern fits. Using the average score of the test computations, we found that the added genes significantly improved the fitting of the Hb and Kr pattern, both for early and mid cycle 14A, with the mid 14A difference being more dramatic. On average, the tests had scores of 128.005 ± 71.703, and the controls had scores of 165.073 ± 32.809.

For the 2-gene model, we find that redundancy serves as a mechanism to find not only better solutions, but also usually to find these solutions faster, in less generations; recruitment significantly raises the efficacy of the evolutionary search. Hence for a small fragment of the network, which could be treated as an "ancestral" primitive primary gene ensemble, redundancy via co-option could substantially facilitate evolutionary searches. To improve pattern, evolution causes such a primitive network to enlarge.

We wondered what kind of spatial patterns are generally made by the recruits, particularly whether they tended to mimic the missing two members of the real 4-gene gap network, *gt* and *kni*. We found that the patterns of the co-opted genes are usually reminiscent of anterior Hb or Gt domains; that is, simple S-shaped patterns, but often with reversed orientation (Fig. 3.3A). We also found cases where recruits had Kr-like patterns (Fig. 3.3B).

We found several cases when one of the co-opted genes formed pattern similar to anterior Gt (Fig. 3.3C). We also saw more complicated patterns, reminiscent of real two-domain gap patterns (Fig. 3.3D). It could be that the evolutionary search is tending to fill in the missing gap patterns to generate the structure of the real, complete gap network. However, these two-domain *gt*-like patterns were relatively rare, and we did not find any *kni*-like patterns.

In summary, we have found that for the case of small fragments of gene ensembles, the co-option of new genes really does facilitate the evolutionary search.

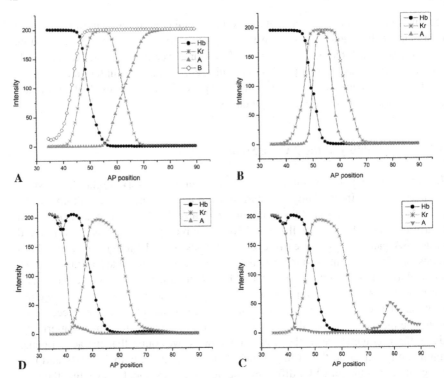

Fig. 3.3 Representative examples of 2-gene models with two recruits. **A**. Recruit patterns are similar to Gt and Hb (reverse orientation; Cf. Fig. 3.1A). **B**. Recruit pattern is similar to Kr. **C**. Recruit pattern looks like anterior Gt (Cf. Fig. 3.1A). **D**. Recruit pattern is reminiscent of real Gt, with anterior and posterior domains

We can speculate that similar mechanisms acted during early evolution of primitive ancestral gene ensembles, while for evolutionarily more mature and larger gene networks this tendency has become less pronounced.

3.3.4 Redundancy and Robustness of Gene Networks

Above, we have shown that our models of evolution, both the 2-gene and 4-gene ones, do account for recruitment of new genes and the selection of redundant networks. Here, we investigate the influence of redundancy on network robustness. A case of robustness that has received much attention in *Drosophila* segmentation is the robustness to variability in the shape of the Bcd morphogen gradient [16–21]. We can use our GA model to study this kind of robustness. The networks in the previous sections were selected on an averaged Bcd gradient (average profile of the real Bcd gradients in the FlyEx database [26]). If we take one of these networks, and now run it on the individual, and varying, Bcd gradients in our database (Fig. 3.4B),

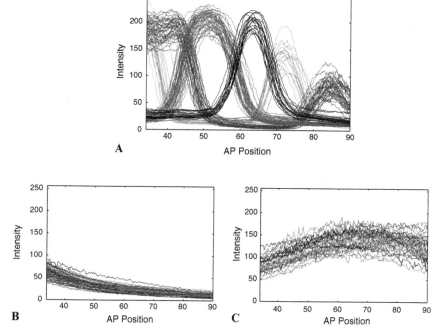

Fig. 3.4 The real, biological variability in the gap gene patterns and the maternal inputs (See [20, 25]; Fly Ex DB). **A.** The between-embryo variability of the gap gene patterns (gt, hb, Kr, kni; Cf. with Fig. 3.1B) for mid cleavage cycle 14A. Much of this variability is probably due to between-embryo variability in the maternal gradients, such as Bcd **B**, and Cad **C. B.** The between-embryo variability of the maternal morphogenetic gradient Bcd, 89 embryos, 13th cleavage cycle. **C.** The between embryo variability of the maternal factor Cad, 38 embryos, 13th cleavage cycle

we get a picture of how robust the network's gap gene patterning is, and how this compares with the observed biological robustness (Fig. 3.4A). We can compare network robustness for the starting 2-gene and 4-gene models, as well as for the evolved redundant models.

3.3.4.1 Robustness of 2-Gene Networks

We start our investigation of robustness on the 2-gene model described in Section 3.3.2, with the same test and control simulations described there. With this simplified model, the effect of variability in Bcd input on robustness is especially evident.

First, we found that the 2-gene solutions can be very robust to Bcd variability. Some solutions are substantially more robust than the robustness level observed for real *Drosophila* segmentation genes (Fig. 3.5A). However, good or even very good solutions (according to fitting score) can show no robustness to Bcd variability.

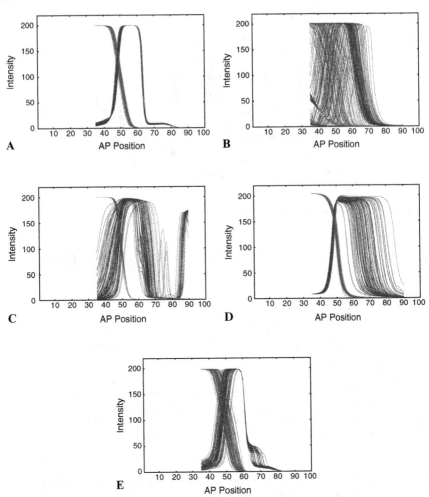

Fig. 3.5 2-gene model: robustness to Bcd variability. Bcd – green; Hb – red; Kr – blue. **A**. Highly robust. **B**. Not robust. **C**. Hb is robust, but Kr is not robust. **D**. All borders are robust, except for posterior Kr. **E**. The posterior Kr border is the most robust

These non-robust solutions can give Hb and Kr variability as high as that for Bcd (Fig. 3.5B), in contradiction to the several-fold drop in variability seen in the data (Fig. 3.4; [20]). The best-fit solutions span from highly robust, capable of filtering out Bcd variability nearly completely, to solutions unable to filter variability at all.

It is biologically established that the position of each domain border of each gap gene pattern is under the control of different combinations of regulatory inputs from the other members of the segmentation ensemble. In the case of the 2-gene model, we have one border for Hb and two borders (anterior & posterior) for Kr. Even for good-scoring solutions, there are cases when Hb is robust but Kr is less robust, or even non-robust (Fig. 3.5C). In many cases, the anterior Kr border is more robust then the posterior one (Fig. 3.5D), but we also saw some cases with the

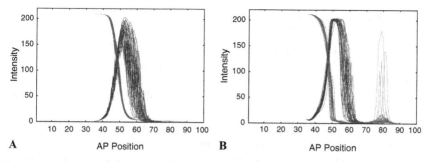

Fig. 3.6 2-gene model: Cad noise filtration. Hb – red; Kr – blue. **A**. Hb and anterior Kr borders are very precise, but posterior Kr is not. **B**. Cad variability can cause, or amplify, the second, posterior domain of Hb

opposite (Fig. 3.5E). Our results show that robustness can evolve relatively independently at each border. Hence, the positional error for each border can be relatively independent.

Detailed analysis of the dynamics underlying this robustness to Bcd variability in the networks, both for the *hb-Kr* pair and the rest of gap ensemble, will be presented in another paper [29].

Because the 2-gene model is under the control of not one, but three external inputs (Bcd, Cad & Tll), we could also study its robustness to the variability of these other factors. Cad displays an even higher variability than Bcd (Fig. 3.4C; Cf. [20]). We have found a set of solutions that can filter this Cad variability to a degree comparable to Bcd filtering. This small set of solutions can filter Cad variability substantially better than real embryos can (data not shown). A typical result from this set has very precise Hb and anterior Kr borders, but un-precise posterior Kr (Fig. 3.6A). This situation is not unexpected, because the more posterior the domain position, the higher Cad intensity level and the higher the Cad positional (horizontal) variability (Fig. 3.4C; Cf. [20]).

We found cases where Cad variability induced not only quantitative, but qualitative changes in the Hb and Kr profiles. For instance, Cad variability can cause or at least highly amplify a second, posterior domain of Hb (Fig. 3.6B). Interestingly, this new posterior domain really does form in embryos during cycle 14, but later than the early stage patterns we used to fit the model.

In summary, the simple 2-gene model, an elementary module of the gap network, shows all possible combinations of robust versus non-robust behavior, including a significant subset of solutions which are very robust to upstream variability. We also found that robustness of the three domain borders can be independently controlled.

We did not find any significant correlation between the fitting scores of solutions and their robustness, either for the control 2-gene networks or redundant solutions with one or two recruits. Robust solutions constitute about 10% or less of the total for 2-gene model, and non-robust solutions constitute a similar proportion.

3.3.4.2 Robustness of the 4-Gene Networks

To see if the external noise filtration we found in the elementary 2-gene model works
in more complicated gene networks, we tested the robustness of 4-gene network solutions to biological variability in a similar way. In these cases, the model has ten
domain borders, which makes systematic analysis more difficult. However, we again
found that there is (1) no evident correlation between fitting scores and robustness,
(2) different borders of different domains can display quite different levels of positional precision, (3), both robust and non-robust solutions are relatively rare, and (4)
there were no evident differences in robustness between the control 4-gene networks
and redundant networks evolved from these.

By using our extended 4-gene model, which includes control by all (or at least
most) of the known maternal factors, Bcd, Hb_{mat}, Cad and Tll, we can investigate
the networks' abilities to filter a more complete set of external variabilities, or to
test their robustness to combinations of these factors.

We performed a series of runs to fit this extended version of the 4-gene model,
for both control conditions and test runs with recruitment of new members to the
core 4-gene ensemble. This extended 4-gene model is several times more computationally intensive, so for this case we have obtained several dozen networks with an
appropriate level of fitting to the experimental data.

We performed a detailed analysis of the 18 best-fit solutions (control and test
runs) for robustness to variability in all four external factors (Bcd, Hb_{mat}, Cad and
Tll), one by one, and in pairs of the factors. For Bcd variability alone, we found that
the behavior of the extended gene network is similar to that observed for 2-gene and
minimal 4-gene models. In most cases (Fig. 3.7A; control runs shown in Fig. 3.7, but
test runs give the same qualitative results), Hb and Gt tend to be highly robust, while
Kr and kni are less robust (but they are comparable to the biologically observed
robustness). In other cases, all the gap domains (with the exception of posterior Hb)
can show similar, and relatively high, levels of positional variability (not shown).
Finally, we see some cases of autonomy in robustness to Bcd variability: the Kr
domain can show very high precision, while the other genes do not (not shown).

With variable Cad input, we have not found robust solutions (Fig. 3.7B). The only
precise domain in the case of Fig. 3.7B is the most anterior Gt one. Robustness of
this domain can be expected because it is chiefly under control of Bcd and relatively
independent of Cad regulation. With variable Tll (also a posterior gradient), the
picture is similar (Fig. 3.7C); the extended 4-gene model is largely not robust to
this. Only the most anterior borders of Gt, Hb and Kr are robust, and again these are
largely under Bcd control and are relatively independent of Tll.

Looking at pairs of external factors, the most interesting case was the pair of Bcd
and Hb_{mat}. It is biologically established that this pair of anteroposterior gradients
cooperatively control patterning of gap genes in the anterior half of the *Drosophila*
embryo [30]. We found that the extended 4-gene model is capable of decreasing
not only Bcd or Hb_{mat} variability separately, but can filter both these variabilities
together (Fig. 3.7D).

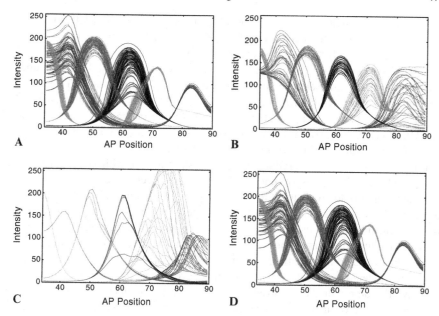

Fig. 3.7 Extended 4-gene model (Bcd, Cad, Tll maternal control). Colors as in Fig. 3.4. **A**. Robustness to Bcd variability. **B**. Non-robustness to Cad variability. **C**. Non-robustness to Tll variability. **D**. Robustness to double Hb + Bcd noise

3.4 Conclusions

In this work, we have presented the results of a computational simulation of evolution of the segmentation gene network, controlling spatial patterning in early fly embryonic development. We used Genetic Algorithms (GA) methods to evolve the parameters of a differential equation model for the segmentation proteins, tested against fitness for matching the biological data for the protein patterns.

We simulated recruitment (co-option) of new genes to existing networks, by introducing gene-addition and gene-removal operators, on top of standard GA techniques. We found that recruitment occurred in all our simulations, even for those in which only point mutations were operating. Crossover aided recruitment, but was not necessary. The recruited genes were either ubiquitous or formed spatial patterns, many of which were similar to real, biological gene patterns, including patterns for genes not in our core starting networks.

With our generated networks, we tested whether recruitment affected evolvability or robustness to variability in external factors. We found the evolvability was especially aided in a 2-gene subnetwork, possibly representing the process by which the ancestral short-germ band segmentation mechanism evolved into the long-germ band mechanism of flies. For robustness, we tested the networks for their ability to filter variability in upstream, regulatory maternal factors. This apparent filtration has been the subject of a great deal of attention in developmental biology in recent years, and we find that a significant subset of our evolved networks (with and without recruitment) have the capacity to filter this maternal variability, i.e. generate domain boundaries with greater precision than the maternal regulatory gradients. Robust networks display a variety of behaviors, demonstrating that domain boundaries can be regulated independently with respect to spatial precision. The model was very successful for filtering variability in anterior gradients, such as Bcd and Hb_{mat} (individually, or as a pair), but less so with posterior gradients, such as Cad and Tll. There is no apparent correlation between the quality of a network's fit to the data and its capacity for robustness to maternal variability.

It has been suggested that redundancy, either structural, in which new genes can substitute for existing genes, or functional, in which new genes create compensatory pathways, can provide robustness to networks [13]. We do not find that recruitment aids robustness to maternal variability, but our recruited genes are not strictly structurally or functionally redundant. Rather, removal of recruited genes results in fitting scores lower than the starting network: genes can not be freely disposed once they have been integrated into the functionality of the network.

Our work demonstrates that relatively simple evolutionary operators can account for network outgrowth. The evolved networks display a number of features of the biological system of interest, such as recruitment of genes from ancestral modules, and robustness to regulatory variability, shedding light on the evolutionary and functional dynamics of this developmental network.

Acknowledgements This work was supported by the Joint NSF/NIGMS BioMath Program grant R01-GM072022. The authors thank an anonymous reviewer for helpful comments on an earlier version of the manuscript.

References

1. Wilkins, A.S. (2002). *The Evolution of Developmental Pathways*, Sinauer, Sunderland, MA.
2. Duboule, D. and Wilkins, A. (1998). The evolution of bricolage, *Trends Genet.* 14:54–59.
3. True, J.R. and Carroll, S.B. (2002). Gene co-option in physiological and morphological evolution, *Annu. Rev. Cell Dev. Biol.* 18:53–80.
4. Carroll, S.B., Grenier, J.K., and Weatherbee, S.D. (2001). *From DNA of Diversity: Molecular Genetics and the Evolution of Animal Design*, Blackwell, Malden, MA.
5. Carroll, S.B. (2005). Evolution at two levels: on genes and form, *PLoS Biol.* 3(7):e245.
6. Davidson, E.H. (2001). *Genomic Regulatory Systems: Development and Evolution*, Academic, San Diego, CA.
7. Goodwin, B.C. and Kauffman, S.A. (1990). Spatial harmonics and pattern specification in early Drosophila development. Part I. Bifurcation sequences and gene expression, *J. Theor. Biol.* 144:303–19.
8. Reinitz, J. and Sharp, D.H. (1995). Mechanism of formation of eve stripes, *Mech. Develop.* 49:133–158.
9. Jaeger, J., et al. (2004). Dynamic control of positional information in the early Drosophila blastoderm, *Nature* 430:368–371.
10. Hunding, A., Kauffman, S.A., and Goodwin, B.C. (1990) Drosophila. segmentation: supercomputer simulation of prepattern hierarchy, *J. Theor. Biol.* 145:369–384.

11. Burstein, Z. (1995). A network model of the developmental gene hierarchy, *J. Theor. Biol.* 174:1–11.
12. Sánchez, L. and Thieffry, D. (2001). A logical analysis of the Drosophila gapgene system, *J. Theor. Biol.* 211:115–141.
13. Jaeger, J., et al. (2004). Dynamical analysis of regulatory interactions in the gap gene system of Drosophila melanogaster, *Genetics* 167:1721–1737.
14. Wagner, A. (2005). Distributed robustness versus redundancy as causes of mutational robustness, *BioEssays* 27:176–188.
15. Carroll, R.L. (2002). Evolution of the capacity to evolve, *J. Evol. Biol.* 15:911–921.
16. Houchmandzadeh, B., Weischaus, E., and Leibler, S. (2002). Establishment of developmental precision and proportions in the early Drosophila embryo, *Nature* 415:798–802.
17. Holloway, D.M., Reinitz, J., Spirov, A.V., and Vanario-Alonso, C.E. (2002). Sharp borders from fuzzy gradients, *Trends Genet.* 18:385–387.
18. Spirov, A.V. and Holloway, D.M. (2003). Making the body plan: precision in the genetic hierarchy of Drosophila embryo segmentation, *In Silico Biol.* 3:89–100.
19. Houchmandzadeh, B., Wieschaus, E., and Leibler, S. (2005). Precise domain specification in the developing Drosophila embryo, *Phys. Rev. E* 72:061920.
20. Holloway, D.M., Harrison, L.G., Kosman, D., Vanario-Alonso, C.E., and Spirov, A.V. (2006). Analysis of pattern precision shows that Drosophila segmentation develops substantial independence from gradients of maternal gene products, *Dev. Dynamics* 235:2949–2960.
21. Gregor, T., Tank, D.W., Wieschaus, E.F., and Bialek, W. (2007). Probing the limits to positional information, *Cell* 130:153–164.
22. Wolpert, L. (1969). Positional information and the spatial pattern of cellular differentiation, *J. Theor. Biol.* 25:1–47.
23. Mjolsness, E., Sharp, D.H., and Reinitz, J. (1991). A connectionist model of development, *J. Theor. Biol.* 152:429–453.
24. Reinitz, J. and Sharp, D.H. (1995). Mechanism of eve stripe formation, *Mech. Develop.* 49:133–158.
25. Surkova, S., Kosman, D., Kozlov, K., Manu, Myasnikova, E., Samsonova, A.A., Spirov, A., Vanario-Alonso, C.E., Samsonova, M., and Reinitz, J. (2008). Characterization of the Drosophila segment determination morphome. *Develop. Biol.* 313:844–862.
26. Poustelnikova, E., Pisarev, A., Blagov, M., Samsonova, M., and Reinitz, J. (2004). A database for management of gene expression data in situ. *Bioinformatics* 20:2212–2221.
27. Press, W.H., Flannery, B.P., Teukolsky, S.A., and Vetterling, W.T. (1988). *Numerical Recipes*, Cambridge University Press, Cambridge.
28. Schwefel, H.-P. (1981). *Numerical Optimization of Computer Models*, Wiley, Chichester.
29. Manu, Surkova, S., Spirov, A.V., Gursky, V., Janssens, H., Kim, A., Radulescu, O., Vanario-Alonso, C.E., Sharp, D.H., Samsonova, M., and Reinitz, J. (submitted). Canalization of gene expression in the Drosophila blastoderm by dynamical attractors. *Nature*.
30. Simpson-Brose, M., Treisman, J., and Desplan, C. (1994) Synergy between the hunchback and bicoid morphogens is required for anterior patterning in Drosophila, *Cell* 78:855–865.

Chapter 4
Comprehensive Genetic Database of Expressed Sequence Tags for Coccolithophorids

Mohammad Ranji and Ahmad R. Hadaegh

Abstract Coccolithophorids are unicellular, marine, golden-brown, single-celled algae (Haptophyta) commonly found in near-surface waters in patchy distributions. They belong to the Phytoplankton family that is known to be responsible for much of the earth reproduction. Phytoplankton, just like plants live based on the energy obtained by Photosynthesis which produces oxygen. Substantial amount of oxygen in the earth's atmosphere is produced by Phytoplankton through Photosynthesis. The single-celled Emiliana Huxleyi is the most commonly known specie of Coccolithophorids and is known for extracting bicarbonate ($HCO3$) from its environment and producing calcium carbonate to form Coccoliths. Coccolithophorids are one of the world's primary producers, contributing about 15% of the average oceanic phytoplankton biomass to the oceans. They produce elaborate, minute calcite platelets (Coccoliths), covering the cell to form a Coccosphere and supplying up to 60% of the bulk pelagic calcite deposited on the sea floors. In order to understand the genetics of Coccolithophorid and the complexities of their biochemical reactions, we decided to build a database to store a complete profile of these organisms' genomes. Although a variety of such databases currently exist, (http://www.geneservice.co.uk/home/) none have yet been developed to comprehensively address the sequencing efforts underway by the Coccolithophorid research community. This database is called CocooExpress and is available to public (http://bioinfo.csusm.edu) for both data queries and sequence contribution.

Keywords Marine alga · Coccolithophorids · Emiliana Huxleyi · Expressed Sequence Tags · Relational database · Entity Relationship Diagram

4.1 Introduction

Within the past decade many organisms have been examined and their DNA sequences were decoded. These DNA sequences are stored in databases and are used to solve many of the biological mysteries such as determining genes that code for

M. Ranji and A.R. Hadaegh (✉)
Department of Computer Science at California State University San Marcos
e-mail: ahadaegh@csusm.edu

S.-I. Ao et al. (eds.), *Advances in Computational Algorithms and Data Analysis,*
Lecture Notes in Electrical Engineering 14,

51

proteins. There are several tools developed every year to contribute to these efforts and make break through where not possible before due to limitation of technology.

CoccoExpress is a project dedicated to creating such one tool to serve as database and search engine of genetic data of Coccolithophorids, a family of species known for their unique ability to produce calcium carbonate platelets. Through extensive research, increasing number of sequences of this family is extracted. This project was initiated to create a tool to maintain this data and allow a meaningful representation of it by building relational structure among different sets of data.

CoccoExpress contains over 120,000 ESTs from Emiliania huxleyi. Emiliania Huxleyi (Ehux) is the most commonly known species of Coccolithophorids and was selected for whole genome sequencing by the Joint Genome Institute (JGI). The sequences are obtained through a number of experiments and collected by the researchers at California State University San Marcos. As explained in [1–3], each EST sequence is a single pass read from a randomly selected cDNA clones. ESTs can be assembled into either overlapping (contigs) or non-overlapping (singletons or singlets). Cluster analysis reveals the most likely arrangement of these fragments. The products of cluster analysis are known as consensus sequences, representing overlapping stretches of cDNA in which each string position is filled by the most likely nucleic acid for that position. By comparing each fragment to large, publicly available databases of known genes, the identification of the consensus sequences within the genome can be partially established. The consensus sequences are blasted against the NCBI non-redundant sequence database, and the top matches are stored for later analysis [1–4]. This is still an ongoing research at California State University San Marcos. The primary goal of this exercise is to reduce the EST dataset into a biological meaningful set of sequences which can be readily maintained, manipulated and queried in a database.

In order to be able to profile expression patterns under specific conditions and to determine the portion of the genome that is transcriptionally active, it is necessary to build an EST gene database to help analyze and organize the EST sequence data. While the processes used in developing an EST database vary, the primary goal of the exercise is to reduce the EST dataset into a biological meaningful set of sequences which can then be readily queried.

"CoccoExpress" was designed to assist in the gene annotation, and is the first known databases built to store and organize data for Coccolithophorids. "CoccoExpress" serves both as a repository for ongoing sequencing efforts and facilitates the public dissemination of sequence information as it relates to this organism. In brief, this research explained in this paper makes a number of contributions to the work underway on the genome discovery of Coccolithophorids. The major contributions of this paper are:

- Creating a database schema to ensure that data anomalies and redundancies are minimized and information are correctly linked within the well-structured tables.
- Design of a set of backend tools that is cost efficient, optimizes the throughput of the system, and facilitates the storage, and maintenance of data.
- Creating a simple user-friendly front-end user-interface that allows researchers to query, upload or download the data.

The rest of this paper is organized as follows. Section 4.2 gives a general overview of our database design. Section 4.3 describes the main components of the Entity Relationship Diagram (ERD) that is our database schema. In addition, it details the implementation of the front and back end tools that reflect the hardware and software components used in this project. Finally, future work and concluding remarks are presented in the last sections of this paper.

4.2 Relates Work

Thus, "Cocco Express" was designed to assist in the gene annotation of Coccolithophorids, and is the first known database built to store and organize data for Coccolithophorids. Using the search engines in CoccoExpress, researchers can profile expression patterns under specific conditions to determine the portion of the Coccolithophorids genome that is transcriptionally active. CoccoExpress serves both as a repository for ongoing sequencing efforts and facilitates the public dissemination of sequence information of Coccolithophorids.

Some other similar EST tools and databases developed for retrieval and gene analysis of different species include [5–9]. Compared to these existing similar databases, CoccoExpress has several new and unique features. First, its dynamic website is developed based on the most current tools and programs available today in the market. Thus CoccoExpress has relatively better manipulation speed and less overhead. Second, many of the exiting databases have very limited search capabilities. Built-in rigid queries limit the user's choices in retrieval of the information. Occasionally, researches may have to send special request to the administrators of the database and ask for specific data that could not be easily queried directly from the provided web interface. CoccoExpress relaxes this constraint by providing a very simple customized search. Researchers can set any valid criteria that can be acceptable by the database server on any field without having the knowledge of structural query programming.

In addition, many of the exiting databases have developed their own packages to secure their data, provide help, or administrate their information. In general, nowadays, developing these tools for databases are waste of time. CoccooExpress takes advantage of the open source tools such as phpAdmin, WikiPedia, and allstat. and integrate them with the database.

4.3 Planning

4.3.1 Technology Choice

Technology used for implementing CoccoExpress was chosen by taking several factors into consideration. Main objective of this engine was to provide simple yet efficient platform to be used by both scientists and bio-info engineers. For platform's operating system, Linux was picked for several reasons:

- Exceptional features and security provided for multi-user projects.
- Unix-based operating systems (OS) are the primary OS for bio-informatics tools and research.
- Minimal overall cost of ownership and scalability.

The overall structure of CoccoExpress can be divided to three main sections: Front-end, Back-end, and Database. Each section was carefully planned to accommodate the objectives of building simple and efficient system.

4.3.2 Front-End Interface

This search engine is subject to be used by scholars from all around the globe. To provide a complete, useable, and easily system accessible, a web interface was chosen for front-end functionality. With several hundred thousands of records in database, primary challenge was to create an interface that is easily expandable to include new records yet is easily maintainable. In an effort to meet these requirements, we chose to create a dynamic front-end display that could adjust dynamically to reflect the changes of database.

Some of the challenges during planning of front-end were:

- Search engine friendly so scholars from around the world could find CoccoExpress by searching the main search engines
- Dynamically maintained and redesigned using forms and database instead of static html editing
- Several methods of searching the database for simple and complex searches
- Scalability to dynamically adopt the new species added to the database
- User friendliness to provide robust navigation
- Help and documentation to match the front-end and be accessible throughout the entire site for given elements

4.3.3 Backend Engine

In the above, we addressed that front-end of CoccoExpress relays heavily on dynamic content. To provide a bridge between database and front-end we needed a backend engine that could provide such dynamic requirements. After sufficient research PHP programming language was adopted for constructing the backend. This decision was made after evaluating several criteria such as:

- Naturally optimized for Linux servers
- Database access layer support

- Web development friendliness
- Development speed
- Runtime performance

Meanwhile we decided to make use of Perl and C++ for some of the back-end routine functionality such as backup rolling and data parsing. Generally speaking PHP was picked for web development and Perl/C++ was reserved for Batch/Cron processes.

4.3.4 Database

Back-end database was indeed the core of this project. This database must have ability to provide relational structure to house several thousands of records and possibly grow to several millions over years. In addition, database engine to be used for CoccoExpress must have good indexing capabilities to provide several different indices to be employed to speed up heavy cpu/disk intensive queries over hundreds of thousands of records.

Most importantly, database should be capable of running stable on Linux platforms. Due to structure and relationship of our data, several relational database systems were researched to find the proper engine. MySQL, Oracle and were nominated to perform as our backend database.

Although both Oracle and MySQL both were well qualified for housing the CoccoExpress database, MySQL was picked primarily for lower cost of ownership, better compatibility with PHP, and ease of scalability.

4.4 Implementation

4.4.1 Database Schema

This section describes the architecture of the database schema of the CoccoExpress. The schema is shown in Fig. 4.1. The main entities of this schema are Library, Clone, EST, Consensus, Submitter, and the Blast_Results of Coccolithophorids.

4.4.2 Front-End

4.4.2.1 Front-End and user Interface

Several different technologies were employed to construct the front-end. Each technology was used to better enhance the usability of CoccoExpress and achieve great user satisfaction.

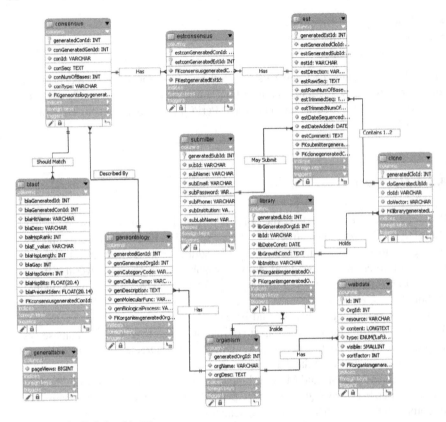

Fig. 4.1 Entity Relationship Diagram

Front-end is primarily built on Hyper Text Markup Language, commonly known as *HTML*. Since *HTML* is the common standard used for constructing web pages, there seemed to be no better solution for presentation layer. However, *HTML* is static and doesn't reflect dynamic changes or factors such as time, user, query, etc. In order to overcome this limitation, we used a server side scripting language to generate and stream *HTML* dynamically to user's browser. For better client experience we employed different client side techniques such as *Cascading Style Sheets (CSS)*, *JavaScript* and *Asynchronous JavaScript and XML* also known as *AJAX*. Once *HTML* content is sent over to user's browser, user will be required to navigate to other pages from that page. This is usually the expected method of navigation. However, this can become time consuming when dealing with search results page, in which client is forced to navigate back and forth between search form and search result to find the desired information. We employed *AJAX* to dynamically change the search result upon users request for faster and more user friendly search experience.

Front-end was developed primarily with one objective in mind. (Dynamic manipulation of pages). With that in mind, all menus for different species were constructed by dynamically querying database. Species home pages were also pulled from database to provide constant information across the platform. This would provide robust

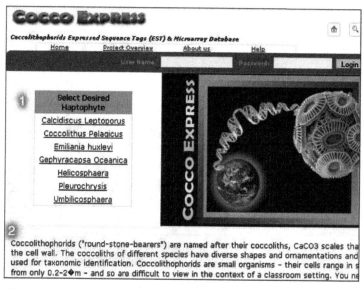

Fig. 4.2 CoccoExpress menu page

yet scalable system that could grow to house more species with minimal development efforts.

Figure 4.2 demonstrates the front-page of CoccoExpress. Sections 4.1 and 4.2 in Fig. 4.2 are dynamically built using data pulled from database, based on requested page. Section 4.1 is list of species in database and will dynamically grow/shrink when new specie is added or existing specie is deleted. Section 4.2 is also dynamically generated based on information regarding selected specie. This will allow anybody even with minimum knowledge of web designing, maintain front-end of CoccoExpress.

Figure 4.3 demonstrates the home page of Emiliana Huxleyi. Sections marked as 1 thru 4 in Fig. 4.3 are dynamically built using data pulled from database. Section 4.1 demonstrates user friendly/search engine friendly naming conventions of CoccoExpress pages. This well-formatted URL is picked by search engines and is easy to parse by search engine's back-end parser. It is important to remember that "/Coccolithophorids/Emiliana-Huxleyi/" is not an existing folder on web server, instead it is a dynamically built URL for better structure and search engine optimization. In back-end implementation we will discuss this in more details. Section 4.2 of Fig. 4.3 demonstrates the dynamically built menu for the given specie. Section 4.3 and 4 are also pulled dynamically from database based on selected specie.

4.4.2.2 Quick Search

As the name describes, Quick Search was implemented to provide a fast method of searching the database for different records of specie.

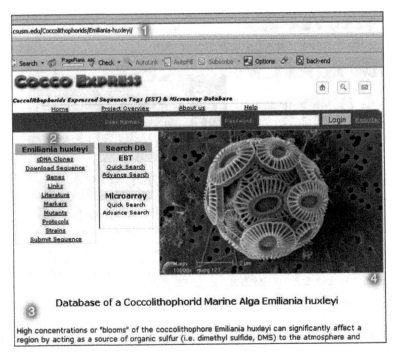

Fig. 4.3 Emiliania Huxleyi menu page

Figure 4.4 demonstrates the quick search tool for Emiliana Huxleyi (this tool is also available for all other species in the database). Users can search by different criteria as shown in the figure. Help button is also provided to show help documentation on data type of any given criteria.

Figure 4.5 demonstrates the results of a quick search submitted with "F2" as criteria for library name. Once this result is sent to client's browser, client can easily sort the result by desired column in desired order. As it is demonstrated in the figure, client can click on the record to see more detail about any given column. This request for more details is sent to backend system using Ajax, once the answer to request is available, results page is updated accordingly. This will allow fast access to several records in different tables without leaving the page.

4.4.2.3 Advance Search

Figure 4.6 demonstrates the interface for advance search. As shown in Section 4.1 in this figure, fields for selected tables are pulled dynamically from database to construct series of windows with fields of each table. If a field is added/removed from a table in database, this form is updated automatically to reflect the changes. By using advance search, user is capable of selecting desired fields to be pulled from

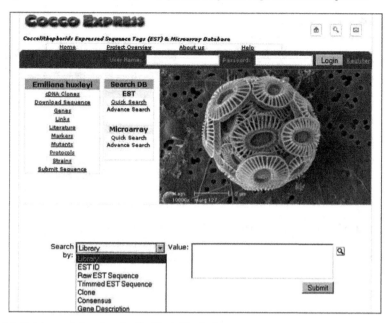

Fig. 4.4 Quick search interface for Emiliania Huxleyi

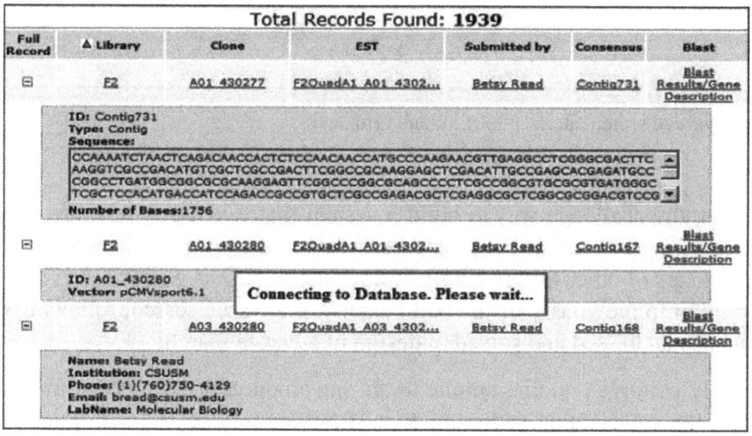

Fig. 4.5 Result of quick search for Emiliania Huxleyi

database (Section 4.2 in the figure). In addition, users can specify desired criteria for any given field. (Sections 4.3 and 4.4 in the figure). The query behind quick search is constructed automatically using back-end engine written in PHP.

Quick search was implemented with a very straight forward sequence flow in mind. While it can provide a fast and easy access to records, it has several limitations by design. These limitations (i.e. lack of ability to select other fields or criteria versus predefined options) led to implementation of a more sophisticated search system.

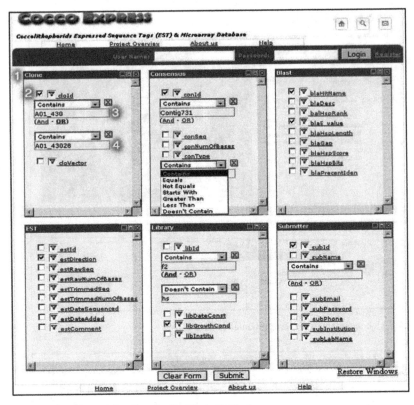

Fig. 4.6 Advanced search interface for Emiliania Huxleyi

Once again, the challenge was to build a system that can satisfy advance queries with an easy to use interface that could be used not just by computer scientists but also by scholars of other fields, mainly biologists.

In addition to previously mentioned benefits of advance search, following concerns are also addressed and solved in design of advance search:

- Manually writing of queries similar to the one produced by advance search could be very time consuming and requires in-depth knowledge of our backend database.
- User written queries are usually written and tweaked several times before desired results is achieved, wasting both user's time and server's resources.
- User created queries are often not well-optimized and do not use proper indices, therefore they are slow and take up a lot of resources to run.
- To write such queries, knowledge of database and SQL is required.
- Returning results in manually written queries are in a text format and are hard to manipulate, where in advance search results are returned as spread sheet.

Each window in advance search represents a table of CoccoExpress database. These windows are constructed with ability to shrink or grow in size based on available

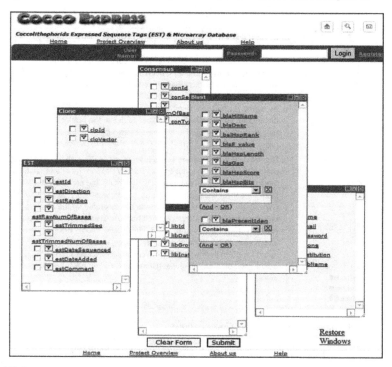

Fig. 4.7 Dynamic feature of advanced search

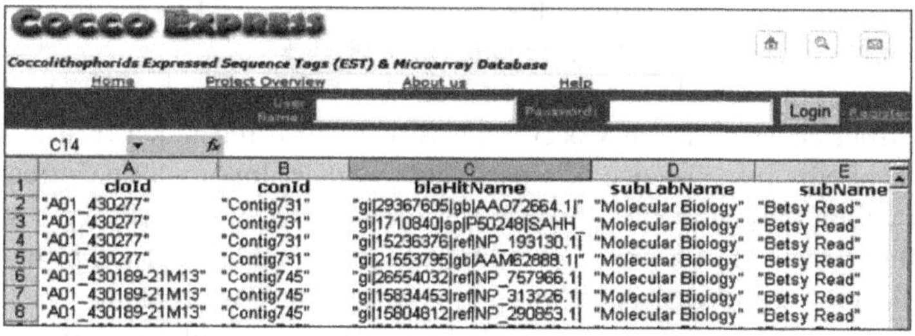

Fig. 4.8 Result of advance search for Emiliania Huxleyi

screen size and total number of tables and fields. As demonstrated in Fig. 4.7, for user comfort, these windows can be moved around and resized manually to achieve desired length and width just like any other window based desktop application.

Figure 4.8 demonstrates the results returned by an advance search. Results are provided in spreadsheet format so they are easier to modify and work with for end users. Columns in spreadsheet are built based on selected fields by user.

A centralized help system was an absolute necessity since CoccoExpress consists of several technical terms and has a unique structure. After some research, we decided to adopt an already developed platform rather than reinventing the wheel.

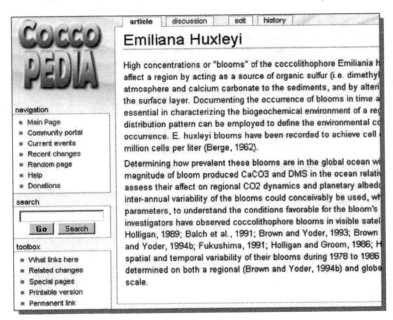

Fig. 4.9 CoccoPedia help integrated with CoccoExpress

This help system was created by taking advantage of open-source software also used by Wikipedia. This help system was to be linked to from different places of CoccoExpress to provide definition and help on different technical terms. Figure 4.9 is snapshot of CoccoPedia entry for Emiliana Huxleyi.

While we could manually link to CoccoPedia's articles and definitions from anywhere in front-end website of CoccoExpress, it seemed too much of a hassle to check CoccoPedia for existence of a page and manually linking all occurrences of that term within the entire site. It was even more hassle to manually go back and update all occurrences of a term every time a page was added, deleted or modified in CoccoPedia. CoccoPediaLinker was born to overcome this problem.

CoccoPediaLinker is a parser system that is enabled by default for all the pages in CoccoExpress website; it can be turned off manually per page. It is responsible for parsing the page content and examining the content against available articles of CoccoPedia. If a term is found in CoccoPedia engine, CoccoPediaLinker will link the term to CoccoPedia entry automatically. This will provide a well documented front-end for CoccoExpress. More importantly, documentation of CoccoExpress pages will get richer as more articles are added to CoccoPedia. Figure 4.10 Demonstrates the term "Coccolithophorids" that was picked up by CoccoPediaLinker and was linked to proper CoccoPedia article page.

The links built by CoccoPediaLinker are underlined with dashed line to be extinguished from normal links of the site. CoccoPediaLinker was built with some intelligence to avoid un-wanted link creations such as modifying an already linked keyword or modifying keywords within html tags and attributes.

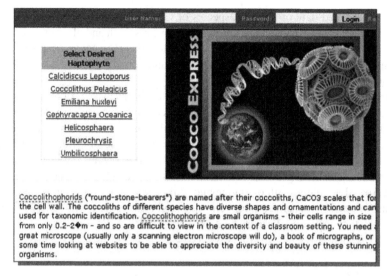

Fig. 4.10 CoccoPedia linker feature of CoccoExpress

4.4.3 Backend and Database

Back-end system was written primarily in PHP. In most cases PHP was used to generate the HTML and Apache was responsible for running PHP and streaming HTML over to user's browser. We previously mentioned that MySQL was picked as our back-end database system. This allowed us to take advantage of compatibility between PHP and MySQL. MySQL connection was established at the beginning of each page by including a centralized header. Later this connection was used within the PHP script to access the database for various reasons such as dynamically building the navigation menu section. We deployed two primary databases, one to be used by CoccoExpress and one to serve as data warehouse for CoccoPedia.

CoccoExpress database structure is consist of 11 tables, out of which 9 were used primarily to house genomic data and two were used to control CoccoExpress front-end interface. To easily access and work with the database, phpMyAdmin was deployed on the server. phpMyAdmin is an open source software that will create a web-based interface for MySQL database. With phpMyAdmin, database administrators can simply make modifications to database such as inserting new record or deleting an existing record with minimum knowledge of database administration.

4.5 Future Work and Conclusion

CoccoExpress was designed to be scalable and grow dynamically to house several different species of Coccolithophorid. At this time majority of data in our database belongs to Emiliana Huxleyi, (Ehux). Future research and work can be done to place the analyzed data of different species of Coccolithophorid family in Cocco-Express. There is also an extensive amount of active research and planning has been conducted in developing the Microarray database [4]. Once Microarray database is completed, it can be linked to EST database and some possible cross-database functionalities could be added.

In brief, CoccoExpress was designed to serve as the primary tool to access and manage EST database of Coccolithophorid. Project was approached with two primary objectives of advance usability and easy maintenance/usage. It was originally designed as Ehux Express [10]. Due to requirements for several fundamental enhancements, it was redesigned and built from ground up to better accommodate the requirements of the fast growing data set. As further research is conducted, we will benefit from dynamic growth of engine and will have a better understanding of the mechanism causing carbon fixation. That shall lead us toward new discoveries about these important species and life matters.

References

1. T. M. Wahlund, A. R Hadaegh, R. Clark, M. Fanelli, and B. A. Read, Analysis of Expressed Sequence Tags in Emiliana Huxleyi, *Nature*, 2003, 377:320–323.
2. T. M. Wahlund, X. Zhang, and B.A. Read, EST Expression Profiles from Calcifying and Non-Calcifying Cultures of Emiliana Huxleyi, *Micropaleontology*, 2004, 50(Suppl. 1):145–155.
3. B. Nguyen, R. M. Bowers, T. Wahlund, and B. A. Read, Suppressive Subtractive Hybridization and Differences in Gene Expression Content of Calcifying and Noncalcifying Cultures of Emiliania Huxleyi Strain 1516, *Applied and Environmental Microbiology*, 2005, 71(5):2564–2575.
4. P. Quinn, R. M. Bowers, X. Zhang, T. M. Wahlund, M. A. Fanelli, D. Olszova, and B. A. Read, cDNA Microarrays as a Tool for Identification of Biomineralization Proteins in the Coccolithophorid Emiliania Huxleyi (Haptophyta), *Applied Environmental Microbiology*, 2006, 72(8):5512–5526.
5. A. Caprera, B. Lazzari, A. Stella, I. Merelli, A. Caetano, and P. Mariani, GoSh: A Web-Based Database for Goat and Sheep EST Sequences, *Bioinformatics*, 2007, 23(8):1043–1045.
6. N. D'Agostino, M. Aversano, L. Frusciabte, and L. Chiusano, TomatEST Database: In Silico Exploitation of EST Data to Explore Expression Patterns in Tomato Species, *Nucleic Acids Research*, 2007, 35(Database issue):D901–D905.
7. M. Hiller, S. Nikolajewa, K. Huse, K. Saransk, P. Rosentiel, S. Schuster, R. Backofen, and M. Platzer, TassDB: A Database of Alternative Tandem Splice Sites, *Nucleic Acids Research*, 2007, 35(Database issue):D188–D192.
8. E. A. O'Brien, L. B. Koski, Y. Zhang, L. Yang, E. Wang, M. W. Gray, G. Burger, and B. F. Lang, TBestDB: A Taxonomically Broad Database of Expressed Sequence Tags (ESTs), *Nucleic Acids Research*, 2007, 35(Database issue):D445–D451.
9. N. Pavy, J. Johnson, J. Crow, C. Paule, T. Junau, J. Mackay, and E. Retzel, ForestTreeDB: A Database Dedicated to the Mining of Tree Transcriptomes, *Nucleic Acids Research*, 2007, 35(Database issue):D888–D894.
10. M. Ebert, A EhuxExpress: Novel Genetic Database for the Marine Alga Emiliania Huxleyi, 2004, Master project, California State University, San Marcos, TX.

Chapter 5
Hybrid Intelligent Regressions with Neural Network and Fuzzy Clustering

Sio-Iong Ao

Abstract A hybrid intelligent algorithm of neural network regression and unsupervised fuzzy clustering is proposed for clustering datasets of nonparametric regression models. In the new formulation, (i) the performance function of the neural network regression models is modified such that the fuzzy clustering weightings can be introduced in these network models; (ii) the errors of these network models are feed-backed into the fuzzy clustering process. This hybrid intelligent approach leads to an iterative procedure to formulate neural network regression models with optimal fuzzy membership values for each object such that the overall error of the neural network regression models can be minimized. Our testing results show that this hybrid algorithm NN-FC can handle cases that the K-means and Fuzzy C-means perform poorly. The overall training errors drop down rapidly and converge with only a few iterations. The clustering accuracy in testing period is consistent with these drops of errors and can reach up to about 100% for some problems that the other classical fuzzy clustering algorithms perform poorly with about 60% accuracy only. Our algorithm can also build regression models, which has the advantage of the NN component, being non-parametric and thus more flexible than the fuzzy c-regression.

Keywords Neural network · Regression models · Fuzzy clustering · Fuzzy performance function

5.1 Introduction

Hathaway and Bezdek [1] developed a methodology of switching regression models and fuzzy clustering. Their main idea is to employ a fuzzy clustering for linear or polynomial regression models for data sets. Their approach leads to an iterative

S.-I. Ao
Oxford University Computing Laboratory, University of Oxford, Wolfson Building, Parks Road, Oxford OX1 3QD, UK
e-mail: siao@graduate.hku.hk

S.-I. Ao et al. (eds.), *Advances in Computational Algorithms and Data Analysis*,
Lecture Notes in Electrical Engineering 14,
© Springer Science+Business Media B.V. 2009

procedure to formulate linear or polynomial regression functions with optimal fuzzy membership values for each object such that the overall error of the linear or polynomial regression functions is minimized. Hathaway and Bezdek have pointed out that there are various applications of the switching regression models in economics. And an example from fisheries was illustrated. The sexuality of a fish called halibut is indistinguishable. The mean length of a male halibut depends on its age for a certain range of ages and so does the female halibut. And the problem can be treated as a switching regression problem of two models, one for the male and another for the female. For the example, Hathaway's two models are

$$y = \beta_{11}x + \beta_{12} + \varepsilon_1$$
$$y = \beta_{21}x + \beta_{22} + \varepsilon_2 \tag{5.1}$$

where y = length and x = age. And our proposed approach can give solutions of the form

$$y = f_1(x) + \varepsilon_1$$
$$y = f_2(x) + \varepsilon_2 \tag{5.2}$$

where the functions are to be simulated by two neural network regression models respectively. Menard [2] extended fuzzy clustering and switching regression models using ambiguity and distance rejects. The main drawback of switching regression approach is that the parameters of the formal generating functions must be known. However, this may not be valid in practice.

The general regression models for representing the dependent variable $\vec{Y}$ and independent variable $\vec{X}$ can be written as followed:

$$\vec{Y} = F_i(\vec{X}) + \varepsilon \tag{5.3}$$

for $1 \leq i \leq K$, where ε are the noises of the data type, K is the number of different regression functions, and F_i is the ith regression function between $\vec{X}$ and $\vec{Y}$. The difficulty of solving this problem is that (i) the regression functions are unknown and (ii) the regression functions are not labeled. For (i), it implies that the form of F_i is unknown and that it may be linear or nonlinear. For (ii), it implies that the clustering of the data $\vec{X}$ and $\vec{Y}$ for the regression functions is required. Our aim is to cluster a set of points into two groups, and to construct their corresponding regression functions for this set of points.

In this paper, we propose the nonparametric neural network (NN) regression models to remove such parametric models on regression functions. Our intelligent algorithm can provide non-linear non-parameter solutions to the above applications of the switching regression models. It has been illustrated from the difference between Eqs. (5.2) and (5.3). Furthermore, it can cluster datasets produced by some underlying generated functions.

Many clustering problems will be solved poorly with common clustering methods, for instance K-means clustering and Fuzzy c-means (FCM). Our tests of applying these clustering methods can give us a clustering accuracy of about 60% only.

It is because the K-means and Fuzzy C-means techniques have strict restrictions on the shapes of the clusters being studied (hyper-spherical clusters of equal size for Euclidean distance and hyper-elliptical clusters for Mahalanobis distance) [3]. Even though Gustafson-Kessel and Gath-Geva algorithms have extended the Fuzzy C-means algorithm for shapes of ellipses and ellipsoids [4], there are still strict restrictions on the clusters' shapes. For the clustering problem with two centers, the FCM will cluster the left-hand side data points into one group and the right-hand side data into another group. It is clear that this is a very poor cluster decision for this problem. Fuzzy c-regression can solve the problem partially, only when the parameters of the formal generating functions are given. But, in reality, it is difficult to know in advance about the generating functions.

There are studies [5–8] that combine the fuzzy clustering and the neural network for supervised classification purposes. They have applied the fuzzy clustering to the original data and got the membership values for each object in each cluster. This information can serve as the weighting for the neural network output at the performance level in Sarkar's study. Or it is used during the combining of different NN models' outputs in Ronen's work. In Boca's study, they employ the Fourier analysis and fuzzy clustering to extract the signal features for the supervised neural network classification. Bortolan has applied the fuzzy clustering as a preprocessor for the initialization of the receptive fields of the radial basis function neural network for supervised classification.

It should be noted that these studies have employed the hybrid system in the loose format. That is to pass the variables through each component only once. They have reported satisfactory results for problems that are suitable for supervised learning. But, these studies are restricted to problems that suit the Fuzzy C-means clustering and will perform poorly for other problems. Our study differs from these above methodologies in that we deal with unsupervised learning problems instead of their supervised classification problems. Also, we have developed our algorithm in an iterative manner and achieve the hybrid objectives of unsupervised clustering and neural network regressions with our algorithm.

We have formulated a hybrid iterative methodology of neural network and unsupervised fuzzy clustering so that the clustering and regression components can supplement each other for further improvement. A general solution for these problems can be found with our new NN-FC algorithm, which can give clustering accuracy of about 100% in our testing period.

The outline of this paper is as follows. In Section 5.2, we present the proposed algorithm. In Section 5.3, we illustrate the effectiveness of the proposed method by some examples. Finally, a summary is given in Section 5.4.

5.2 The Proposed Algorithm

Our proposed intelligent algorithm consists of the neural network and the unsupervised fuzzy clustering components. The neural network is for investigating the regression relationships between the dependent and independent variables, while the unsupervised fuzzy clustering is for clustering the objects into different generating functions. Then, NN regressions are implemented to simulate different generating functions with the modified performance functions that each error is weighted by the output from the FC. The respective performance errors are passed back to the FC. FC component will adjust the membership values for each object in each cluster based on the corresponding errors of that object in each generating NN model.

The idea comes from the fact that, if an object fits a particular NN model well, its error in that model will be much lower than its errors in other NN models. At the same time, when the memberships are more and more close to the corrected generated clusters, it can help the NN to have better simulations for each underlying regression function. These two components work in a hybrid way and form a loop until the overall errors do not show any further significant improvement.

5.2.1 Neural Network Regression Models

Neural network is well known for its non-linear capability and is usually employed with the three-layer architecture. The layers are input layer, hidden layer and output layer. The inspiring idea for this structure is to mimic the working of our brain. The mathematical structure for the above neural network structure can be expressed as followed [9]:

$$y = f\left(\sum_{j=1}^{J} w_j^{(2)} f\left(\sum_{i=1}^{I} w_{ji}^{(1)} x_i\right)\right) \tag{5.4}$$

where the function f is the activation function of the network, I denotes the number of inputs, J the number of hidden neurons, x_i the ith input, $w^{(1)}$ the weights between the input and hidden layers, $w^{(2)}$ the weights between the hidden and output layers.

Unlike traditional neural networks, we have employed the fuzzy membership results from the clustering as the weighting for each output error of the network. We have adopted the fuzzy clustering instead of the hard clustering. If the hard clustering of membership values 0 and 1 were employed, the network would be trained with these crisp weightings and then optimized with respect to these crisp weights. When we perform the clustering procedure for such data in the next iteration, the objective function value does not improve. The main reason is that the NN regression models fit the data points very well. It is difficult to adjust the weightings to the optimal membership values when we were restricted with the membership values 0 and 1 only.

As said, we have modified the performance function of a typical network

$$E = \frac{1}{2NM} \sum_{n=1}^{N} \sum_{m=1}^{M} (z_{nm} - t_{nm})^2$$

with our new one:

$$E = \frac{1}{2NM} \sum_{k=1}^{K} \sum_{n=1}^{N} \sum_{m=1}^{M} w_{knm}^{\alpha} (z_{knm} - t_{knm})^2 \tag{5.5}$$

where N is the number of examples in the data set, M the number of outputs of the network, t_{knm} the mth target output for the nth example in the kth cluster, z_{knm} the mth output for the nth example in the kth cluster, K the number of clusters. w_{knm} is the fuzzy membership value for each sample to belong a certain cluster k, and α is the fuzzy weighting exponent. We have developed specific learning laws for this modified NN performance function similar with the neural network part of Sarkar's study and the derivation for a simple three-layer network with logistic transfer function is in Appendix. Sarkar et al. have discussed the motivations and advantages of introducing of this fuzzy mean square error term. In brief, with the introduction of the fuzzy mean square error term, the restriction of an input datum belonging to one and only one cluster/class has been removed. It addresses the situations where the datum may belong to more than one cluster. And the training of the networks can be conceptually viewed as a fuzzy constraint satisfaction problem.

5.2.2 Fuzzy Clustering

K-means and Fuzzy C-means are two conventional clustering methods. The difference of the K-means clustering and the Fuzzy C-means clustering is on the overlapping or not of the boundaries between the clusters. In K-means clustering, the belonging of a datum x to a cluster k or not is crisp, usually donated by a *membership function* $u_k : X \rightarrow \{0,1\}$, where $u_k(x) = 1$ if and only if $x \in k$, and $u_k(x) = 0$ if and only if $x \notin k$. The task of the K-means clustering algorithm is to determine the K cluster centers and the $u_k(x)$ values for every datum and cluster.

In the real life situations, boundaries between the classes may be overlapping [5] and it is uncertain if a datum belongs completely to a certain cluster. This is one of the motivations for our adoption of Fuzzy C-means clustering here. In Fuzzy C-means clustering, the membership function u_k is no longer crisp. Instead, here, it can take any values between 0 and 1, with the constraint

$$\sum_{k=1}^{K} u_k(x) = 1$$

for every datum x and every cluster k.

The objective of applying fuzzy clustering component in our study is to minimize the above performance function (5.5) of the neural network with respect to the w_{knm}, where

$$w_{kmn} \in \{0,1\}, \quad and \sum_{k=1}^{K} w_{kmn} = 1 \forall m, n \tag{5.6}$$

Define

$$E_{knm} = \frac{1}{2NM}(z_{knm} - t_{knm})^2$$

as the dissimilarity measure between the object nm and the k cluster center, we can have

$$E = \sum_{k=1}^{K}\sum_{n=1}^{N}\sum_{m=1}^{M} w_{knm}^{\alpha} E_{knm} \tag{5.7}$$

which can be recognized as a fuzzy clustering problem like [10, 11] etc., and can be solved by taking partial derivative of E with respect to w_{knm}. For $\alpha > 1$, the minimizer $\hat{w}_{knm}$ is given by

$$\hat{w}_{hnm} = \begin{cases} 1, & if E_{hnm} = 0 \\ 0, & if E_{knm} = 0 \text{ } for some other } k \neq h \\ 1/\sum_{k=1}^{K}\left[\frac{E_{hnm}}{E_{knm}}\right]^{1/(\alpha-1)}, & otherwise \end{cases} \tag{5.8}$$

where $1 \leq h \leq K$ and $1 \leq k \leq K$.

5.2.3 Hybrid Neural Network and Fuzzy Clustering (NN-FC)

Instead of clustering the data only once and then passing it to the neural network, our algorithm further utilize the information of the clustering and neural network. It works in a hybrid iterative loop. As said, the motivation is that outputs from each of the two components can improve each other component in the following round. The algorithm is given as follow:

Algorithm 1 The Hybrid NN-FC Algorithm

Step 1. Randomize the fuzzy membership matrix w_{knm};

Step 2. Train the neural network models to minimize E for each cluster, using our modified gradient-descent rules and keeping w_{knm} as constants;

Step 3. Update the fuzzy membership values w_{knm} of every object in each cluster, such that E is minimized with respect to w_{knm}.

Step 4. Repeat the above steps (2) and (3) in iteration until the improvement of the performance function between the successive iterations drops below a certain level.

Then, the accuracies of the clustering results will be checked in our testing period.

5.3 Experimental Results

We have carried out some tests of the NN-FC algorithm on different synthetic datasets that cannot be solved well by the previous studies of NN-FC that are in loose hybrid format [5, 6]. And the results show that the proposed algorithm is capable of clustering the data sets accurately and forming the regression functions accurately.

The dataset comes from two linear generating functions as shown:

$$F_1 : \varepsilon(X)$$
$$F_2 : X/10 + \varepsilon(X) \tag{5.9}$$

where X will have values drawn evenly from the interval studied, ε is the corresponding noise. For the following dataset, one generating function is of second-order and the other of third order, with noises at different levels:

$$F_1 : 260 \times (X - 7.5)^2 + 5 + 0.1 \times \varepsilon(X)$$
$$F_2 : 8 \times X^3 + 0.1 \times \varepsilon(X) \tag{5.10}$$

The following experiments are with the datasets 3 and 4 respectively. The dataset 3 show us a problem from two linear generating functions with intersection:

$$F_1 : -5.6 \times X + 90 + \varepsilon(X)$$
$$F_2 : 6 \times X + \varepsilon(X) \tag{5.11}$$

In the dataset 4, the two generating functions are of second-order and first-order respectively:

$$F_1 : 10 \times (X - 6.5)^2 + 5 + 0.1 \times \varepsilon(x)$$
$$F_2 : -68 \times X + 620 + 0.1 \times \varepsilon(X) \tag{5.12}$$

And, Table 5.1 shows us the clustering results for these datasets. We can see the accuracies of our proposed method are close to 100% for all datasets in the testing period. Sarkar's results are obtained with fuzzy membership assignment. In their methodology, each of the data sets will pass the fuzzy clustering and neural network components once and only once. While the neural network component have

Table 5.1 Clustering accuracies (in percentage) for the datasets in the testing period

Datasets	1	2	3	4
Our method	96%	100%	99%	97%
Sarkar's	61%	50%	61%	77%
K-means	62%	51%	61%	77%
Hierarchical	51%	100%	55%	63%
Quantum	59%	50%	62%	80%

Table 5.2 Clustering accuracies (in percentage) for the datasets in the testing period

Noise levels	0.1	400	1600
Accuracies	96%	91%	96%

added as a classifier, the actual clustering jobs are done with the fuzzy clustering. K-means clustering results are also listed. As said, Sarkar's method has employed the fuzzy clustering technique for the clustering process and we can see that their performances are very similar with the K-means clustering. The hierarchical cluster results are obtained with the single linkage algorithm. The quantum clustering algorithm has been recently suggested by Horn and Gottlieb [12] and has been successfully applied to the genome datasets [13]. The algorithm is based on the solution for the Schrödinger equation.

We have applied different noise levels to the first dataset to check the robustness of our method. The following generating functions are that for dataset 1 with different noise levels controlled by α:

$$F_1 : 260 \times (X - 7.5)^2 + 5 + \alpha \times \varepsilon(X)$$
$$F_2 : 8 \times X^3 + \alpha \times \varepsilon(X) \tag{5.13}$$

The noise level 0.1 is the one we have used and is with noise level α equal to 0.1. And the noise level 400 is with α value equal to 400 and the noise level 16 is with α value equal to 1,600. This is comparable with the magnitude of our datasets. Table 5.2 shows that the clustering results of these different noise levels. We can see that our method is robust over these levels and can maintain its clustering accuracies above 90% while other methods can give only about 60% accuracy for the original dataset.

Three different neural network structures have also been tested to see the effect of different NN architectures. Among the test structures, they are three-layer networks of 5 hidden neurons, 10 hidden neurons and 15 hidden neurons respectively. All of the three networks with sample 2 datasets can produce stable clusters in less than 20 iterations and obtain 100% clustering accuracies. We can see these results in the below figure. This suggests that our hybrid NN-FC algorithm is stable with respect to the network's structure. And we can observe the total training errors decrease rapidly and converge to a small value in just a few epochs.

5.4 Summary and Discussion

In our results, it is shown that our NN-FC is able to handle problems that cannot be properly solved with the K-means, fuzzy c-means, and the previous fuzzy clustering and neural network systems. In the studies [5–8], the fuzzy clustering has been designed as a preprocessing tool for feeding the neural network with better inputs without any feedback to the fuzzy clustering. Here, we have successfully developed a feedback algorithm so that the fuzzy clustering can be further improved for solving unsupervised learning problems. And the steady decreases of the sums of errors confirm with this. Table 5.2 shows us that our algorithm can have stable solutions for a range of different noise levels.

There is still much room for the optimization of this methodology. It is expected that further improvement can be obtained with the optimal design of network structure like the number of hidden neurons, with faster second-order learning laws of fuzzy performance function, and with the tabu-search algorithm. Therefore, the proposed algorithm can explore the solution space beyond local optimality in order to aim at finding a global optimal solution of unsupervised learning problems [14].

Appendix

In a simple three-layer network, the inputs to hidden neuron j can be donated as $u_j = a_{0j} + \sum\limits_{i=1}^{I} a_{ij}x_i$, and its output as $y_j = g(u_j)$, where the function is the Logistic transfer function. The inputs to k output are given as $v_k = b_{0k} + \sum\limits_{j=1}^{J} b_{jk}y_j$, and its output is $z_k = g(v_k)$. For the on-line learning case, the mean squared error is given by:

$$\bar{E} \triangleq \frac{1}{2} \sum_{k=1}^{K} (z_k - t_k)^2$$

Here, we have replaced this measurement of error by our new one E^* with the weightings w_k^* for each output k, where w_k^* is obtained from our fuzzy clustering component,

$$E^* \triangleq \frac{1}{2} \sum_{k=1}^{K} w_k^*(z_k - t_k)^2$$

As a result, the updating procedure for the neural network is different from the typical network and will be derived as followed in a similar way as Sarker et al.

Taking derivative of E^* with respect to z_k,

$$\frac{\partial E^*}{\partial z_k} = w_k^*(z_k - t_k)$$

And we can update $\frac{\partial E^*}{\partial b_{jk}}$ as followed:

$$\frac{\partial E^*}{\partial b_{jk}} = \frac{\partial E^*}{\partial z_k} \frac{\partial z_k}{\partial v_k} \frac{\partial v_k}{\partial b_{jk}} = \begin{cases} P_k^*, for \ j = 0 \\ P_k^* y_j, for \ j = 1,\ldots,J \end{cases}$$

where

$$P_k^* = w_k^*(z_k - t_k)z_k(1 - z_k)$$

And $\frac{\partial E^*}{\partial a_{ij}}$ is given by:

$$\frac{\partial E^*}{\partial a_{ij}} = \left(\sum_{k=1}^{K} \frac{\partial E^*}{\partial z_k} \frac{\partial z_k}{\partial v_k} \frac{\partial v_k}{\partial y_j} \right) \frac{\partial y_j}{\partial u_j} \frac{\partial u_j}{\partial a_{ij}} = \left\{ \begin{array}{l} Q_j^*, for\ i = 0 \\ Q_j^* x_i, for\ i = 1, \ldots, I \end{array} \right.$$

where

$$Q_j^* = \left[\sum_{k=1}^{K} P_k^* b_{jk} \right] y_j (1 - y_j).$$

References

1. R. J. Hathaway and J. C. Bezdek, "Switching Regression Models and Fuzzy Clustering", IEEE Transactions on Fuzzy Systems, 1, 195–204, 1993.
2. M. Menard, "Fuzzy Clustering and Switching Regression Models Using Ambiguity and Distance Rejects", Fuzzy Sets and Systems, 133, 363–399, 2001.
3. K. Jajuga, A. Sokolowski, and H.-H. Bock (eds.), "Classification, Clustering, and Data Analysis: Recent Advances and Application", Springer, 35–42, 2002.
4. F. Hoppner, F. Klawonn, R. Kruse, and T. Runkler, "Fuzzy Cluster Analysis: Methods for Classification, Data Analysis and Image Recognition", Wiley, New York, 1999.
5. M. Sarkar, B. Yegnanarayana, and D. Khemani, "Backpropagation Learning Algorithms for Classification with Fuzzy Mean Square Error", Pattern Recognition Letters, 19, 43–51, 1998.
6. M. Ronen, Y. Shabtai, and H. Guterman, "Rapid Process Modelling-Model Building Methodology Combining Supervised Fuzzy-Clustering and Supervised Neural Networks", Computers Chemical Engineering, 22, S1005–S1008, 1998.
7. A. Del Boca and D. C. Park, "Myoelectric Signal Recognition Using Fuzzy Clustering and Artificial Neural Networks in Real Time", IEEE World Congress on Computational Intelligence, 5, 3098–3103, 1994.
8. G. Bortolan and W. Pedrycz, "Fuzzy Clustering Preprocessing in Neural Classifiers", Kybernetes, 27(8), 900, 1998.
9. J. C. Principe, N. R. Euliano, and W. C. Lefebvre, "Neural and Adaptive Systems: Fundamentals Through Simulations", Wiley, New York, 2000.
10. J. Z. Huang, "Extensions to the k-Means Algorithm for Clustering Large Data Sets with Categorical Values", Data Mining Knowledge Discovery, 2, 283–304, 1998.
11. J. Z. Huang and M. K. Ng, "A Fuzzy k-Modes Algorithm for Clustering Categorical Data", IEEE Transactions on Fuzzy Systems, 7, 446–452, 1999.
12. D. Horn and A. Gottlieb, "Algorithm for Data Clustering in Pattern Recognition Problems Based on Quantum Mechanics", Physical Review Letters, 88, 018702, 2002.
13. D. Horn and I. Axel, "Novel Clustering Algorithm for Microarray Expression Data in a Truncated SVD Space", Bioinformatics, 19(9), 1110–1115, 2003.
14. M. Ng and J. Wong, "Clustering Categorical Data Sets Using Tabu Search Techniques", Pattern Recognition, 35, 2783–2790, 2002.

Chapter 6
Design of DroDeaSys (Drowsy Detection and Alarming System)

Hrishikesh B. Juvale, Anant S. Mahajan, Ashwin A. Bhagwat, Vishal T. Badiger, Ganesh D. Bhutkar, Priyadarshan S. Dhabe, and Manikrao L. Dhore

Abstract The paper discusses the Drowsy Detection & Alarming System that has been developed, using a non-intrusive approach. The system is basically developed to detect drivers dozing at the wheel at night time driving. The system uses a small infra-red night vision camera that points directly towards the driver's face and monitors the driver's eyes in order to detect fatigue. In such a case when fatigue is detected, a warning signal is issued to alert the driver. This paper discusses the algorithms that have been used to detect drowsiness. The decision whether the driver is dozing or not is taken depending on whether the eyes are open for a specific number of frames. If the eyes are found to be closed for a certain number of consecutive frames then the driver is alerted with an alarm.

Keywords Alarm · Detection · Driver · Drowsiness · Fatigue

H.B. Juvale (✉)
Cybage Software Pvt. Ltd., Pune – 411037, India
e-mail: hrishikeshjuvale@gmail.com.

A.S. Mahajan
Accenture, Pune – 411037, India
e-mail: anant_mahajan99@yahoo.com

A.A. Bhagwat
e-mail: ashwin.bhagwat@yahoo.com

V. T. Badiger
Wipro Technologies, Pune – 411037, India
e-mail: vishalbadiger@yahoo.com

G.D. Bhutkar and P.S. Dhabe
Vishwakarma Institute of Technology, Pune – 411037, India
e-mail: ganesh.bhutkar@vit.edu, e-mail: priyadarshan.dhabe@vit.edu

M.L. Dhore
Computer Engineering Department, Vishwakarma Institute of Technology, Pune – 411037, India
e-mail: hodcomp@vit.edu

S.-I. Ao et al. (eds.), *Advances in Computational Algorithms and Data Analysis*,
Lecture Notes in Electrical Engineering 14,
© Springer Science+Business Media B.V. 2009

6.1 Introduction

Driver fatigue is a significant factor in a large number of vehicle accidents. The development of technologies for detecting or preventing drowsiness at the wheel is a major challenge in the field of accident avoidance systems. By monitoring the eyes, it is believed that the symptoms of driver fatigue can be detected early enough to avoid a car accident. Detection of fatigue involves a sequence of images of a face, and the observation of eye movements and blink patterns [1–4].

The eye detection algorithm as well as the drowsy detection procedure has been implemented using a self developed algorithm. The system is developed using image processing fundamentals. The focus of the system is on accurately determining the open or closed state of the eyes. Depending on the state of the eyes it can be said whether the driver is alert or not.

To achieve the result we have used the clustering & slope detection algorithms. The images of the drivers face are acquired from the infra-red night vision camera. The infra red camera illuminates the drivers face at night time. The images obtained are converted to binary images first & then clusters on those images are found out. The slope detection algorithm is used to make the former algorithm more accurate in detecting the state of the eyes. It calculates the slope between each of the clusters & keeps on discarding the clusters as long as we don't get the right clusters as the pupils of the eyes.

This paper discusses these algorithms as well as the flow of the system in greater detail in the following sections.

6.2 Block Diagram

The block diagram shown in Fig. 6.1 describes the dynamic flow of the system wherein a new image/frame is extracted pre-processed, processed and post processed to determine whether the state of drowsiness is reached. If this state is reached then an alert is given to the driver and the process continues until all frames are processed.

6.3 Phases of the System

The system uses a completely software approach & has been broken down into three phases:

1. Pre-processing
2. Processing
3. Post-processing

The processing phase forms the major part of the system & this is where the algorithm to detect the state of the eyes has been implemented.

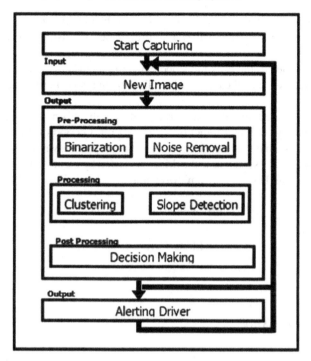

Fig. 6.1 Block diagram of DroDeASys

6.3.1 Pre-processing

The images acquired from the infra red night vision camera are converted into binary images using a specific threshold. Also the image is enhanced by isolating independent pixels.

6.3.2 Processing

The binary image is then input to the clustering algorithm wherein clusters are found out within the binary image. Depending on the illumination from the camera at that instant of time & the skin color of the person there will be different number of clusters that will be found each time. Clusters are nothing but the areas of the face which are turned on after applying a specific threshold. Once the clusters are detected the centers of each of the clusters is found out & distance is calculated. We have tested the algorithm on the samples of a number of different people & found out the approximate distance within which the two pupils lie. To detect the eyes the distance is checked between the clusters & if ever the clusters are found to be within that range then the eyes are detected. One problem with this algorithm is that the same distance can be there between a different set of clusters which are really not the eyes.

To accurately detect the eyes the slope detection algorithm is used to calculate the slope between each of the clusters & it discards the clusters till finally the eyes are detected.

If ever the eyes are found then the driver is alert & there is no need of raising an alarm. But if the eyes are not found or are closed for a period of 3 s continuously then it is safe to assume that the alertness level has dropped down to certain level & the driver is dozing. In such a case the driver is alerted by raising an alarm.

where, (x1,y1) and (x2,y2) are any two points.

$$dist = Sqrt(x2 - x1)2 + (y2 - y1)2) \tag{6.1}$$

Distance formula

$$Slope = dy/dx \tag{6.2}$$

Slope formula

6.3.3 Post Processing

Depending on the state of the eyes found in the previous stage an appropriate decision is made & then displayed on screen (Fig. 6.2).

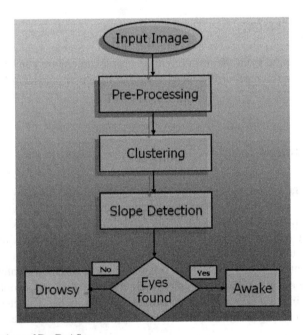

Fig. 6.2 Flowchart of DroDeASys

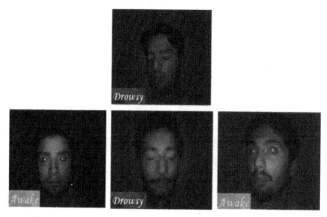

Fig. 6.3 Tested samples

6.4 Samples Tested

Following are the tested samples under conditions mentioned below:
1. Constant distance from camera
2. Tilted to one side
3. Looking straight in front
4. Deep drowsy state
5. Looking at a rear view mirror

6.5 Conclusion

As shown in the tested samples in Fig. 6.3, the system works on a variety of samples. The results of the experiments carried out on around 70 samples indicate a success rate of 90%. Thus this approach will be used to implement the system on a full-fledged basis.

References

1. Qing Ji and Xiaojie Yang, "Real-Time Eye, Gaze, and Face Pose Tracking for Monitoring Driver Vigilance," Real-Time Imaging, Vol. 8, pp. 357–377 (2002).
2. M. Yang, D. Kriegman, and N. Ahuja, "Detecting faces in Images: Survey," IEEE Trans. PAMI, Vol. 24, No. 1, pp. 34–58 (2002).
3. E. Hjelmas, "Face Detection: A Survey," Computer Vis. Image Understanding, 83, pp. 236–274 (2001).
4. Estimation of Face Direction Using Near-Infrared Camera Takuya Akashi, Hironori Nagayama, Minoru Fukumi, and Norio Akamatsu University of Tokushima.

Chapter 7
The Calculation of Axisymmetric Duct Geometries for Incompressible Rotational Flow with Blockage Effects and Body Forces

Vasos Pavlika

Abstract In this paper a numerical algorithm is described for solving the boundary value problem associated with axisymmetric, inviscid, incompressible and irrotational flow with a circumferentially arranged cascade of aerofoils placed in the duct. The governing equations are formulated in terms of the stream function $\psi(x,y)$ and the function $\varphi(x,y)$ as independent variables where for irrotational flow $\varphi(x,y)$ can be recognized as the velocity potential function, for rotational flow $\varphi(x,y)$ ceases being the velocity potential function but does remain orthogonal to the stream lines. A numerical method based on finite differences solving a Poisson type equation on a uniform mesh is employed. The technique described is capable of tackling the so-called inverse problem where the velocity wall distributions are prescribed from which the duct wall shape is calculated, as well as the direct problem where the velocity distribution on the duct walls are calculated from prescribed duct wall shapes. Results for the case of prescribing the radius i.e. the so called Dirichlet boundary conditions are given. A downstream condition is prescribed such that cylindrical flow, that is flow which is independent of the axial coordinate, exists. The numerical results are obtained by using Green's function for the Laplace equation on a rectangle. The presence of the blades has a bearing on the rate of mass flow and thus alters the usual equation of continuity.

7.1 Introduction

Designers of ducts require numerical techniques for calculating wall shapes from a prescribed velocity distribution [1–3]. The objective of the prescribed velocity is typically to avoid boundary layer separation. At inlet a velocity is prescribed to

V. Pavlika
Harrow School of Computer Science University of Westminster, School of Computer Science, Harrow, Middlesex, UK
e-mail: V.L.Pavlika@westminster.ac.uk

S.-I. Ao et al. (eds.), *Advances in Computational Algorithms and Data Analysis*,
Lecture Notes in Electrical Engineering 14,
© Springer Science+Business Media B.V. 2009

allow for a vorticity to be present calculated from $\underline{\omega} = \underline{\nabla} \wedge \underline{v}$ where the $\wedge$ denotes the usual cross product of vectors, $\underline{v}$ is the vorticity vector and v the velocity vector respectively.

The present paper describes a numerical algorithm for solving the boundary value problem that arises when the independent variables are φ and ψ where φ may be identified as the velocity potential function (for irrotational flow only), for flow with vorticity, φ ceases being the velocity potential function but does remain orthogonal to ψ which may be identified as the stream function. The dependent variable y, is the radial coordinate and x is the axial coordinate. The numerical technique is based on the finite difference scheme on a uniform mesh.

7.2 The Design Plane

As shown in Pavlika [4] when the independent variables are $\varphi(x,y)$ and $\psi(x,y)$ where the $\varphi(x,y)$ and $\psi(x,y)$ have been previously defined it can be shown that the governing partial differential equation that the radius satisfies is given by:

$$\frac{\partial}{\partial\varphi}\left(\frac{A}{B}\frac{\partial y}{\partial\varphi}\right) + \frac{\partial}{\partial\psi}\left(\frac{B}{A}\frac{\partial y}{\partial\psi}\right) = 0 \qquad (7.1)$$

with the speed calculated from

$$\frac{1}{q^2} = \frac{1}{A^2}\left(\frac{\partial y}{\partial\psi}\right)^2 + \frac{1}{B^2}\left(\frac{\partial y}{\partial\varphi}\right)^2 \qquad (7.2)$$

and completion of the physical coordinates are provided from

$$dx = \frac{B}{A}\frac{\partial y}{\partial\psi}d\varphi - \frac{A}{B}\frac{\partial y}{\partial\varphi}d\psi$$

where x is the axial coordinate and A and B satisfy their own first order quasi-linear hyperbolic partial differential equations with characteristics parallel to the φ and ψ axes which maps the physical flow field into an infinite strip in the (φ, ψ) plane. In fact the A and B satisfy:

$$\frac{\partial}{\partial\varphi}(\log(A)) = \frac{\eta}{q^2}B \qquad (7.3)$$

and

$$\frac{\partial}{\partial\psi}(\log(B)) = -\frac{\omega_\alpha}{q^2}A \qquad (7.4)$$

Regarding temporarily η, ω_α and q as known functions of φ and ψ the system (7.3) and (7.4) as previously mentioned is quasi-linear hyperbolic with characteristics parallel to the φ and ψ axes which maps the physical flow field into an infinite strip in the (φ, ψ) plane. Bearing in mind the freedom available in the stream wise variation of φ and the cross stream variation of ψ, suitable values of A can be prescribed along one φ characteristic and those of B can be prescribed along one ψ characteristic.

7.3 The Numerical Algorithm in the Design Plane

Rewriting the partial differential equation that y satisfies i.e. Eq. (7.1) as:

$$\frac{\partial}{\partial \phi}\left(C\frac{\partial y}{\partial \phi}\right) + \frac{\partial}{\partial \psi}\left(\frac{1}{C}\frac{\partial y}{\partial \psi}\right) = c \tag{7.5}$$

where $C = \frac{A}{B}$, for problems posed in the design plane $c = 0$, the value of C will vary depending on whether the flow field is irrotational or swirl free etc. Equation (7.5) will be re-written as a Poisson equation that is as:

$$\nabla^2 y = \frac{c}{C} + \left(1 - \frac{1}{C^2}\right)\frac{\partial^2 y}{\partial \psi^2} - \left(\frac{\partial}{\partial \phi}\log_e |C|\right)\frac{\partial y}{\partial \phi}$$
$$- \frac{1}{C}\frac{\partial}{\partial \psi}\left(\frac{1}{C}\right)\frac{\partial y}{\partial \psi} \tag{7.6}$$

where ∇^2 is the usual Laplacian operator

$$\nabla^2 \equiv \frac{\partial^2}{\partial \psi^2} + \frac{\partial^2}{\partial \phi^2}$$

so that

$$\nabla^2 y = g\left(\frac{\partial^2 y}{\partial \psi^2}, \frac{\partial y}{\partial \phi}, \frac{\partial y}{\partial \psi}, C, c\right)$$

where g is a function of the arguments shown as defined by expression (7.6). Writing in finite difference form using central differences gives:

$$Y^{(i-1)} + AY^{(i)} + Y^{(i+1)} = E^{(i)}, \tag{7.7}$$

where

$$\underline{Y}^{(i-1)} = \begin{bmatrix} y_{i-1,1} \\ y_{i-1,2} \\ \cdot \\ \cdot \\ y_{i-1,N} \end{bmatrix}, \underline{Y}^{(i)} = \begin{bmatrix} y_{i,1} \\ y_{i,2} \\ \cdot \\ \cdot \\ y_{i,N} \end{bmatrix}, \underline{Y}^{(i+1)} = \begin{bmatrix} y_{i+1,1} \\ y_{i+1,2} \\ \cdot \\ \cdot \\ y_{i+1,N} \end{bmatrix}$$

$$\underline{E}^{(i)} = h^2 \begin{bmatrix} g_{i,1} - y_{i,0} \\ g_{i,2} \\ \cdot \\ \cdot \\ g_{i,N} - y_{i,N+1} \end{bmatrix}$$

and

$$A = \begin{bmatrix} -4 & 1 & 0 & 0 & \cdot \\ 1 & -4 & 1 & & \\ 0 & 1 & -4 & \cdot & \cdot \\ & & & \cdot & 1 \\ & & & 1 & -4 \end{bmatrix}$$

on a uniform mesh with $\Delta\varphi = \Delta\psi = h$.

7.4 Direct Solution of the Difference Equations

The matrix-vector Eq. (7.7) is

$$Y^{(i-1)} + AY^{(i)} + Y^{(i+1)} = E^{(i)},$$

With all matrices of order (NxN), and column vectors $\underline{Y}^{(i)}$ and $\underline{E}^{(i)}$ of order N. To solve the vector recurrence relation a speculation is made that the $\underline{Y}^{(i-1)}$ vector can be related linearly to the $\underline{Y}^{(i)}$ vector as follows:

$$Y^{(i-1)} = B^{(i)}Y^{(i)} + K^{(i)} \qquad (7.8)$$

where the $B^{(i)}$ and the $K^{(i)}$ are at present unknown matrices and column vectors respectively. Substituting (7.8) into (7.7) gives

$$(W^{(i)}B^{(i)} + A)Y^{(i)} = -W^{(i)}K^{(1)} - E^{(i)}Y^{(i+1)}$$

$$Y^{(i)} = -(W^{(i)}B^{(i)} + A)^{-1}E^{(i)}Y^{(i+1)}$$
$$+ (W^{(i)}B^{(i)} + A)^{-1} - W^{(i)}K^{(i)})$$

but

$$Y^{(i)} = B^{(i+1)}Y^{(i+1)} + K^{(i+1)}$$

Thus equating coefficients implies

$$B^{(i+1)} = -(W(i)B(i) + A)^{-1}E(i) \qquad (7.9)$$

and

$$K^{(i+1)} = (W^{(i)}B^{(i)} + A)^{-1} - W^{(i)}K^{(i)})$$

For i = 0 this gives
$$Y^{(0)} = B^{(1)}Y^{(1)} + K^{(1)} \qquad (7.10)$$

To determine the $K^{(1)}$, if the first iterate $B^{(1)} = 0$ then

$$K^{(1)} = Y^{(0)}$$

The matrix and vector sequences are now defined by Eqs. (7.9) and (7.10) for i = 1 to M. The $Y^{(i)}$ vectors are now calculated starting from right to left (as $Y^{(M+1)}$ is known) using $Y^{(M)} = B^{(M+1)}Y^{(M+1)} + K^{(M+1)}$.

7.5 Axisymmetric Flow in the Presence of Body Forces

So that the algorithm can accommodate a larger class of flow problems as shown in Pavlika [5] equations are given for the general case of axisymmetric, inviscid and rotational flow in the presence of body forces. Numerical results are obtained however without incorporating the model including the body forces. Adopting cylindrical polar coordinates with y being the radial coordinate, a the circumferential and x the axial coordinate, defining velocity components u_y, u_α and u_x with corresponding vorticity components $\omega_y, \omega_\alpha, \omega_x$ in the direction of increasing y, α and x respectively, suppose also that there is a body force resolved into two components, one orthogonal to the flow direction and one parallel to the flow direction, then:

$$\underline{F_1} = F_{1,y}e_y + F_{1,\alpha}e_\alpha + F_{1,x}e_x \text{ and } \underline{F_2} = F_{2,y}e_y + F_{2,\alpha}e_\alpha + F_{2,x}e_x$$

Where e_y, e_α and e_x are unit vectors in the direction of increasing y, α and x respectively. Imposing orthogonality between $\underline{F_1}$ and $\underline{u}$ implies that $\underline{u}.\underline{F_1} = 0 \Rightarrow$ $u_y F_{1,y} + u_\alpha F_{1,\alpha} + u_x F_{1,x} = 0$ with equation of continuity with unit density giving:

$$\frac{D\underline{u}}{Dt} = \underline{F} - \underline{\nabla}.\underline{p} \tag{7.11}$$

Where $\frac{D}{Dt}$ is the material derivative. Equation (7.11) can be written using well known vector identities as:

$$\frac{\partial u_y}{\partial t} + u_x \frac{\partial u_y}{\partial x} + u_y \frac{\partial u_y}{\partial y} - \frac{u_\alpha^2}{y} = F_y - \frac{\partial p}{\partial y}$$

$$\frac{\partial u_\alpha}{\partial t} + u_x \frac{\partial u_\alpha}{\partial x} + u_y \frac{\partial u_\alpha}{\partial y} - \frac{u_\alpha u_y}{y} = F_\alpha$$

$$\frac{\partial u_x}{\partial t} + u_x \frac{\partial u_x}{\partial x} + u_y \frac{\partial u_x}{\partial y} = F_x - \frac{\partial p}{\partial x} \tag{7.12}$$

Furthermore

$$\frac{\partial \underline{u}}{\partial t} + (\underline{u}.\underline{\nabla})\underline{u} = \underline{F} - \underline{\nabla}.\underline{p}$$

can be written (once again using an appropriate vector identity as)

$$\frac{\partial \underline{u}}{\partial t} + (\underline{\omega} \wedge \underline{u}) = -\underline{\nabla}(\underline{p} + \frac{1}{2}q^2) - \underline{F}$$

Thus for steady flow Crocco's form of the equation of motion is obtained, i.e.

$$(\underline{u} \wedge \underline{\omega}) = \underline{\nabla} H - \underline{F} \qquad (7.13)$$

where H is the total head defined by $H = p + \frac{1}{2}q^2$. So that

$$\underline{u}.(\underline{u} \wedge \underline{\omega}) = \underline{u}.\underline{\nabla} H - \underline{u}.\underline{F}_1 - \underline{u}.\underline{F}_2$$
$$\Rightarrow \underline{u}.\underline{\nabla} H = \underline{u}.\underline{F}_2$$

so for axisymmetric flow this becomes

$$u_y \frac{\partial H}{\partial y} + u_x \frac{\partial H}{\partial x} = u_y F_{2,y} + u_\alpha F_{2,\alpha} + u_x F_{2,x}$$

and Eq. (7.12) (axial component) gives

$$u_y \omega_\alpha = \frac{\partial H}{\partial x} + u_\alpha \omega_x - (F_{1,x} + F_{2,x})$$

with

$$\omega_y = -\frac{\partial}{\partial x}(y u_\alpha)$$

now

$$\frac{\partial H}{\partial x} = \frac{\partial \varphi}{\partial x} \frac{\partial H}{\partial \varphi} + \frac{\partial \psi}{\partial x} \frac{\partial H}{\partial \psi}$$

thus

$$\frac{\partial H}{\partial x} = \frac{u_x}{B} \frac{\partial H}{\partial \varphi} - y u_y \frac{\partial H}{\partial \psi}$$
$$\Rightarrow \frac{\partial H}{\partial x} = -\frac{u_y}{u_x} \frac{\partial H}{\partial y} + F_{2,x} + \frac{u_y}{u_x} F_{2,y} + \frac{u_\alpha}{u_x} F_{2,\alpha}$$

the following operator relations are valid

$$\frac{\partial}{\partial x} \equiv \frac{\partial \varphi}{\partial x} \frac{\partial}{\partial \varphi} + \frac{\partial \psi}{\partial x} \frac{\partial}{\partial \psi}$$
$$\frac{\partial}{\partial y} \equiv \frac{\partial \varphi}{\partial y} \frac{\partial}{\partial \varphi} + \frac{\partial \psi}{\partial y} \frac{\partial}{\partial \psi}$$

and

$$\frac{\partial}{\partial \psi} \equiv \frac{\partial x}{\partial \psi} \frac{\partial}{\partial x} + \frac{\partial y}{\partial \psi} \frac{\partial}{\partial y}$$
$$\frac{\partial}{\partial \varphi} \equiv \frac{\partial x}{\partial \varphi} \frac{\partial}{\partial x} + \frac{\partial y}{\partial \varphi} \frac{\partial}{\partial y}$$

so that

$$\frac{\partial H}{\partial x} = -\frac{u_y^2}{B u_x} \frac{\partial H}{\partial \varphi} - y u_y \frac{\partial H}{\partial \psi} + F_{2,x} + \frac{u_y}{u_x} F_{2,y} + \frac{u_\alpha}{u_x} F_{2,\alpha}$$

and hence

$$\frac{\omega_\alpha}{y} = -\frac{u_y}{Byu_x}\frac{\partial H}{\partial \varphi} - \frac{\partial H}{\partial \psi} + \frac{F_{2,y}}{yu_x} + \frac{u_\alpha}{u_xu_{yy}}F_{2,\alpha} + \frac{u_\alpha u_x}{y^2 u_y B}\frac{\partial}{\partial \varphi}(yu_\alpha) + \frac{u_\alpha}{y}\frac{\partial}{\partial \psi}(yu_\alpha)$$

(7.14)

and for axisymmetric flow the vorticity vector $\underline{\omega}$ becomes

$$\underline{\omega} = \underline{\nabla}\wedge\underline{u} = \left\{-\frac{\partial u_\alpha}{\partial x}\right\} + \left\{\frac{\partial u_y}{\partial x} - \frac{\partial u_x}{\partial y}\right\}\underline{\alpha} + \left\{\frac{1}{y}\frac{\partial(yu_\alpha)}{\partial y}\right\}\underline{x}$$

(7.15)

furthermore the axial velocity profile is constant given by $u_x(y) = \beta$, where β, is a constant and the swirl velocity $u_\alpha(y)$, will be of the form $u_\alpha(y) = \frac{l}{y}$ where l is a constant and the term l/y represents the so-called free vortex term.

7.6 The Blockage Effect: Deriving the Additional Flow Equation Due to the Circumferentially Arranged Aerofoils

In deriving the additional flow equation the effect of the circumferentially arranged blades placed in the duct must be considered. The blades effect the rate of mass flow η, considering Fig. 7.1, with $k \equiv k(x,y)$ representing the blockage effect, the mass flow into and out of the fluid element is:

Face A:

$$\left(2u_y + \frac{\partial u_y}{\partial x}\delta x\right)\delta x\left(2k + \frac{\partial k}{\partial x}\delta x\right) + O((\delta x)^2)$$

Face B:

$$-\left(2u_x + 2\frac{\partial u_x}{\partial x}\delta x + \frac{\partial u_x}{\partial y}\delta y\right)*$$
$$\delta y\left(2k + 2\frac{\partial k}{\partial x}\delta x + \frac{\partial k}{\partial y}\delta y\right)$$
$$+ O((\delta x)^2) + O((\delta y)^2)$$

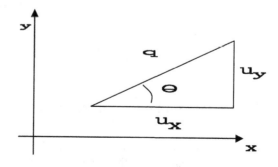

Fig. 7.1 The meridional plane

Face C:

$$-\left(2u_y+\frac{\partial u_y}{\partial x}\delta x+2\frac{\partial u_y}{\partial y}\delta y\right)*$$

$$\delta x\left(2k+2\frac{\partial k}{\partial y}\delta y+\frac{\partial k}{\partial x}\delta x\right)$$

$$+O((\delta x)^2)+O((\delta y)^2)$$

Face D:

$$\left(2u_x+\frac{\partial u_x}{\partial y}\delta y\right)\delta y\left(2k+\frac{\partial k}{\partial y}\delta y\right)$$

$$+O((\delta y)^2)$$

summing these terms, using the principle of conservation of mass and taking the limit as $\delta x \to 0, \delta y \to 0$ gives:

$$\lim_{\substack{\delta x \to 0 \\ \delta y \to 0}}\left(\frac{massflow}{4\delta x\delta y}\right)=\frac{\partial(ku_x)}{\partial x}+\frac{\partial(ku_y)}{\partial y}=0$$

which may be identified as the continuity equation for two dimensional compressible flow with the density term being replaced by the blockage factor k.

7.7 The Blockage Function k(x,y)

The blockage function k(x,y) is defined to be of the form $k(x,y) = 1 - \frac{\lambda(x)}{T(y)}$ where the function $\lambda(x)$ represents the contour shape of the aerofoil and the term T(y) is a scaling factor given by $T(y) = \frac{2\pi}{N}y$, where N is the number of blades (arbitrary). If the axial span of the aerofoil is x_1 then the function $\lambda(x)$ is defined to have a maximum at $x_1/5$ and $\lambda(x_1/5) = x_1/10$. Furthermore $\lambda(x)$ is chosen to vanish at $x = 0$ and $x = x_1$. Choosing $\lambda(x)$ to be of the form $\lambda(x) = cx^\alpha(x^\beta - x_1^\beta)$, where c is a constant. With $\beta = 1(arbitrary) \Rightarrow \alpha = \frac{1}{4}$, applying these conditions gives

$$c=-\frac{1}{8}\left(\frac{5}{x_1}\right)^{1/4} \text{ and } \lambda(x)=-\frac{1}{8}\left(\frac{5x}{x_1}\right)^{1/4}(x-x_1).$$

7.8 The Design Plane Counterparts

In order to compute numerical solutions in the design plane (with no body forces), expressions are required for the terms A, B and ω_α, thus

$$\frac{\partial u_x}{\partial x} + \frac{\partial u_y}{\partial y} = -\frac{1}{y}\left(u_x \frac{\partial y}{\partial x} + u_y\right)$$

$$= -q\frac{\partial}{\partial s}(\log(y))$$

or

$$\eta = -\frac{q^2}{B}\frac{\partial}{\partial \varphi}(\log(y)),$$

but

$$\eta = \frac{q^2}{B}\frac{\partial}{\partial \varphi}(\log(A))$$

thus $Ay = f(\psi)$, that is $\frac{\partial \psi}{\partial n} = \frac{yq}{f(\psi)}$. The arbitrary function $f(\psi)$ represents the freedom in the cross stream distribution of ψ and choosing $f(\psi)$ to be unity everywhere ψ can be identified as the usual Stokes stream function given by

$$\frac{\partial \psi}{\partial x} = -yu_y; \frac{\partial \psi}{\partial y} = yu_x$$

Equation (7.12), (circumferential component) gives

$$0 = u_x\frac{\partial(yu\alpha)}{\partial x} + u_y\frac{\partial(yu\alpha)}{\partial y}$$

Referring to the meridional plane Fig. 7.2, it may be deduced that

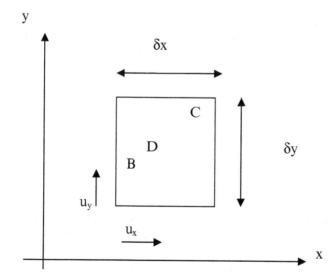

Fig. 7.2 A fluid element

$$u_x = q\frac{\partial x}{\partial s}; u_y = q\frac{\partial y}{\partial s}$$

$$\Rightarrow \frac{\partial}{\partial s}(yu_\alpha) = 0$$

$$\therefore yu_\alpha = C(\psi)$$

where $q = \frac{ds}{dt}$. In terms of $C(\psi)$ the vorticity vector (Eq. (7.15)) becomes

$$\underline{\omega} = \underline{\nabla} \wedge \underline{u} = \left\{-\frac{1}{y}\frac{\partial C}{\partial x}\right\}\underline{y} + \left\{\frac{\partial u_y}{\partial x} - \frac{\partial u_x}{\partial y}\right\}\underline{\alpha}$$

$$+ \left\{\frac{1}{y}\frac{\partial C}{\partial y}\right\}\underline{x}$$

$$= \omega_y \underline{y} + \omega_\alpha \underline{\alpha} + \omega_x \underline{x}, \text{ by definition.}$$

An expression for ω_α is required as this appears in the expression for B, so using the radial component of Eq. (7.14) (with no body forces) gives

$$\omega_\alpha = \frac{u_x}{u_\alpha}\left(\frac{1}{y}\frac{\partial C}{\partial y}\right) - \frac{1}{u_x}\frac{\partial H}{\partial y}$$

using the Stokes' stream function this becomes

$$\omega_\alpha = \frac{C(\psi)}{y}\left(\frac{dC}{d\psi}\right) - y\frac{dH}{d\psi}$$

which is the required expression to be used in calculation of B according to definition (7.4). If far upstream the flow is assumed to be cylindrical so that all quantities are independent of x, then with unit density the equation of motion and the Stokes' Stream function give

$$u_y = 0; \frac{\partial p}{\partial x} = 0; \frac{\partial p}{\partial y} = \frac{u_\alpha^2}{y}; \frac{\partial \psi}{\partial x} = 0; \frac{\partial \psi}{\partial y} = yu_x$$

giving

$$\omega_\alpha = \frac{C(\psi)}{y}\left(\frac{dC}{d\psi}\right) - \frac{y}{2}\frac{d}{d\psi}(u_x^2 + u_\alpha^2) - \frac{u_\alpha^2}{u_x y}$$

With $u_x(y) = \beta$ and $u_\alpha(y) = \frac{l}{y}$ as previously defined. Once $\frac{dH}{d\psi}$ has been calculated upstream it takes this value throughout the (ϕ, ψ) since as is self evident the expression is independent of ψ. This last expression for ω_α is required in the calculation of B and numerical coupling with Eq. (7.1) gives the numerical solution in the design plane.

7.9 Downstream Conditions

Downstream a cylindrical flow condition as discussed below will be prescribed. Defining the pressure function $H(\psi)$ and the function $C(\psi)$ as

$$H(\psi) = \frac{1}{2}(u_x^2 + u_\alpha^2) + \frac{p}{\rho} \text{ and } C(\psi) = yu_\alpha$$

for cylindrical flow radial equilibrium (from Eq. (7.12)) (with no body forces) radial component gives:

$$\frac{1}{\rho}\frac{dp}{dy} = \frac{u_\alpha^2}{y}$$

Integrating gives

$$\frac{1}{\rho}(p - p_{y-inner}) = \int_{y-inner} \frac{u_\alpha^2}{y} dy = \int_{y-inner} \frac{C^2(\psi)}{y^3} dy$$

Which gives $H(\psi)$ as

$$H(\psi) = \frac{1}{2}(u_x^2 + u_\alpha^2) + \frac{p_{y-inner}}{\rho}$$
$$+ \int_{y-inner} \frac{C^2(\psi)}{y^3} dy$$

Now $\int_{y-inner} \frac{C^2(\psi)}{y^3} dy = -\frac{1}{2} \int_{y-inner} C^2 d(1/y^2)$

$$= -\frac{1}{2}\left[\frac{C^2}{y^2} - \left(\frac{C^2}{y^2}\right)_{y-inner}\right] + \frac{1}{2}\int_{y-inner} \frac{1}{y^2}\frac{dC^2}{dy} dy$$

Therefore

$$H(\psi) = \frac{1}{2}u_x^2 + \frac{p_{y-inner}}{\rho} + \frac{1}{2}(u_\alpha^2)_{y-inner}$$
$$+ \int_{\psi=0} \frac{1}{y^2}\frac{dC^2}{d\psi} d\psi$$

Suppose $u_{x,1} = u_{x,1}(\psi)$ and $u_{\alpha,1} = u_{\alpha,1}(\psi)$, where the subscript 1 denotes upstream conditions, then $u_{x,2} = u_{x,2}(\psi)$ and $u_{\alpha,2} = u_{\alpha,2}(\psi)$ are required as functions of ψ, where the subscript 2 similarly denoting downstream conditions, so that

$$\frac{1}{2}u_{x,2}^2 = H(\psi) - \frac{p_{2,inner}}{\rho} - \frac{1}{2}(u_{\alpha,2}^2)_{inner}$$

$$- \frac{1}{2}\int_{\psi=0} \frac{1}{y_1^2}\frac{dC^2}{d\psi}d\psi \qquad (7.16)$$

and

$$\int_{\psi=0} \frac{d\psi}{u_{x,2}}d\psi = \frac{1}{2}\left(y_2^2 - y_{2,inner}^2\right)$$

Furthermore $C(\psi) = y_1 u_{\alpha,1} = y_2 u_{\alpha,2}$, and Eq. (7.16) now give

$$\frac{1}{2}u_{x,2}^2 = \frac{1}{2}u_{x,1}^2 + \frac{p_{1,inner}}{\rho} - \frac{p_{2,inner}}{\rho}$$

$$+ \frac{1}{2}((u_{\alpha,1}^2)_{inner} - (u_{\alpha,2}^2)_{inner})$$

$$+ \frac{1}{2}\int_{\psi=0} \left(\frac{1}{y_1^2} - \frac{1}{y_2^2}\right)d(C^2)$$

or

$$u_{x,2}^2 = u_{x,1}^2 + K + \int_{\psi=0} \left(\frac{1}{y_1^2} - \frac{1}{y_2^2}\right)d(C^2) \qquad (7.17)$$

where

$$K = 2\left(\frac{p_{1,inner}}{\rho} - \frac{p_{2,inner}}{\rho}\right) + (u_{\alpha,1}^2)_{inner} - (u_{\alpha,2}^2)_{inner}$$

$$\text{and } y_2^2 = y_{2,inner}^2 + 2\int_{\psi=0} \frac{d\psi}{u_{x,2}} \qquad (7.18)$$

with $u_{x,2}$ in this case given by (7.17).

7.10 Calculation Procedure

The calculation of the downstream radii $y_2(\psi)$ follow from Eq. (7.18) with $u_{x,2}$ given by Eq. (7.17), which can be written as

$$u_{x,2}^2 = g(\psi) + K, \text{ where } g(\psi) = u_{x,1}^2 + \int_{\psi=0} \left(\frac{1}{y_1^2} - \frac{1}{y_2^2}\right)\frac{d(C^2)}{d\psi}d\psi \qquad (7.19)$$

In order to calculate the $(n+1)^{th}$ iterate it is known that:

$$\frac{\partial}{\partial K}(y_{2,outer}^2) = 2\int_{\psi=0} \frac{\partial}{\partial K}\left(\frac{d\psi}{\sqrt{g(\psi)+K}}\right) = -\int_{\psi=0}^{\Psi} \frac{d\psi}{\left(u_{x,2}^3\right)^{(n)}}$$

but

$$\left(\frac{\partial}{\partial K}(y^2_{2,outer})\right)^{(n)} = \frac{(y^2_{2,outer})^{(n+1)} - (y^2_{2,outer})^{(n)}}{K^{(n+1)} - K^{(n)}} \tag{7.20}$$

from which as can be seen from Eq. (7.20) the $K^{(n)}$ must be calculated iteratively with $K^{(0)} = 0$, Once the $K^{(n+1)}$ has been calculated it is introduced into Eq. (7.19), giving rise to a new $(u^2_{x,2})^{(n+1)}$ which in turn gives a new $(y^2_{x,2})^{(n+1)}$ from Eq. (7.18) and the process repeated until some convergence criteria is satisfied.

7.11 Prescription of Wall Geometries

In this paper the Dirichlet boundary conditions will be prescribed on the wall boundaries so that it is the radii values, y that are given as a function of φ on the boundaries. The function chosen to give a y distribution is based on the hyperbolic tangent, choosing $y(\varphi) = Ct\ anh(a\varphi + b) + k$ where C, a, b and k are constants, applying the conditions that $y = y_u$ at $\varphi = 0$ and $y = y_d$ at $\varphi = \Phi$ taking $a\Phi + b = 3$(arbitrary) and $b = -3$, so that $\tanh(a\Phi + b) \approx 1$ and $\tanh(b) \approx -1$, then it follows that

$$y(\varphi) = \left(\frac{y_d - y_u}{2}\right)\tanh(a\varphi + b) + \left(\frac{y_d + y_u}{2}\right) \tag{7.21}$$

replacing φ, by x in Eq. (7.21) gives a y(x) distribution. The inner radius is prescribed to be given by a hyperbolic tangent function in this paper. The geometries produced are shown in Figs. 7.3–7.5 respectively.

7.12 Alternative Solution Using an Integral Formula Based on Green's Theorem

Here a second method of solution is derived using an integral formula. Commencing with the generalized form of Green's theorem for the self adjoint elliptic operator E(t) in normal form given by:

$$\int\int_R vE(t) - tE^{(A)}(v)d\varphi d\psi = \oint_C t\frac{\partial v}{\partial n} - v\frac{\partial t}{\partial n}ds$$

where $t = y^2$, where $E(u) = E^{(A)}(u)$ is the adjoint of E and v is the fundamental solution to the adjoint equation. In this case the adjoint equation is given by $\nabla^2 v = 0$ and $E(t) = g$ as defined by Eq. (7.6). The contour C bounding the surface R is traversed in the counter clockwise sense. For a doubly connected region introducing a singularity at the point (φ_0, ψ_0) (inside or on the contour C) and assuming $v(\varphi, \psi) =$

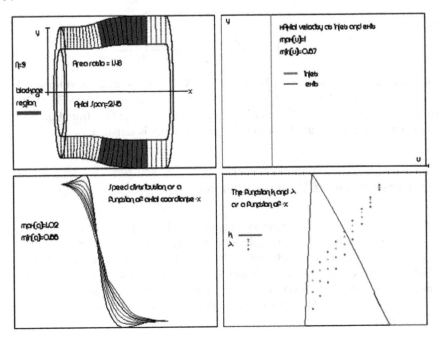

Fig. 7.3 The geometry and speed distribution produced with N = 3

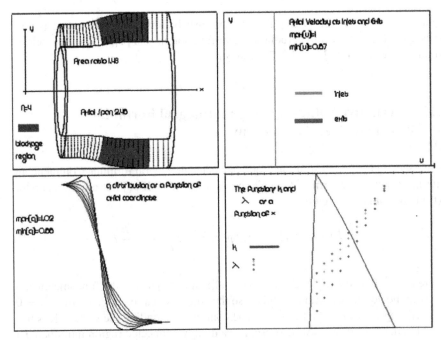

Fig. 7.4 The geometry and speed distribution produced with N = 4

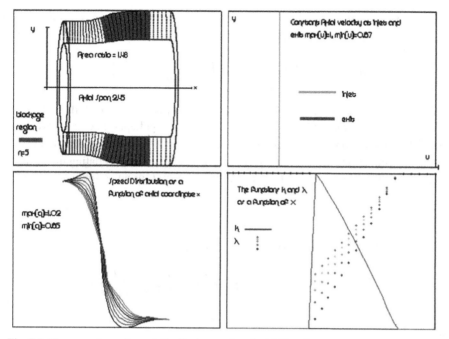

Fig. 7.5 The geometry and speed distribution produced with N = 5

$F(\varphi, \psi) \log_e |r|$ so that the distance r is given by: $r = \left((\varphi - \varphi_0)^2 + (\psi - \psi_0)^2 \right)^{1/2}$ with $F(\varphi, \psi)$ analytic, then it can be shown that

$$m\pi t(\varphi_0, \psi_0) F(\varphi_0, \psi_0) = \oint_C t \frac{\partial v}{\partial n} - v \frac{\partial t}{\partial n} ds \qquad (7.22)$$

$$- \iint_R vg \left(\frac{\partial^2 t}{\partial \varphi^2}, \frac{\partial^2 L}{\partial \varphi^2} \right) d\varphi d\psi, \quad m = 1, 2 \qquad (7.23)$$

with $L = \log_e t$. Now m = 2 if (φ_0, ψ_0) is within C and m = 1 if (φ_0, ψ_0) is on C (the m = 1 case can be shown using the appropriate Plemelj formulae or by indenting the contour at (φ_0, ψ_0)). For the Dirichlet case of boundary condition of $t(\varphi, \psi)$ he requirement is that $v(\varphi, \psi) = 0$ on C in addition to $v(\varphi, \psi)$ being harmonic and for the Neumann conditions on t, the requirement is that $\frac{\partial v}{\partial n} = 0$ and $v(\varphi, \psi)$ once again satisfying Laplace's equation. Much literature is available for the Green's function for the Laplace equation (see Williams [6]) and need not be mentioned here. Hence for the Dirichlet problem setting $F(\varphi, \psi) = 1 \forall \varphi, \psi$ without loss of generality and for interior points:

$$2\pi t(\varphi_0, \psi_0) = - \oint_C t \frac{\partial v_D}{\partial n} ds - \iint_R v_{Dg} \left(\frac{\partial^2 t}{\partial \varphi^2}, \frac{\partial^2 L}{\partial \varphi^2} \right) d\varphi d\psi \qquad (7.24)$$

i.e., the Green's function v_D satisfies Laplace's equation on $\partial\Omega$ where $\partial\Omega$ is defined by: $\varphi_0 \leq \varphi \leq \varphi_{M+1}$, $\psi_0 \leq \psi \leq \psi_{N+1}$ and vanishes on C. For the Neumann problem

$$2\pi t(\varphi_0, \psi_0) = -\oint_C v_N \frac{\partial t}{\partial n} ds - \int\int_R v_N g\left(\frac{\partial^2 t}{\partial\varphi^2}, \frac{\partial^2 L}{\partial\varphi^2}\right) d\varphi d\psi$$

Which give integral formulae for the square of the radius t, from which the radius y can be determined, above Green's function v_N satisfies the Laplace equation on $\partial\Omega$ with $\frac{\partial v_N}{\partial n}$ vanishing on C. Knowledge of the derivatives $\frac{\partial t}{\partial\varphi}$ and $\frac{\partial t}{\partial\psi}$ are also required for the determination of the speed q given by Eq. (2) hence differentiating under the integral sign above with respect to φ and ψ gives integral formulae for both $\frac{\partial t}{\partial\varphi}$ and $\frac{\partial t}{\partial\psi}$, such that:

$$2\pi\frac{\partial}{\partial\psi}t(\varphi_0, \psi_0) = \oint_C \frac{\partial t}{\partial\psi}\cdot\frac{\partial v_D}{\partial n} + t\frac{\partial^2 v_D}{\partial\psi\partial n} ds$$

$$- \int\int_R v_D\frac{\partial}{\partial\psi}\left(g\left(\frac{\partial^2 t}{\partial\varphi^2}, \frac{\partial^2 L}{\partial\varphi^2}\right)\right) + \frac{\partial v_D}{\partial\psi}\cdot g\left(\frac{\partial^2 t}{\partial\varphi^2}, \frac{\partial^2 L}{\partial\varphi^2}\right) d\varphi d\psi$$

and similarly for $\frac{\partial t}{\partial\varphi}$

$$2\pi\frac{\partial}{\partial\varphi}t(\varphi_0, \psi_0) = \oint_C \frac{\partial t}{\partial\varphi}\cdot\frac{\partial v_D}{\partial n} + t\frac{\partial^2 v_D}{\partial\varphi\partial n} ds$$

$$- \int\int_R v_D\frac{\partial}{\partial\varphi}\left(g\left(\frac{\partial^2 t}{\partial\varphi^2}, \frac{\partial^2 L}{\partial\varphi^2}\right)\right) + \frac{\partial v_D}{\partial\varphi}\cdot g\left(\frac{\partial^2 t}{\partial\varphi^2}, \frac{\partial^2 L}{\partial\varphi^2}\right) d\varphi d\psi$$

7.13 Iterative Solution

To convert formula (7.24) to a system of linear algebraic equations the point $t(\varphi_0, \psi_0)$ inside C is related to its boundary values on C. To obtain the first iterates $t_i^{(1)}(\varphi_0, \psi_0)$, $g_i^{(0)}$ is set equal to zero, so that

$$2\pi t_i^{(1)} = \sum_j^{2N+2M+4} \left(\frac{\partial v_D}{\partial n}\right)_j t_j \Delta s_j \quad \iota = 0, 1, 2, \ldots 2N+2M+4$$

Using the trapezoidal rule

$$\pi t_i^{(1)} = \sum_j^{2N+2M+4} \frac{1}{4}\left(\frac{\partial v_D}{\partial n}\right)_j (s_{j+1} - s_{j-1}) t_j$$

$$\iota = 0, 1, 2, \ldots 2N+2M+4$$

Where $K(v_D, s) = \frac{1}{4\pi} \left(\frac{\partial v_D}{\partial n} \right) (s_{j+1} - s_{j-1})$

Using this method there is a simple self-consistency check. i.e. the t_j are known upstream and downstream for $j = 0, 1, 2, \ldots, N+1$ and $j = N+M+3, N+M+4, \ldots .2N+M+3$, hence the first iteration may be written as:

$$
\begin{bmatrix}
1 - K_{N+2} & -K_{N+3} & \cdot & \cdot & -K_{2N+2M+4} \\
-K_{N+2} & 1 - K_{N+3} & -K_{N+4} & \cdot & -K_{2N+2M+4} \\
\cdot & & \cdot & \cdot & \cdot \\
& & & \cdot & \cdot \\
-K_{N+2} & \cdot & \cdot & \cdot & -K_{2N+2M+4}
\end{bmatrix}
\begin{bmatrix}
t_{N+2} \\
t_{N+3} \\
\cdot \\
\cdot \\
t_{2N+2M+4}
\end{bmatrix}
$$

$$
=
\begin{bmatrix}
\sum_j K_j t_j \\
\sum_j K_j t_j \\
\cdot \\
\cdot \\
\sum_j K_j t_j
\end{bmatrix}
$$

so that $A^{(1)} \underline{t}^{(1)} = \underline{b}^{(i)}$ (7.25)

where the summations on the right hand side are performed over $j = 0, 1, \ldots N+1$ and $j = M+N+3, M+N+4 \ldots, 2M+N+3$. Once the first iterate t_j has been calculated the field integral containing g is then computed, where the central difference approximation to the second derivative is used, this is then introduced into the right hand side of Eq. (7.25) and compute the second iterate $\underline{t}^{(2)}$. The procedure is repeated until some convergence criteria is satisfied e.g. $\| t_i^{(k)} - t_i^{k-1} \|_p < \varepsilon$, where ε is a constant and the p denotes the p-norm ($p = 1, 2$ or ∞).

7.14 Conclusions

As shown, geometries have been produced subject to given upstream and downstream conditions with prescribed Dirichlet boundary conditions. In this case vorticity at inlet has been specified by defining the axial velocity to be of the form $u_x(y) = \beta$ and the swirl velocity of the form $u_\alpha(y) = \frac{l}{y}$, where l is a constant, defining the so-called free vortex whirl respectively. The downstream conditions where such that: cylindrical flow was present, Dirichlet boundary conditions were prescribed, however the case with Neumann conditions can be accommodated using the algorithm, in addition so can the case with Robin boundary condition. Further examples of the algorithm with a combination of boundary condition is given in Pavlika [5]. It was found that at most eight iterations were required to achieve an acceptable level of convergence, with the technique accelerated using Aitken's Method.

References

1. Cousins. J,M., Special Computational Problems Associated with Axisymmetric Flow in Turbo-machines, Ph.D. thesis (CNAA), 1976.
2. Curle, N and Davies, H.J., Modern Fluid Dynamics, van Nostrand Reinhold, New York, 1971 Chapter .
3. Klier, M., Aerodynamic Design of Annular Ducts, Ph.D. thesis (CNAA), 1990 Chapter 1.
4. Pavlika, V., Vector Field Methods and the Hydrodynamic Design of Annular Ducts, Ph.D thesis, University of North London, Chapter VI, 1995.
5. Pavlika, V., Vector Field Methods and the Hydrodynamic Design of Annular Ducts, Ph.D. thesis, University of North London, Chapter VIII, 1995.
6. Williams, W.E., Partial Differential Equations, Clarendon Press, Oxford, 1980.

Chapter 8
Fault Tolerant Cache Schemes

H.-yu. Tu and Sarah Tasneem

Abstract Most of modern microprocessors employ on-chip cache memories to meet the memory bandwidth demand. These caches are now occupying a greater real estate of chip area. Also, continuous down scaling of transistors increases the possibility of defects in the cache area which already starts to occupies more than 50% of chip area. For this reason, various techniques have been proposed to tolerate defects in cache blocks. These techniques can be classified into three different categories, namely, cache line disabling, replacement with spare block, and decoder reconfiguration without spare blocks. This chapter examines each of those fault tolerant techniques with a fixed typical size and organization of L1 cache, through extended simulation using SPEC2000 benchmark on individual techniques. The design and characteristics of each technique are summarized with a view to evaluate the scheme. We then present our simulation results and comparative study of the three different methods.

Keywords Fault tolerant · Cache block disabling · Spare cache · PADded

8.1 Introduction

High-performance VLSI processors results increasing demand of memory bandwidth which the memory technology never be able to satisfy [1]. Thus, extensive use of on-chip cache memories became essential to sustain memory bandwidth demand of the CPU. Meanwhile, advances in semiconductor technology and continuous down scaling of feature size creates extra-space for additional functionality on a single chip. The most popular way to make use of this extra-space is integrating bigger size of cache so that a microprocessor is able to gain higher performance. We can observe that the area occupied by cache in current processor already exceeded 50%

H.-yu. Tu (✉) and S. Tasneem
Eastern Connecticut State University, Willimantic, CT 06226, USA
e-mail: tuH@easternct.edu

S.-I. Ao et al. (eds.), *Advances in Computational Algorithms and Data Analysis*,
Lecture Notes in Electrical Engineering 14,
© Springer Science+Business Media B.V. 2009

of total area of CPU die. However, an increase in the circuit density is closely coupled with an increase in probability of defect. Furthermore, the increased defects can mostly be in the on-chip cache area since the area occupied by cache grows larger and larger. Consequently, the defect level of the cache has a significant impact on the defect level of overall CPU. Therefore, the first fault tolerant cache design was proposed in [2] for the purpose of enhancement in yield of micro-processors.

To design fault tolerant cache, we first need to observe that cache is a redundant structure which is employed to enhance the performance of CPU. Thus, the CPU can work correctly without cache-memory. Among many components in microprocessor, redundant structures, such as cache is called *non-critical component* and defects in that structure is called *non-critical defects* [2]. The *non-critical defects* can be easily tolerated by simply disabling *non-critical component* which contains defect. Thus, disabling defective part of caches were investigated in [2–4]. However, simply disabling the defective part of cache will result degradation of overall performance of CPU. Thus, use of redundancy to tolerate defect in cache memory is studied in the literature. The redundancy techniques that are used for RAM can easily applied to cache. Using a SEC-DED code [5] codes can mask out defective bits in cache as well as main memory. However, a detailed investigation in [2] showed that employing SEC-DED code for on-chip cache is not appropriate due to the delay introduced by SEC-DED hardware. Using redundancy and reconfiguration logic is another method to tolerate faults in cache by providing spare cache blocks [6, 7]. The defective block is switched to spare block by reconfiguration mechanism. The reconfiguration can be done either electrical or laser fuses to permanently replace defective blocks [6]. In [7], instead of permanent replacement, they employed small fully associative cache to dynamically replace the faulty block.

Yet another technique which called *PADded cache* [8] is presented recently. This new technique can sooth the degradation of cache performance without spare cache block. Instead of using explicit spare blocks, the physical or logical neighborhood blocks play a role of spare block.

In this chapter, we examine and compare three different fault tolerant schemes, namely, *cache line disabling* [2–4], spare cache [7], and *PADded cache* [8]. In Section 8.2, a brief overview of the organizations of the different schemes as well as summary of previous results are presented. In Section 8.3, we evaluate each technique with realistic, unbiased setup for fair comparison. Also, the results of our comparisons of the schemes are reported.

8.2 Fault Tolerant Caches

8.2.1 Cache Block Disabling

In general, redundancy, which is explicitly provided, is used as the main technique to tolerate defect/fault. However, cache is provided not for the fault tolerance but for the performance enhancement. Thus, disabling the faulty cache will not affect the correct operation of CPU. However, disabling entire cache will significantly degrade computer performance. Thus, one may consider disabling some portion of cache, for example, disable one unit (way) of set associative cache [3]. However, purging the entire way is wasteful since all the other fault free blocks in the same unit cannot be utilized. The opposite extreme solution is disabling faulty byte or word in data array to maximize the utilization of fault free bits. However, most cache implementation fetches data from the main memory by the size of multiple words (*i.e. transfer block size*), instead of by a byte or a word, which is usually the same as block size of cache. Hence, disabling a single block containing fault/defect was investigated in [2, 4], where a single bit is used to indicate the presence of fault in a block. This indicator bit is called the *availability bit* [2], the *purge bit*, or the *second valid bit* [4]. In the present chapter, we will call this indicator bit as *faulty-bit*.

The faulty blocks (*i.e. blocks containing defect/s*) can be identified either by (a) manufacturing test to enhance the yield of micro-processors [2, 4] or (b) error detection code to tolerate permanent fault occurred even during normal operation [5, 9]. The *faulty-bit* of a faulty block will be set once the block is identified as faulty. When access to the certain address of cache occurs, the cache control logic makes use of the *faulty-bit* by treating the access as a miss and also excludes the block from cache replacement algorithm. Figure 8.1 illustrates simplified block diagram of this scheme. The effect of disabling faulty block is first presented in [2] and mathematically extended in [4].

According to their result and analysis, the performance degradation incurred by disabling defective block is sensitive to the cache organization. The associativity of cache, especially, has the biggest impact on this sensitivity. In case of direct mapped cache, all the memory blocks that are mapped onto defective block will be excluded from the cache. Thus, one can expect the linear degradation of performance on fraction of faulty block. On the other hand, set associative cache has

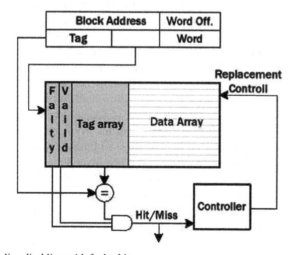

Fig. 8.1 Cache line disabling with faulty bit

less degradation ratio to the fraction of faulty block. Suppose M-way set associative cache. If one block among M-way in a same set is defective, the remaining $(M - 1)$ healthy blocks are still able to accommodate the corresponding memory blocks that are mapped onto the set. However, the replacement rate on a faulty set will increase due to higher probability of conflict miss by decreased number of ways in that set. A fully associative cache always allows every memory block to be cached in every cache blocks. Therefore, the degradation of cache performance would solely depend on the probability of conflict miss. To overcome this, the idea of using small fully associative spare cache has been evolved which is described in the next subsection.

The cache block disabling method is the most primitive and therefore, has little overhead (single bit for each block). Many other fault-tolerant cache schemes as explained in following sub-sections are based on this method.

8.2.2 Replacement Using Small Spare Cache

The crux of fault-tolerant technique is the use of redundancy and replacement. Even though the cache itself is a redundant component of main memory and simply disabling them would guarantee correct operation of CPU, the performance degradation can be significant.

To recover the performance loss due to disabling faulty blocks, a replacement scheme called the *Memory Reliability Enhancement Peripheral* (MREP) is discussed in [6]. The main idea is to provide extra words which can replace any faulty words in memory. Using this scheme on cache memory, there is no performance loss if the number of faulty block is less than the number of spare words. However, number of spare block will limit the capability of recovery. A similar method is used in [7]. While MREP replaces a faulty block with the a spare which is dedicated for the specific faulty block, Vergos and Nikolos [7] used small fully associative cache as a spare for the faulty blocks of direct mapped primary cache. Since any blocks in fully associative spare cache – hereafter called *spare cache* – can store data of all possible indexes that is used for its primary cache, it can temporary replace more faulty blocks than the number of spare blocks in *spare cache*.

The organization of spare cache scheme is illustrated in Fig. 8.2. When cache is accessed, the primary cache and *spare cache* is accessed with the same address simultaneously. In case of read operation, cache hit from *spare cache* will override hit/miss signal from primary cache regardless of faulty status of primary cache block. On the other hand, write operation should be treated more carefully. If write access occurs to the primary cache, the controller will check if the addressed block is faulty or not. In case of write operation on faulty block, the controller redirects write-access to the *spare cache*. The replacement will also be controlled by the status of *faulty-bit* of replacement candidate in primary cache.

Use of *spare cache* for direct-mapped cache and its performance recovery are extensively studied in [7]. In order to limit our scope of study on various fault-tolerant

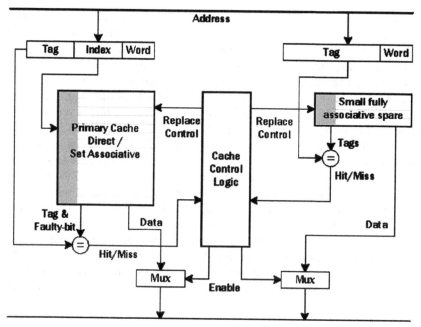

Fig. 8.2 Block diagram of a fully associative spare cache

cache schemes, it is worth to make some observation from their study. First, they defined a matrix called MR (*Miss-recovery Rate*) to measure the effectiveness of *spare cache* which is:

$$MR = \frac{Misses\ removed\ by\ spare\ cache}{Misses\ caused\ by\ faults} \times 100\% \qquad (8.1)$$

Based on MR, their study on *spare cache* can be summarized as follows:

1. MR decreases as number of faulty block increases.
2. For a constant block size and number of faulty blocks, MR decreases as total cache size decreases.
3. Block size of 16 or 32 maximizes MR.
4. One or two blocks are sufficient for a small number of faulty blocks.

It is obvious to see that the first and second results are due to the increased conflict miss in *spare cache*. As the number of faulty block increases, more blocks will contend to occupy limited number of blocks in *spare cache* resulting more conflict misses in *spare cache*. Also, as the size of the primary cache decreases, the number of addresses which is mapped onto faulty block will increase and eventually will increase the conflict miss in spare cache. The third observation (*i.e. the relation of MR to the cache's block size*) is more complicated. For the block size of less than 16, MR increases as the cache's block size increases because the misses caused by fault is greater for caches with larger block size and thus the spare caches with same block size will cover those misses (*i.e. the denominator and the numerator in Eq. (8.1) will increase by the same amount*). However, if block size becomes too big, the addressable space in primary cache will decrease and results into more conflict misses in *spare cache*. The fourth observation (*i.e. required number of spare blocks*) is related to the temporal locality [1] on the access to the faulty blocks in primary cache. Due to the presence of temporal locality, the number of spare blocks does not have to be proportional to the number of faulty blocks in primary cache to achieve reasonable MR. Based on the above observations, we re-examined the use of the *spare cache* scheme from various aspects. First, we investigate the effectiveness of *spare cache* for bigger size of primary cache and spare cache, since previous simulation results and conclusion are outdated, regarding the size of caches. It might require more than one or two spare blocks to achieve reasonable MR. Moreover, we examine the effectiveness of spare cache on set-associative cache. This is valuable to study because most of today's micro-processors employ set associative cache to increase hit rate. Furthermore, since cache line disabling on set associative cache already has acceptable degradation for fault tolerance with a very little hardware overhead, it is worthy to investigate the effectiveness of *spare cache* with set-associative primary cache. The new simulation results and analysis on extended study is presented in next section.

8.2.3 Programmable Decoder (PADded Cache)

As mentioned before, caches have an intrinsic redundancy since the purpose of caches is to improve performance; it should not be relative to the preciseness of operation. Many architectures can work without any cache at the cost of degraded performance. Therefore, adding extra redundancy, using spare blocks, could be inefficient. There is a phenomenon to cut this second redundancy. Because of the spatial and temporal locality of memory references not all of the sets in a cache are hot at the same time. Thus, there must be some cache sets which can substitute the *spare blocks*. A special *Programmable Address Decoder* (PAD) [4] is introduced to exploit this nature.

When a memory reference happens, a decoder maps it to the appropriate block. A *PAD* is a decoder which has programmable mapping function for the fault tolerance. Once a faulty block is identified, a *PAD* automatically redirects access to that block to a healthy block in the same primary cache. For example, let's consider a *PAD*ded cache, which is equipped with *PAD*. If *PAD*ded cache has n cache blocks and one of the blocks is faulty, the cache will work as if it has $n-1$ cache blocks. *PAD* re-configures the mapping function so that a healthy block acts as a spare block. The method to find suitable defect-free block is predefined and implemented in hardware. Figure 8.3 shows one implementation of *PAD* where $S0 = f'0 \cdot a'0 + f1$ and $S1 = f'1 \cdot a0 + f0$ ($a0$ is the least significant bit of the index). In the case of where *block 0* is faulty, $S0$ will be always on. That is, index for both *block 1* and *block 0*

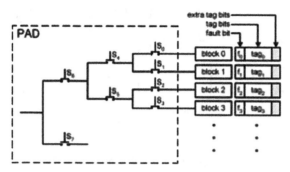

Fig. 8.3 One-level programmable address decoder (PAD)

will be directed to *block 1*. However, during the remapping process, one bit of index information will be lost. Therefore, an extra tag bit (shaded boxed in Fig. 8.3) is required to determine whether the contents in block 1 is its own or not. Otherwise, bogus hits can be generated in *block 1*. For instance, suppose that memory address *000* is originally mapped to *block 0* but redirected to *block 1* because *block 0* is faulty. If after the reference to *block 0*, memory address *001* is given, it will be a hit even though the contents come from memory address *000*.

For better reliability, multiple levels can be utilized by applying above scheme recursively. *S*4 and *S*5 can be modified to reflect the faults among *block 0* to *block 1* for the case when both *block 0* and *block 1* are faulty. As expected, corresponding number of extra tag bits will be required. For the multiple-level *PAD*s, the orders of input to *PAD*s affect their result because caches exploit the special locality of memory access space. Suppose faulty block and re-mapped block are too close to each other in address space. Then access to the consecutive memory address will break the spatial locality between those consecutive sets. However, just reversing the order of index can reduce that effect. It is simple and works well since the order of input is independent to the function of *PAD*s. Indeed, the simulation result showed that reverse order indexed *PAD*s work better than normal ordered *PAD*s. Besides the level-expansion, applying *PAD*s for each way enables to utilize set-associative caches. For instance, each way of a four-way associative cache can have separate *PAD*s. Figure 8.4 illustrates how the *PAD*s work. The Shaded block in *Set 1* substitutes the faulty block in *Set 0*. The shared block belongs to both of *Set 0* and *Set 1*; the faulty block does not belong to any set. That is, *Set 0* and *Set 1* have three their own blocks and they share one block.

To evaluate *PAD*ded caches, they compared this technique with the cache block disabling method for several configurations: cache size (2, 4, 8, 16 and 32 KB), block size (8, 16, 32 bytes) and associativity (1, 2 and 4). Many sets of traces were used: ATUM traces, traces from SPEC92 and SPEC95, and the IBS traces. All *PAD*s were assumed that they are programmable for all levels and reverse ordered. The results of their simulation showed that the miss rate of *PAD*ded caches stay relatively flat and increase slowly towards the end. Simulations of caches with different sizes show that when half of the blocks are faulty, the miss rate of a *PAD*ded cache is

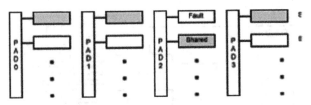

Fig. 8.4 Four-way associative cache employing *PAD*

almost the same as a healthy cache of half the size. The authors claimed that the full capacity of healthy blocks is utilized and the performance decreases with almost minimum possible degradation. Hardware overhead of *PAD*ded cache is also estimated to be 11%, in terms of area [8]. According to their calculation, *PAD*s cost 3%, extra tag bits cost 7%, and the remaining 1% is due to faulty bits.

8.3 Evaluation and Comparisons

In this section, we first discuss our simulation set up to evaluate above surveyed techniques and present the results. Results are divided into two sub-sections. Each fault tolerant schemes are evaluated separately and their results are given in Section 8.3.1 and the comparisons of different schemes are presented in Section 8.3.2.

We modified the *SimpleScalar* [1] execution-driven architecture simulator to simulate each technique. Five benchmark programs (i.e. *bzip2, gcc, mcf, vortex,* and *parser*) from SPEC2000 benchmark suite were simulated. For each benchmark, 5 million instructions were simulated (total 25 million). Moreover, caches were flushed on every system calls to mimic realistic operating system environment.

We total cache size and block size was fixed to 32 KB and 16 byte, respectively, which is a reasonable configuration for most L1 cache of modern days. Although the total size of cache and the size of block can affect the performance of each technique, the manageable amount of result should be used to focus on the comparison of each technique. However, we varied associativity (1, 2 and 4) since the performance of surveyed techniques were significantly sensitive to the associativity. All simulated caches used write-back, allocate-on-write-miss, and LRU replacement algorithm. No pre-fetching or sub-blocking was used. Also, unified instruction/data cache was used since simulation on separated caches showed no significant difference with unified caches' results.

The *random spot defect model* [10] was assumed to inject permanent faults in random locations in the cache. We assume that power up BIST identifies faulty block with 100% coverage. The *fraction of faulty block* (hereafter called *FFB*) was set to range from 0% to 100% for all simulations. The *FFB* lower than 15% was examined more closely for realistic evaluations.

Miss rate is used, to measure effectiveness of individual techniques. The MR (*miss recovery rate, Eq. (8.1)*) is used to compare relative performance of spare-cache and PADded cache to cache block disabling method.

8.3.1 Evaluations

8.3.1.1 Cache Line Disabling

Cache line disabling is examined first since it is the most primitive way of tolerating fault in cache and will be used as the basis of comparing effectiveness of other techniques. In addition to direct mapped, a two-way and four-way set associative cache, and a fully associative cache is also simulated to get the lower bound of degradation for cache block disabling. Total size of cache and block size is fixed to 32 KB and 16 bytes, respectively.

Figure 8.5 depicts the cache miss ratio versus *FFB*. For the direct mapped cache organization, the miss ratio increases almost linearly with *FFB*. Each block in direct mapped cache is mapped to its congruent memory blocks. In other words, a specific address can be mapped to no more than one block. Thus, the memory blocks those are mapped to disabled block cannot be cached. All the references to these blocks will be missed in cache, resulting linear degradation.

On the other hand, disabling faulty blocks in fully associative cache shows 0.07 increase only, even with 90% of faulty blocks. As discussed in previous section, blocks in fully associative cache can accommodate every possible memory blocks and the miss rate depends only on the increased replacement rate. Two-way and four-way set associative cache showed their degradation somewhere in between direct mapped cache and fully associative cache.

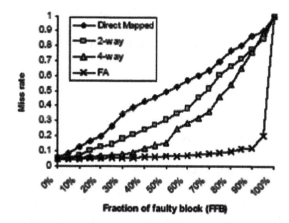

Fig. 8.5 Miss rate of cache line deletion (32 KB)

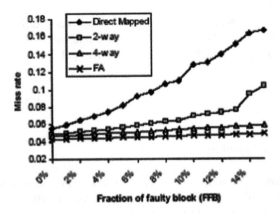

Fig. 8.6 Miss rate of cache line deletion (32 KB)

Our result shows that the fully associative cache can be the best solution for cache line disabling. However, the implementation of fully associative cache is prohibitively large and impractical.

Although there is a large probability for cache to be faulty, we believe that considering more than 15% of *FFB* is impractical. Thus, we closely examined the degradation of each cache with *FFB* lower than 15% and the results are plotted in Fig. 8.6. While the relative degradation between two- or four-way set associative cache and fully associative cache is significant for large number of faulty blocks, there was only up to 0.01 difference in miss rate between four-way and fully associative cache when FFB is less than 15%. Furthermore, there was only up to 0.015 miss rate difference between two-way associative cache and fully associative cache with FFS less then 15%.

In summary, the cache block disabling can be more effective when the associativity of cache is larger. However, for reasonable value of FFB and feasibility, two- or four-way set associativity is sufficient to get the advantage of associative organization of cache. The results presented for cache block disabling will be used in subsequent subsections as the basis for comparing effectiveness of other schemes.

8.3.1.2 Spare Cache

Extensive simulation on *spare cache* for the direct mapped primary cache is done in [7] and their results are summarized in previous section. However, the simulation set up is obsolete since the simulated cache size was too small (less than 16 KB) and the effectiveness of spare cache in the conjunction with set associative primary cache is not considered. Thus, we extended our simulation for larger cache size (32 KB). Moreover, a two- and four-way set associative primary cache as well as a direct mapped cache was simulated. First, the effects of varying the spare cache size for three different primary caches are considered. Figures 8.7–8.9 plots miss rate versus *FFB* for Direct-mapped, two- and four-way set associative cache, respectively. For

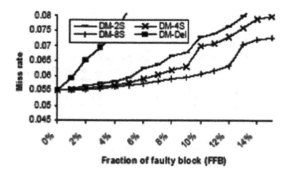

Fig. 8.7 Spare cache with direct-mapped cache

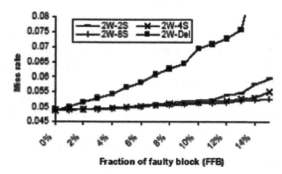

Fig. 8.8 Spare cache with two-way set associative cache

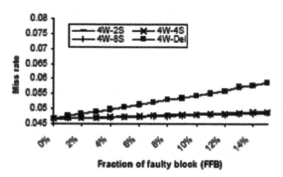

Fig. 8.9 Spare cache with four-way set associative cache

each of the cases a 2, 4, and 8-block sized spare cache was considered. In those figures, DM, 2W, 4W denotes direct-mapped cache, two-way set associative, and four-way set associative cache, respectively, "-Del" means cache line disabling or deletion, and "-xS" denotes *spare cache* of x blocks.

For the entire primary cache configuration, the sensitivity to the size of spare cache is noticeable. The direct mapped cache is more sensitive to spare cache size

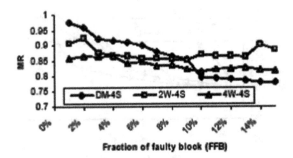

Fig. 8.10 Miss recovery rate

as opposed to both a two- and four-way set associative caches. While there is more than 0.05 miss rate difference between 2-block spare and 8-block spare with direct-mapped primary cache at 15% of *FFB*, two-way and four-way set associative caches showed less than only 0.005 miss rate difference between 2-block and 8-block spare cache. Especially, the number of blocks in spare cache showed negligible miss rate difference for the four-way set associative cache.

In case of direct-mapped cache, we notice that employing spare cache improves the miss rate significantly as compared to set associative caches. While the miss rate of cache line deletion already exceeds 0.08 with only 5% of FFB, spare cache suppress the miss rate under 0.08 for the *FFB* of up to 14%. Similar but less improvement over cache line disabling can be observed in two-way set associative cache. On the other hand, there was no significant miss rate improvement (less than 1% difference of miss rate) for *FFB below*15% in four-way set associative cache. This is because the degradation due to cache line disabling on set-associative cache is already small compared to that of direct-mapped cache. This result brings up new question: Is spare cache effective only for a direct mapped or a small associativity cache? To answer this question, we need to compare MR (*Miss-recovery Rate*) for each of the primary cache organizations. Figure 8.10 compares the MR for three different primary cache organizations for a fixed spare cache size of 4 blocks. The size of spare cache is fixed to solely compare the effectiveness of employed spare cache on primary caches with different associativity. For small *FFB*, the misses recovered by 4-block spare cache is higher for lower associativity. However, the MR for direct mapped cache quickly drops when *FFB* exceeds 8% resulting less *MR as* compared to four-way associative caches. The similar drop of *MR* for two-way set associative was also observed.

Thus, the following observation can be made from our results:

- Primary cache with larger associativity is less sensitive to the size of spare cache. This is because the associativity of primary cache can already be able to reduce the misses caused by faulty block access. If misses caused by faulty block access is lower, the contention in the spare cache will be lower, therefore, a very small number (2 or 4-block) spare cache size is enough to tolerate *FFB* of less than 15%.
- The *MR* is higher for smaller associative cache for a lower value of *FFB*, then quickly drops below the *MR* of higher associative primary cache when *FFB*

increases further. The reason is that there are more misses caused by fault in smaller associativity cache than larger associativity cache, while the quick drop occurs when the contention in the spare cache causes more replacement in *spare cache.*

8.3.1.3 PADded Cache

In this subsection we present simulation results of *PAD*ded cache, the last scheme. In [8], *PAD*ded caches are simulated extensively on the range of . . . . . . % to 100% to emphasize the effectiveness of *PAD*ded cache in case of large *FFB*. Although *PAD*ded cache has this nice property, we believe that considering more than 15% of *FFB* would be impractical. Figure 8.11 plots miss rate versus *FFB* for Direct-mapped cache from 0% to 15% of *FFB*. Cache block disabling is added to compare the effectiveness. As shown in Fig. 8.11, *PAD*ded caches show very low and flat miss rate for one level of *PAD* is used,. However, there is no significant change of miss rate after level two, which implies employing more than two levels could be extravagant for small range of *FFB*. Therefore, we have used only level two *PAD*ded caches to evaluate *PAD*ded caches, although other levels have been simulated to verify our result.

From observation on Fig. 8.11, it seems that *PAD*ded caches successfully dissolve spare blocks in themselves as they are supposed. However, there is another point. Inserting spare blocks into caches cause the change of locality in the caches. For example, there is a two-way associative *PAD*ded cache. It has a faulty block, *block A,* so that the references to *block A* are redirected to an adjacent healthy *block B.* Then, it is hard to exploit the spatial locality of the set which contains *block b.* As expected, this phenomenon can be reduced by different ordering of address bits in the *PAD*. The simulation results confirm that reverse order indexed *PAD*s have better performance than normal ordered *PAD*s. Thus, the result presented in this report is based on reverse-ordered index *PAD*.

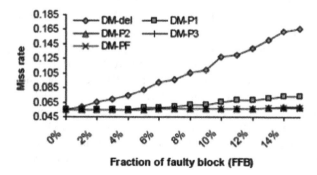

Fig. 8.11 Miss rates for different *PAD*ded levels

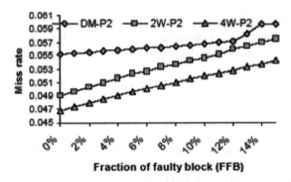

Fig. 8.12 Miss rates for different associativities

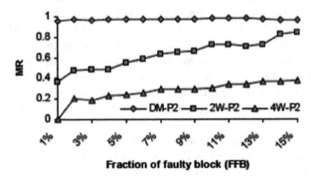

Fig. 8.13 MRs for different associativities

To assess the effectiveness of *PAD*ded caches, we also simulated two-way and four-way associated caches. Figure 8.12 depicts the variation of MR with FFB for both two-way and four-way associated caches. In Fig. 8.12, *PAD*ded caches seems that it takes full advantage of associative caches. Figure 8.13 shows MR vs. FFB for three different schemes. The results shows that the *PAD*ded cache has less effectiveness when it is used for set-associative cache as opposed to direct mapped cache, *MR* decreases as associativity increases. The conflict of these two figures comes from another locality issue. Suppose there are two sets of cache block: *A* and *B*. Set *A* has a faulty block and its references are re-mapped to set *B*. Even though set *B* is far away from set *A* in address space, this redirection impedes set *B* to exploit the temporal locality. Increasing associativity give *PAD*ded caches more chances to break the temporal locality. As a result, *PAD*ded caches do not take full advantage of associative cache, which indicates that the performance improvement, as shown in Fig. 8.12, is due to mainly because of primary cache's associativity. *PAD*ded caches have the less portion of contribution to improved performance for the higher associativity. In an extreme case of fully associated cache, *PAD*ded caches have obviously no effect.

For direct-mapped caches, MR stays close to 1 for the entire range of *FFB*, indicating that *PAD*s recovers all most all of misses due to faulty cache blocks. Since there is no other redundancy to accommodate the misses, if some of them are recovered, it must be done by *PAD*s. For associative caches, MR decreases as *FFB* increases. For low *FFB*, associativity contributes more than *PAD*s. However, as *FFB* increases, *PAD*s start to surpass associativity since *PAD*s have more candidates for faulty blocks than associativity. This result seems not in accordance with that of spare caches. Nonetheless, it is quite in accord. The number of spare blocks is very small compared to the number of total cache blocks. Opposite to spare, *PAD*s have huge amount of spare blocks up to 50% of the total number of cache blocks depend on *PAD* level. That is, the capacity of *PAD*s is much larger than that of the spare the cache method. The following is the summary of the simulation results of the *PAD*ded caches:

- Two-level is enough for *PAD*ded caches from a practical view point. For small percentage of faults, small number of levels is actually utilized. Therefore, surplus *PAD* level is not desirable; high level of *PAD*s is expensive.
- The order of inputs to *PAD*s is important to its performance. Which affects the spatial locality of caches.

*PAD*ded caches contribute to the performance of direct-mapped caches more than that of associativity caches. Since PADs influence the temporal locality of caches, high associativity caches suffer from second conflict miss due to *PAD*s.

8.3.2 Comparisons

In this sub-section we will compare individually examined schemes together in terms of their characteristics, advantages, effectiveness and hardware overhead.

We compare the simulation results of the cache line disabling, 2-level *PAD*ded cache, and 4-block spare cache for each of direct-mapped and four-way set associative caches. Since 2-level *PAD*ded cache showed close result to its ideal case (i.e. full-level PAD) and there was no significant difference between 4-block and 8-block spare cache, these two configurations of each scheme would be good candidates for the comparison. Direct-mapped and four-way set associative cache is chosen to clearly compare the characteristics of each scheme when they are applied to the primary cache with different associativities.

In Fig. 8.14, miss rate for each configuration is compared for varying FFB. First of all, we can easily observe that all the techniques with direct-mapped primary cache has higher miss rate than those with four-way set associative cache. Even cache block disabling method has better performance than the best case of direct-mapped primary cache. Thus, one might conclude that using higher set associativity cache would be the best solution when the hardware cost is the main issue. However, if the latency of cache is the primary consideration, there is situation where direct mapped cache is preferred [11] as well as fault tolerant features. In this case, *PAD*ded cache seems to be the best solution since the miss rate of 2-level PADded cache has much less degradation for more than 10% of *FFB* in Fig. 8.14. Further-

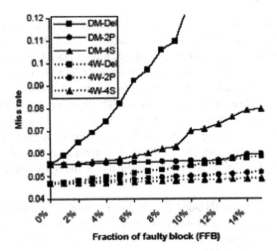

Fig. 8.14 Comparison of disabling, spare cache, and four-way set associative cache

Table 8.1 Comparison on three different fault tolerant cache techniques

	Cache line disabling	Spare cache	PADded cache
Suitable primary cache	Large associative cache	Any	Direct mapped or two way set associative cache
Hardware overhead	Lowest	High	Low for direct-mapped Higher for set associative cache

more, the hardware cost for *PAD*ded cache can be minimized when it is applied for direct-mapped cache. On the other hand, the spare cache scheme can achieve the minimum degradation with four-way set associative cache. Although there is a slight difference between the 2-level *PAD*ded cache and the 4-block *spare cache*, *PAD*ded cache may not be a good solution for four-way set associative cache since PADded cache will require four separate decoders for each ways of set-associative primary cache.

8.4 Conclusion

As VLSI technology and performance of micro-processor advances, on-chip cache memory becomes essential and continues to grow in size. This trend results more chance of defect in cache area. Consequently, many fault tolerance scheme had been presented in literature.

In this chapter, we present the results of extensive simulation study to investigate and compare three different fault tolerant cache schemes.Our simulation results for individual technique expose their characteristics and indicate ways to achieve low degradation in system performance. In addition, the result demonstrates that each of fault tolerant techniques has its own advantages and there is no one scheme which is better than the other in all the situations considered as shown in Table 8.1. However, more thorough investigation on hardware cost of each technique should be done to obtain more precise comparison.

References

1. Hennessy, J.L. and D.A. Patterson. Computer Architecture: A Quantitative Approach, 2nd edition, Morgan Kaufmann, San Mateo, CA, 1996.
2. Sohi, G.S., "Cache Memory Organization to Enhance the Yield of High-Performance VLSI Processors," IEEE Trans. Comp., Vol. 38, No. 4, pp. 484–492, April 1989.
3. Ooi, Y., M. Kashimura, H. Takeuchi, and E. Kawamura, "Fault-Tolerant Architecture in a Cache Memory Control LSI," IEEE J. Solid-State. Circ., Vol. 27, No. 4, pp. 507–514, April 1992.
4. Pour, A.F. and M.D. Hill, "Performance Implications of Tolerating Cache Faults," IEEE Trans. Comp., Vol. 42, No. 3, pp. 257–267, March 1993.
5. Turgeon, P.R., A.R. Stell, and M.R. Charlebois, "Two Approaches to Array Fault Tolerance in the IBM Enterprise System/9000 Type 9121 Processor," IBM J. Res. Develop., Vol. 35, No. 3, pp. 382–389, May 1991.
6. Lucente, M.A., C.H. Harris, and R.M. Muir, "Memory System Reliability Improvement Through Associative Cache Redundancy," Proc. IEEE Custom Integr. Circ. Conf., pp. 19.6.1–19.6.4, Boston, MA, May 1990.
7. Vergos, H.T. and D. Nikolos, "Performance Recovery in Direct-Mapped Faulty Caches via the Use of a Very Small Fully Associative Spare Cache," Proc. Intl. Comp. Perform. Dependability Symp., pp. 326–332, April 1995.
8. Shirvant, P.P. and E.J. McCluskey, "PADded Cache: A New Fault-Tolerance Technique for Cache Memories," IEEE VLSI Test Symp., pp. 440–445, April 1999.
9. O'Leary, B.J. and A.J. Sutton, "Dynamic Cache Line Delete," IBM Tech. Disclosure Bull., Vol. 32, No. 6A, p. 439, Nov. 1989.
10. Stapper, C.H., F.M. Armstrong, and K. Saji, "Integrated Circuit Yield Statistics," Proc. IEEE, Vol. 71, pp. 453–470, April 1983.
11. Hill, M.D., "A Case for Direct-Mapped Caches," IEEE Computer Vol. 21, No. 12 Micro, pp. 25–40, Dec. 1988.

Chapter 9
Reversible Binary Coded Decimal Adders using Toffoli Gates

Rekha K. James, K. Poulose Jacob, and Sreela Sasi

Abstract Reversibility plays a fundamental role when computations with minimal energy dissipation are considered. This research describes Toffoli Gate (TG) implementations of conventional Binary Coded Decimal (BCD) adders, adders for Quick Addition of Decimals (QAD), and carry select BCD adders suitable for multi-digit addition. For an N-digit fast adder, **partial parallel processing is done on all digits in the decimal domain. Such high-speed BCD adders find application in real-time processors and internet-based computing**. An analysis of delay normalized to a TG and quantum cost of BCD adders is presented. Implementations using TGs and Fredkin Gates (FRGs) are compared based on quantum cost, number of gates, garbage count and delay, and the results are tabulated.

Keywords Reversible implementation · Quantum cost · High speed arithmetic · Toffoli gates

9.1 Introduction

Currently, fast decimal arithmetic is gaining popularity in the computing community due to the growing importance of commercial, financial, and internet-based applications which normally process decimal data. Low power designs with high performance are given prime importance by researchers as power has become a first-order design consideration. While efforts are being made to reduce power dissipation due to leakage currents, alternate circuit design considerations are also gaining

R.K. James (✉) and K.P. Jacob
Cochin University of Science and Technology, Cochin, Kerala, India
e-mail: rekhajames@cusat.ac.in

S. Sasi
Gannon University, Erie, PA, USA

S.-I. Ao et al. (eds.), *Advances in Computational Algorithms and Data Analysis*,
Lecture Notes in Electrical Engineering 14,
© Springer Science+Business Media B.V. 2009

importance. In recent years, reversible logic has been in demand for high-speed power aware circuits. Classical logic gates such as AND, OR and XOR are not reversible. These gates dissipate heat and may reduce the life of the circuit. So, reversible logic has emerged as one of the most important approaches for power optimization with its application in nanotechnology, low power CMOS, and quantum computing.

Landauer's [1] principle states that a heat equivalent to $kT*ln2$ is generated for every bit of information lost, where k is the Boltzmann's constant and T is the temperature. At room temperature T, though the amount of heat generated may be small it cannot be neglected for low power designs. The amount of energy dissipated in a system bears a direct relationship with the number of bits erased during computation. Bennett [2] showed that energy dissipation would not occur if the computations were carried out using reversible circuits since these circuits do not lose information. A completely specified n-input, n-output Boolean function is called reversible if it maps each input vector to a unique output vector and vice versa. There is a significant difference in the synthesis of logic circuits using classical logic gates and reversible gates. While constructing reversible circuits, the fan-out of each output must be _1_ without any feedback loops. As the number of inputs and outputs are made equal there may be a number of unutilized outputs in certain reversible implementations called _garbage_. It is defined as the number of outputs added to make an n-input, k-output Boolean function reversible. For example, a single output function of _n_ variables will require at least _n-1_ garbage outputs. An important aspect for evaluating reversible circuits is the garbage count. Hence, one of the major issues in designing a reversible circuit is in garbage minimization.

A reversible conventional BCD adder is proposed by Babu [3] using NG (New Gate) and NTG (New Toffoli Gate) reversible gates. Even though the implementation is modified by Thapliyal [4] using TSG reversible gates, this approach does not take care of the fan-out restriction of reversible circuits, and hence it is only a near-reversible implementation. An improved reversible implementation of decimal adder with reduced number of garbage outputs is proposed by James [5]. Further reduction in number of logical computations was achieved by using HNG gates in the implementation by Haghparast [6]. All these implementations are for conventional BCD adders. These are relatively slow, and are implemented using different types of reversible gates. In this research, Toffoli Gate (TG) **reversible implementations of conventional and fast decimal adders are done.** Implementations using TG and that using Fredkin Gates (FRG) [7, 8] are compared based on quantum cost (QC), number of gates, garbage count and delay, and the results are tabulated. **Quantum cost analysis** is done to compare the equivalent number of _two-qubit quantum gates_ required for the implementation.

The organization of this paper is as follows. Decimal adders such as conventional BCD adder, QAD [9] and carry select BCD adders [8] are described, followed by the implementations using only generalized reversible Toffoli gates. An analysis of delay normalized to a TG and quantum cost of different BCD adders is presented. Finally, a comparison in terms of delay, number of gates and garbage count is done for implementations of different BCD adders using TGs and FRGs.

9.2 BCD Adders

This section gives a brief overview of the different BCD adders compared.

9.2.1 Conventional Decimal Adder

The conventional BCD adder shown in Fig. 9.1 has three blocks: 4-bit binary adder, 6-correction circuit and a final adder, which is a modified special adder. The final adder is a 3-bit adder with two half adders and one full adder. Six-correction circuit generates 'L' bit as given in Eq. (9.1).

$$L = C_{out} + S_3(S_1 + S_2) \qquad (9.1)$$

The total (worst case) delay of an N-digit conventional BCD adder $(T_{dsum(conventional)})$ given in Eq. (9.2) is the sum of 'N' times the 'carry delay' through one digit and 'sum delay' $(T_{sum\text{-}digit})$ of the last digit.

$$T_{dsum\ (conventional)} = NT_{dcout} + T_{sum\text{-}digit} \qquad (9.2)$$

9.2.2 Quick Decimal Adder

A BCD adder for quick addition of decimals, QAD [9] is shown in Fig. 9.2. It consists of a 4-bit binary adder, a 6-correction circuit, and a special adder along

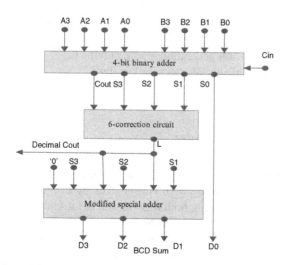

Fig. 9.1 Conventional BCD adder

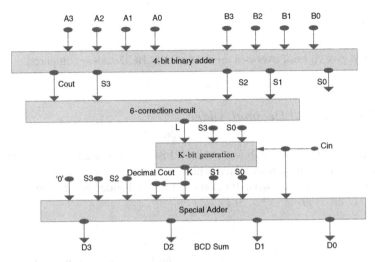

Fig. 9.2 BCD adder for quick addition of decimals (QAD)

with a K-bit generation circuit to generate decimal carry out (d_{cout}). The d_{cout} (K-bit) is generated using the Eq. (9.3).

$$K = S_3 S_0 C_{in} + L \tag{9.3}$$

An N-digit QAD will have a total (worst case) delay ($T_{dsum(QAD)}$) equal to the sum of the 'carry delay' through the first digit (T_{dcout}), the K-bit generation delays through the next (N-1) digits, and the 'sum delay' through the last digit ($T_{sum\text{-}digit}$). This is given in Eq. (9.4).

$$T_{dsum(QAD)} = T_{dcout} + (N-1)T_K + T_{sum\text{-}digit} \tag{9.4}$$

T_{dcout} is the delay to generate d_{cout} from the BCD inputs for the first digit.

T_K is the delay for a K-bit generation after receiving C_{in}.
$T_{sum\text{-}digit}$ is the delay of special adder for last digit.

9.2.3 Carry Select BCD Adder

If the carry select technique is adopted for K-bit generation then k_1 denotes the K-bit with $C_{in} = 1$ and k_0 with $C_{in} = 0$. This is given by $k_1 = S_3 S_0 + L$ and $k_0 = L$ (from Eq. (9.3)). After computing both bits (k_1 and k_0) a selection is done using a 2:1 multiplexer as shown in Fig. 9.3. An N-digit carry select adder will have a total (worst case) delay ($T_{dsum(carryselect)}$) equal to the sum of the 'carry delay' through the first digit (T_{dcout}), the carry select delays through next (N-1) digits, and 'sum delay' through the last digit ($T_{sum\text{-}digit}$). This is given in Eq. (9.5).

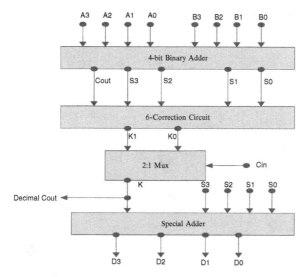

Fig. 9.3 Carry select BCD adder

$$T_{\text{dsum(carry select)}} = T_{\text{dcout}} + (N-1)T_{\text{mux}} + T_{\text{sum-digit}} \qquad (9.5)$$

T_{dcout} is the delay to generate K-bit from the BCD inputs for the first digit.

T_{mux} is the delay of a 2:1 multiplexer.
$T_{\text{sum-digit}}$ is the delay of special adder for last digit.

9.3 Reversible Implementations Using Toffoli Gates

The proposed reversible implementations of the BCD adders are done using the generalized Toffoli gates. Since the design uses only one type of gate (Toffoli), it is suitable for VLSI implementation. Figure 9.4 shows a 3*3 Toffoli Gate (TG) [7]. A TG can be used to generate an AND, and XOR functions. If C = '0', then R = AB and if B = '1' then $R = A \oplus C$. TG can be considered as a universal gate since any Boolean function in Positive Polarity Reed Muller (PPRM) form can be realized using this gate. This is also referred as a T3 (3-input TG). Figure 9.5 shows a Feynman Gate (FG) [10]. FG can be used as a copying gate. Since a fanout greater than one is not allowed, this gate is useful for duplication of the required outputs. If B= '0', then P = A and Q = A. This is referred as a 2-input TG (T2).

Figures 9.6 and 9.7 show the implementation of a half adder and a full adder using Toffoli gates. The implementation of half adder makes use of two Toffoli gates: 3-input Toffoli (T3) and 2-input Toffoli (T2) with one garbage.

Full adder implementation makes use of four Toffoli gates (two T3, two T2) and results in two garbage outputs. These are optimum solutions in terms of number of

Fig. 9.4 Three-input Toffoli
gate (T3)

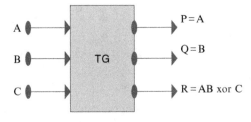

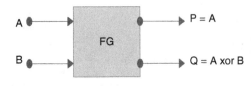

Fig. 9.5 Two-input Toffoli
gate (T2)

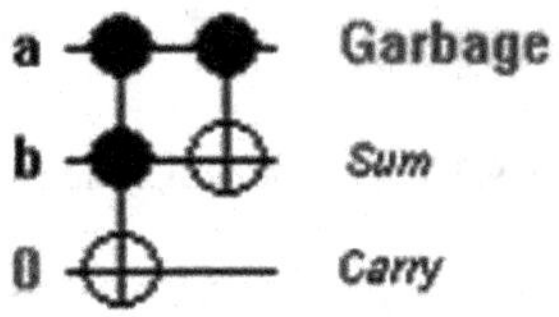

Fig. 9.6 Half adder

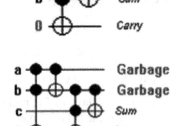

Fig. 9.7 Full adder

garbage outputs. The circuits were simulated in RC Viewer, which gives the quantum cost of the full and half adders as 8 and 4 respectively.

9.3.1 Reversible Conventional BCD Adder

A 4-bit binary adder for the conventional BCD adder realized using 4 full adders is shown in Fig. 9.8.

The least significant bit requires a path delay of three TGs to generate C_0(carry) from the addends. Carry ripples through the subsequent full adders with a path delay of one TG per bit. This is because the first two TGs of all full adders work in parallel in an n-bit binary adder. The 'Sum' is generated after one more TG delay after generating Carry. The delay to generate 'Sum' in the n-bit binary adder is given in Eq. (9.6).

$$T_{\text{sum}-\text{ripple(conventional)}} = 4 + (n - 1) \tag{9.6}$$

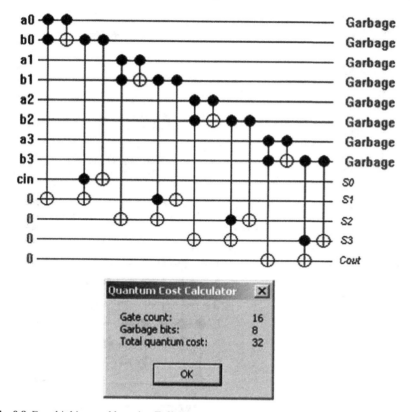

Fig. 9.8 Four-bit binary adder using Toffoli gates

For a BCD adder this delay is the delay with n = 4 for each digit. The implementation gives a gate count of 16, with eight garbage outputs at a quantum cost of 32.

The Toffoli reversible implementation of the 6-correction circuit is shown in Fig. 9.9. The implementation requires four TGs to generate the 'L' output, with one garbage output at a quantum cost of 12. This circuit takes only one more delay after generating the 'S$_3$' to generate the 'L' bit and is given in Eq.(9.7).

$$T_{L(conventional)} = 5 \mid (n-1) = 8(\text{with } n = 4) \tag{9.7}$$

Special adder shown in Fig. 9.10 requires eight TGs to generate the BCD$_{sum}$, d$_{sum}$. The first T2 is used to duplicate 'L' bit or Decimal C$_{out}$, d$_{cout}$. So, the total delay in terms of one TG delay for generation of d$_{cout}$ for an N digit conventional BCD adder is given in Eq. (9.8).

$$T_{dcout(conventional)} = 9N \tag{9.8}$$

Fig. 9.9 Six-correction circuit
using Toffoli gates

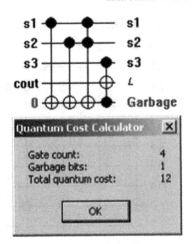

Fig. 9.10 Modified special
adder of conventional BCD
adder

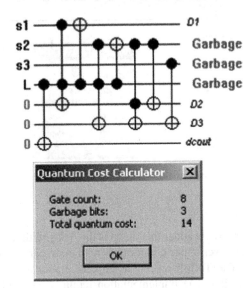

The modified special adder gives an additional delay of six TGs to generate BCD_{sum}.
Figure 9.11 shows the schematic of the reversible circuit for the conventional BCD
adder implemented using Toffoli gates given by the RC Viewer. The circuit uses
28 gates and results in 12 garbage outputs. The quantum cost of the implementation is 58.

The total delay for generating the BCD_{sum}, d_{sum} from the inputs in terms of TG
delay for N-digit BCD addition is given in Eq. (9.9).

$$T_{d\text{-sum(conventional)}} = 6 + 9N \tag{9.9}$$

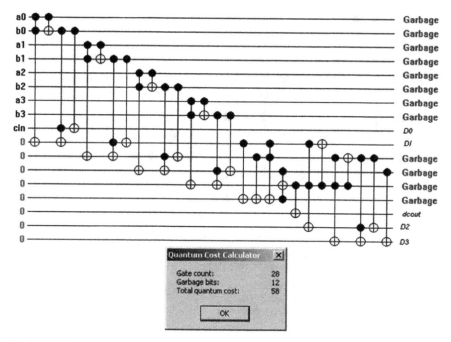

Fig. 9.11 Toffoli gate implementation of conventional BCD adder

9.3.2 Reversible QAD Adder

Reversible implementation of QAD is also done using TGs. A 4-bit binary adder realized using one half adder and 3 full adders is shown in Fig. 9.12.

The least significant bit requires a path delay of two TGs only. The carry ripples through the subsequent full adders with a path delay of one TG per bit. This is because the first two TGs of all full adders work in parallel with the least significant bit half adder. 'Sum' is generated after one more TG delay subsequent to generating the final carry. The delay to generate 'Sum' in the 4-bit binary adder in QAD is given in Eq. (9.10).

$$T_{\text{sum-rippleQAD}} = 3 + (n-1) = 6(\text{with } n = 4) \tag{9.10}$$

The delay to generate 'L' bit from the BCD inputs in a QAD adder is given in Eq. (9.11). As in conventional BCD adder the circuit takes only one more delay after generating the 'S$_3$' to generate the 'L' bit.

$$T_{\text{L(QAD)}} = 4 + (n-1) = 7(\text{with } n = 4) \tag{9.11}$$

Figure 9.13 shows the reversible implementation for generating Decimal C_{out} (d_{cout}) or K-bit in a QAD. The design makes use of four TGs. It is seen that $S_3 S_0$ is

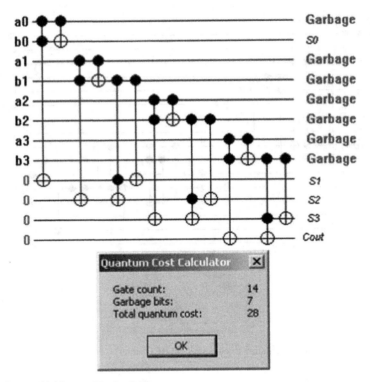

Fig. 9.12 Four-bit binary adder for QAD

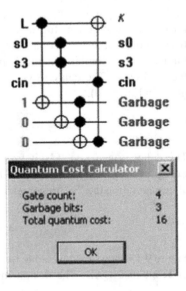

Fig. 9.13 K-bit generation
using Toffoli gates

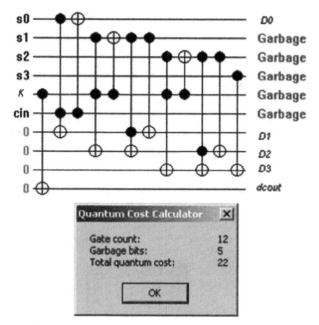

Fig. 9.14 Toffoli implementation of special adder for QAD

generated after three TG delays after the sum output 'S' is produced. So, the additional delay after receiving C_{in} is due to one TG in each stage.

Special adder for QAD requires 12 TGs to generate the BCD_{sum} d_{3-0} as shown in Fig. 9.14.

The first T2 is used to duplicate K-bit for d_{cout}. For an N-digit BCD addition the delay for generation of K-bit or Decimal C_{out} from the BCD inputs for QAD adder is as given in Eq. (9.12).

$$T_{d\text{-cout(QAD)}} = 7 + 3 + 1 + N = 11 + N \qquad (9.12)$$

The 'C_{in}' input or K-bit passes through a maximum of seven TGs to generate the BCD sum d_{3-0}. So the special adder gives an additional delay of seven gates. The total delay for generating the BCD_{sum}, d_{sum} from the inputs in terms of TG delay is

$$T_{d\text{-sum(QAD)}} = 11 + 7 + N = 18 + N \qquad (9.13)$$

Fig. 9.15 shows the schematic of the reversible circuit for the QAD implemented using TGs given by the RC Viewer. The circuit uses 33 gates and results in 16 garbage outputs. The quantum cost of the implementation is 81.

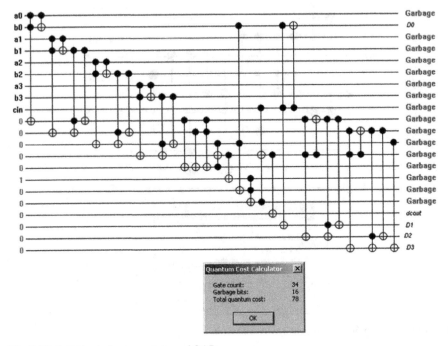

Fig. 9.15 Toffoli gate implementation of QAD

9.3.3 Reversible Carry Select BCD Adder

Reversible implementation of carry select BCD adder will differ from QAD implementation only in the generation of K-bit from L-bit and C_{in}. Figure 9.16 shows the generation of K-bit using k_1 and k_0 for a carry select BCD adder. The generation of k_1 and k_0 takes the delay of four TGs after receiving 'L' bit as shown in Fig. 9.16. On receiving C_{in}, K-bit is computed after a delay of one TG. So, the additional delay in each stage to generate K-bit on receiving C_{in} is due to only one TG. One more T2 is used to duplicate K-bit for d_{cout}. Now, the delay gets modified as shown in Eqs. (9.14) and (9.15).

$$T_{d\text{-cout(carryselect)}} = 7 + 4 + 1 + N = 12 + N \tag{9.14}$$

$$T_{d\text{-sum(carryselect)}} = 12 + 7 + N = 19 + N \tag{9.15}$$

Similar analysis is done on reversible implementations of conventional, QAD and carry select BCD adders using FRGs by James [8]. Table 9.1 shows a comparison of implementations of different BCD adders using FRGs and TGs in terms of quantum cost, number of reversible gates, garbage outputs and delay. The percentage reduction attained in quantum cost is computed as [QC(FRG) – QC(TG)]/QC(FRG). It is seen that TG is superior to FRG implementation in terms of quantum cost,

Fig. 9.16 K-bit generation of Carry select BCD Adder

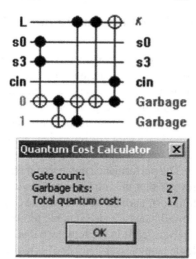

Table 9.1 Comparative analysis of reversible BCD adders implemented using Toffoli and Fredkin gates

Reversible	BCD adders	Conventional BCD adder	QAD	Carry select BCD adder
Toffoli implementation	Quantum cost	58	78	79
	No: of gates	28	34	35
	Garbage count	12	16	15
	Delay for Decimal an N-C_{out}	9N	11 + N	12 + N
	digit BCD sum adder	6 + 9N	18 + N	19 + N
Fredkin implementation	Quantum cost	105	126	126
	No: of gates	35	42	42
	Garbage count	36	45	44
	Delay for Decimal an N-C_{out}	11N	10 +2N	11 + N
	digit BCD sum adder	6 + 11N	15 + 2N	17 + N

garbage count and number of gates. Conventional BCD adder implemented using TGs shows 45% reduction in quantum cost, 67% reduction in garbage count and 18% reduction in gate count compared to FRG implementation. QAD gives a corresponding reduction of 38%, 63% and 15%. For carry select BCD adder the respective reduction factors are 37%, 64% and 12%. The implementation using TGs gives a reduction in delay for conventional BCD adder and QAD compared to FRG implementation.

9.4 Conclusion and Future Work

This paper describes reversible implementations of several BCD adders using only Toffoli gate. The architectures of the decimal adders are specially designed to make them suitable for reversible logic implementation. It is demonstrated that the design is highly optimized in terms of number of reversible gates and garbage outputs. The design strategy is to reduce the number of garbage outputs, which is the most important factor for reversible circuit cost. This approach also gives suitability for VLSI implementation due to the use of only one type of building block.

The performance comparison of VLSI implementations of different BCD adders reveals that the implementations using Toffoli gates are superior in terms of quantum cost, garbage count and gate count, compared to Fredkin gate implementations. Toffoli gates are also suitable for implementations of Reed Muller expressions. This analysis leads to the inference that decimal adders can be implemented in reversible logic using lesser number of gates and garbage count when the logic is expressed in Reed Muller form. Investigations for determining alternate implementations can be done using logic synthesis methods given by researchers [11, 12]. Implementations using other standard reversible gates such as TSG or HNG gates can also be tried. New families of 'n-input' – 'n-output' reversible gates that can be used for regular structures can be investigated. This research is an initial step towards building complex reversible systems, which can execute more complicated functions.

References

1. R. Landauer, "Irreversibility and Heat Generation in the Computational Process", IBM Journal of Research Development, 5, 1961, 183–191.
2. C. Bennett, "Logical Reversibility of Computation", IBM Journal of Research and Development, 17, 1973, 525–532.
3. Md. Hafiz Hasan Babu and A. R. Chowdhury, "Design of a Reversible Binary Coded Decimal Adder by Using Reversible 4-Bit Parallel Adder", VLSI Design 2005, Jan. 2005, Kolkata, India, pp. 255–260.
4. H. Thapliyal, S. Kotiyal, and M. B Srinivas, "Novel BCD Adders and Their Reversible Logic Implementation for IEEE 754r Format", 19th VLSI Design 2006, Jan. 2006, Hyderabad, India, pp. 387–392.
5. R. K. James, T. K. Shahana, K. P. Jacob, and S. Sasi, "Improved Reversible Logic Implementation of Decimal Adder", IEEE 11th VDAT Symposium, Aug. 2007, Kolkata, India.
6. M. Haghparast and K. Navi, "A Novel Reversible BCD Adder for Nanotechnology Based Systems", American Journal of Applied Sciences, 5(3), 2008, 282–288, ISSN 1546–9239.
7. E. Fredkin and T. Toffoli, "Conservative Logic", International Journal of Theoretical Physics, 21, 1982, 219–253.
8. R. K. James, T. K. Shahana, K. P. Jacob, and S. Sasi, "Performance Analysis of Reversible Fast Decimal Adders", International Conference on Computer Science and Applications, Oct. 2007, San Francisco, USA, pp. 234–239.
9. R. K. James, T. K. Shahana, K. P. Jacob, and S. Sasi, "Quick Addition of Decimals Using Reversible Conservative Logic", 15th International Conference on Advanced Computing & Communication ADCOM 2007 18–21 December 2007, IIT Guwahati, India, pp. 191–196.

10. R. Feynman, "Quantum Mechanical Computers", Optical News, 1985, Claydon, Ipswich, pp. 11–20.
11. D. Maslov, "Reversible Logic Synthesis", Ph.D. Dissertation, Computer Science Department, University of New Brunswick, Canada, Oct. 2003.
12. P. Gupta, A. Agrawal, and N. K. Jha, "An Algorithm for Synthesis of Reversible Logic Circuits", IEEE Transactions on Computer Aided Design of Integrated Circuits and Systems, 25(11), Nov. 2006, 2317–2330.

Chapter 10
Sparse Matrix Computational Techniques in Concept Decomposition Matrix Approximation

Chi Shen and Duran Williams

Abstract Recently the concept decomposition based on document clustering strategies has drawn researchers' attention. These decompositions are obtained by taking the least-squares approximation onto the linear subspace spanned by all the concept vectors. In this chapter, a new class of numerical matrix computation methods has been developed in computing the approximate decomposition matrix in concept decomposition technique. These methods utilize the knowledge of matrix sparsity pattern techniques in preconditioning field. An important advantage of these approaches is that they are computationally more efficient, fast in computing the ranking vector and require much less memory than the least-squares based approach while maintaining retrieval accuracy.

Keywords Term-document matrix · Concept vectors · Concept decomposition matrix · Sparse matrix approximation · Least-squares

10.1 Introduction

Concept decomposition for text data has been first proposed by [1]. In this approach, data are modeled as a term-document matrix and the large document dataset is first partitioned into tightly structured clusters. The centroid (average) of the documents in a cluster (group) is computed and normalized as the *concept vector* of that cluster, as it represents the general meaning of that group of tightly structured homogeneous documents. The term-document matrix projected on the concept vectors is compared favorably to the truncated Singular Value Decomposition (SVD) [2] of the original term-document matrix. Gao and Zhang [3] have indicated that the retrieval accuracy from the concept decomposition can be comparable to that from SVD.

C. Shen (✉) and D. Williams
Kentucky State University, Frankfort, KY 40601, USA
e-mail: chi.shen@kysu.edu

S.-I. Ao et al. (eds.), *Advances in Computational Algorithms and Data Analysis*,
Lecture Notes in Electrical Engineering 14,
© Springer Science+Business Media B.V. 2009

However, the numerical computation based on the straightforward implementation of the concept matrix decomposition is expensive, as there is an inverse matrix of the normal matrix formed by the concept vector matrix is required for each query computation.

In this chapter, we develop numerical matrix computation methods to compute the matrix associated with the concept matrix decomposition. The matrix sparse pattern techniques based on Chow's SAI [4] and Grote and Huckle's SPAI [5] have been proposed in computing the approximate decomposition matrix. These computations are related to weighted least squares optimization and approximate pseudo matrix inversion. Our experimental results show that these approaches greatly reduce the time and memory space required to compute the decomposition matrix and perform document retrieval.

This chapter is organized as follows. Section 10.2 introduces document clustering and concept decomposition. Section 10.3 discusses the approximate least squares based approaches, the key part of the numerical computation in this chapter. Numerical experiments are given in Section 10.4. We summarize this chapter in Section 10.5.

10.2 Document Clustering and Concept Vectors

A large dataset can be divided into a few smaller ones, each contains data that are close in some sense. One of the best known clustering algorithms is the k-means, with many variants [6–8]. In our study, we use a k-means algorithm to cluster our document collection into a few tightly structured ones. Due to the high dimensionality and low sparsity of the text data, the sub-clusters usually have a certain "self-similar" behavior, i.e., documents of the similar classes are grouped into the same cluster. The centroid vector (defined below) of a tightly structured cluster can usually capture the general description of documents in that cluster. An ideal cluster contains homogeneous documents that are relevant to each other.

10.2.1 Document Clustering

Let the set of document vectors be $A = [a_1, a_2, \ldots, a_j, \ldots, a_n]$, where a_j is the jth document in the collection. We partition the documents into k sub-collections $\{\pi_j\}_{j=1}^k$ such that

$$\bigcup_{j=1}^k \pi_j = \{a_1, a_2, \ldots, a_n\} \text{ and } \pi_j \cap \pi_i = \phi \text{ if } j \neq i.$$

For each fixed $1 \leq j \leq k$, the centroid vector of each cluster is defined as

$$\tilde{c}_j = \frac{1}{n_j} \sum_{a_i \in \pi_j} a_i,$$

where $n_j = |\pi_j|$ is the number of documents in π_j. We normalize the centroid vectors such that

$$c_j = \frac{\tilde{c}_j}{\| \tilde{c}_j \|}, \quad j = 1, 2, \ldots, k.$$

An intuitive definition of the clusters is that, if

$$a_i \in \pi_j,$$

then

$$a_i^T c_j > a_i^T c_l \quad \text{for} \quad l = 1, 2, \ldots, k, \quad l \neq j,$$

i.e., documents in π_j are closer to its centroid than to the other centroids. If the clustering is good enough and each cluster is compact enough, the centroid vector may represent the abstract concept of the cluster. So they are also called *concept vectors* [1]. Define the *concept matrix* as an $m \times k$ matrix such that, for $1 \leq j \leq k$, the jth column of the matrix is the concept vector c_j, i.e., the concept matrix is

$$C_k = [c_1, c_2, \ldots, c_k].$$

The concept matrix C_k is still a sparse matrix, as each column c_j is sparse. The degree of sparsity of C_k is inversely proportional to the number of clusters generated. If we assume that the concept vectors are linearly independent, the concept matrix has rank k.

For any partitioning of the document vectors, we can define the corresponding concept decomposition A_k of the term-document matrix A as the least squares approximation of A onto the column space of the concept matrix C_k. We can write the concept decomposition as an $m \times n$ matrix

$$A_k = C_k \tilde{M},$$

where $\tilde{M}$ is a $k \times n$ matrix that is to be determined by solving the following least squares problem

$$\tilde{M} = \arg\min_M \|A - C_k M\|_F^2. \tag{10.1}$$

Here the norm $\| \cdot \|_F$ is the Frobenius norm of a matrix. It is well-known that problem (10.1) has a closed-form solution,

$$\tilde{M} = (C_k^T C_k)^{-1} C_k^T A.$$

The concept matrix and concept decomposition have been studied extensively by Dhillon and Modha [1]. It is shown that the concept decomposition is very close to the SVD decomposition for some term-document matrices. The concept matrix is much more sparse than the SVD matrix of a term-document matrix.

10.2.2 Retrieval Procedure and Research Strategies

For convenience, we ignore the subscript of C_k and write the concept matrix as C.

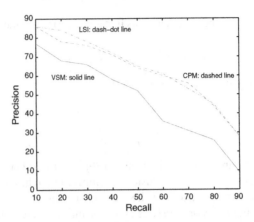

Fig. 10.1 Precision-recall curve comparison of three different retrieval methods on the Medline dataset

Given a query vector q, the retrieval with the concept projection matrix (CPM) A_k can be computed as

$$r^T = q^T A_k = q^T C (C^T C)^{-1} C^T A, \tag{10.2}$$

where r^T is the ranking vector. After sorting the entries of r^T in a descending order, we have a ranking list of documents which are related to the query (listed from the most relevant to the least relevant). The retrieval accuracy based on CPM is quite good, compared to the *Latent Semantic Indexing* (LSI) technique based on SVD. Figure 10.1 shows a comparison of the plain *vector space model* (VSM), LSI, and CPM strategies on the well-known Medline database which contains 5,831 terms and 1,033 documents. It can be seen that the LSI and CPM techniques are comparable, both are much better than the plain VSM.

However, the retrieval procedure in Eq. (10.2) is not efficient in terms of numerical computation and storage. It needs to compute $(C^T C)^{-1}$, the inverse of the normal matrix of the concept matrix, which is likely to be a dense matrix. Thus if the CPM based retrieval is computed as in Eq. (10.2), this technique may not be competitive, compared to the LSI based on SVD.

To make the numerical computation procedure and storage cost more competitive, a strategy which utilizes the knowledges of both preconditioning techniques and the computational theorems have been studied. This strategy is to compute a sparse matrix to approximate the projection matrix directly, i.e., we compute a matrix M to solve the least squares problem (10.1) approximately. The matrix M computed must be sparse in order for the strategy to be competitive against the inverse of the normal matrix of the concept matrix approach (CPM).

10.3 Approximate Least Squares Based Strategies

Note that all we want is to find a *sparse* matrix M such that the functional $f(M) = \|A - CM\|_F^2$ is minimized, or more precisely, *approximately minimized*. A similar problem has been studied recently in the preconditioning community to compute a sparse approximate inverse matrix M for a nonsingular coefficient matrix A to solve a sparse linear system $Ax = b$. The sparse matrix M is computed by minimizing

$$f(M) = \min_{M \in \mathcal{G}} \|I - AM\|_F^2, \tag{10.3}$$

subject to certain constraints on the sparsity pattern of M. Here the sparsity pattern of M is restricted to a (usually unknown *a priori*) subset $\mathcal{G}$.

So, taking the sparse approximate inverse computation in mind, for a term-document matrix A, we can minimize the functional

$$f(M) = \min_{M \in \mathcal{G}} \|A - CM\|_F^2 \tag{10.4}$$

with a constraint such that M is sparse. The most important part of this minimization procedure is to determine the sparsity pattern constraint set $\mathcal{G}$, that gives the sparsity pattern of M.

10.3.1 Computational Procedure

For a moment, we suppose that a sparsity pattern set $\mathcal{G}$ for M is given somehow, the minimization problem (10.3) is decoupled into n independent subproblems as

$$\|A - CM\|_F^2 = \sum_{j=1}^{n} \|(A - CM)e_j\|_2^2 = \sum_{j=1}^{n} \|a_j - Cm_j\|_2^2, \tag{10.5}$$

where a_j and m_j are the jth column of the matrices A and M, respectively. (e_j is the jth unit vector.) It follows that the minimization problem (10.5) is equivalent to minimizing the individual functions

$$\|Cm_j - a_j\|_2, \qquad j = 1, 2, \ldots, n, \tag{10.6}$$

with certain restrictions placed on the sparsity pattern of m_j. In other words, each column of M can be computed independently. This certainly opens the possibility for parallel implementation. Since we assume the sparsity pattern of m_j (and M) is given, i.e., a few, say n_2, entries of m_j at certain locations are allowed to be nonzero, the rest of the entries of m_j are forced to be zero. Denote the n_2 nonzero entries of m_j by $\bar{m}_j$ and the n_2 columns of C corresponding to $\bar{m}_j$ by C_j. Since C is sparse, the submatrix C_j has many rows that are identically zero. After removing the zero rows of C_j, we have a reduced matrix $\bar{C}_j$ with n_1 rows. The individual minimization problem (10.6) is reduced to a much smaller least squares problem of order $n_1 \times n_2$

$$\|\bar{C}_j \bar{m}_j - \bar{a}_j\|_2, \qquad j = 1, 2, \ldots, n, \tag{10.7}$$

in which $\bar{a}_j$ consists of the entries of a_j corresponding to the remaining columns of C_j. We note that the matrix $\bar{C}_j$ is now a very small rectangular matrix. It has full rank if the matrix C does.

There are a variety of methods available to solve the small least squares problem (10.7). Assume that $\bar{C}_j$ has full rank. Since $\bar{C}_j$ is small, the easiest way is probably to perform a QR factorization on $\bar{C}_j$ as

$$\bar{C}_j = Q_j \begin{pmatrix} R_j \\ 0 \end{pmatrix}, \tag{10.8}$$

where R_j is a nonsingular upper triangular $n_2 \times n_2$ matrix. Q_j is an $n_1 \times n_1$ orthogonal matrix, such that $Q_j^{-1} = Q_j^T$. The least squares problem (10.7) is solved by first computing $\bar{c}_j = Q_j^T \bar{a}_j$ and then obtaining the solution as $\bar{m}_j = R_j^{-1} \bar{c}_j (1 : n_2)$. In this way, $\bar{m}_j$ can be computed for each $j = 1, 2, \ldots, n$, independently. This yields an approximate decomposition matrix M, which minimizes $\|CM - A\|_F$ for the given sparsity pattern.

The remaining problem for constructing a sparse approximate decomposition matrix M is choosing or deciding a good sparsity pattern for M. Here we introduce static sparsity pattern (SSP) and dynamic sparsity pattern (DSP) approaches. They are based on the similar strategies proposed in the preconditioning community [5, 9–11]. The difference between the static and dynamic strategies lies in that the static sparsity patterns are decided before the matrix construction phase (*a priori*) and unchanged during the computation while dynamic sparsity patterns are adjusted adaptively in the approximate decomposition matrix construction phase.

10.3.2 Static Sparsity Pattern (SSP)

In preconditioning field, there are some heuristic strategies developed for choosing suitable sparsity patterns for M. A particularly useful and effective strategy is to use the sparsity pattern of the coefficient matrix C or C^T. Chow [4] offers the strategy of using sparsity patterns of C as the sparsity pattern for M. The difficulty for choosing a static sparsity pattern in information retrieval lies in the fact that there is no known study that has been done to find a suitable sparsity pattern, to the best of our knowledge. This work ventures into a non-traditional application of computational numerical linear algebra with approximate decomposition matrix computation in information retrieval.

Once a good sparsity pattern is chosen or found, the static sparsity pattern algorithms are relatively easier than the dynamic sparsity pattern (to be discussed later) to implement [4, 12, 13].

The knowledge from the preconditioning field can be exploited for choosing a suitable sparsity pattern for our application. Note that the concept matrix C describes the relationship between the term vectors and the concept vectors. If a term is related to a concept vector, this relationship may be maintained in the approximate decomposition matrix in some sense. However, this line of reasoning is much more difficult than that in the preconditioning field. This is because of the fact that the dimensions of the matrix C and those of M do not match. The dimensions of C are $m \times k$ and those of M are $k \times n$. To make such an approach practically useful, several auxiliary strategies based on the sparsity pattern of C and entry values of both C and A are proposed.

- Our first strategy is based on the numerical computation, vector-vector product. That is $c_i m_j = a_{ij}$, where c_i is the ith row of C, m_j is the jth column of M, and a_{ij} is the entry at the ith row, jth column of A. The sparsity pattern of m_j is given in this way: If a_{ij} is the largest entry in the jth column of A, the sparsity pattern of the jth column of M, m_j, is the same as that of the ith row of C, i.e. c_i. Here we use small matrices to illustrate our ideas. Suppose we have three matrices: $C_{4\times3}$, $M_{3\times5}$, and $A_{4\times5}$. The pattern of $CM = A$ is depicted by the Eq. (10.9).

$$\begin{pmatrix} x & 0 & x \\ 0 & x & 0 \\ x & x & 0 \\ 0 & x & 0 \end{pmatrix}_{4\times3} \begin{pmatrix} - & - & - & - & - \\ - & - & - & - & - \\ - & - & - & - & - \end{pmatrix}_{3\times5} = \begin{pmatrix} 0 & x & 0 & x & 0 \\ 0 & 0 & 0 & x & x \\ x & x & 0 & 0 & 0 \\ x & 0 & x & 0 & 0 \end{pmatrix}_{4\times5} \quad (10.9)$$

Here, "x" denotes nonzero entry, "-" denotes undefined pattern. We determine the sparsity pattern of M column by column. First, find the largest entry in each column of A, suppose they are a_{31}, a_{12}, a_{43}, a_{14}, and a_{25} in Eq. (10.9). Then the sparsity pattern of m_1, the first column of M, is the same as that of the third row of C, c_3 and the sparsity pattern of m_2 is the same as that of c_1, m_3 and m_4 have the same sparsity pattern of c_4 and c_1 respectively. Finally we have the sparsity pattern of M like this: $\begin{pmatrix} x & x & 0 & x & 0 \\ x & 0 & x & 0 & x \\ 0 & x & 0 & x & 0 \end{pmatrix}$.

Since there may be more than one largest entries in each column of term-document matrix A, the following rules may be applied to choosing the largest term.

- Start the above procedure from the column of A that has the smallest number of nonzero entries.
- Do not use the same row's sparsity pattern of C if possible.

Based on this strategy, the sparsity ratio of M is almost the same as that of C.
- In order to improve the accuracy and robustness, the first strategy can be applied again to those second largest entries in each column of A. For example the second largest entry in the second column is a_{32}. Comparing the pattern of the third row of C, (x x 0), with that of the second column of M, (x 0 x), we simply fill in more nonzero entries in the second column of M based on the nonzero positions in the third row of C. Now the sparsity pattern of the second column of M is

$(x\ x\ x)$. This strategy can be repeated couple of times as needed. The matrix M may be more dense. However it might be more accurate and robust. We can also control the number of fill-ins like that in preconditioning techniques.

10.3.3 Dynamic Sparsity Pattern (DSP)

The dynamic sparsity pattern strategies first compute an approximate decomposition matrix by solving the least squares problem (10.4) with respect to an initial sparsity pattern guess. Then this sparsity pattern is updated according to some rules and is used as the new sparsity pattern guess for solving (10.4). The approximate decomposition matrix computation may be repeated several times until some stopping criteria are satisfied. Different update rules lead to different dynamic sparsity pattern strategies. One useful rule, suggested by Grote and Huckle [5] in computing sparse approximate inverse preconditioners, adds the candidate indices into the sparsity pattern of the current approximation S that can most effectively reduce the residual $g = C(\cdot, S)\bar{m}_j - a_j$.

The candidate indices are chosen from the set

$$\beta = \{j \notin S \mid C(\alpha, j) \neq 0\},$$

where $\alpha = \{ i \mid g(i) \neq 0\}$. This is a one-dimensional minimization problem

$$\min_{u_j} \|g + u_j(Ce_j - a_j)\|_2, \quad j \in \beta,$$

which e_j is the jth unit vector. Denote $l_j = Ce_j - a_j$, the above minimization problem has the solution

$$u_j = -\frac{g^T l_j}{\|l_j\|_2^2}.$$

For each j, we compute the two-norm of the new residual as

$$\rho_j = \|g\|^2 - \frac{(g^T l_j)^2}{\|l_j\|_2^2}.$$

Then we can choose the most profitable indices j which lead to the smallest new residual norm ρ_j. The procedure to augment the sparsity structure S is as follows.

Dynamic Sparsity Pattern Construction Algorithm

Given the maximum number of update steps $ns > 0$,
*   stopping tolerance ε, integers $\mu > 0$,*
*   and an initial diagonal sparsity pattern S;*
Loop:
*   compute ρ_j for all indices $j \in \beta$;*
*   Compute the mean λ of $\{\rho_j\}$;*
*   At most μ indices with $\rho_j < \lambda$ will be added into S;*
*   Until $\|r\|_2 < \varepsilon$ or exceed the steps ns.*

Barnard et al. released an MPI implementation, SPAI_3.0, [14] for computing the sparse approximate inverse preconditioner of a nonsingular matrix. We modified the code for our computation of the sparse approximate decomposition matrix.

10.4 Numerical Experiments

To evaluate the performance of the proposed matrix approximation approaches SSP and DSP for the concept matrix decomposition, we apply them to three popular text databases: CRAN, MED and CISI and compare them with the concept project matrix method (CPM). The databases are downloaded from http://www.cs.utk.edu//lsi/ [15]. The information about the three databases are given in Table 10.1.

A standard way to evaluate the performance of an information retrieval system is to compute precision and recall values. The precision is the proportion of the relevant documents in the set returned to the user; the recall is the proportion of all relevant documents in the collection that are retrieved by the system. We average the precision of all queries at fixed recall values as $10\%, 20\%, \ldots, 90\%$. The clustering, query and precision evaluation codes in matlab are acquired from [15]. In addition of precision-recall tests, we compare their storage costs and CPU time required for query procedure and approximate matrix computation for each of the three approaches. The precision computation and query time are carried out in UNIX system using matlab. The sparse matrix computation and matrix inverse are carried out in IBM Power/Intel(Xeon) hybrid system at University of Kentucky using C.

As presented in [15], the better number of clusters k for three databases are around 200 and 500. Therefore in all of the following tests, we use $k = 256$ and $k = 500$ for CISI and CRAN databases; $k = 200$ and $k = 500$ for MED database. For DSP approach, three parameters are used in algorithm 10.3.3. "ε" is used to control the quality of the approximation matrix. For all the tests, we choose $\varepsilon = 1$. "ns" is maximum number of improvement steps per row in DSP. "μ" is maximum number of new nonzero candidates per step. Higher values of ns, μ lead to more work, more fill-in, and usually more accurate matrix.

We first test their query precisions. The precision test results for all three databases are given in Figs. 10.2–10.4. We are not surprised that CPM has better query results for every database. This is because CPM is much more dense (memory costs will be given in other figures) and hence more accurate than that of SSP and DSP.

Table 10.1 The information of three databases

Database	Matrix size	Number of queries	Source
CISI	5609×1460	112	Fluid dynamics
CRAN	4612×1398	225	Science indices
MED	5831×1033	30	Medical documents

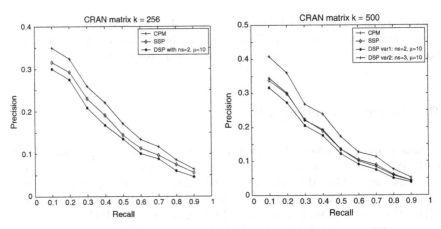

Fig. 10.2 Left panel: CRAN database, k = 256. Right panel: CRAN database, k = 500

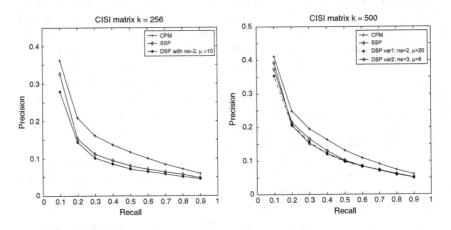

Fig. 10.3 Left panel: CISI database, k = 256. Right panel: CISI database, k = 500

We then compare their storage costs required in previous tests by listing the number of nonzeros of the approximate matrix M and CPM matrix $(C^T C)^{-1} C^T A$. The test results presented in the left and right panel of the Fig. 10.5 are corresponding to that in the left and right panel of the Figs. 10.2–10.4 respectively. From this test we see that SSP and DSP use much less memory space than CPM. Considering the fact that with about 90% less memory costs than CPM, SSP and DSP suffer only 6% precisions lost. They may be more attractive if storage cost is a bottle-neck. We also tried to sparsify the CPM matrix by dropping small entries. By reducing 20% of the memory storages, the precision is lost more than 40% for CISI with $k = 500$ test.

In previous precision-recall tests, SSP looks slightly better than DSP, especially when the number of cluster k is small. However, SSP matrix has more than double number of nonzeros than that of DSP matrix. In order to see which sparsity pattern is good, we compare SSP and DSP by increasing the density of the DSP matrix.

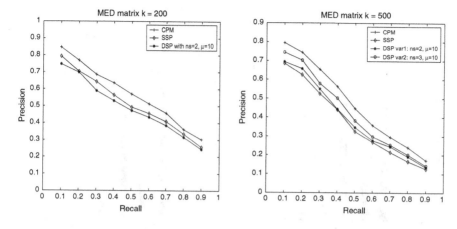

Fig. 10.4 Left panel: MED database, k = 200. Right panel: MED database, k = 500

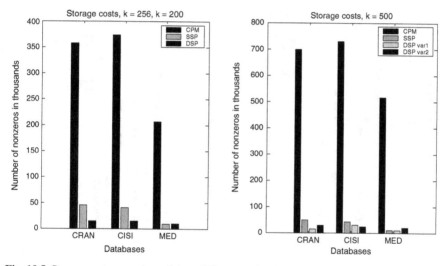

Fig. 10.5 Storage cost comparison of three different retrieval methods.

In the tests of all three databases with $k = 500$, we either increase the update steps or the number of fill-ins at each step to compute DSP matrix M for accurately. In the right panel of the Figs. 10.2–10.4, we see the precision-recall curves for DSP corresponding to DSP var2 are better or very closer to that of SSP while still uses smaller memory space compared with SSP. That means, given the same storage constraint, dynamic sparsity pattern strategy more accurately computes sparse approximate matrix than the static sparsity pattern strategy. Figure 10.5 shows the storage cost comparison of three different retrieval methods.

Finally we give the query time and total CPU time required for the query procedures and CPM, SSP, and DSP three matrices computation. The query time is one

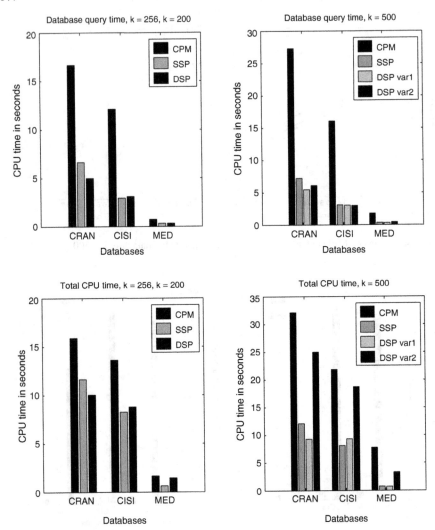

Fig. 10.6 CPU time for query procedure and matrix construction

of the most important aspects in information retrieval system. It affects the retrieval performance. We see from the upper panel of the Fig. 10.6 that SSP and DSP greatly reduce the query time required by CPM approach due to their very sparse matrices. The total CPU time for matrix construction and query procedure is given in the bottom panel of the Fig. 10.6. If $k = 256$, or 200, the size of the $C^T C$ is either 256 by 256 or 200 by 200, which is small, the inverse of $C^T C$ takes less time than SSP and DSP matrix construction time. However, for $k = 500$, CPM takes much more time compared with SSP. SSP also is much faster than DSP if the sparse matrix M is constructed with the same density.

10.5 Summary

We have developed numerical matrix computation methods based on static sparse pattern (SSP) and dynamic sparse pattern (DSP) to compute the approximate matrix to the concept matrix decomposition. We tested and compared with CPM based retrieval schema. In our numerical experiments, the sparsity pattern based approaches SSP and DSP turn out to be more competitive in terms of query precision, computational costs and memory space.

With the comparison of SSP and DSP sparsity pattern strategies, DSP displays some of the following advantages: First is that it computes a more accurate approximate matrix than SSP given the same density constraint. Second is that the accuracy of the approximate decomposition matrix can be controlled easily. By allowing more update steps, we can compute a more accurate approximate decomposition matrix. Third, as the information retrieval database may be updated periodically to accommodate new documents or to remove some out-of-dated documents, the computed sparsity pattern may be used as a static sparsity pattern in some intermediate database update. But the dynamic sparsity pattern strategy is more expensive in terms of computational cost. Note that in information retrieval, these computations are called *pre-processing* computation to prepare the database for the purpose of retrieval. Thus, an expensive one-time cost can be allowed if the prepared database enables more accurate and faster retrieval. If the storage costs and query time are the bottle-neck during the query procedure, the dynamic sparsity pattern strategy looks more attractive.

References

1. I. S. Dhillon and D. S. Modha. Concept decompositions for large sparse text data using clustering. *Mach. Learn.*, 42(1):143–175, 2001.
2. G. H. Golub and C. F. van Loan. *Matrix Computations*. The Johns Hopkins University Press, Baltimore, MD, 3rd edition, 1996.
3. J. Gao and J. Zhang. Clustered SVD strategies in latent semantic indexing. *Inform. Process. Manage.*, 41(5):1051–1063, 2005.
4. E. Chow. A priori sparsity patterns for parallel sparse approximate inverse preconditioners. *SIAM J. Sci. Comput.*, 21(5):1804–1822, 2000.
5. M. Grote and T. Huckle. Parallel preconditioning with sparse approximate inverses. *SIAM J. Sci. Comput.*, 18:838–853, 1997.
6. J. Hartigan. *Clustering Algorithms*. Wiley, New York, 1975.
7. J. Hartigan and M. Wong. Algorithm AS136: A k-means clustering algorithm. *Appl. Stat.*, 28:100–108, 1979.
8. A. K. Jain, M. N. Murty, and P. J. Flynn. Data clustering: a review. *ACM Comput. Surv.*, 31(3):264–323, 1999.
9. E. Chow and Y. Saad. Approximate inverse preconditioners via sparse-sparse iterations. *SIAM J. Sci. Comput.*, 19(3):995–1023, 1998.
10. J. D. F. Cosgrove, J. C. Diaz, and A. Griewank. Approximate inverse preconditionings for sparse linear systems. *Int. J. Comput. Math.*, 44:91–110, 1992.
11. L. Y. Kolotilina. Explicit preconditioning of systems of linear algebraic equations with dense matrices. *J. Soviet Math.*, 43:2566–2573, 1988.

12. K. Wang, J. Zhang, and C. Shen. A class of parallel multilevel sparse approximate inverse preconditioners for sparse linear systems. *J. Scalable Comput. Pract. Exper.*, 7:93–106, 2006.
13. K. Wang and J. Zhang. MSP: a class of parallel multistep successive sparse approximate inverse preconditioning strategies. *SIAM J. Sci. Comput.*, 24(4):1141–1156, 2003.
14. S. T. Barnard, L. M. Bernardo, and H. D. Simon. An MPI implementation of the SPAI preconditioner on the T3E. *Int. J. High Perform Comput. Appl.*, 13:107–128, 1999.
15. J. Gao and J. Zhang. Text retrieval using sparsified concept decomposition matrix. Technical Report No. 412-04, Department of Computer Science, University of Kentucky, KY, 2004.

Chapter 11
Transferable E-cheques: An Application of Forward-Secure Serial Multi-signatures

Nagarajaiah R. Sunitha, Bharat B.R. Amberker,
and Prashant Koulgi

Abstract With the modern world going on-line for all businesses, we need to transact with various business organizations all over the world using different modes of payment obtained through various Financial Institutions. We have considered the following bank transaction. A person X having an account in bank C issued a cheque for certain amount in favor of a person Y having an account in bank B. But, Y wants to issue a cheque for the same amount favoring Z. In normal course, Y need to deposit the cheque in bank B, wait for clearance and then issue a cheque in favor of Z. This consumes time as bank B sends the cheque to bank C for clearance.

In this paper, we propose a scheme in which a cheque is transferable. That is, the cheque in favor of Y can be reissued for the same amount to Z without presenting in the bank. Person Z can deposit the cheque directly in his bank, thus saving time and work load on banks. Further, the scheme can be used to transfer the cheque to any number of persons and only the last person deposits in the bank. In view of this, e-cheque can be used as e-cash to a limited extent. Our scheme is based on multi-signatures. We have augmented the multi-signature scheme to provide forward security. This guarantees the security of cheques signed in the past even if the signer's secret key is exposed today. We propose two flavors of Forward-secure multi-signatures, one based on Elgamal signatures and the other based on Digital Signature Algorithm.

Keywords E-cheque · Transferable · Multi-signature · Forward-secure · Elgamal signature · Digital signature

N.R. Sunitha (✉) and P. Koulgi
Department of Computer Science & Engineering, Siddaganga Institute of Technology,
Tumkur, Karnataka, India
e-mail: nrsunitha@gmail.com

B.B.R. Amberker
Department of Computer Science & Engineering, National Institute of Technology,
Warangal, Andhra Pradesh, India

S.-I. Ao et al. (eds.), *Advances in Computational Algorithms and Data Analysis,*
Lecture Notes in Electrical Engineering 14,
© Springer Science+Business Media B.V. 2009

11.1 Introduction

A fundamental requirement of E-commerce protocol [1, 2] is efficient schemes for payments. E-cheques are a mode of electronic payments. This technology was developed couple of years ago and has been promoted by many of the financial institutions. E-cheques work the same way as paper cheques and are a legally binding promise to pay. Rather than handwritten or machine-stamped signatures, however, e-cheques are affixed with digital signatures. The payer writes an e-cheque by structuring an electronic document with the information legally required to be in a cheque and digitally signs it. The payee receives the e-cheque over email or web, verifies the payer's digital signature, writes out a deposit and digitally signs it. The payee's bank verifies the payer's and payee's digital signatures and forwards the cheque for clearing and settlement. The payer's bank verifies the payer's digital signature and debits the payer's account.

The standard notion of digital signature [3–5] security is extremely vulnerable to leakage of the secret key which over the lifetime of the scheme may be quite a realistic threat. Indeed if the secret key is compromised any message can be forged. Forward-secure signature schemes, first proposed by Anderson in [4] and formalised by Bellare and Miner in [5] are intended to address the above limitation. A forward-secure digital signature scheme [5–8] is a method for creating digital signatures signed with secret keys changing with time periods, all of which can nevertheless be verified by the verifier using the same public key. An adversary with access to this public key and the secret key of some time period, will be unable to forge signatures for an earlier time period.

When a signature depends on more than one signer we call it a multi-signature. A multi-signature scheme [9–12] enables a group of signers to produce a compact, joint signature on a common document. As many applications require multiple signers to sign the same document, we propose to apply the concept of forward-security to multi-signatures. Using Forward-secure multi-signatures all signers of the document can guarantee the security of document signed in the past even if their secret key is exposed today. An adversary will not be able to forge such a multi-signature unless the secret key of all the signers are compromised in the same time period, which is practically not possible.

Cheques once issued to a customer must be deposited in a bank for further processing. Generally there is no provision for a cheque to be transferred among customers i.e. if a customer has a cheque in hand for a specified amount, he cannot give the same cheque to another customer for the same specified amount. He must first deposit the cheque in his bank and issue another cheque to the customer. We have come up with a proposal through e-cheques, to provide an option for a customer to transfer a cheque to another customer without depositing it in the bank. Each customer receiving the cheque is convinced that the cheque is from the intended sender. Only the last receiver of the e-cheque deposits it in the bank. All previous customers transfer the e-cheque off-line. This reduces the work load on bank to clear the e-cheques. Also e-cheques can be used like hard cash to some extent. We use the concept of Serial Forward-Secure multi-signatures [13] which ensure forward-security of the document and allow signers to sign the same document serially.

In Section 11.2, we discuss how to make the basic schemes like ElGamal and DSA signature schemes forward-secure. In Section 11.3, we apply forward-secure concept for a group of signers who need to sign the same document. Here we discuss Forward-secure serial multi-signatures which ensure forward-security of the document and allow signers to sign the same document serially. We propose schemes based on both Elgamal signatures and DSA. We also explain the model to use the concept of forward-secure serial multi-signature to transfer e-cheques among customers. In Section 11.4, we give the security analysis of our scheme by considering the possible attacks against the multi-signature scheme and in Section 11.5, we discuss the forward-security of our scheme. Lastly in Section 11.6, we conclude.

11.2 Basic Schemes made Forward-Secure

To specify a forward-secure signature scheme, we need to (i) give a rule for updating the secret key (ii) specify the public key and (iii) specify the signing and the verification algorithms. In saying that our forward-secure scheme is based on a basic signature scheme, we mean that, given a message and the secret key of a time period, the signing algorithm is the same as in the basic signature scheme. As for the other specifications we use the idea of Bellare and Miner in [5]: Given the secret key of some time period, the secret key of the subsequent period will be calculated just as the public key would be if this were the secret key in the (underlying) basic signature scheme. The public key for the forward-secure signature scheme is the key obtained on running T times the update rule for secret keys.

Now, we need to be able to write a verification equation relating the public key and the signature (and incorporating the time period of the signature) from which the claim of forward security can be deduced.

In the following sections we give the algorithms to make the basic schemes like ElGamal Signature scheme and the Digital signature scheme forward-secure.

11.2.1 Forward Secure ElGamal Signature Scheme

Following are the details of the algorithm:

1. **Secret Key Updation:** Let p be a large prime.
   Let $\phi(p-1) = p_1^{r_1} \ldots p_k^{r_k}$ where $p_1 < p_2 < \ldots < p_k$.
   Choose α such that

$$\gcd(\alpha, p) = 1, \gcd(\alpha, \phi(p)) = 1, \gcd(\alpha, \phi^2(p)) = 1, \ldots,$$
$$\gcd(\alpha, \phi^{T-1}(p)) = 1$$

where $\phi(p)$ is the totient function and $\phi^{T-i}(p) = \phi(\phi^{T-i-1}(p))$ for $1 \le i \le T-1$ with $\phi^0(p) = p$. It may be noted that a prime α chosen in the range $p_k < \alpha < p$ satisfies the above condition. The base secret key a_0 (this is the initialisation for the secret key updation) is chosen randomly in the range $1 < a_0 < p-1$.

The secret key a_i in any time period i is derived as a function of a_{i-1}, the secret key in the time period $i-1$, as follows:

$$a_i = \alpha^{a_{i-1} \bmod \phi^{T-i+1}(p)} \bmod \phi^{T-i}(p)$$

for $1 \le i < T$. Once the new secret key a_i is generated for time period i, the previous secret key a_{i-1} is deleted. Thus an attacker breaking in period i will get a_i but cannot compute $a_0, \ldots, a_{i-1}$, because of difficulty of computing discrete logarithms.

2. **Public Key Generation:** In Bellare–Miner scheme, the public key is obtained by updating the base secret key $T+1$ times. However, we obtain the public key by executing the Secret Key Updation Algorithm T times as follows:

$$\beta = \alpha^{a_{T-1}} \bmod p = a_T \bmod p \tag{11.1}$$

3. **Signature Generation:** The signature generated in any time period i is $\langle y_{1,i}, y_{2,i} \rangle$. The computation of $y_{1,i}$ is

$$y_{1,i} = \alpha^k \bmod p$$

where k is a random number chosen such that $0 < k < p$ and $gcd(k, (p-1)) = 1$. The computation of $y_{2,i}$ is

$$y_{2,i} = (H(m\|i) - (\mathcal{A}(\alpha, T-i-1, a_i) \cdot y_{1,i}))k^{-1} \bmod (p-1) \tag{11.2}$$

where H is a collision-resistant hash function. While hashing, i is concatenated with m to indicate the time period in which the message is signed.

By the notation $\mathcal{A}(\alpha, u, v) = \alpha \ldots \alpha v$ we mean that there are u number of α 's in the tower and the topmost α is raised to v, i.e. in the above equation there are $(T-i-1)$ number of α's in the tower and the topmost α is raised to a_i. Notice that the public key β can also be given in terms of a_i as,

$$\beta = \mathcal{A}(\alpha, T-i, a_i) \bmod p,$$

This relation gets employed in the verification of validity of the signature.

4. **Signature Verification:** As for verification, a claimed signature $\langle y_{1,i}, y_{2,i} \rangle$ for the message m in time period i is accepted if

$$\alpha^{H(m\|i)} = \beta^{y_{1,i}} \, y_{1,i}^{y_{2,i}} \bmod p \tag{11.3}$$

else rejected.

11.2.2 Forward Secure DSA Signature Scheme

As the key generation algorithm is same for both ElGamal and DSA signature schemes, the secret key updation and public key generation algorithms discussed under Forward-Secure ElGamal Signature scheme can be used even for Forward-Secure DSA scheme.

1. **Signature Generation:** The signature generated in any time period i is $\langle r, s, i \rangle$. The computation of r is

$$r = (\alpha^k \mod p) \mod q$$

where k is a random number chosen such that $0 < k < p$ and $gcd(k, (p-1)) = 1$. The computation of s is

$$s = k^{-1}(SHA(m||i) + (\mathcal{A}(\alpha, T - i - 1, a_i) * r)) \mod q$$

where SHA is a collision-resistant hash function.

2. **Signature Verification:** $w = (si)^{-1}$; $u1 = SHA(m||i) * w$; $u2 = r * w$; $v = \alpha^{u1} * y^{u2}$. As for verification, a claimed signature $\langle r, s, i \rangle$ for the message m in time period i is accepted if

$$v^i = r$$

else rejected.

11.3 The Forward-Secure Serial Multi-signature Scheme

We use the Forward-secure Elgamal and DSA signature schemes discussed in previous section to design Forward-Secure Serial Multi-signature scheme [12].

The Forward-secure Serial Multi-signature Scheme ensures forward-security of the document and allows multiple signers to sign the same document serially i.e. one after the other. Here signing order need not be predetermined. During this process each signer verifies the signature of his/her predecessor's and then signs the document by creating a partial multi-signature. The signature generated by the last signer will be the multi-signature which can be verified by any verifier with a single public key.

11.3.1 Forward-Secure Serial Multi-signature Scheme based on Forward-Secure Elgamal Signatures

Following is the protocol to create Forward-secure Serial Multi-signature Scheme using Forward-secure Elgamal Signatures.

11.3.1.1 Partial Multi-signature Generation and Verification

Any signer $U_j(2 \le j \le n)$ computes $(y_j, y'_{j,i})$ as follows,

$$y_j = \alpha^{k_j} \mod p$$

where k_j is a random number chosen such that $0 < k_j < p - 1$ and $gcd(k_j, p-1) = 1$.

$$y'_{j,i} = (H(m||i) - ((\mathcal{A}(\alpha, T - i - 1, a_{j,i}) \cdot y_j))k_j^{-1} \mod (p-1) \qquad (11.4)$$

where H is a collision-resistant hash function.

The signer signs the message m by creating the partial multi-signature, $\langle((\sigma_{j,1}, \sigma_{j,2}), m)\rangle$ where

$$\sigma_{j,1} = \sigma_{j-1,1} \cdot y_j^{y'_{j,i} \cdot y_1^{-1}} \mod p$$

$$\sigma_{j,2} = (\sigma_{j-1,2} + y_j^{-1}) \cdot H(\sigma_{j,1}) \mod p.$$

This partial multi-signature is sent to the next signer U_{j+1}. Any partial multi-signature received by a signer $U_j(2 \le j \le n)$ is verified using the following equation:

$$\alpha^{H(m||i) \cdot \sigma_{j,2}} = (\beta_{1\ldots(j)} \cdot \sigma_{j,1})^{H(\sigma_{j,1})}$$

where the public key $\beta_{1\ldots(j)}$ is computed as the product of public keys of previous signers.

The partial multi-signature generated by the last signer is the Forward-secure Serial Multi-signature of n signers which can be verified by any external verifier. The verification equation for the external verifier is

$$\alpha^{H(m||i) \cdot \sigma_{n,2}} = (\beta_{1\ldots n} \cdot \sigma_{n,1})^{H(\sigma_{n,1})}.$$

11.3.2 Forward-Secure Serial Multi-signature Scheme based on Forward-Secure DSA Signatures

Following is the protocol to create Forward-secure Serial Multi-signature Scheme using Forward-secure DSA Signatures.

11.3.2.1 Partial Multi-signature Generation and Verification

Any signer $U_j(2 \le j \le n)$ computes $\langle(rj, sj)m, i\rangle$. The computation of rj is

$$rj = (\alpha^{kj} \mod p) \mod q$$

where kj is a random number chosen such that $0 < kj < p$ and $gcd(kj,(p-1)) = 1$. The computation of sj is

$$sj = kj^{-1}(SHA(m||i) + (\mathcal{A}(\alpha, T - i - 1, a_{j,i}) * rj)) \mod q \qquad (11.5)$$

He creates the partial multi-signature, $\langle((\sigma_{j,1}, \sigma_{j,2}), m)\rangle$ where

$$\sigma_{j,1} = \sigma_{j-1,1}.r_j^{r_j^{-1}.sj}$$

$$\sigma_{j,2} = (\sigma_{j-1,2} + r_j^{-1}).H(\sigma_{j,1}).$$

This partial multi-signature is sent to the next signer U_{j+1}. Any partial multi-signature received by a signer $U_j (2 \le j \le n)$ is verified using the following equation:

$$\alpha^{SHA(m||i).\sigma_{j-1,2}} = (\beta^{-1}.\sigma_{j-1,1})^{H(\sigma_{j-1,1})}$$

where the public key β is computed as the product of public keys of previous signers. The verification equation for the external verifier is

$$\alpha^{SHA(m||i).\sigma_{n,2}} = (\beta^{-1}.\sigma_{n,1})^{H(\sigma_{n,1})}$$

11.3.3 Our Model

We propose a scheme in which a cheque is transferable. That is, the cheque in favor of Y can be reissued for the same amount to Z without presenting in the bank. Person Z can deposit the cheque directly in his bank, thus saving time and work load on banks. Further, the scheme can be used to transfer the cheque to any number of persons and only the last person deposits in the bank. In view of this, e-cheque can be used as e-cash to a limited extent. Our scheme is based on multi-signatures. We have augmented the multi-signature scheme to provide forward security. This guarantees the security of cheques signed in the past even if the signer's secret key is exposed today.

We assume that the customers issue/transfer e-cheques only if sufficient balance exists in their account. As seen in Fig. 11.1, the payer1 requests his bank (FI C as this financial institution clears the e-cheque later) for e-cheque leaves. On storing the e-cheque details like e-cheque no, security parameters and so on, the FI-C sends the e-cheque leaves to payer1. The payer1 enters the e-cheque amount, details of the payee and signs it using the forward secure ElGamal/DSA signature as discussed in Section 11.3 and generates the partial multi-signature as discussed in Section 11.4. Payer1 is the initiator. This e-cheque is sent to the payee. The payee verifies the multi-signature as discussed in Section 11.3. If it is verified the payee can deposit in his bank (FI-S – as this financial institution submits the e-cheque to FI-C) or can transfer the same e-cheque to another payee. If he is transferring the

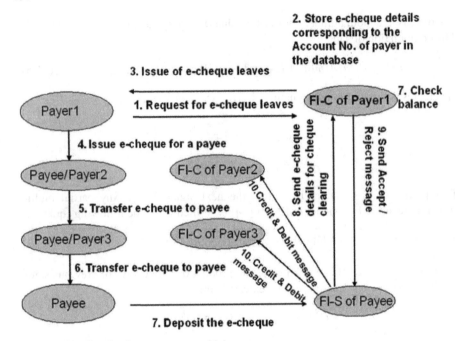

Fig. 11.1 Transfer of e-cheques among multiple customers

e-cheque, he becomes payer2. Payer2 must sign as done by payer1 using the forward secure ElGamal/DSA signature and generate partial multi-signature as discussed in Section 11.3. The payee can verify the partial multi-signature and either submit it in his FI-S or transfer it to another payee. This can continue for any number of customers. Once the payee verifies the partial multi-signature, he is convinced that he has received from the intended sender. The e-cheque is given a validity period before which it has to be submitted in a bank. When the last person deposits the e-cheque in his FI-S, the e-cheque is cleared by the FI-C of the first payer. Thus the transfer of e-cheques is done off-line. Also, the e-cheque needs to be cleared only once. The FI-S of the last customer just sends a message to all FI-C of signers of e-cheque to credit and debit the e-cheque amount. This information is required to keep track of all the transactions of a customer.

11.4 Security Analysis

In this section we analyze the possible attacks against our forward-secure multi-signature scheme:

11.4.1 Attacks Aiming to get Private Keys

1. **Recover Secret Key from Public Key:** The public key β for the Forward-Secure Serial Multi-signature is computed as the product of public keys of individual signers: Recovering $a_{j,i}$ from β_j is equivalent to solving discrete log problem which is computationally not possible.

2. **Determining Secret Key from a Set of Signatures:** There are n equations of the form (4/5), but $(n+1)$ unknowns (since each signature uses different secret k_j). The system of equations cannot be solved and the private key $a_{j,i}$ is secure.

3. **Recovering k_j and then Determine $a_{j,i}$:** If an adversary is able to get k_j, he can determine $a_{j,i}$. But recovering k_j from y_j is equivalent to solving discrete log problem.

4. When the private keys of one or more users are lost and if the intruder holds this secret information and intend to get private keys of other users, he must break the security as mentioned above(1, 2&3).

11.4.2 Attacks for Forging Multi-signatures

1. **The Substitution Attack:** This attack is prevented by the use of one-way hash functions (see Eq. (11.4)/Eq. (11.5)).

2. Any signer $U_j (2 \leq j \leq n)$ may want to forge a multi-signature for a message m and then declare that m is signed by $U_1, \ldots, U_{j-1}$ and U_j itself. By this j signer is making all the previous $j-1$ signers responsible for the forged message. This is once again prevented by the use of one-way hash functions.

11.5 Forward Security of the Proposed Scheme

Here we prove that, given a secret key a_i of some time period i an adversary cannot find the secret key for some period $j < i$. We show that in Eq. (11.2), finding secret key using public key (as public key is obtained by updating the base secret key T times) is equivalent to solving discrete log problem.

Let P_1 be the discrete log problem where given α and B we want to find A in

$$B \equiv \alpha^A \bmod \phi^{T-j-1}(p) \tag{11.6}$$

This problem is believed to be computationally hard.

Let P_2 be a problem where given α and a_i we need to find the secret key a_j in

$$a_i \equiv \mathcal{A}(\alpha, i - j, a_j) \bmod \phi^{T-i}(p) \tag{11.7}$$

We claim that if P_1 is hard, then P_2 is also hard. Thus, if we can find a_j from a_i in Eq. (11.7) we can find A from B in Eq. (11.6). We prove this by contradiction.

Proof. Let us assume that P_2 is not hard. We will show that P_1 is also not hard. Set

$$a_i \equiv \mathcal{A}(\alpha, i - j - 1, \beta) \bmod \phi^{T-i}(p)$$

then a_j obtained from solving P_2 satisfies

$$\alpha^{a_j} \equiv \beta \bmod \phi^{T-j-1}(p)$$

By setting $a = a_j$ we have obtained a solution in P_1 which is a contradiction.

11.6 Conclusion

Many applications require multiple signers to sign the same document. A multi-signature scheme enables a group of signers to produce a compact, joint signature on a common document. We have come up with a scheme to provide an option for a customer to transfer a cheque to another customer without depositing it in the bank. Each customer receiving the e-cheque is convinced that the cheque is from the intended sender. Only the last receiver of the e-cheque deposits it in the bank. All previous customers transfer the e-cheque off-line. This reduces the work load on bank to clear the e-cheques. We use the concept of Serial Forward-Secure multi-signatures. The schemes work with Elgmal and DSA signatures and ensure forward-security of the messages. The signatures can be verified using a single public key though multiple signers are involved.

References

1. Bruce, S., *Applied Cryptography*, Wiley (2001).
2. Graff, J.C., *Cryptography and E-Commerce*, Wiley (2001).
3. Taher, E., *A Public Cryptosystem and a Signature Scheme Based on Discrete Logarithms*, IEEE Transactions on Information Theory, Vol. IT-31, No. 4, pp. 469–472 (1985).
4. Anderson, R., *Invited Lecture*, Fourth Annual Conference on Computer and Communications Security, Zurich, Switzerland ACM (1997).
5. Bellare, M. and Miner, S., *A Forward-Secure Digital Signature Scheme*. In: Wiener, M. (Ed.): Advances in Cryptology-Crypto 99 proceedings, LNCS, Vol. 1666, Springer (1999).
6. Abdalla, M. and Reyzin, L., *A New Forward-Secure Digital Signature Scheme*. In: ASI-ACRYPT 2000, LNCS, Vol. 1976, pp. 116–129 Springer (2000).
7. Kozlov, A. and Reyzin, L., *Forward-Secure Signatures with Fast Key Update*. In: Security in Communication Networks (SCN 2002), LNCS, Vol. 2576, pp. 241–256, Springer (2002).

8. Malkin, T., Miccianco, D., and Miner, S., *Efficient Generic Forward Secure Signatures with an Unbounded Number of Time Periods*. In: Proceedings of EuroCrypt 2002, LNCS, Vol. 2332, pp. 400–417, Springer (2002).
9. Boyd, C., *Digital Multi-signatures*. In: Cryptography and Coding, Oxford University Press, pp. 241–246 (1989).
10. Itakura, K. and Nakamura, K., *A Public Key Cryptosystem Suitable for Digital Multi-Signatures*, NEC Research and Development, Vol. 71, pp. 1–8 (1983).
11. Micali, S., Ohta, K., and Reyzin, L., *Accountable Subgroup Multi-Signatures,* In: ACM Conference on Computer and Communications Security, pp. 245–254 Philadelphia, Pennsylvania, USA (2001).
12. Shieh, S.-P., Lin, C.-T., Yang, W.-B., and Sun, H.-M., *Digital Multi-Signature Schemes for Authenticating Delegates in Mobile Code Systems*, IEEE Transactions on Vehicular Technology, Vol. 49, No. 4 pp. 1464–1473 (July 2000).
13. Sunitha, N.R., Amberker, B.B., Prashant Koulgi, *Transferable E-Cheques Using Forward-Secure Multi-Signature Scheme,* In: The World Congress on Engineering and Computer Science 2007, San Francisco, CA, USA, 24–26 October, 2007.

Chapter 12
A Hidden Markov Model based Speech Recognition Approach to Automated Cryptanalysis of Two Time Pads

Liaqat Ali Khan and M.S. Baig

Abstract Although keystream reuse in stream ciphers and one time pads has been a well known problem in stream ciphers for several decades, yet the threat to real systems has still been underestimated. The keystream reuse in case of textual data has been the focus of cryptanalysts for quite some time now. In this chapter, we present the use of hidden Markov models based speech recognition approach to cryptanalysis of encrypted digitized speech signals in a keystream reuse situation, also known as the two time pad. We show that how an adversary can automatically recover the digitized speech signals encrypted under the same keystream provided the language (e.g. English) and digital encoding scheme (e.g. linear predictive coding) of the underlying speech signals are known. The technique is flexible enough to incorporate all modern speech coding schemes and all languages for which the speech recognition techniques exist. The technique is simple and efficient and can be practically employed with the existing HMM based probabilistic speech recognition techniques with some modification in the training (pre-computation) and/or the maximum likelihood decoding procedure. The simulation experiments showed promising initial results by recognizing around 80% correct phoneme pairs encrypted by the same keystream.

Keywords Cryptanalysis · Hidden Markov model · Keystream reuse · One time pad

L.A. Khan (✉)
College of Telecommunications, National University of Science and Technology, Rawalpindi, Pakistan
e-mail: liaquatalikhan@gmail.com

M.S. Baig
Centre for Cyber Technology and Spectrum Management, NUST, Islamabad, Pakistan

S.-I. Ao et al. (eds.), *Advances in Computational Algorithms and Data Analysis,*
Lecture Notes in Electrical Engineering 14,

12.1 Introduction

In a stream cipher, the plaintext p is exclusive ORed with a keystream k to produce the ciphertext c i.e. $p \oplus k = c$. If the keystream k is purely random then the stream cipher becomes provably unbreakable [1] and the cipher is known as the one time pad. The security of the stream cipher rests on never reusing the keystream. If two different plaintexts p_1 and p_2 are encrypted with the same keystream k then their results $p_1 \oplus k$ and $p_2 \oplus k$ can be XORed to neutralize the effect of the keystream, thereby obtaining $p_1 \oplus p_2$. The key reuse problem in stream ciphers has been studied since long. It has recently been mentioned in the literature as the "two time pad" problem [2]. The vulnerability of keystream reuse exists with many practical systems which are still in use. The practical systems which are vulnerable to such type of attacks include Microsoft Office [2, 3], 802.11 Wired Equivalent Privacy [4], WinZip [5] and the point to point tunneling protocol [6] used in virtual private networks (VPNs). This problem is predicted to remain there for quite some time in the future also as mentioned by Mason et al. [2]: "We do not expect that this problem will disappear any time soon: indeed, since NIST has endorsed the counter mode for AES, effectively turning a block cipher into a stream cipher, future systems that might otherwise have used CBC with a constant IV may instead reuse keystreams."

Hidden Markov Models (HMMs) are very rich in mathematical structure and form the theoretical basis for use in a wide range of applications particularly in machine recognition of speech [7]. Most modern speech recognizers are based on HMMs. Digitization, compression and encryption of speech communications between two parties have been important areas of communication. Encryption schemes particularly designed for speech, starting from the old aged analog speech inverters to the modern aged partial speech encryption techniques, have been the focus of security professionals since long. With the advancement in the speech digitization and compression techniques such as low rate Vocoders which use parameter encoding, the speech signal is now treated as an ordinary data stream of bits as far as encryption is concerned. The properties of the digitized speech signals exploited by the speech recognition equipment especially in case of speech recognition from Codec bitstreams [8] and automatic transcription of telephone conversations [9] have encouraged us to look at their characteristics from the cryptanalytic point of view in the keystream reuse scenario. It has enabled us to extend the natural language approach from automated cryptanalysis of encrypted text based data to the digital data extracted from underlying verbal conversation between two parties. An interesting by-product of our attack is that it would not only decipher the information but would automatically transcribe it during the process of speech recognition.

This chapter is organized as follows: In Section 12.2, we discuss the previous and related work on the keystream reuse problem as well as the use of HMMs in various cryptanalytic procedures. Section 12.3 presents details regarding our approach and method of attack. In Section 12.4, we present the implementation procedure which we adopted and the details of the tools used for launching the attack along with experimental results. Section 12.5, concludes the paper and gives directions for future work on the topic.

12.2 Previous and Related Work

The previous and related work in this domain can be divided into two categories: the prior work related to the keystream reuse or the two time pads and secondly, the previous work with respect to the use of HMMs in cryptanalysis.

12.2.1 Keystream Reuse

Key stream reuse vulnerability of stream ciphers has been studied for quite some time, starting from National Security Agency's VENONA project [10, 11] which started in 1943, going through the work of Rubin [12], Dawson and Neilson [13] and finally to the automated cryptanalysis of two time pads by Mason and coauthors [2]. Mostly the keystream reuse problem discussed previously is with respect to the textual data and mainly based on heuristic rules for obtaining the two plaintexts p_1 and p_2 from $p_1 \oplus p_2$ except for the work of Mason et al. [2] which uses statistical finite states language models and natural language approach. Previous works also exist on automated cryptanalysis of analog speech signals [14], but no previous work exists on the use of modern speech recognition techniques based on hidden Markov models being used for cryptanalysis of the two time pad problem for the digitally encoded and/or compressed speech signals. Hence, based on the available literature on this topic, a user can safely reuse keystreams if the underlying plaintext data is speech. Our work prove it the other way round that just like text based data, speech signals can also be reconstructed from their plaintext XORs. Moreover, most of the reconstruction techniques for separately identifying two text based data files from their bitwise XOR fail if compression is applied on the plaintext files before encryption. Since speech recognition can also be carried out from the Codec bitstreams [15–17], even for highly compressed speech data, therefore, the chances of applying the same techniques for cryptanalysis of two time pads even in the compressed environment has a fair chance of producing encouraging results. This task may be taken up as a future work in this area.

12.2.2 Use of HMMs in Cryptanalysis

As regards to the use of hidden Markov models in cryptography and cryptanalysis, these have recently been used for several problems in these areas. The most prominent ones are the works of Narayanan and Shamtikov [18] who used hidden Markov models for improving fast dictionary attacks on human memorable passwords; Song et al. [19] who used HMMs for timing attacks on SSH; substitution deciphering of compressed documents using HMMS by Lee [20]; and Zhuang et al. [21] works on keyboard acoustic emanations with the help of HMMs; Karlof and Wagner [22] who modeled countermeasures against side channel cryptanalysis as HMMs; and finally the most relevant work of Mason and coauthors [2] who used the viterbi beam search for finding the most probable plaintext pairs from their XOR in case of textual data. It is worth mentioning here that most of the work involving cryptanalysis with the aid of HMMs relate to text based data with no or very little attention to digitized encrypted speech. Our algorithm for cryptanalysis of the plaintext XOR of the digitized speech signals using hidden Markov model based speech recognition techniques is the first of its kind and has showed encouraging preliminary results.

12.3 Proposed Approach

Our method of cryptanalyzing the speech signals being encrypted with the same key is based on the hidden Markov model based automatic speech recognition (ASR) techniques. All modern speech recognition equipment use this technique because of its robustness, flexibility and efficiency. The goal of any ASR system is to find the most probable sequence of words $W = (w_1, w_2, w_3, \ldots \ldots)$ given an acoustic observation $O = (o_1, o_2, o_3, \ldots . o_T)$. Mathematically,

$$\widehat{W} = \arg\max_{i \in L} P(w_i / O)$$

where L indicates the phonetic units in a language model. This cannot be calculated directly but using Baye's Rule, the above equation can be modified as

$$\widehat{W} = \arg\max_{i \in L} \frac{P(O/w_i) P(w_i)}{P(O)}$$

or it can also be written as

$$\widehat{W} = \arg\max_{i \in L} P(O/w_i) P(w_i)$$

here, $P(O/w_i)$ is calculated using HMM based acoustic models, whereas $P(w_i)$ is determined from the language model. In the keystream reuse case the unknown observation O is not a plain acoustic observation sequence but bitwise XOR of two observation sequences O_1 and O_2. The job of the cryptanalyst is to separate these two sequences from their XOR. Keeping in view the techniques of speech recognition, there could be two approaches to tackle this problem. One approach would be that we train our HMMs corresponding to phoneme pairs (phoneme$_1$ xor phoneme$_2$) and then try to recognize the phoneme pair which best fits the observation sequence which in fact is a bitwise XOR of two observation sequences. In this approach, we mainly modify the training phase of the speech recognition procedure. The second approach would be to find out the phoneme pair which gives the best joint probability given the observation sequence. Here, a modification in the decoding phase of the recognition procedure is required. The succeeding paragraphs discuss the pros and cons of the two approaches in this regards.

12.3.1 HMM Training Phase Modification

In the conventional speech recognition techniques the hidden Markov models are trained for complete words in case of isolated word recognizers and for phonemes in case of continuous speech recognizers. In our case, for isolated word recognition, we have to first list down all the possible combination of words and hence the HMMs required to be trained will increase from n to n^2. Since the list of words is generally very large, therefore this approach of training the HMMs would be very computational intensive and may be impractical. A better and more efficient approach is to train the HMMs for the XORed pairs of all the possible phonemes in the language under test. In this case the list of phonemes is generally not high and it is computationally feasible to train our HMMs for the phonemes instead of complete words. For example in English language there are about *40* to *50* phonemes and hence the number of HMMs to be trained in this case would be at the most 50^2 which is not high as regards to the computational resources available to a normal user these days. Using this approach would not require any modification in the conventional speech recognition procedure, except that the decoded phonemes would not be the actual speech signals but the bitwise XORed of two phonemes. In this case it is only in the training phase that, instead of training the HMMs for individual phonemes, we train them for the XORed pairs of phonemes for all combinations.

12.3.2 Recognition Phase Modification

Another approach in this case is to keep the training part of the HMMs unchanged but modify the recognition algorithm at the decoding stage. Using this approach is simple and elegant in case of textual data like it is applied by Mason and coauthors [2]. In this case, assuming x to be the known XOR of the two ciphertexts c_1 and c_2, a feasible solution to the two time pad problem is a string pair (p_1, p_2) such that $p_1 \oplus p_2 = x$. If p_1 and p_2 are drawn independently from known probability distributions Pr_1 and Pr_2 respectively then we look for the most probable of the feasible solutions i.e. the (p_1, p_2) pair that maximizes the joint probability $Pr_1(p_1).Pr_2(p_2)$. In case of speech, there are a lot of gray areas which need to be clarified, before we get some conclusive results. In case of textual data, since the plaintext XOR x $(p_1 \oplus p_2)$ is known (equal to $c_1 \oplus c_2$) and the probability of x given p_1 and p_2 i.e. $P[x/(p_1, p_2)]$ is simply 1. In case of speech signals the XORed observation vector available to the cryptanalyst is the combination of two vectors which in turn are probabilistic functions of phonemes, thereby making the situation more complex as compared to the textual data. In this case we have to pair each phoneme with every other phoneme at each stage of the recognition network including itself and then try to find the most probable path. This will lead to state explosion even at the early stage of the recognition network and in return would require exorbitant amount of computations. We postpone further details and complexity analysis of this approach for future work and restrict this chapter to the modification at the training phase level.

12.4 Simulation

The simulation part of our attack involves a pre-computation phase and then the actual attack phase. Both the phases are interrelated and interdependent and the accuracy of the attack is greatly dependent on how well these two parts of the attack are carefully employed and joined. For both these phases, we used *HTK* which is an open source toolkit based on *C* language available for building hidden Markov models [23]. It is primarily designed for building HMM based speech processing tools, particularly speech recognizers. *HTK* training and recognition tools can be efficiently employed to break the two time pads of encrypted digitized speech.

12.4.1 Pre-computation Phase

The pre computation phase corresponds to the training part of the HMMs and is done once for a particular language and specific speech encoding procedure. The accuracy of the attack is greatly dependent on how well the selected HMMs are trained with diversified nature of speech signals. For training data, we selected the *Switchboard* Corpus [24] which is a selection of telephone bandwidth conversational speech data collected from T1 Lines. The speech files are fully transcribed. In order to simulate the keystream reuse scenario, we selected 256 speech files and XORed them with each other. The XORed speech files were transcribed using the *HTK* tool *HSLab*. Figure 12.1 shows the waveform and phoneme level transcription of the words *Clothes* and *Glitters* and their corresponding bitwise XOR. The acoustical events were modeled with 588 HMMs with each HMM corresponding to one pair of phonemes. Since all the possible phonemes do not occur always hence the actual number of phoneme pairs (588) is quite less than the total possible number (2,500). The basic design of the HMM we used in this case for all the models is as shown in Fig. 12.2.

The speech recognition tools cannot process speech waveforms directly. These have to be represented in a more compact and efficient way like the case in all

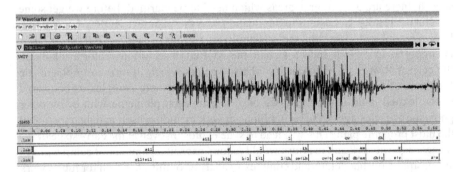

Fig. 12.1 Waveform and transcription of the words "*Clothes*" xor "*Glitters*"

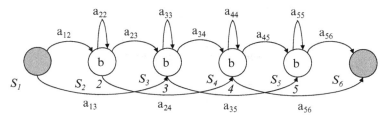

Fig. 12.2 Basic topology of a six state HMM

modern speech digitizers and encoders. Different acoustic feature representations were used for recognition purposes. For all the representations, we used frame length of 25 milliseconds with 10 milliseconds frame periodicity. The parameters which were to be estimated for each HMM during the training phase were transitional probabilities a_{ij} and the single Gaussian observation function for each emitting state which is described by a mean vector and variance vector (the diagonal elements of the autocorrelation matrix). The *HTK* tools *HInit*, *HCompV*, and *HRest* can be used for this purpose.

The different acoustic features extracted for recognition purposes includes Linear predictive coefficients, reflection coefficients, Cepstral coefficients and Mel Frequency Cepstral coefficients along with delta and reflection coefficients. In addition to the training of each HMM, there is one more important task which has to be completed in the pre computation phase before we go on to the decoding phase. Before using our HMMs we have to define the basic architecture of our recognizer. In actual case this depends on the language and the syntactic rules of the underlying task for which the recognizer is used. We assume that these things like the language of the speakers and the digital encoding procedures are known to the cryptanalyst before hand. *HTK* like most speech recognizers work on the concept of recognition network which are to be prepared in advance, and the performance of the recognizer is greatly dependent on how well the recognition network maps the actual task of recognition. In addition to the recognition network, we need to have a task dictionary which explains how the recognizer has to respond once a particular HMM is identified. At this stage, our speech recognition task completely defined by its network, its task dictionary and its HMM set is ready for use. The recognition network in our case is shown in Fig. 12.3. The *HParse* tool of *HTK* can be used for this purpose. *HSGen* can be used for its verification.

12.4.2 Decoding Phase

Once the pre-computation phase has carefully been completed, the decoding process becomes pretty simple and elegant. An input speech signal comprising of n observation vectors, which in our case are the XOR of two unknown sequences of vectors, is then fed as input to the recognizer. Every path from the start node to the end node in the recognition which passes through exactly n emitting states is a potential

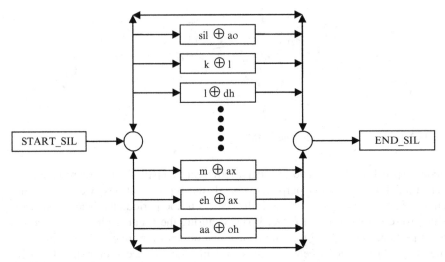

Fig. 12.3 Recognition network

recognition hypothesis. Each of these paths has a log probability which is computed by summing the log probability of individual transition in the path and the log probability of each emitting state generating the corresponding XORed vector. Within the model, transitions are determined from the model parameters (a_{ij}), between two models the transitions are regarded as constant and in case of large recognition networks the transition between end words are determined by language models likelihoods attached to the word level networks. The job of the decoder is to find those paths through the network which have the highest log probability. *HTK* tools *HVite* can be used for this purpose.

12.4.3 Experimental Results

The performance analysis of the recognizer can be done using the *HResults* tool of *HTK*. It reads in a set of label files output by the recognition tool (*HVite* in our case) and compares them with the corresponding reference transcription files. For the analysis of speech recognition output the comparison is based on a dynamic programming based string alignment procedure. The experimental results with respect to different acoustic features extraction mechanisms are depicted in Table 12.1. The best accuracy results were presented by the Mel Frequency Cepstral Coefficients (MFCC) with delta and acceleration coefficients-a 39 dimensional vector comprising of 12 first MFCC coefficients, the null MFCC coefficient which is proportional to the total energy in the frame, 13 Delta coefficients estimating the first order derivative of MFCC coefficients and 13 acceleration coefficients estimating the second order derivatives. This is perfectly inline with the conventional speech recognition accuracies with respect to the acoustic features.

Table 12.1 Recognition accuracies of different acoustic features

SNo	Feature extraction mechanism	Recognition accuracy (%)
1.	Linear predictive coefficients	65.93
2.	Linear predictive reflection coefficients	69.06
3.	Linear predictive cepstral coefficients	72.09
4.	Mel frequency cepstral coefficients (MFCC)	74.72
5.	Linear predictive cepstral + delta coefficients	77.15
6.	Mel frequency cepstral + delta + acceleration coefficients	79.96

12.5 Future Work and Conclusion

This chapter presents the implementation of plaintext XORs of the key reuse problem of stream ciphers for speech signals encoded with modern encoding techniques. The text based plain text XORs have been discussed in the literature for quite some time now and the techniques have matured quite well. Conventional speech recognition tools such as *HTK* can effectively be employed for the automated cryptanalysis of two time pads in case of speech signals. Two main approaches for achieving this have been discussed while experimental results for the training part modification have been presented. The decoding part modification can be taken up as a future work. Detailed complexity analysis of the training and decoding parts of the recognition technique also needs to be looked into in the future assignment.

References

1. Shannon, C.E., *A mathematical theory of communication*. Bell System Technical Journal, 27, 379–423, July, 1948.
2. Mason, J., Watkins, K., Eisner, J., and Stubblefield, A., *A natural language approach to automated cryptanalysis of two time pads*. In 13th ACM Conference on Computer and Communications Security, November, 2006, Alexandria, Virginia, USA.
3. Wu, H., *The misuse of RC4 in Microsoft Word and Excel,* Cryptology ePrint Archive, Report 2005/007, 2005. http://eprint.iacr.org.
4. Borisov, N., Goldberg, I., and Wagner, D., *Intercepting mobile communications: The insecurity of 802.11*, MOBICOM 2001, 2001.
5. Kohno, T., *Attacking and repairing the WinZip encryption scheme*, In 11th ACM Conference on Computer and Communications Security, pp. 72–81, October, 2004.
6. Schneier, B., Mudge, B., and Wagner, D., *Cryptanalysis of Microsoft PPTP Authentication Extensions (ms-chapv2)*. CQRE'99, 1999.
7. Rabiner, L.R., *A tutorial on hidden Markov models and selected applications in speech recognition*, Proceedings of the IEEE, 77(2), 257–286, February, 1989.
8. Raj, B., Migdal, J., and Singh, R., *Distributed speech recognition with codec parameters,* IEEE Automatic Speech Recognition and Understanding 2001, Cambridge, MA, USA, December, 2001.

9. Gales, M.J.F., Jia, B., Liu, X., Sim, K.C., Woodland, P.C., and Yu, K., *Development of the CUHTK 2004 RT04F Mandarin conversational telephone speech transcription system*. Proceedings of ICASSP 2005, I, 841–844, March, 2005.

10. Benson, R.L. and Warner, M., *VENONA: Soviet espionage and the American response 1939–1957*. Central Intelligence Agency, Washington, DC, 1996.

11. Wright, P., *Spy Catcher*. Viking, New York, NY,1987.

12. Rubin, R., *Computer methods for decrypting random stream ciphers*. Cryptologia, 2(3), 215–231, July, 1978.

13. Dawson, E. and Nielsen, L., *Automated cryptanalysis of XOR plaintext strings*. Cryptologia, 20(2), 165–181, April, 1996.

14. Goldburg, B., Dawson, E., and Sridharan, S., *The automated cryptanalysis of analog speech scramblers*, EUROCRYPT'91, Springer LNCS 457, pp. 422, Germany, April, 1991.

15. Carmen P.M., Ascension G.A., Diego F.G.C., and Fernando D.M., *A comparison of front-ends for bitstream-based ASR over IP*, Signal Processing, 86, 2006.

16. Choi, S.H., Kim, H.K., and Lee, H.S., *Speech recognition using quantized LSP parameters and their transformations in digital communications*, Speech Communication, April, 2000.

17. Kim, H.K., Cox, R.V. and Rose, R.C., *Performance improvement of a bitstream-based front-end for wireless speech recognition in adverse environments*, IEEE Transactions on Speech and Audio Processing, August, 2002.

18. Narayanan, A. and Shmatikov, V., *Fast dictionary attacks on human-memorable passwords using time-space trade-off*. In 12th ACM Conference on Computer and Communications Security, pp. 364–372, Washington, DC, November, 2005.

19. Song, D.X., Wagner, D., and Tian, X., *Timing analysis of keystrokes and timing attack on SSH*. In 10th USENIX Security Symposium, Washington, D.C., USA, August, 2001.

20. Lee, D., *Substitution deciphering based on HMMs with application to compressed document processing*. IEEE Transactions on Pattern Analysis and Machine Intelligence, 24(12), 1661–1666, December, 2002.

21. Zhuang, L., Zhou, F., and Tygar, J.D., *Keyboard acoustic emanations revisited*. In 12th ACM Conference on Computer and Communications Security, pp. 373–382, Washington, DC, November, 2005.

22. Karlof, C. and Wagner, D., *Hidden markov models cryptanalysis*. Cryptographic Hardware and Embedded Systems – CHES'03, Springer LNCS 2779, pp. 17–34, 2003.

23. Young, S.J., Evermann, G., Hain, T., Kershaw, D., Moore, G.L., Odell, J. J., Ollason, D., Povey, D., Valtchev, V., and Woodland, P.C., *The HTK Book*. Cambridge University, Cambridge, 2003. http://htk.eng.cam.ac.uk.

24. Godfrey, J.J., Holliman, E.C., and McDaniel J., *SWITCHBOARD: Telephone speech corpus for research and development*, Proceedings of ICASSP, San Francisco, 1992.

Chapter 13
A Reconfigurable and Modular Open Architecture Controller: The New Frontiers

Muhammad Farooq, Dao Bo Wang, and N.U. Dar

Abstract A novel and flexible open architecture controller platform is presented for PUMA Robot system. The original structure of the PUMA robot has been retained. All computational units are removed from the existing PUMA controller, and the PC assumes the role of computing the control strategy. By assembling the controller from off-the-shell hardware and software components, the benefits of reduced cost and improved robustness have been realized. An Intel Pentium IV industrial computer is used as the central controller. The control software has been implemented using VC ++ programming language. The trajectory tracking results show the validity of the new PC based controller.

Keywords PUMA robot · Computed torque control (CTC) · Open architecture · Amplifier · Graphical user interface (GUI)

13.1 Introduction

Robots form an essential part of mechatronics and computer integrated manufacturing (CIM) systems. Robots are generally controlled by dedicated controllers. As upgrades become costly and interfacing becomes complex due to hardware and software conflicts, the flexibility of the robotic manipulators is reduced. Dedicated hardware and proprietary software which normally allows only high level programming by the users are costly and difficult to understand.

M. Farooq (✉) and D.B. Wang
College of Automation Engineering, Nanjing University of Aeronautics & Astronautics, Nanjing, P.R. China, 210016
e-mail: muh_farooq1974@yahoo.com

N.U. Dar
College of Mechanical Engineering, University of Engineering and Technology, Taxila, Pakistan

S.-I. Ao et al. (eds.), *Advances in Computational Algorithms and Data Analysis,*
Lecture Notes in Electrical Engineering 14,
© Springer Science+Business Media B.V. 2009

Since the early years of 1980s many projects have been carried out to develop an open architecture controller such as NGC [1] GISC [2], ROBLINE [3] and so on. They try to solve a problem of realization of an open architecture controller, and several prototype systems have been developed. However, they are not widely accepted due to the overly restrictive definitions and special standards.

The Unimate PUMA (Programmable Universal Machine for Assembly) 500 series Robots mainly uses DEC LSI 11 processor running VAL robot control software [4]. Methods of bypassing VAL (Variable Assembly Language) are discussed in literature, including Unimation technical reports [5]. However, most of these procedures have been confined to replacing the LSI 11 with another DEC computer, leaving peripheral hardware intact. A much refined open structure architecture for industrial robot was discussed in [6]. However, it is mainly based on Common Object Request Broker Architecture (COBRA), leaving scope to simplify the hardware and software work. A hardware retrofit for Puma 560 robot is discussed in [7] but still it relies on special-purpose TRC041 cards installed on the backplane of Mark II controller.

The shift towards personal computer open architecture robot controller and the impact of using these newer controllers for system integration is discussed in [8–12]. However, they offer problem-specific solutions instead of presenting a generalized open architecture approach. In fact, it is far more cost effective to develop new hardware using less specific interfaces. In our paper a flexible, modular hardware is developed for the puma robot, incorporating a personal computer, in-house as well as specialized hardware. Some technical problems in the previous designs [13] for velocity test profile of joints 1, 2 and 4 have also been addressed. The joints position tracking error at high velocities is also minimized in our design.

13.2 Original Unimate PUMA 500

The Unimation Mark II is an industrial robot controller. It consists of mainly a DEC LS11 computer with ADAC parallel interface board, DLV11-J serial interface board; CMOS board and EPROM board; servo interface board and Six digital servo boards [12].

The original system used a large number of operational amplifiers and discrete components for conditioning of shaft encoder signals and amplification of analog control voltages. This leaves considerable scope to simplify and compact the controller design by substitution of more modern components.

13.3 New Hardware Configuration

The PUMA 512 robot used for work is described in Fig. 13.1, is a member of the Unimate PUMA 500 series of Robots, having six joints. Each of PUMA 512 joints is driven by a gear train with a permanent magnet DC servo motor which incorporates

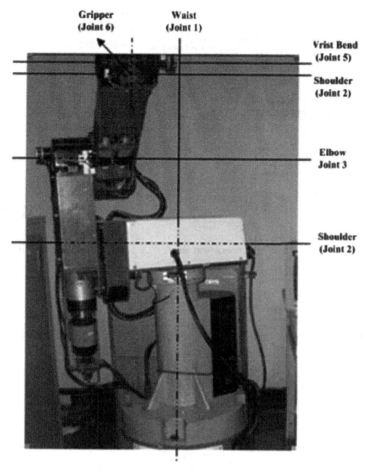

Fig. 13.1 Puma 512 Robot joints identification

a rotary shaft encoder, a tachometer and a potentiometer. The potentiometers & shaft encoders use a common 5 V supply.

The new system's block diagram is shown in the Fig. 13.2. It mainly consists of PWM amplifiers box, control unit, power supply box and an industrial computer.

The PWM amplifier box contains six in-house built amplifiers employing SA01. The SA01 amplifier is a pulse width modulation amplifier that can supply 2 kW to the load. It operates at 40 Vdc control supply and ±15 Vdc servo supply. This form of amplifier technology provides particular benefit in the high power ranges where operating efficiencies as high as 90% can be achieved to dramatically reduce heat sinking requirements.

The control box includes an internally designed digital conditioner card for shaft encoder signals and an analog conditioner card for potentiometer and tachometer.

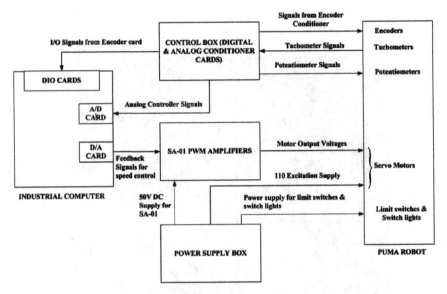

Fig. 13.2 Schematic diagram of new robot hardware

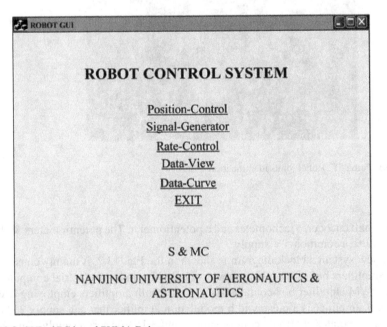

Fig. 13.3 GUI of PC based PUMA Robot

The in-house built encoder conditioner card uses ALTERA MAX 7256AETC100-10 CPLD [14] as shown in Fig. 13.3. It belongs to MAX 7000A programmable device family.

The robotic arm needs two digital conditioner cards. The CPLDs are programmed using VHDL language. The signals A+, A−, B+, B−, Z+ and Z−, VCC and DGND are the eight signals from rotary shaft encoder which are interfaced to the CPLD via a differential line receiver MC3486. The 24 signals D0_waist to D23_waist go to six channels 722 DIO card. The other five joints' shaft encoders are connected to digital conditioner card in the same way.

A significant advantage of CPLDs is that they provide simple design through re-programming. Unlike the commercially available encoder conditioner cards, the developed CPLD card is re-configurable. The power supply unit incorporates power supplies for PWM amplifiers units, signal conditioner cards and an excitation 110 V power supply for six servo motors. It has one six channel 722 DIO card, 16 bits 816 A/D and 6126 D/A cards. The analog signals from tachometers and potentiometers are fed into a six-channels analog conditioner card. The card was designed at Simulation & Machine Control lab (S & MC). After conditioning the signals, they are fed to industrial PC (A/D card).

13.3.1 Advantages for New PC based PUMA Robot

There were three main advantages of development of a new PC-based PUMA robot manipulator control system: complexity, flexibility and cost. The hardware and software complexity of the Unimate Mark II robot controller is extensively reduced as discussed in Section 13.3. Flexibility refers to the ability to implement arbitrary control strategies which can easily integrate sensor information into low level control. Flexibility also refers to the ability to easily use wide variety of sensors in the trajectory generator. The suggested PC based platform has the ability to integrate sensors easily such as sonars, ranging lasers, cameras etc. that would allow the user to implement more complex control strategies (e.g. vision based control). With the current digital servo boards in the Mark II, none of these advanced modes are feasible.

13.4 Description of the Control Scheme

In this work, the reference torque for each joint of the arm is calculated using computed torque control [15]. This technique is used to remove the nonlinearities of the PUMA by employing feedback linearization. The arms dynamics are given by

$$M(q)\ddot{q} + N(q, \dot{q}) + \tau_d = \tau \qquad (13.1)$$

where $q(t) \in \Re^6$ is a vector of joint variables, $\tau(t) \in \Re^6$ the control torque, $\tau_d(t) \in \Re^6$ is a disturbance, $M(q)$ is the inertia matrix, $N(q \cdot \dot{q})$ represents nonlinear terms including coriolis/centripetal effects, friction and gravity.

Suppose that a reference trajectory $q_d(t)$ has been chosen for the arm motion. The tracking error is defined as:

$$e(t) = q_d(t) - q(t) \tag{13.2}$$

If the tracking error is differentiated twice,
then

$$\ddot{e} = \ddot{q}_d + M^{-1}(N + \tau_d - \tau) \tag{13.3}$$

The feedback input linearizing function may be defined as:

$$u = \ddot{q}_d + M^{-1}(N - \tau) \tag{13.4}$$

and the disturbance function as

$$w = M^{-1}\tau_d \tag{13.5}$$

Then the tracking error dynamics can be expressed as

$$\frac{d}{dt}\begin{bmatrix} e \\ \dot{e} \end{bmatrix} = \begin{bmatrix} 0 & I_{6\times6} \\ 0 & 0 \end{bmatrix}\begin{bmatrix} e \\ \dot{e} \end{bmatrix} + \begin{bmatrix} 0 \\ I_{6\times6} \end{bmatrix}u + \begin{bmatrix} 0 \\ I_{6\times6} \end{bmatrix}w \tag{13.6}$$

Hence, as a result of using the feedback linearization transformation (13.4), the tracking error dynamics are given by a linear state equation with constant coefficients in (13.6).

The feedback linearization transformation can be inverted to give

$$\tau = M(q)(\ddot{q})_d - u + N(q, \dot{q}) \tag{13.7}$$

This is the computed torque control law. An outer loop controller is often used. The role of the outer loop controller is to provide the input u. In this paper, PD computed torque controller has been used as an outer controller. The tuning of the CTC controller for Puma robot using (3.12) leads to the gains as shown in the Table 13.1. The computed torque control technique is known to perform well when the robotic arm parameters are known fairly accurately. Fortunately, the dynamics of PUMA 560 manipulator are well known and reported. The inverse dynamics and Denavit-Hartenberg arm parameters employed in this work are those reported in [17].

Table 13.1 Controller gains

	1	2	3	4	5	6
K_P	18,000	10,000	8,900	6,200	10,500	5,500
K_D	19	16	16	14	18	16

Table 13.2 Static friction for each joint

Joint	Static friction, Nm
1	68.5
2	54.9
3	52.2
4	3.85
5	2.9
6	1.8

13.4.1 Effect of Large Gear-Ratios & Friction

There are two main barriers for industrial robots to successfully interact with the environment: large gear ratio and friction. Puma robot manipulators typically have a large gear ratio for handling large payloads. These linkages are usually heavy to provide rigidity. Large joint friction is caused by large gear ratios and heavy linkages. The main disadvantage of the large gear ratios is the backlash problem which may cause higher position tracking errors.

Static friction was identified by increasing the torque command gradually and recording the joint velocity. The torque command was increased 0.01 N every 0.05 second. The torque command was increased 0.01 N every 0.05 second.

The static friction modeling of joint 2 and joint 3 is more complicated than other joint because of gravity loading. The estimated gravity compensation force used in this experiment was 173 Nm for joint 2 and 70 Nm for joint 2. Table 13.2 shows the static friction of each joint.

13.4.2 Viscous Friction and Inertia

Identification of link inertias is needed for the controller implementation and improved trajectory tracking performance was assumed that the motor inertias, reflected through the gear reductions, resulted in an inertia tensor that was primarily constant and diagonal. The inertia of each link was estimated by moving the robot in sine-wave motion and recording the joint velocity and joint torque simultaneously.

By observing the magnitudes and relative phase of two signals, the friction and joint inertia can be estimated. To make the estimation feasible and according to the fact that both dynamic coulomb and viscous frictions are in phase with velocity, the total friction was identified as though it were solely viscous. The results from the experiments confirmed that the simplification yielded adequate model accuracy. The equations for this analysis are:

$$\dot{\theta}_{cmd} = V_{cmd} \sin \omega t \tag{13.8}$$

$$\dot{\theta}_{act} = V_p \sin(\omega t + \phi) \tag{13.9}$$

$$\ddot{\theta}\phi_{act} = V_p \cos(\omega t + \phi) \tag{13.10}$$

$$\tau = \tau_p \sin(\omega t + \phi t) \tag{13.11}$$

$$\tau = I\ddot{\theta}_{act} + K_{vis}\dot{\theta}_{act} \tag{13.12}$$

Where

$\dot{\theta}_{cmd}$: Commanded joint velocity.
V_p : Peak value of actual joint velocity.
V_{cmd} : Pak value of the commanded joint velocity.
ω : Desired angular velocity.
$\dot{\theta}_{act}$: Actual joint velocity.
$\ddot{\theta}_{act}$: Actual joint acceleration.
ϕ : Phase lag of the actual velocity.
ϕ_t : Phase lag of the joint torque.
I : Link inertia.
K_{vis} : Viscous friction.
τ_p : Peak value of joint torque

From Eqs. (3.8)–(3.12), the relationship between joint torque, viscous friction and inertia can be written as:

$$\begin{bmatrix} \tau_p \cos \phi t \\ \tau_p \sin \phi t \end{bmatrix} = \begin{bmatrix} V_p \cos \phi & -Vp\omega \sin \phi \\ V_p \sin \phi & Vp\omega \cos \phi \end{bmatrix} \begin{bmatrix} K_{vis} \\ I \end{bmatrix} \tag{13.13}$$

By using least square fitting to estimate τ_p, V_p, ϕ_t and ϕ, the link inertia and viscous friction can be computed from Eq. (13.13). Table 13.3 summarizes the estimated values of link inertias and viscous frictions for each joint.

Table 13.3 Inertia and viscous friction parameters for each joint

Joint	Inertia, Kg m^2	Viscous friction Nm (rad/second)
1	9.3	125.4
2	18.2	219.7
3	6.61	106.7
4	0.0309	1.8
5	0.0406	2.1
6	0.0021	0.29

13.5 Software Design for the Controller

To implement the control algorithms developed in Section 13.3, a real time software was developed using $C++/VC++$ [17]. The graphical user interfaces developed for robot are shown partially in Figs. 13.3 and 13.4.

Figure 13.3 shows different options for robot control. The "Position-Control" and "Rate- Control" are used to control the robot 6 joints' position and speed respectively. The "Signal-Generator" is designed mainly for testing the robot position-trajectory performance. The "Data-View" and "data-Curve" display the joints position and speed data. Figure 13.4 demonstrates "Position-Control" window only. The software has several levels:

- System initialization and self diagnosis: which initializes custom boards, configures the robot and diagnoses each block of the system.
- System coordination and safety check: which works with the safety device to monitor robot status and stop operations in case of errors or emergency.
- Basic I/O routines for feedback information and output control signals: which reads joint encoders, position signals, estimate velocities and convert digital control signals into analog ones.
- Kinematic & dynamic routines: which includes forward and inverse kinematics for path planning as well dynamic routines.

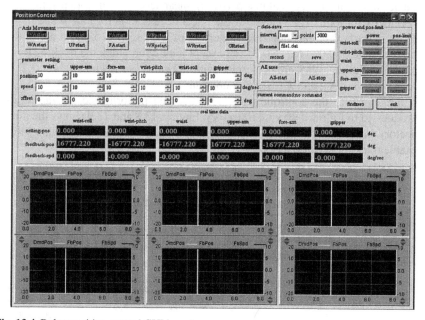

Fig. 13.4 Robot position-control GUI layout

- User interface: that provides users with control buttons to properly operate the robot, convenient means of planning experiments and post-processing experimental data.

13.6 Results

The suggested system was developed and applied to Puma 560 robot. The original proprietary controller was replaced with it. Figure 13.5 depicts the experimental layout of the developed system. The PC runs the Windows 2000 operating system.

The PC-based controller is evaluated from two different aspects. The first aspect is to examine how easy the system integrations and modifications are. The second is to examine the performance of the control. The first aspect is obvious. The suggested PC-based hardware and software ensure that the extensibility and scalability are available. The second aspect is evaluated by the trajectory-tracking experiment.

To verify the effectiveness of the new controller, experiments were performed to test the tracking control of the robot manipulator. Firstly, each joint is separately requested to follow a desired trajectory. In this test, each joint is asked to move to a specified destination while following a predetermined path. Figures 13.6 and 13.7 show the desired position trajectories and position tracking errors respectively for six joints.

The desired profile consists of two parts: (a) linear acceleration from rest to maximum velocity for joints 1, 2, 3 and 5 while linear deceleration for joint 4 and 6; (b) decelerate linearly. The position tracking errors of joint 1, 3, 4, 5 and 6 are in the acceptable range of 0.005 rad to 0.01 rad except joint 2 for which the tracking error is 0.02 rad. All these tests were performed with PD gains as shown in Table 13.2. A careful selection of K_D and K_p is very important.

To further verify the controller trajectory tracking performance, all the joints were requested to follow a varying trajectory. Figures 13.8 and 13.9 shows the joint 1 trajectory tracking performance and error respectively. All the other joints show acceptable tracking performances.

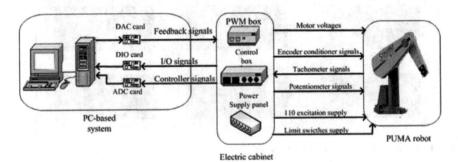

Fig. 13.5 The experimental platform for implementing open architecture controller

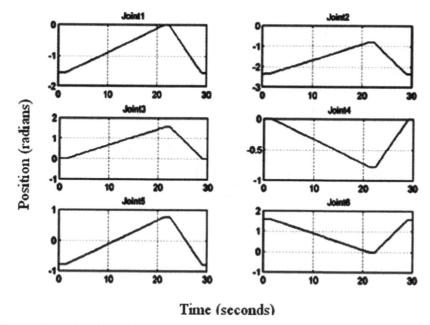

Fig. 13.6 Desired position trajectories

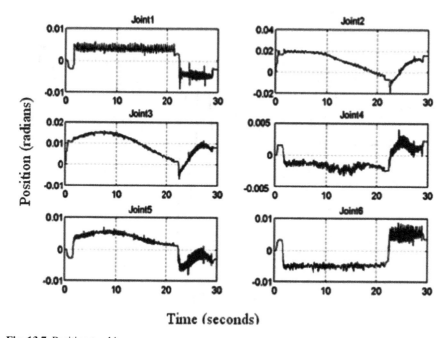

Fig. 13.7 Position tracking error

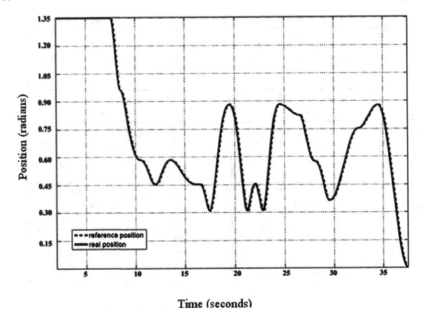

Fig. 13.8 Comparison between planned and followed trajectory of joint1

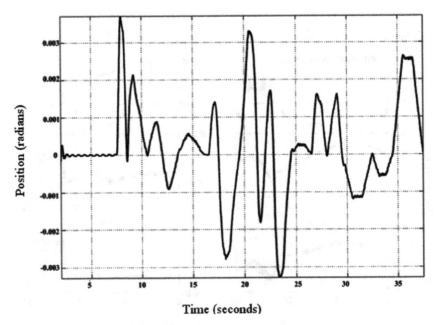

Fig. 13.9 Trajectory tracking error for joint1

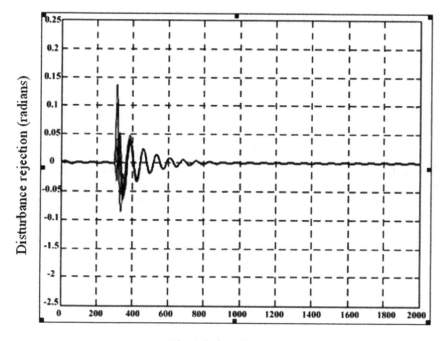

Time (seconds)

Fig. 13.10 Disturbance rejection performance of the controller

To check the robustness of the proposed controller, an unknown output disturbance was applied at 0.3 seconds. Figure 13.10 shows the disturbance rejection
performance or robustness of the controller. Two further tests were performed to
examine the simultaneous joints movement: (1) the maximum velocity was set at
15,000 counts/second. To test the simultaneous joints movement, all six joints were
asked to move at their fastest respective speeds and (2) the linear interpolation mode
test in which the individual joints should arrive at their respective destinations at the
same time.

The same tests were performed with varying joints' velocities and accelerations.
Joint 1 and joint 3 showed higher position tracking errors at higher velocities and
accelerations, however, all the remaining joints showed satisfactory performance.

The control method, CTC, used in this paper, is a scheme for canceling the nonlinearities in the dynamics to yield a linear error system. It works well if all the
parameters of the robotic arm are known exactly. One of the reasons in position
tracking errors may be some variance in those parameters as the PUMA robot in S
& MC lab, which is used in this experiment, may have some divergence from the
standard parameters because of aging.

13.7 Discussion & Conclusions

In this paper, the preliminary results are achieved in the development and implementation of a new simple PC based replacement controller for PUMA 512 robot. Although the experiments were performed at educational and research institution, the research is oriented towards industrial applications.

Though, some technical problems were faced while performing the tests at higher velocities for joint 1 and joint 3, the new designed hardware and software works very well and overcomes the problems in the previous PC based design for PUMA robot. All the joints show satisfactory performance at low velocity as well as they exhibit low position tracking error while following a velocity profile at high speeds.

The software graphical user interface for the robot was developed using VC++. It encompasses all the features needed to control an industrial robot. The experimental results showed that it is feasible to implement modern control methods for PUMA 500 series robots through software routines running on a PC.

References

1. Sorenson, S., Overview of Modular, Industry Standard Based Open Architecture Controller, *Proceedings of the International Conference of Robots and Vision Automation*, Detroit, MI.
2. Miller, D.J., Lenox, R.C., An Object Oriented Environment for Robot System Architecture, *Control System Magazine, IEEE*, Vol. 11, No. 2 (1991), pp. 14–23.
3. Leahy, M.B. Jr., Petroski, S.B., "Unified Telerobotic Architecture Project Status Report", *IEEE International Conference Systems, Man and Cybernetics*, San Antonio, TX, Vol. 1 (1994), pp. 249–253.
4. Unimation Robotics. User's Guide to VAL 398P2A: A Robot Programming and Control System, *Unimation Inc.*, Danbury, CT (1983).
5. Vistness, R., Breaking Away from VAL. Technical Report, *Unimation Inc.*, Danbury, CT (1982).
6. Pan Laingdong, Huang Xinhan., "Implementation of a PC-Based Robot Controller with Open Architecture", *Proceedings of the IEEE International Conference on Robotics and Biometrics* (2004), pp. 790–794.
7. Becerra, V.M., Cage, C.N.J., Harwin, W.S., Sharkey, P.M., "Hardware Retrofit and Computed Torque Control of a Puma 560 Robot Updating an Industrial Manipulator", *IEEE Control Systems Magazine*, Vol. 24, No. 5 (2004), pp. 78–82.
8. Fiedler, P., Schlib, C., "Open Architecture Robot Controllers and Workcell Integration", *Robotics Today*, Vol. 11, No. 4 (1998), pp. 1–4.
9. Rajsuman, R., Noriyuki, M., "Open Architecture Test System: System Architecture and Design", *IEEE International Test Conference* (2004), pp. 403–412.
10. Wang, S., Shin, K.G., "Reconfigurable Software for Open Architecture Controllers" *Proceedings – IEEE International Conference on Robotics and Automation*, Vol. 4 (2001), pp. 4090–4095.
11. Tang Qiangping, Wu Yu, Qin Guorong, "A Middleware for Open CNC Architecture". *IEEE International Conference on Automation Science and Engineering* (2006), pp. 558–561
12. Tatsuno, K., "An Example of Open Robot Controller Architecture", *Proceedings of the 2005 International Symposium on Micro-NanoMechatronics and Human Science*. The Eighth Symposium Micro- and Nano-Mechatronics for Information-Based Society (IEEE Cat. No. 05TH8845) (2005), pp. 35–40

13. Katupitiya, J., Radajewski, R., Sanderson, J., Tordon, M., "Implementation of a PC Based Controller for a PUMA Robot", *Proceedings of the IEEE Conference on Mechatronics and Machine Vision in Practice*, Australia (1997), pp. 14–19.
14. Chartrand, L., Advanced Digital Systems and Concepts with CPLD's (1st ed), *Thomson Delmar Learning*, CENGAGE Delmar Learning, New York, USA (2004).
15. Lewis, F.L., Abdallah, C.T., Dawson, D.M., *Robot Manipulator Control, Theory and Practice* (2nd ed), Marcel Dekker, New York (2004).

Chapter 14
An Adaptive Machine Vision System for Parts Assembly Inspection

Jun Sun, Qiao Sun, and Brian Surgenor

Abstract This paper presents an intelligent visual inspection methodology that addresses the need for an improved adaptability of a visual inspection system for parts verification in assembly lines. The proposed system is able to adapt to changing inspection tasks and environmental conditions through an efficient online learning process without excessive off-line retraining or retuning. The system consists of three major modules: region localization, defect detection, and online learning. An edge-based geometric pattern-matching technique is used to locate the region of verification that contains the subject of inspection within the acquired image. Principal component analysis technique is employed to implement the online learning and defect detection modules. Case studies using field data from a fasteners assembly line are conducted to validate the proposed methodology.

Keywords Visual inspection system · Adaptability · Online learning · Defect detection · Principal component analysis · Parts assembly

14.1 Introduction

In mass-production manufacturing, the automation of product inspection processes by utilizing machine vision technology can improve product quality and increase productivity. Various automated visual inspection systems have been used for quality

J. Sun (✉) and Q. Sun
Department of Mechanical and Manufacturing Engineering, University of Calgary, Calgary, Alberta, Canada T2N 1N4
e-mail: jun.sun@ucalgary.ca

B. Surgenor
Department of Mechanical and Materials Engineering, Queen's University, Kingston, Ontario, Canada K7L 3N6

S.-I. Ao et al. (eds.), *Advances in Computational Algorithms and Data Analysis,*
Lecture Notes in Electrical Engineering 14,

assurance in parts assembly lines. Instead of human inspectors, a visual inspection system can automatically perform parts verification tasks to assure that parts are properly installed and reject improper assemblies. However, most of the existing visual inspection systems are hand-engineered to provide an *ad-hoc* solution for inspecting a specific assembly part under specific environmental conditions [1–3]. Such system does not work properly when changes take place in the assembly line. For instance,

1. *Changing products in response to market demands. This requires changing the inspection algorithms of the existing system to deal with a new assembly part.*
2. Changing environmental conditions, for example, lighting conditions, camera characteristics, and system fixation after a certain period of system operating time. This may render the existing system obsolete, and thus require adjusting the original inspection algorithms for new environmental conditions.

Developing a new algorithm or adjusting the original inspection algorithm is not a trivial task. It is time consuming and expensive. Therefore, an ideal visual inspection system is required to quickly adapt to the changes in the assembly line.

For the past two decades, researchers have attempted to apply supervised machine learning-based strategies to improve the adaptability of visual inspection systems [4, 5]. Conventional approaches typically use pixel-based matching templates or feature-based matching patterns that are pre-defined manually through a trial and error procedure. In contrast, a learning-based approach allows the system to establish inspection functions based on the recognition patterns learned from the training samples. The role of human inspectors is to label the training samples based on the quality standard for a particular inspection task. As such, the system is flexible to be trained to handle different inspection problems in variable environmental conditions.

Machine learning can be conducted through off-line or online processes. In off-line learning, all training data are previously obtained, and the system does not change the learned recognition knowledge (e.g., functions, models or patterns) after the initial training phase. In online learning, the training data are presented incrementally during the system operation and the system updates the learned knowledge if necessary.

Popular off-line learning techniques used in visual inspection system include probabilistic methods, multi-layer perceptron (MLP) neural networks, adaptive neuron-fuzzy inference systems (ANFIS), and a decision tree (e.g., C4.5). These learning techniques have been used to build the recognition and classification functions in a visual inspection system [4–9]. However, it is often raised as a concern by the end user that the performance of an off-line learning based system relies heavily on the quality of the training data. In many situations it may be difficult or even impossible to collect sufficient representative, defective and non-defective, samples for training the system over a limited period of time. To solve this problem, both Beck et al. [4] and Hata et al. [10] developed defect image generation units for surface inspection systems to increase the number of training samples. Defect images were generated by artificially manipulating the defect geometrical characteristics, such as size, intensity, location, orientation, and edge sharpness. Evidently, this approach may not be practicable in the assembly parts inspection due to a large variation and complexity of assembly parts.

Recently there has been an increasing interest to apply online learning techniques to the development of an adaptive visual inspection system. By learning the inspection patterns incrementally during operation, the system does not require an excessive off-line training process when it faces the situations of changing inspection tasks or environmental conditions. The first systematic study on such a system was published by Abramovich et al. [1]. They proposed a novel online part inspection method based on a developmental learning architecture that used the incremental hierarchical discriminant regression (IHDR) tree algorithm. The method was capable of adapting to the variation of parts and environmental properties incrementally by updating rather than reprogramming inspection algorithms. Although the aforementioned research has shown promising results in utilizing online learning techniques to realize an adaptive visual inspection, there is still a need of more systematic studies with the support of practical experiments. Particularly, the following desire and objectives led to this research:

- Investigating potential online learning techniques for optimal feature extraction and effective recognition or classification
- Improving the efficiency of an online learning process by minimizing human operator involvement in the supervised learning process
- Developing an effective method to evaluate the sufficiency of the inspection function/model that is established thought online learning

This paper presents an intelligent visual inspection system that addresses the need for an improved adaptability of a visual inspection system for parts verification in assembly lines. The remaining of this paper is organized as follows. Section 14.2 describes the architecture of the adaptive visual inspection system. Sections 14.3–14.5 introduce the three major modules of the proposed system, i.e., region localization, defect detection, and online learning. Section 14.6 provides a case study to illustrate the performance of the proposed system. Section 14.7 summarizes this research.

14.2 Architecture of an Adaptive Visual Inspection System

In general, an adaptive visual inspection system can be designed in such a way that the system is capable of defect detection using a recognition model and updating the recognition model online if necessary. As illustrated in Fig. 14.1, the proposed adaptive visual inspection system consists of three major modules:

1. Region Localization Module

This module locates the region of interest (ROI) containing the installation assembly area within the acquired image. It then extracts the corresponding region of verification (ROV) containing the assembly part to be inspected.

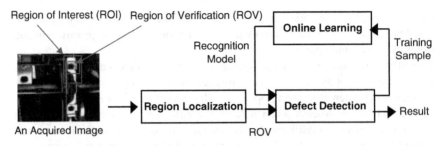

Fig. 14.1 Architecture of an adaptive visual inspection system

2. Defect Detection Module

This module checks whether the inspection case appears as a defective assembly case, e.g., part missing or improper installation, based on the located ROV. A recognition model is used in the inspection process. In our system, the recognition model characterizes non-defective assembly cases, thus an inspection case is considered defective if it derivates from the recognition model.

3. Online Learning Module

This module builds and updates the recognition model with each newly arrived inspection case if it is selected as a training sample. Here, the training samples for the supervised learning are actual non-defective assembly cases. Evidently, it would be impracticable and inefficient to require human inspectors to label non-defective cases during the system operation (i.e., online). In order to minimize the human inspector's involvement, the proposed system also makes use of the defect detection module to effectively select the training samples that require manually checking/labeling for updating the recognition model. The training strategy will be described in details in Section 14.5.

14.3 Localization of ROV

Region localization is required so that the amount of data processing can be reduced by processing the region of verification (ROV) instead of the whole acquired image. In the proposed system, the region of interest (ROI) within an acquired image contains two sub-regions, as shown in Fig. 14.2.

1. *Region of Verification (ROV)*: It must include the inspection subject, that is, the assembly part being inspected. Appearance verification of this region may indicate part missing or improper installation.
2. *Region of Matching (ROM)*: This region contains features that are invariant to both non-defective and defective cases. Therefore, the appearance pattern of this region can be used as a matching reference/template to search and locate

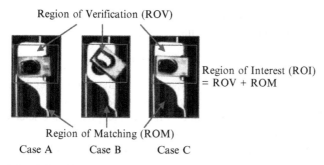

Fig. 14.2 Defining ROI, ROV, and ROM

the ROI and its corresponding ROV. The ROM provides efficient and effective identification of ROV. The ROM can be defined manually during the system setting-up and tuning stage.

An edge-based pattern-matching technique is employed in the region localization module. Instead of comparing pixels of the whole image, the edge based technique compares edge pixels with the template. It offers several advantages over the pixel-to-pixel correlation method. For example, it offers reliable pattern identification when part of an object is obstructed, as long as about 60% of its edges remain visible. Since only edge information is used, this technique can rotate and scale edge data to find an object, regardless of its orientation or size. In addition, this technique can provide good results with a greater tolerance of lighting variations. In this research, the region localization module is implemented using an edge-based geometric pattern-matching function (i.e., Geometric Model Finder) in Matrox® Imaging Library (MIL-version 8), a commercial software provided by Matrox® Imaging Inc. More background on this function can be obtained in Matrox® MIL8 – User Guide (2005).

14.4 Defect Detection using a Principal Component Analysis Technique

The system performs the defect detection of part assembly based on a recognition model that is built using a principal component analysis (PCA) technique. Principal component analysis involves a mathematical procedure that allows optimal representation of images with a reduced order of basis vectors called eigenpictures. The eigenpictures are generated from a set of training images. The projection of an image onto the subspace of eigenpictures is a more efficient representation of the image. Many works on image recognition and reconstruction have adopted the idea of the PCA based image representation and decomposition.

Given a set of N training images, $X = [x_1, x_2, \ldots, x_i, \ldots, x_N]$, each image consists of w by h pixels. In PCA, an image is represented by a vector of size $w*h$, $x_i = [p_1, p_2, \ldots, p_i, \ldots, p_{w*h}]^T$ where p_i denotes the intensity value of pixel i.

The set of eigenpictures $U = [a_1, a_2, \ldots, a_i, \ldots, a_{w*h}]$, with $a_i = [v_1, v_2, \ldots, v_{w*h}]^T$, can be obtained by solving the following equation:

$$U^T (X - m)(X - m)^T U = \Lambda \tag{14.1}$$

where vector $\mu = [m_1, m_2, \ldots, m_{w*h}]^T$ is the mean of the training set X, and Λ is a diagonal matrix of eigenvalues: $\Lambda = diag(\lambda_1, \lambda_2, \ldots, \lambda_{w*}h)$.

Corresponding to l largest eigenvalues, l major eigenpictures are selected to form a subspace of the eigenpictures $U_l = [a_1, a_2, \ldots, a_l]$, By projecting onto the subspace of U_l, a newly acquired image x can be represented by the a set of projection coefficients $c = [c_1, c_2, \ldots, c_l]^T$:

$$c = U_l^T (x - m) \tag{14.2}$$

These coefficients can be used to reconstruct the original images within certain error tolerance. The image being reconstructed based on its projection coefficients in Eq. (14.2) can be represented as:

$$y = U_l c + m \tag{14.3}$$

The image reconstruction error measures the difference between the original and the reconstructed images. For a newly acquired image x, the reconstruction error can be defined as the sum of the residual squares between the original image x and the reconstructed image y:

$$Q = (x - y)^T (x - y) \tag{14.4}$$

The reconstruction errors can be used to detect abnormality or novelty in image recognition [11, 12]. In this research, this concept is adopted for defect detection as follows.

Let θ_1, θ_2, and θ_3 denote the summations of first order, second order, and third order of eigenvalues going from $l + 1$ to K, respectively. That is:

$$\theta_1 = \sum_{i=l+1}^{K} \lambda_i, \quad \theta_2 = \sum_{i=l+1}^{K} \lambda_i^2, \quad \theta_3 = \sum_{i=l+1}^{K} \lambda_i^3, \tag{14.5}$$

where l and K denote the number of major eigenvalues and the total number of non-zero eigenvalues, respectively.

Assuming that all training images are non-defective cases of part assembly, a Gaussian approximation for the distribution of the reconstruction error, Q, can be represented as [11, 13]:

$$q = (Q/\theta_1)^{h_0} \sim N[\mu, \sigma^2]$$

$$h_0 = 1 - \frac{2\theta_1 \theta_3}{3\theta_2^2}, \quad \mu = 1 + \frac{\theta_2 h_0 (h_0 - 1)}{\theta_1^2}, \quad \sigma = \sqrt{\frac{2\theta_2 h_0^2}{\theta_1^2}} \tag{14.6}$$

That is, q is obtained from the normalizing transformation of Q and follows a normal distribution with the mean value m and the standard deviation σ. h_0 is the so-called joint moment being used in the transformation function.

In the defect detection module of this system, the PCA based recognition model is used to detect defective assembly cases. The recognition model is built on non-defective cases, thus an inspection case may be considered defective if it appears as an outlier of the recognition model with a certain confidence interval α. For instance, 99.74% confidence level translates to $\alpha = 3$ for a normal distribution. A newly arrived case is considered defective if

$$q \not\subset [\mu - \alpha * \sigma, \mu + \alpha * \sigma] \tag{14.7}$$

where $\mu - \alpha^* \sigma$ and $\mu + \alpha^* \sigma$ are the lower and upper thresholds of non-defective cases, respectively. Equation (14.7) can be used as the defect detection criteria.

14.5 Online Learning of PCA based Recognition Model

In the existing literature, the principal component analysis technique (PCA) is used in off-line learning mode that requires all the training data to be available in beforehand. It is unsuitable to applications that demand online updates to the recognition model. This research proposes an efficient online training algorithm that can build and update the recognition model as new training samples emerge.

A major challenge to perform a supervised online learning is that all training samples need to be verified and labeled manually by human inspectors. The costly human involvement affects the efficiency of an online learning process. To address this issue, the following two learning strategies are proposed in this system, as illustrated in Fig. 14.3.

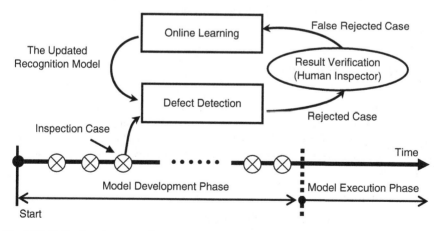

Fig. 14.3 Online learning strategies

1. The PCA based recognition model is built and updated on the false rejected cases. In an inspection process, there are two types of cases that could be incorrectly classified, i.e., false rejected case and false accepted case:

   - A *rejected case* is the one that is classified as a defective case by the existing system. Thus, a *false rejected case* refers to an actual non-defective case being incorrectly classified as a defective case.
   - An *accepted case* is the one that is classified as a non-defective case by the existing system. Thus, a *false accepted case* refers to an actual defective case being incorrectly classified as a non-defective case.

   Since the system attempts to build the recognition model to characterize non-defective cases, each false rejected case can be considered to contain a new no-defective pattern that needs to be incorporated in the recognition model. Therefore, the system only requires the human inspector to do manual verification of every rejected case to generate non-defective training samples. Also, the online learning module updates the recognition model only when a false rejected case is identified and added into the training samples. Consequently, the efficiency of the online learning process is improved with less training samples and human involvements.

2. The system only performs the online learning during the model development phase. Two different system operation phases are defined to address the training sufficiency. In the model development phase, the system builds and updates the recognition model using false rejected cases that are identified by human inspectors. Once the recognition model is stabilized by modeling error convergence (See Section 14.5.2), a model execution phase begins. In the model execution phase, the system employs the constructed recognition model without requiring a human inspector to identify the false rejected cases for the model updating. Based on the two aforementioned strategies, an efficient supervised online learning algorithm is proposed in this research as follows.

14.5.1 Model Development Phase

At a given time index i, the PCA based recognition model is denoted by $MP(i)$, which contains the generated eigenpictures $U_l(i)$ and the mean of training samples $\mu(i)$. It also contains the normal distribution function (Eq. (14.6)) of reconstruction errors based on the existing training samples. The model development phase consists of four steps:

1. Model Initialization

   (a) Initialize a PCA based recognition model, i.e., $MP(i)$ at $i = 0$:

   $$MP(0) = \{U_l(0) = 0_{l*(w*h)}, m(0) = 0_{1*(w*h)}\}.$$

2. Defect Detection

    (a) For $i > 0$, compute the projection coefficients of the acquired ROV image $x(i)$ of each newly arrived inspection case by using Eq. (14.2). Reconstruct the image of the ROV and then calculate the reconstruction error by using Eqs. (14.3), (14.4), based on the currently existing recognition model $MP(i-1)$.

    (b) Conclude on whether the newly arrived case appears as a rejected case according to Eq. (14.7).

3. Result Verification

    (a) Request a human inspector to verify the inspection result obtained in the step 2b.

    (b) Add the newly arrived case $x(i)$ into the training set, if $x(i)$ is identified as a false rejected case. Otherwise, return to the step 2 to process the next inspection case and model remains unchanged: $MP(i) = MP(i-1)$.

4. Model Update

    (a) Build a new PCA based recognition model $MP(i)$ using the updated set of training samples. If the newly arriving case $x(i)$ is the first rejected one, the recognition model is set to be $MP(i) = \{U_l(i) = 0_{l*(w*h)}, m(i) = x(i)\}$. Otherwise update the model using Eq. (14.1) and the corresponding distribution function of reconstruction errors using Eq. (14.6).

    (b) Return to the step 2 to process the next inspection case.

14.5.2 Model Execution Phase

The model development phase is considered completed when the recognition model $MP(i)$ has stabilized. The condition for the stabilized model is that both the mean $\mu(i)$ and the standard deviation $\sigma(i)$ of the reconstruction error have converged to stead-state values. Subsequently, the model execution phase begins and the latest constructed recognition model can be employed to detect defective cases by using Eqs. (14.2)–(14.7) without requiring further updating until needed.

14.6 Case Study

In the case study, field data were collected from an existing fastener inspection system for a truck cross-car beam assembly. The visual inspection system examines a total of 46 metal clips inserted by assembly robots for their proper installation. The existing system works well after excessive amount of manual tuning. Improving the system adaptability to changes has been a top priority. Figure 14.4 shows two types of clips on the cross-car beam assembly.

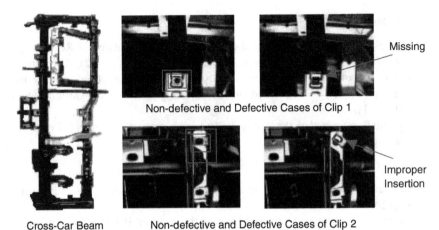

Missing

Non-defective and Defective Cases of Clip 1

Improper
Insertion

Cross-Car Beam   Non-defective and Defective Cases of Clip 2

Fig. 14.4 Examples of clips on the cross-bar beam assembly

14.6.1 System Performance

In order to simulate an online operation, the proposed adaptive inspection system
was applied to a dataset of images that were acquired sequentially in 24 h. This
section presents the experimental results of inspecting Clip 1 as shown in Fig. 14.4.
The dataset for this type of clip includes 1,295 non-defective cases and 3 defective
cases. The observations from the experiment are summarized as follows:

- In the model development phase which is during the first 700 cases, the adaptive system updated the PCA based recognition model each time when a false
  rejected case was identified. As shown in Figs. 14.5 and 14.6, there were 40
  cases whose reconstruction errors were beyond the thresholds of non-defective
  range. Manual verification confirmed 1 defective case and hence 39 false rejected cases. These 39 false rejected cases were incrementally incorporated into
  the PCA based recognition model for updates.
- The PCA based recognition model became stabilized after learning the 39 false
  rejected cases that are identified from a total of 700 case images in the model development phase. As illustrated in Fig. 14.7a, the convergence of both the mean
  and standard deviation of the reconstruction errors indicated a stabilized recognition model. Figure 14.7b shows the distribution of reconstruction errors on the
  training samples approximately followed a normal distribution with a mean value
  of $\mu = 0.9968$ and a standard deviation of $\sigma = 0.0209$. The 99.74% confidence
  interval with $\alpha = 3$ was used to define the thresholds.
- In the model execution phase, the constructed recognition model were able to
  efficiently identify all defective cases without generating false rejected cases and
  false accepted cases, as illustrated in Fig. 14.8. The PCA based recognition model
  contained eight major eigenpictures corresponding to the eight largest eigenvalues that contributed 80% of the total of eigenvalues.

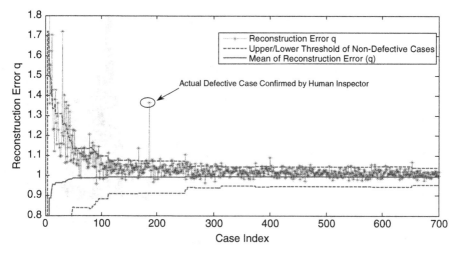

Fig. 14.5 Reconstruction errors of inspection cases in the model development phase

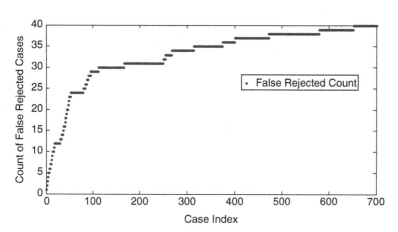

Fig. 14.6 Count of false rejected cases in the model development phase

14.6.2 Comparison with Off-Line Approach

To illustrate the efficiency of the adaptive system, this section provides a comparison of the system performance between online and off-line learning. We used the same sequence of inspection cases as used in Section 14.6.1. Using the off-line approach, the system built four PCA based recognition models with training samples ranging from 0 to 100, 0 to 300, 0 to 600, and 0 to 700, respectively. The training samples in the off-line mode required manual labeling (as defective or non-defective) in beforehand. Using the proposed online approaches, four recognition models were built and updated on the identified false rejected cases after processing cases 100,

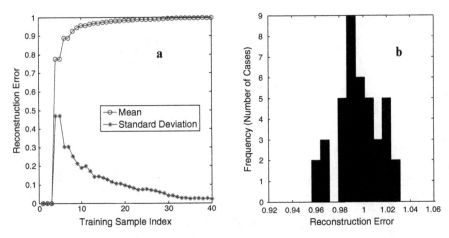

Fig. 14.7 Stabilization of the PCA based recognition model

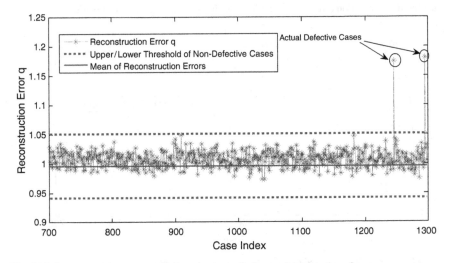

Fig. 14.8 Reconstruction errors of inspection cases in the model execution phase

300, 600, and 700, respectively. The system accuracy was evaluated with a same set of inspection cases, ranging from 701 to 1,295, for both online and off-line learning approaches.

The comparison results are listed in Table 14.1. Observations can be made from the comparison:

- For both off-line and online approaches, the system accuracy could be improved by increasing the size of training samples. In the off-line model, the total number of false rejected cases dropped from 27 to 2 when the number of training samples was increased from 100 to 700. In the online mode, the total number of false rejected cases was decreased from 33 to 0 when the number of inspection cases

Table 14.1 A comparison between the online learning and off-line learning approaches

Method	System training		System testing case processed: 701–1,295, total no. = 595			
	Cases processed	Cases labeled	Total accepted	False accepted	Total rejected	False rejected
Off-line-I	0–100	100	566	0	29	27
Off-line-II	0–300	300	587	0	8	6
Off-line-III	0–600	600	591	0	4	2
Off-line-IV	0–700	700	591	0	4	2
Online-I	0–100	30	560	0	35	33
Online-II	0–300	35	589	0	6	4
Online-III	0–600	39	592	0	3	1
Online-IV	0–700	40	592	0	2	0

was increased from 100 to 700. No false accepted cases were generated for the system running in both off-line and online modes.
- Comparing with the off-line approach, the online approach provided the system with a higher training efficiency. In other words, the proposed adaptive system was able to achieve a higher inspection accuracy using a relatively fewer number of training samples. Consequently, it reduced the amount of human effort required for checking and labeling training samples.

14.6.3 Experiments on Changing Conditions

In order to demonstrate the system performances on different scenarios with changing inspection part and environmental conditions, the proposed system was applied respectively on two different datasets. The two datasets were acquired for inspecting two types of clips from two different periods, in which the lighting conditions and camera settings were different. The experimental results showed that the proposed system had similar performances as presented in Section 14.6.1 for the two different datasets.

14.7 Conclusion

This paper presented an adaptive machine vision system for assembly parts inspection. The experimental results were promising, indicating that the proposed system was able to adapt to changing inspection tasks and environmental conditions through an efficient online learning process without excessive off-line retraining or retuning.

The future work will focus on further development which includes (*a*) validating the proposed system with a greater range of data reflecting variations in both inspection tasks and environmental conditions; (*b*) investigating alternative online learning algorithms that can build classification functions or models for identifying different types of defective patterns. The expected outcome of this research will be beneficial to the application of intelligent visual inspection in the manufacturing industry.

Acknowledgements This work was partially supported by the Auto21 Network of Centers of Excellence, Canada. Van Rob Stampings Inc. provided experimental data for the case study.

References

1. Abramovich, G., Weng, J., and Dutta, D., 2005, Adaptive part inspection through developmental vision, *Journal of Manufacturing Science and Engineering* **127**:1–11.
2. Garcia, H., Villalobos, R., and Runger, G. C., 2006, An automated feature selection method for visual inspection systems, *IEEE Transactions on Automation Science and Engineering* **3(4)**:394–406.
3. Newman, S. and Jian, A. K., 1995, A survey of automated visual inspection, *Computer Vision and Image Understanding* **61(2)**:231–262.
4. Beck, H., McDonald, D., and Brzakovic, D., 1989, A self-training visual inspection system with a neural network classifier, *Proceedings of International Joint Conference on Neural Networks* **I**:307–311.
5. Perkins, W. A., 1983, INSPECTOR: A computer vision system that learns to inspect parts, *IEEE Transactions on Pattern Analysis and Machine Intelligence* **5(6)**:584–592.
6. Cho, T. H. and Conners, R. W., 1991, A neural network approach to machine vision systems for automated industrial inspection, *International Joint Conference on Neural Networks (Seattle, USA)* **I**:205–208.
7. Killing, J., Surgenor, B. W. C., and Mechefske, K., 2006, A neuro-fuzzy approach to a machine vision based parts inspection, *Proceedings of North American Fuzzy Information Processing Society Annual Conference (Montreal, Canada)*, pp. 696–701.
8. Piccardi, M., Cucchiara, R., Bariani, M., and Mello, P., 1997, Exploiting symbolic learning in visual inspection, *Lecture Notes on Computer Science* **1280**:223–236.
9. Su, J. C. and Tarng, Y. S., 2006, Automated visual inspection for surface appearance defects of varistors using an adaptive neuron-fuzzy inference system, *International Journal of Advanced Manufacturing Technology* **35**:789–802.
10. Hata, S., Matsukubo, T., Shigeyama, Y., and Nakamura, A., 2004, Neural network visual system with human collaborated learning system, *Proceedings of IEEE International Conference on Industrial Technology* **1**:214–128.
11. Jackson, J. E. and Mudholkar, G. S., 1997, Control procedures for residuals associated with principle component analysis, *Technometrics* **21(3)**:341–349.
12. Vieira Neto, H. and Nehmzow, U., 2005, Incremental PCA: An alternative approach for novelty detection, *Proceedings of Conference on Towards Autonomous Robotic Systems* (London, UK), pp. 227–233.
13. Jensen, D. R. and Solomon, H., 1972, A Gaussian approximation for the distribution of definite quadratic forms, *Journal of the American Statistical Association* **67**:898–902.

Chapter 15
Tactile Sensing-based Control System for Dexterous Robot Manipulation

Hanafiah Yussof, Masahiro Ohka, Hirofumi Suzuki, and Nobuyuki Morisawa

Abstract Object manipulation is one crucial task in robotics. This chapter presents a precision control scheme of a multi-fingered humanoid robot arm based on tactile sensing information to perform object manipulation tasks. We developed a novel optical three-axis tactile sensor system mounted on the fingertips of a humanoid robot to enhance its ability to recognize and manipulate objects. This tactile sensor can acquire normal and shearing forces. Trajectory generation based on kinematical solutions at the arm and fingers, combined with control system structure and a sensing principle of a tactile sensor system, is presented. Object manipulation experiments are conducted using hard and soft objects. Experimental results revealed that the proposed control scheme enables the finger system to recognize low force interactions based on tactile sensing information to grasp the object surface and manipulate it without damaging the object or the sensor elements.

Keywords Tactile sensing-based control algorithm · Object manipulation · Optical three-axis tactile sensor · Multi-fingered humanoid robot arm

15.1 Introduction

Robot manipulation fundamentally relies on contact interaction between the robot and the world [1]. As blind people convincingly demonstrate, the sense of touch or tactile sensing alone can support extremely sophisticated manipulation. Tactile

---

H. Yussof (✉) and M. Ohka
Graduate School of Information Science, Nagoya University
e-mail: hanafiah@nuem.nagoya-u.ac.jp

H. Suzuki and N. Morisawa
Graduate School of Engineering, Nagoya University, Furo-cho, Chikusa-ku, Nagoya, 464-8601, Japan

S.-I. Ao et al. (eds.), *Advances in Computational Algorithms and Data Analysis,*
Lecture Notes in Electrical Engineering 14,
© Springer Science+Business Media B.V. 2009

sensing in robotics is defined as the process of determining physical properties and events by contact with objects in the world. A tactile sensor system is essential as a sensory device to support robot control system [2–4], particularly in object manipulation tasks. This tactile sensor can sense normal force, shearing force, and slippage, thus offering exciting possibilities for applications in the field of robotics for determining object shape, texture, hardness, etc. However, tactile sensors are a particularly appropriate sensing device that has too often been neglected in favor of vision-based approaches. To date, even though much research has been applied to the development of sensors, especially visual and auditory sensors, comparatively little progress has been made with on sensors that translate the sense of touch [4].

Recent research in robot manipulation has been focusing in development of new tactile sensor that take advantage of advances in materials [5], microelectromechanical systems (MEMS) [6], and semiconductor technology [7]. Unfortunately, many traditional tactile sensing technologies do not fit the requirements of robot manipulation in human environments due to lack of sensitivity, dynamic range and material strength. Therefore, besides the development of novel tactile sensor system, the development of precision control algorithm based on the tactile sensing information in the robot control system is inevitably important.

In this research, we developed and analyzed the performance of a novel optical three-axis tactile sensor system mounted on the robotic fingers of a humanoid robot arm to conduct object manipulation tasks. Figure 15.1 shows the structure of the multi-fingered arm of humanoid robot Bonten-Maru II used in this research. It consists of two robotic fingers, and tactile sensors are mounted on each fingertip. This multi-fingered arm system is developed for experimental evaluations of the tactile sensor system toward future applications in actual humanoid robots. In this report, first we present the development of the optical three-axis tactile sensor. Second we explain the trajectory generation and kinematical solutions of the multi-fingered arm, in which the tactile sensor is mounted on each fingertip. Next, we explain the control algorithm of the robotic fingers based on tactile sensing information. Finally, we present the experimental results of object manipulation tasks.

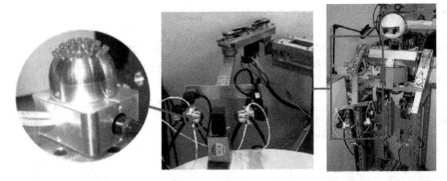

Fig. 15.1 Multi-fingered humanoid robot arm mounted with optical three-axis tactile sensors at fingertips

15.2 Optical Three-Axis Tactile Sensor

Research on tactile sensors is basically motivated by the tactile sensing system of the human skin. Our tactile sense is very accurate and sensitive. The most important distinction between the sensor types concern static and dynamics, in the sense that dynamics do not react to constant pressure. Dynamics sensors have special importance for actively checking surface texture as well as such properties as roughness, flatness, etc. Meanwhile, static sensors are improvements of imperfect grips. In humans, the skin structure provides a mechanism to sense static and dynamics pressure simultaneously with extremely high accuracy. On the other hand, most tactile sensors currently being developed are capable of detecting both static and dynamic pressure. Although accuracy and consistency remain significant problems, tactile sensing characteristics offer exciting possibilities for applications in the field of robotics, especially in robotic finger or gripper systems that perform object manipulation tasks. Basically, to effectively perform object manipulation tasks, the robotic systems required at least two types of tactile information: contact sense and slippage. Therefore, the tactile sensor system must be able to measure force in the direction of three axes. The contact sense is normally defined by measuring normal force by static tactile sensing and slippage is defined by measuring shearing force by dynamic tactile sensing.

A tactile sensor can measure a given property of an object or contact event through physical contact between the sensor and the object. To date, several basic sensing principles are commonly used in tactile sensor: capacitive, piezoelectrical, inductive, optoelectrical, and piezoresistive. In this research, to establish object manipulation ability in an actual humanoid robot, we developed an optical three-axis tactile sensor capable of acquiring normal and shearing forces to be mounted on the fingertips of a humanoid robot arm. This tactile sensor uses an optical waveguide transduction method and applies image processing techniques [8]. Since such a sensing principle provides comparatively greater sensing accuracy to detect contact phenomena from the acquisition of the three axial directions of forces, normal and shearing forces can be measured simultaneously [9]. The proposed three-axis tactile sensor has higher potential than ordinary tactile sensors for fitting a dextrose robotic arm.

Our optical three-axis tactile sensor has a hemispherical dome shape that consists of an array of sensing elements. This shape is to mimics the structure of human fingertips for easy compliance with objects of various shapes. The hardware novelty consists of an acrylic hemispherical dome, an array of 41 pieces of sensing elements made from silicon rubber, a light source, an optical fiber scope, and a CCD camera (Fig. 15.2). The silicone rubber sensing element, which is also shown in this figure, is comprised of one columnar feeler and eight conical feelers. The eight conical feelers remain in contact with the acrylic surface while the columnar feeler's tip touches an object. The sensing elements are arranged on the hemispherical acrylic dome in a concentric configuration with 41 sub-regions (Fig. 15.3). Referring to Fig. 15.2, the light emitted from the light source is directed toward the edge of the hemispherical acrylic dome through optical fibers. When an object contacts the columnar feelers,

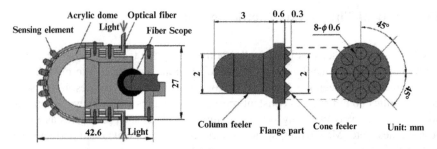

Fig. 15.2 Structure of hemispherical optical three-axis tactile sensor and sensing element

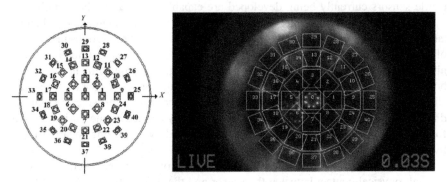

Fig. 15.3 Arrangement of sensing elements on fingertip and CCD camera-captured image of contact phenomenon in the tactile sensor

resulting in contact pressure, the feelers collapse. At the points where the conical feelers collapse, light is diffusely reflected from the reverse surface of the acrylic surface because the rubber has a higher reflective index.

The contact phenomena consisting of bright spots caused by the collapse of the feelers are observed as image data, which are retrieved by the optical fiber scope connected to the CCD camera and transmitted to the computer. Figure 15.3 shows the image data acquired by the CCD camera. The dividing procedure, digital filtering, integrated grayscale value, and centroid displacement are controlled on the PC using an auto-analysis program applying image analysis software, Cosmos32.

In this situation, the normal forces of Fx, Fy, and Fz values are calculated using integrated grayscale value G, while shearing force is based on horizontal centroid displacement. The displacement of grayscale distribution u is defined in (15.1), where i and j are orthogonal base vectors of the $x-$ and $y-$axes of a Cartesian coordinate, respectively:

$$u = u_x i + u_y j. \tag{15.1}$$

This equation reflects calibration experiments, and material functions are identified with piecewise approximate curves such as bilinear and curves [10]. Consequently, each force component is defined in (15.2):

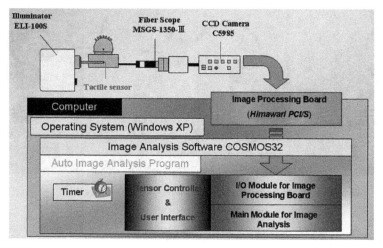

Fig. 15.4 Control system structure of optical three-axis tactile sensor

$$F_x = f(u_x), F_y = f(u_y), F_z = g(G). \tag{15.2}$$

Figure 15.4 shows the layout of the tactile sensor controller. Based on image data captured by the CCD camera, in the tactile sensor controller, an image processing board Himawari PCI/S (Library Corp.) functions as a PCI bus and selects the image and sends it to an internal buffer created inside the PC main memory. Sampling time for this process is 1/30 s. We used a PC with Windows XP OS installed with Microsoft Visual C + + . The image data are then sent to the image analysis module, where Cosmos32 software controls the dividing procedure, the digital filtering, the calculation of integrated grayscale values, and centroid displacement. Since the image was warped due to projection from a hemispherical surface, software Cosmos32 with an auto image analysis program installed in the computer modifies the warped image data and calculates G, u_x and u_y to obtain the three-axis force applied to the tip of the sensing element with Eq. (15.2). These control schemes enable the finger controller to perform force-position control to adjust the grasp pressure of the two fingers on the given object.

15.3 Trajectory Generation of Multi-Fingered Humanoid Robot Arm

The humanoid robot arm developed in this research consists of three dofs: two dofs (pitch and roll) at the shoulder and one dof (roll) at the elbow. The arm structure and control system was designed based on Bonten-Maru II. However, refinement and new modules have been added to it to comply with the robotic finger and tactile sensor systems. The trajectory generation of the arm is generated by the determination

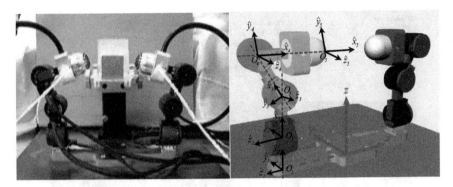

Fig. 15.5 Two robotic fingers mounted with optical three-axis tactile sensors

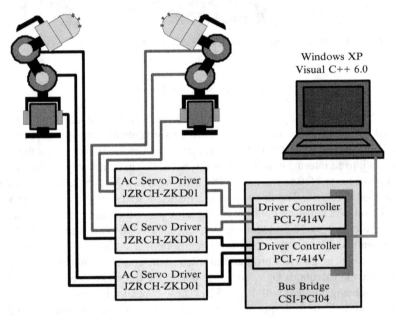

Fig. 15.6 Layout of robotic fingers system structure

of the forward and inverse kinematics solutions [11]. To describe the translation and rotational relationship between adjacent joint links, we employ a matrix method proposed by Denavit-Hartenberg [12] that systematically establishes a coordinate system for each link of an articulated chain.

Meanwhile, the robotic finger system is comprised of two articulated fingers (Fig. 15.5). Each finger has three dofs with microactuators (YR-KA01-A000, Yasukawa) in each joint. The microactuator consists of a micro AC servomotor, a harmonic gear (reduction ratio: 1/80, maximum torque: 0.7 Nm), and a digital encoder. An optical three-axis tactile sensor is mounted on each fingertip. The robotic finger's hardware system structure is shown in Fig. 15.6. Each joint is connected to a PC by a motor driver and motor control board. The PC is installed with a Windows OS and a Visual C++ compiler. In this research, the integrated robotic fingers and the tactile sensor system were designed to comply with the humanoid robot arm in terms of mechanical design and control structure.

The trajectory generation of the fingers is defined by kinematical solutions derived by the same convention as the humanoid robot arm. Based on Denavit-Hartenberg notation, a model of the finger consists of configurations and the orientation of each joint coordinate and the five sets of joint-coordinates frames are shown in Fig. 15.5. The actuator angular velocity is derived by kinematics-based resolved motion rate control, which is a common algorithm for solving path-tracking problems in robotic control.

At this point, since the finger's joint angles are defined by the kinematics solution as $\theta = [\theta_1, \theta_2, \theta_3]^T$ and the fingertip moving velocity in global coordinate space as $\dot{r} = [\dot{x}, \dot{y}, \dot{z}]^T$, the joint rotation velocity at the finger is defined as following:

$$\dot{\theta} = J(\theta)^{-1} \dot{r}. \tag{15.3}$$

Here, an inverse Jacobian matrix was employed to solve the joint angle velocity, which consequently satisfies specified velocity vector $\dot{r}$ of the fingertip in the global coordinate plane. The Jacobian matrix was initially defined in (15.4).

$$J(\theta) = \begin{bmatrix} -R_{13}(l_2 + l_3 c\theta_2 + l_4 c\theta_{23}) & l_3(R_{11}s\theta_3 + R_{12}c\theta_3)R_{12}l_4 + R_{12}l_4 & R_{12}l_4 \\ -R_{23}(l_2 + l_3 c\theta_2 + l_4 c\theta_{23}) & l_3(R_{32}s\theta_3 + R_{22}c\theta_3)R_{22}l_4 + R_{22}l_4 & R_{22}l_4 \\ -R_{33}(l_2 + l_3 c\theta_2 + l_4 c\theta_{23}) & l_3(R_{31}s\theta_3 + R_{32}c\theta_3)R_{32}l_4 + R_{32}l_4 & R_{32}l_4 \end{bmatrix} \tag{15.4}$$

Meanwhile, the rotational transformation from the local frame of the fingertip, where the tactile sensor is attached, to the frame of the workspace (see Fig. 15.5) is calculated as follows:

$$\begin{aligned} {}^{0}_{s}R &= \begin{bmatrix} R_{11} & R_{12} & R_{13} \\ R_{21} & R_{22} & R_{23} \\ R_{31} & R_{32} & R_{33} \end{bmatrix} \begin{bmatrix} \cos 90° & 0 & \sin 90° \\ 0 & 1 & 0 \\ -\sin 90° & 0 & \cos 90° \end{bmatrix} \\ &= \begin{bmatrix} -R_{13} & R_{12} & R_{11} \\ -R_{23} & R_{22} & R_{21} \\ -R_{33} & R_{32} & R_{31} \end{bmatrix}. \end{aligned} \tag{15.5}$$

A direction cosine is obtained to estimate the slippage direction of the grasped object while handling it. Referring to the definition of the sensor element coordinate position at the hemispherical shape tactile sensor (Fig. 15.7), the direction cosine of the k-th sensing element in the local frame of tactile sensor ($\alpha_k, \beta_k, \gamma_k$) is defined in (15.6):

$$\begin{bmatrix} \alpha_k \\ \beta_k \\ \gamma_k \end{bmatrix} = \begin{bmatrix} \sin \theta_k \cos \phi_k \\ \sin \theta_k \sin \phi_k \\ \cos \theta_k \end{bmatrix}. \tag{15.6}$$

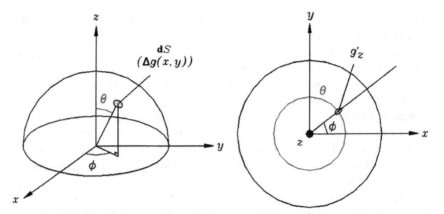

Fig. 15.7 Coordination of sensing position in hemispherical dome tactile sensor

Hence, the direction cosine in the frame of workspace $(\alpha_{Gk}, \beta_{Gk}, \gamma_{Gk})$ is calculated as follows:

$$\begin{bmatrix} \alpha_{Gk} \\ \beta_{Gk} \\ \gamma_{Gk} \end{bmatrix} = \begin{bmatrix} -R_{13} & R_{12} & R_{11} \\ -R_{23} & R_{22} & R_{21} \\ -R_{33} & R_{32} & R_{31} \end{bmatrix} \begin{bmatrix} \sin\theta_k \cos\phi_k \\ \sin\theta_k \sin\phi_k \\ \cos\theta_k \end{bmatrix}. \tag{15.7}$$

15.4 Control System Structure

Figure 15.8 shows the control system structure of the multi-fingered humanoid robot arm with robotic fingers and an optical three-axis tactile sensor. This system is comprised of three main controllers: arm, finger, and a tactile sensor. All controllers are connected to each other using TCP/IP protocols by the internet. The arm controller consists of two main modules: robot controller and motion instructor. Shared memory is used to connect these two modules. The control system architecture for the robot finger controller, based on tactile sensing, is shown in Fig. 15.9. It comprised of three modules: a Connection Module, Thinking Routines, and a Hand/Finger Control Module. The architecture is connected to the tactile sensor controller by the connection module using TCP/IP protocols. In addition, to obtain low force interactions of the fingers while exploring the object surface without causing damage, the rotation velocity at each joint is precisely defined based on the joint angle obtained in the kinematics calculations, whereby force-position controls were performed.

The following are the most important considerations for controlling finger motions while performing the object manipulation tasks: what kind of information is to be acquired from the tactile sensor, how to translate and utilize this information, and how to send commands to the robot finger to properly control the velocity of the finger motion. These processes are performed inside the Thinking Routines Module. As shown in Fig. 15.9, inside the Thinking Routines Module, there is a Thinking Routine Chooser that consists of a Pin Status Analyzer and a Velocity

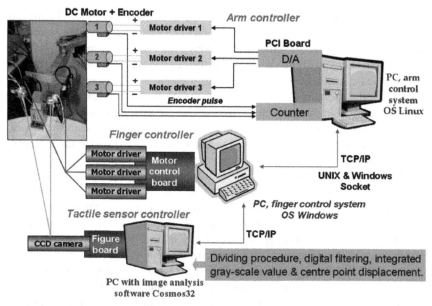

Fig. 15.8 Control system structure of multi-fingered humanoid robot arm with optical three-axis tactile sensor

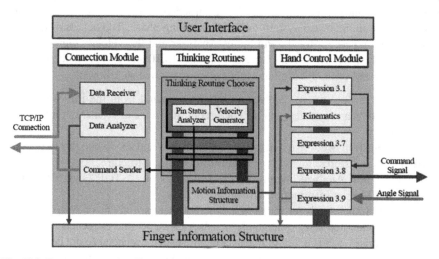

Fig. 15.9 System conception diagram of finger controller

Generator. Moreover, there is a Motion Information Structure that connects to both the Pin Status Analyzer and the Velocity Generator. The Pin Status Analyzer module receives information from the tactile sensor about the condition of the sensing elements and uses this information to determine a suitable motion mode. Then it sends a list of sensing elements to the Connection Module to acquire tactile sensing information. Meanwhile, the Velocity Generator Module determines finger velocity based on the Finger Information and Motion Information Structure. The Motion Information Structure consists of initial velocity and motion flag modes, which are used to control the finger movement. Meanwhile, the Finger Information Structure provides connections to all modules so that they can share finger orientation, joint angle, and tactile sensing data from each sensor element. A User Interface was designed so that the operator can provide commands to the finger control system. The finger control module controls finger motions by calculating joint velocities and angles. In fact, this module can move the finger without using sensing feedback. The Thinking Routines module received tactile sensing data from the tactile sensor and used them to calculate the fingertip velocity.

To deepen understanding about the data communication process in the finger controller, we present a simple case study where a finger touched an object and then avoided or evaded it by moving in a reverse direction. First, the finger moving velocity to search for the object was defined as V_0. Next, we fixed thresholds of normal forces F_1 and F_2. During the searching process, when any of the sensor elements touched an object, and if detected normal force F_n exceeded normal force threshold F_1[N], the finger stopped moving. If detected normal force F_n exceeded threshold F_2[N], the finger moved away from the sensing element that detected the highest force. At this moment the reverse velocity is defined as $|V_{re}|$. The parameters values of V_0, F_1, F_2 and $|V_{re}|$ are kept inside the Motion Information Structure. The thresholds F_1 and F_2 are also sent to the sensor controller. When the finger starts moving, commands to request the status of each sensor element are sent to the sensor system based on the finger system's control sampling phase. The details of the data communication process for the Pin Status Analyzer and the Velocity Generator for this case study are shown in the flowchart of Figs. 15.10 and 15.11, respectively.

The processes of the Pin Status Analyzer are explained as follows:

1. When the sensor system receives a request command from the Pin Status Analyzer, it returns the status flag of each requested sensing element condition as feedback.
2. The Connection Module receives the feedback data and sends it to the Pin Status Analyzer, and stores it inside the Finger Information Structure.
3. The Pin Status Analyzer then resets the finger motion ("STOP" and "EVADE") inside the Motion Information Structure.
4. If the Pin Status Analyzer receives a data flag that exceeds F_1 or F_2 (or both) it lists the concerned sensor elements. If sensor elements that exceeded F_1 are detected, it raises a "STOP" flag. If sensor elements that exceeded F_2 are detected, it raises an "EVADE" flag. These data lists are then sent to the Connection Module.
5. The Connection Module creates a command to request the normal force data of the related sensor elements and sends them to the sensor system.
6. When the sensor system receives this request command, it feedbacks the normal force data of the requested sensor element to the Connection Module.
7. The Connection module receives the feedback of the normal force data and sends them it to the Finger Information Structure. Based on these data, the Velocity Generator module will decides the finger velocity.

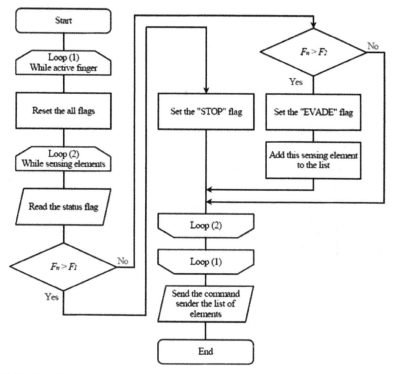

Fig. 15.10 Algorithm flowchart at the Pin Status Analyzer for case study

The processes of the Velocity Generator are explained as follows:

1. If no "STOP" or "EVADE" flags are raised, the finger moves based on initial velocity V_0.
2. If the "STOP" flag is raised, finger velocity becomes 0.
3. If the "EVADE" flag is raised, the finger moves in the reverse direction of the sensing element that detected the highest normal force value. To decide finger velocity, when the finger evading velocity is described as $V_r = (V_{rx}, V_{ry}, V_{rz})$, the direction cosine in the frame of workspace $(\alpha_{Gk}, \beta_{Gk}, \gamma_{Gk})$ is calculated as Eq. (15.8).

$$V_r = -|V_{re}| \begin{bmatrix} \alpha_{Gk} \\ \beta_{Gk} \\ \gamma_{Gk} \end{bmatrix} \tag{15.8}$$

Here, this generation of velocity is basically sent to the Hand/Finger Control Module to solve joint rotation velocity at finger. Therefore, finger controls based on tactile sensing information are conducted.

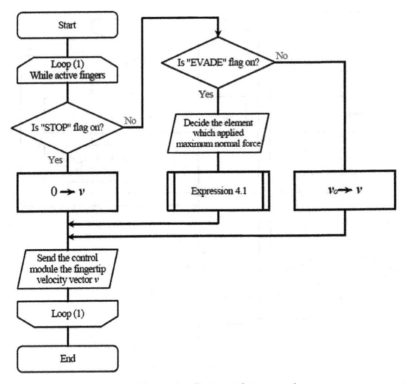

Fig. 15.11 Algorithm flowchart at the Motion Generator for case study

15.5 Experiment and Results

The ability to sense hardness and/or softness will be particularly important in future applications of humanoid robots. Therefore, we conducted a set of experiments to recognize and manipulate hard and soft objects. We used a cubic-shaped wood block as a hard object, and a cubic-shaped thin paper box as a soft object. Since the parameters of the human hand and fingers involved in sensing an object's hardness and/or softness of an object have not been fully researched, we conducted calibration tests with our tactile sensor system to grasp objects with different hardnesses to obtain the basic data to estimate optimum grasping. From the test results we estimate the force parameters, as shown in Table 15.1, which also indicate the parameters of the finger control system.

Figure 15.12 shows photographs of the robot arm performing object manipulation with a wood block and a paper box. In both experiments, the fingers managed to grasp the objects within optimum grasp pressure and lift them up. In this experiment, first the two fingers grasp the object to define the optimum gripping pressure. At this moment, the grasp pressure is controlled by the parameters of normal force thresholds. Then both fingers lift the object in the z-axis direction while maintaining the optimum grasp pressure. During this motion, both normal pressure and slip-

Table 15.1 Parameters in manipulation experiments with hard and soft objects

Category		Parameter
Interval for sampling	Sensor	100 ms
	Finger	25 ms
Threshold of normal force	F_1	0.5 N
	F_2	1.8 N
Threshold of centroid change	dr	0.004 mm
Velocity of repush	v_p	2 mm/s
Increment of normal force	ΔF	0.08 N
Progress time	Δt	0.1 s

Fig. 15.12 Experiments of object manipulation with wood block and paper box

page are measured. Therefore the finger controller utilized the parameters of normal force and centroid change thresholds. Here, when shearing force exceeds the centroid change threshold, the finger's velocity that reinforces the grasping pressure is calculated using (15.9), and the vector velocity of finger $v + \Delta v$ is defined by the finger control module in the finger controller:

$$\Delta v = \left| v_p \right| \begin{bmatrix} \alpha_{Gk} \\ \beta_{Gk} \\ \gamma_{Gk} \end{bmatrix}. \tag{15.9}$$

Figure 15.13 shows an example of normal and shearing forces detected and its relation with fingertip movement at the x and z-axes in the experiment with the paper box. The proposed control scheme managed to recognize low force interactions to grasp the object surface within optimum grasp pressure without damaging the object and the sensor elements. In addition, the formulations and algorithm applied in this system enable precise control of the fingertips from the determination of joint rotation angles and velocity.

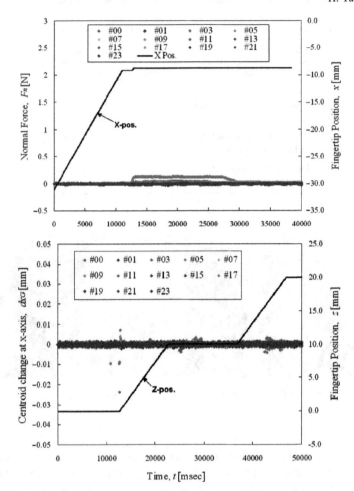

Fig. 15.13 Data graphs of experiment with paper box at right finger: (*Top*) Relation between normal force and *x*-directional fingertip position. (*Bottom*) Amount of *x*-directional centroid change and *z*-directional fingertip position

15.6 Conclusions

We proposed a new control scheme for object manipulation tasks based on tactile sensing in a multi-fingered humanoid robot arm for applications in real humanoid robots. We developed a novel optical three-axis tactile sensor system capable of acquiring normal and shearing forces. To evaluate the proposed system's performance, we conducted object manipulation experiments with hard and soft objects. Experimental results revealed that the proposed system managed to recognize low force interactions to grasp and manipulate objects of different hardnesses. We anticipate that this system's technology will help advance the evolution of humanoid robots working in real life.

References

1. A. Edsinger and C. Kemp, Manipulation in Human Environments, in *Proceedings of IEEE/RSJ International Conference on Humanoid Robotics (Humanoids06)*, Italy (2006).
2. S. Omata, Y. Murayama, and C.E. Constantinou, Real Time Robotic Tactile Sensor System for Determination of the Physical Properties of Biomaterials, *Sensors and Actuators A*, 112(2–3), 278–285 (2004).
3. O. Kerpa, K. Weiss, and H. Worn, Development of a Flexible Tactile Sensor System for a Humanoid Robot, in *Proceedings of IROS2003*, Vol. 1, Las Vegas, NV, pp. 1–6 (2003).
4. M.H. Lee and H.R. Nicholls, Tactile Sensing for Mechatronics: A State of the Art Survey, *Journal of Mechatronics*, 9(1), 1–31 (1999).
5. Y. Ohmura, Y. Kuniyoshi, and A. Nagakubo, Conformable and Scalable Tactile Sensor Skin for Curved Surfaces, in *Proceedings of ICRA2006*, Orlando, FL, pp. 1348–1353 (2006).
6. L. Natale and E. Torres-Jara, A Sensitive Approach to Grasping, in *Proceedings of the 6th International Conference on Epigenetic Robotics*, France (2006).
7. P.A. Schmidt, E. Mael, and R.P. Wurtz, A Sensor for Dynamic Tactile Information with Applications in Human-Robot Interaction and Object Exploration, *Robotics & Autonomous Systems*, 54(12), 1005–1014 (2006).
8. M. Ohka, H. Kobayashi, and Y. Mitsuya, Sensing Precision of an Optical Three-Axis Tactile Sensor for a Robotic Finger, in *Proceedings of 15th RO-MAN06*, pp. 220–225 (2006).
9. Y. Hanafiah, M. Ohka, H. Kobayashi, J. Takata, M. Yamano, and Y. Nasu, Contribution to the Development of Contact Interaction-Based Humanoid Robot Navigation System: Application of an Optical Three-Axis Tactile Sensor, in *Proceedings of 3rd ICARA06*, New Zealand, pp. 63–68 (2006).
10. M. Ohka, Y. Mitsuya, Y. Matsunaga, and S. Takeuchi, Sensing Characteristics of an Optical Three-Axis Tactile Sensor Under Combined Loading, *Robotica*, 22, 213–221 (2004).
11. Y. Hanafiah, M. Yamano, Y. Nasu, and M. Ohka, Trajectory Generation in Groping Locomotion of a 21-dof Humanoid Robot, *Journal of Simulation: System, Science & Technology*, 7(8), 55–63 (2006).
12. J. Denavit and S. Hartenberg, A Kinematic Notation for Lower-Pair Mechanisms Based upon Matrices, *Journal of Applied Mechanics*, 77, 215–221 (1955).

Chapter 16
A Novel Kinematic Model for Rough Terrain Robots

Joseph Auchter, Carl A. Moore, and Ashitava Ghosal

Abstract We describe in detail a novel kinematic simulation of a three-wheeled mobile robot moving on extremely uneven terrain. The purpose of this simulation is to test a new concept, called Passive Variable Camber (PVC), for reducing undesirable wheel slip. PVC adds an extra degree of freedom at each wheel/platform joint, thereby allowing the wheel to tilt laterally. This extra motion allows the vehicle to better adapt to uneven terrain and reduces wheel slip, which is harmful to vehicle efficiency and performance.

In order to precisely model the way that three dimensional wheels roll over uneven ground, we adapt concepts developed for modeling dextrous robot manipulators. The resulting equations can tell us the instantaneous mobility (number of degrees of freedom) of the robot/ground system. We also showed a way of specifying joint velocity inputs which are compatible with system constraints. Our modeling technique is adaptable to vehicles of arbitrary number of wheels and joints.

Based on our simulation results, PVC has the potential to greatly improve the motion performance of wheeled mobile robots or any wheeled vehicle which moves outdoors on rough terrain by reducing wheel slip.

Keywords Kinematics · Mobile robots · Uneven terrain · Dextrous manipulation

16.1 Introduction

Wheeled mobile robots (WMRs) were first developed for indoor use. As such, traditional kinematic models reflect assumptions that can be made about the structured

J. Auchter (✉) and C.A. Moore
Department of Mechanical Engineering, FAMU/FSU College of Engineering in Tallahassee, Florida, USA
e-mail: auchtjo@eng.fsu.edu

A. Ghosal
Department of Mechanical Engineering, Indian Institute of Science in Bangalore, India

S.-I. Ao et al. (eds.), *Advances in Computational Algorithms and Data Analysis,*
Lecture Notes in Electrical Engineering 14,

environment in which the robot operates [1]. For example, the robot is assumed to move on a planar surface. The wheels are modeled as thin disks and the velocity of each wheel center, v, in terms of its angular speed ω is calculated as $v = \omega R$. As a result of these assumptions, the non-holonomic constraints of wheel rolling without slip at the wheel/ground contacts are simple trigonometric relationships.

Recently there have been many attempts to extend the operating range of WMRs to outdoor environments [2–4]. As a result the kinematic modeling process becomes very complex, mainly because the robot is now moving in a three-dimensional world instead of a two-dimensional one. On uneven terrain the contact point can vary along the surface of the wheel in both lateral and longitudinal directions. Therefore it no longer justifiable to model the wheel as a thin disk. Furthremore, the non-holonomic constraints can no longer be determined by simple geometry. Despite these difficulties, kinematic modeling is a crucial process since it is used for control and path planning [5,6] and as a stepping stone to a dynamic model.

16.1.1 Previous Work

There have been several recent efforts to model the kinematic motion of WMRs on uneven terrains. However, none of them provide a complete model for the motion of the wheel rolling over the uneven ground. Capturing this motion precisely is of utmost importance when studying wheel slip, power efficiency, climbing ability, and path planning for outdoor robots.

In reference [7], the authors provide a detailed kinematic model for the Rocky 7 Mars rover, but assume a 2-D wheel and do not provide a model of how the contact point moves along the surface of the wheel as it rolls on an uneven ground. In the reference [8], the authors develop a similar kinematic model for their CEDRA robot. However, they assume that certain characteristics of the wheel motion on the terrain are known without providing any equations describing the motion. The kinematic model in reference [9] places a coordinate frame at the wheel/ground contact point, but no explanation is provided as to where the contact point is or how the motion is influenced by the terrain shape. Grand and co-workers [10] perform a velocity analysis on their hybrid wheel-legged robot Hylos. They identify the contact point for each wheel and an associated frame, but make no mention of how these frames evolve as the vehicle moves. Sreenivasan and Nanua [11] explore first- and second-order kinematic characteristics of wheeled vehicles on uneven terrain in order to determine vehicle mobility. For general terrains, their method is inefficient and involves manual determination of free joint rates to avoid interdependencies.

16.1.2 Kinematic Slip

The unstructured outdoor environment can cause unexpected problems for wheeled mobile robots. In addition to dynamic slippage due to terrain deformation or insufficient friction, a WMR is affected by kinematic slip [4, 12, 13]. Kinematic slip occurs when there is no instantaneous axis of rotation compatible with all of the robot's wheels. For automobiles this can lead to tire scrubbing and is generally avoided using Ackermann steering geometry. This works properly only on flat ground, however. On uneven terrain kinematic slip occurs generally with a standard vehicle since the location of the wheel/ground contact point varies laterally and longitudinally over the wheel surface due to the terrain shape and robot configuration.

Wheel slip causes several problems: first, power is wasted [4, 12], and second, wheel slip reduces the ability of the robot to self-localize because position estimates from wheel encoder data accumulate unbounded error over time [14]. Researchers have reported that reducing slip improves the climbing performance and accuracy of the odometry for a six-wheeled off-road rover [15]. Accurate kinematic models are needed to test robot designs which will potentially reduce this costly kinematic slip. In reference [11], the authors used screw theory to explore the phenomenon of kinematic slip in wheeled vehicle systems moving on uneven terrain. Their analysis showed that kinematic slip can be avoided if the distance between the wheel/ground contact points is allowed to vary for two wheels joined on an axle. The authors of that work suggest the use of a Variable Length Axle (VLA) with a prismatic joint to achieve the necessary motion. The VLA is difficult to implement because it requires a complex wheel axle design.

As a more practical alternative to the VLA, the authors in reference [12] introduced the idea of adding an extra degree of freedom (DOF) at the wheel/axle joint, allowing the wheel to tilt laterally relative to the axle. This new capability, herein named Passive Variable Camber (PVC), permits the distance between the wheel/ground contact points to change without any prismatic joints. On a real vehicle the PVC joints would be actuated by lateral forces at the wheel/ground interface arising from interactions between the two surfaces; therefore, the joint would be "passive" (requiring no extra energy expenditure by the robot). Figure 16.1 shows an example of an axle and two wheels equipped with PVC.

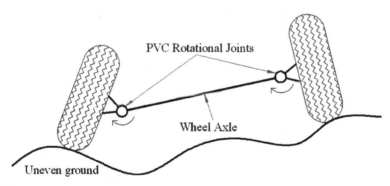

Fig. 16.1 Two tires on uneven ground attached to an axle equipped with Passive Variable Camber. The axis of rotation of each PVC joint is perpendicular to the page

16.1.3 Contribution of this Work

Traditional methods are not suitable for kinematic modeling of outdoor WMRs due to the complex nature of the terrain/robot system. More recent efforts to model outdoor vehicle motion lack convincing descriptions of how a realistic wheel rolls over an arbitrary uneven terrain.

This paper describes in detail a novel kinematic simulation of a three-wheeled mobile robot equipped with Passive Variable Camber and moving on uneven terrain. In order to precisely model the way that three dimensional wheels roll over uneven ground, we adapted concepts developed for modeling dextrous robot manipulators. In the reference [12] the authors began this task by using dextrous manipulation contact kinematics to model how a torus-shaped wheel rolls over an arbitrarily-shaped smooth terrain. In their WMR model they introduced an extra degree of freedom at the wheel which allowed for lateral tilting in order to prevent kinematic slip. In this paper we extend that work by completing the analogy between dextrous manipulators and wheeled vehicles using additional equations which give more insight into the structure of the system, including the number of degrees of freedom and the interdependencies among joint rates.

To the best of our knowledge, the union of the worlds of WMR modeling and dextrous manipulator modeling is novel and does not suffer from many of the assumptions inherent in other modeling techniques. Our method provides a concise and manageable description of the kinematics and is easily adaptable to other vehicle designs of arbitrary complexity.

The purpose of our simulation is to verify that a WMR equipped with PVC can traverse uneven terrain without kinematic slip. Based on the simulation results, PVC has the potential to greatly improve the motion performance of wheeled mobile robots, or any wheeled vehicle which moves outdoors on rough terrain.

16.2 Analogy Between WMRs and Dextrous Manipulators

In this work a kinematic model of the WMR/ground system is developed using techniques from the field of dextrous manipulation. This is extremely useful for the WMR community because the kinematics of dextrous manipulation provide an ideal description of the way wheels roll over uneven terrain. To our knowledge, the analogy between a robotic hand manipulating an object and a WMR traversing a three-dimensional terrain had never been made before the work of Chakraborty and Ghosal [12]. A WMR in contact with uneven ground is analogous to a multi-fingered robotic "hand" (the WMR) grasping an "object" (the ground). Table 16.1 summarizes the analogies between robotic hands and WMRs.

Table 16.1 Relationships between manipulators and WMRs

Manipulators	Mobile robots
Multi-fingered hand	Wheeled mobile robot
Grasped object	Ground
Fingers	Wheels
Palm	Robot platform

16.3 Off-Road Wheeled Mobile Robot Kinematic Model

In this paper we model a three-wheeled mobile robot (one front and two rear wheels). The front wheel is steerable and the two rear wheels have Passive Variable Camber (PVC) joints. The wheels are torus-shaped, which is more realistic than the typical thin-disk model. We adapt techniques from the field of dextrous manipulator modeling to show that the PVC-equipped wheeled mobile robot is able to negotiate extreme terrains without kinematic wheel slip. This is desirable because rolling motion is more controlled than sliding motion. Our simulation and evaluation process involves the following steps:

1. Write kinematic differential equations which describe the system, including the wheel/ground contact
2. Constrain the wheels to roll by suitable modification of the contact equations
3. Run simulations on various types of uneven terrain and
4. Monitor the level of constraint violation

As will be shown, the rolling constraints are well-satisfied for all of the simulations that we have attempted. This means that PVC has the potential to dramatically reduce undesirable wheel slip for a real vehicle operating on rough terrains.

16.3.1 Wheel/Ground Contact Model

At the heart of our novel WMR modeling concept is the use of dextrous manipulation equations to describe how the wheel rolls over the ground surface in response to the robot's velocity inputs. These equations are powerful because they were originally formulated to show how a robotic finger rolls/slides over an object of any shape, provided that both finger and object are smooth surfaces. Therefore, the equations can easily be applied to the special case of smooth wheels rolling over an uneven ground.

Montana [16] was the first to develop kinematic contact equations which describe how two arbitrarily-shaped smooth surfaces roll/slide against each other. In our case the two surfaces are the wheel and ground. In this section we will develop the tools that we need in order to make use of the contact equations. For a good overview of dextrous manipulation and the associated mathematics, see [17].

Fig. 16.2 Wheel surface
parameterization

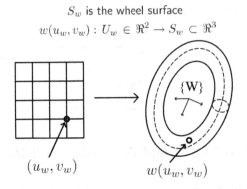

S_w is the wheel surface

$$w(u_w, v_w) : U_w \in \Re^2 \to S_w \subset \Re^3$$

(u_w, v_w) $w(u_w, v_w)$

Fig. 16.3 Ground surface
parameterization

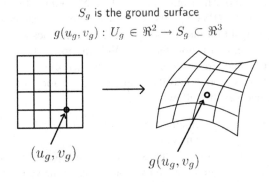

S_g is the ground surface

$$g(u_g, v_g) : U_g \in \Re^2 \to S_g \subset \Re^3$$

(u_g, v_g) $g(u_g, v_g)$

Figures 16.2 and 16.3 show the surface parameterizations that we will use.
The surface of the wheel is parameterized relative to its frame $\{W\}$ by the right-handed orthogonal coordinate chart:

$$w(u,v) : U \in \Re^2 \to S_w \subset \Re^3$$

In other words, specifying two parameters u_i and v_i will locate a unique point on
the surface of wheel i. Similarly, the ground surface is parameterized relative to its
frame $\{G\}$ by the chart $g(x,y)$, meaning that any parameters x and y will locate a
unique point $(x,y,g(x,y))$ or (x,y,z) on the ground surface.

Montana's equations describe the motion of the point of contact across the surfaces in response to a relative motion between the wheel and the ground. This motion has five degrees of freedom (DOFs), the one constraint being that there be no
translational component of motion along their common surface normal and contact
is maintained. These five DOFs have the following interpretation: two DOFs each
for the position of the contact point on the two surfaces (wheel and ground), and
one DOF for rotation about the surface normal. The parameters that describe these
five DOFs for wheel i are

Fig. 16.4 Coordinate frames
of the wheel and ground

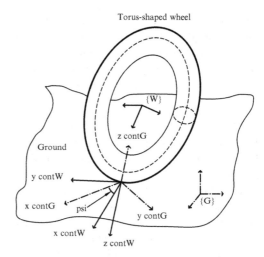

Fig. 16.5 Coordinate frames
of the robot platform

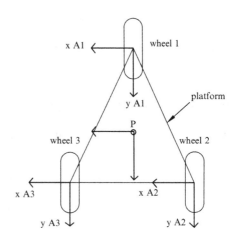

$$\eta_i = [u_i \ v_i \ x_i \ y_i \ \psi_i]^T, \quad i = 1, 2, 3$$

where ψ_i is the angle of rotation about the common surface normal. They are grouped for all three wheels as: $\eta = [\eta_1^T \ \eta_2^T \ \eta_3^T]^T$.

Figures 16.4 and 16.5 show the coordinate frames which are used to develop the WMR kinematic model. The frames in Fig. 16.4 follow the conventions of Montana [16, 18] and Murray et al. [17].

In the above figures frame $\{G\}$ is the ground inertial reference frame and frame $\{contG_i\}$ is the ground contact frame for wheel i. The z-axis of $\{contG_i\}$ is the outward normal to the ground surface at the contact point. Frame $\{P\}$ is the robot platform reference frame, $\{A_i\}$ is the frame at the point of attachment of the wheel i to the platform, and $\{W_i\}$ is the reference frame of wheel i. The frame $\{contW_i\}$ is the contact frame relative to wheel i. Its z-axis is the outward pointing normal from the torus-shaped wheel, which is collinear with the z-axis of $\{contG_i\}$. The variable ψ_i (described above) is the angle between the x-axes of frames $\{contG_i\}$ and $\{contW_i\}$. Note that the origin of $\{contG_i\}$ is the point $g(x_i, y_i)$, and the origin of $\{contW_i\}$ is the point $f(u_i, v_i)$ as described above.

Also important are the velocities of the wheel relative to the ground:

$$contW V_{GW} = V_c = [v_x \ v_y \ v_z \ \omega_x \ \omega_y \ \omega_z]^T \tag{16.1}$$

The leading superscript indicates that the vector is resolved in the $\{contW\}$ frame. The subscript GW means that these velocities are of the $\{W\}$ frame relative to the $\{G\}$ frame.

The purpose of our simulation is to show that Passive Variable Camber (PVC) can reduce wheel slip and allow the vehicle to move with a controlled rolling motion over uneven terrains. For our model, this means that the $\{contW\}$ and $\{contG\}$) frames do not translate relative to each other and only relative rolling is permitted. Mathematically, these conditions are expressed as constraints on the velocities V_c:

$$\tilde{V}_c = \bar{B} V_c \tag{16.2}$$

where $\bar{B} = [0_{3\times3} \ I_{3\times3}]$. $\tilde{V}_c$, a subset of V_c (from (16.1)), are called the allowable contact velocities. For a wheel rolling without slip $\tilde{V}_c = [\omega_x \ \omega_y \ \omega_z]^T$.

16.3.1.1 Kinematic Contact Equations

We are now ready to introduce the contact equations. In terms of the metric (M), curvature (K) and torsion (T) forms, the equations for rolling contact are [16]:

$$\begin{aligned}
(\dot{u}, \dot{v})^T &= M_w^{-1} (K_w + K^*)^{-1} (-\omega_y, \omega_x)^T \\
(\dot{x}, \dot{y})^T &= M_g^{-1} R_\psi (K_w + K^*)^{-1} (-\omega_y, \omega_x)^T \\
\dot{\psi} &= \omega_z + T_w M_w (\dot{u}, \dot{v})^T + T_g M_g (\dot{x}, \dot{y})^T.
\end{aligned} \tag{16.3}$$

where subscript w indicates the wheel and g indicates the ground. The inputs to these equations are the allowable wheel contact velocities $\tilde{V}_c$, and the outputs are $\dot{\eta}$, so we abbreviate the equations by:

$$\dot{\eta} = [CK] \ \tilde{V}_c, \tag{16.4}$$

where $[CK]$ stands for "Contact Kinematics". These are the non-holonomic constraints of the robot/ground system by which rolling contact is enforced. During the simulation if these equations are satisfied then the vehicle is moving without slip.

16.3.2 Wheeled Mobile Robot Kinematic Model

In this section we develop the kinematic model of the three-wheeled mobile robot, making use of the contact equations from the previous section to model the

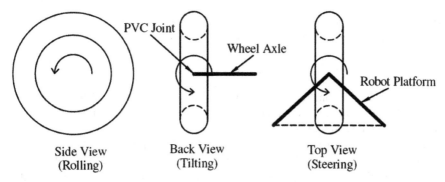

Fig. 16.6 Robot wheel joint velocities

wheel/ground contacts. The inputs and outputs for the forward kinematics are desired wheel and joint velocities $\dot{\theta}$ and position and velocity of robot platform, respectively.

16.3.2.1 Robot Joint Velocities

The vector of joint velocities is:

$$\dot{\theta} = \begin{bmatrix} \dot{\phi}_1 & \dot{\alpha}_1 & \dot{\gamma}_2 & \dot{\alpha}_2 & \dot{\gamma}_3 & \dot{\alpha}_3 \end{bmatrix}^T$$

where $\dot{\alpha}_i$ is the driving rate of wheel i, $\dot{\phi}_1$ is the steering rate of wheel 1, $\dot{\gamma}_i$ is the rate of tilt of the wheel about the PVC joint of wheel i (for $i = 2, 3$). Figure 16.6 graphically illustrates these variables.

16.3.2.2 Choice of Inputs $\dot{\theta}$

The inputs to our forward kinematics simulation are $\dot{\theta}$, the joint velocities (steering rate, driving rate, and PVC tilting rate). Because the robot/ground system has complex non-holonomic constraints (Eq. (16.3)) which depend on the terrain geometry, we cannot choose these inputs arbitrarily: they are interrelated by the structure of the system. In this section we adapt dextrous manipulator equations developed by Han and Trinkle [19] in order to get more insight as to how the joint velocities of our system relate to one another. This will ultimately allow us to calculate $\dot{\theta}$ velocities which are as close as possible (in the least squares sense) to a set of desired velocities.

For the forward kinematics simulation we are interested in calculating the velocities of the robot platform frame $\{P\}$ relative to the ground inertial frame $\{G\}$, which we will call V_{PG}. Following Han and Trinkle [19], we group the platform/ground relative velocities V_{PG} and contact velocities $\tilde{V}_c$ (see Eq. (16.1)) together in $V_{GC} = \left[V_{PG}^T \; \tilde{V}_c^T \right]^T$. Jacobian matrices can be formed such that:

$$J_{GC} \, V_{GC} = J_R \, \dot{\theta}. \tag{16.5}$$

Equation (16.5) are constraints which relate the joint velocities $\dot{\theta}$ to the relative ground/platform and contact velocities V_{GC}. In the general case neither J_{GC} nor J_R are square and thus are not invertible.

Because of the constraints (16.5), we cannot freely choose our inputs $\dot{\theta}$. However, we can calculate inputs consistent with (16.5) which are as close as possible (in the least-squares sense) to a vector of desired inputs. This is done as follows. The QR decomposition [20] of matrix J_{GC} is:

$$J_{GC} = Q \, R. \tag{16.6}$$

Let r denote $rank(J_{GC})$ and c be the number of columns of J_{GC}. Split Q into $[Q_1 \; Q_2]$, where $Q_2 \in \Re^{c \times (c-r)}$. [The matrix] Q_2 forms a orthonormal basis for the null space of J_{GC}^T, meaning $J_{GC}^T \, Q_2 = 0$ or $Q_2^T \, J_{GC} = 0$. Pre-multiplying both sides of (16.5) by Q_2^T yields $Q_2^T \, J_{GC} \, V_{GC} = Q_2^T \, J_R \, \dot{\theta}$, or

$$Q_2^T \, J_R \, \dot{\theta} = 0. \tag{16.7}$$

Equation (16.7) is a set of constraint equations for the inputs $\dot{\theta}$. To make use of these equations, let $C_\theta = (Q_2^T \, J_R) \in \Re^{p \times q}$ where $p = rank(C_\theta)$. The QR decomposition of C_θ^T is:

$$C_\theta^T = [Q_{C1} \; Q_{C2}] \, R_C,$$

where $Q_{C2} \in \Re^{q \times p}$. Then $C_\theta \, Q_{C2} = 0$, meaning Q_{C2} is an orthonormal basis for the null space of C_θ. At this point, we can choose independent generalized velocity inputs $\dot{\theta}_g$ such that

$$\dot{\theta} = Q_{C2} \, \dot{\theta}_g. \tag{16.8}$$

However, since neither C_θ nor Q_{C2} are unique and both change as the robot configuration changes, the generalized inputs $\dot{\theta}_g$ likely have no physical interpretation and their relationship with the actual joint velocities $\dot{\theta}$ is unclear.

Since (16.8) is of limited use, we take another step. We want our actual joint velocities $\dot{\theta}$ to match some desired joint velocities $\dot{\theta}_d$, or $\dot{\theta} \approx \dot{\theta}_d$. Combining this with (16.8), we have:

$$Q_{C2} \, \dot{\theta}_g \approx \dot{\theta}_d.$$

To get as close as possible in the least squares sense to $\dot{\theta}_d$, we use the pseudo-inverse [21] of Q_{C2}:

$$\dot{\theta}_g = Q_{C2}^+ \, \dot{\theta}_d = (Q_{C2}^T Q_{C2})^{-1} Q_{C2}^T \, \dot{\theta}_d. \tag{16.9}$$

Since the columns of Q_{C2} are orthonormal, Q_{C2}^{+} reduces to Q_{C2}^{T}. Noticing that $\dot{\theta} = Q_{C2}\,\dot{\theta}_g$, we can pre-multiply both sides of (16.9) by Q_{C2} to get $Q_{C2}\dot{\theta}_g = Q_{C2}Q_{C2}^{T}\dot{\theta}_d$, or

$$\dot{\theta} = Q_{C2}Q_{C2}^{T}\dot{\theta}_d = J_{in}\dot{\theta}_d. \tag{16.10}$$

The matrix J_{in} can be thought of as a transformation that takes the desired velocities $\dot{\theta}_d$, which can be arbitrary, and transforms them such that $\dot{\theta}$ satisfy the constraints (16.5) while remaining as close as possible to $\dot{\theta}_d$.

Equation (16.10) is a highly useful result for our simulation. First, it eliminates the need to deal with independent generalized velocities $\dot{\theta}_g$, which have no physical meaning. We can instead directly specify a desired set of joint velocity inputs $\dot{\theta}_d$ and get a set of actual inputs $\dot{\theta}$ which satisfies the constraints (16.5) of the robot/ground system. Second, $\dot{\theta}$ is guaranteed to be as close as possible to $\dot{\theta}_d$ in the least squares sense. Third, (16.10) gives us control over the type of motion we want: for instance, if we want a motion trajectory that minimizes the PVC joint angles $\gamma_{2,3}$ then we set $\dot{\gamma}_{2d} = \dot{\gamma}_{3d} = 0$. The actual γ values will then remain as close to 0 as the system constraints permit, given the desired steering and driving inputs.

16.3.2.3 System Degrees of Freedom

The constraint equation (16.5) and Eq. (16.8) are further useful because they provide a way to determine the number of degrees of freedom (DOFs) of the complex robot/ground system. The size of matrix Q_{C2} explicitly tells us the number of system DOFs. For example, for our system Q_{C2} is 6×3, meaning three generalized inputs and therefore three degrees of freedom. Note that this does not mean we can choose *any* three inputs from $\dot{\theta}$; we can however arbitrarily choose the generalized inputs $\dot{\theta}_g$ and therefore can make use of Eq. (16.10). Also note that the size of Q_{C2} depends on the rank of the original Jacobian matrix J_{GC}. Since J_{GC} is a function of the system configuration, its rank might change for certain singular configurations and therefore the instantaneous number of DOFs would change. In our simulations, however, we have not encountered such a situation.

16.3.3 Holonomic Constraints and Stabilization

The robot/ground system is modeled as a hybrid series-parallel mechanism. Each wheel is itself a kinematic chain between the platform and the ground, and there are three such chains in parallel. Figure 16.7 illustrates the idea. Three chains of coordinate transformations each start at frame $\{G\}$ and end at frame $\{P\}$.

The holonomic *closure constraints* (as opposed to the non-holonomic rolling constraints) for the parallel mechanism specify that each kinematic chain must end at the same frame (in this case, $\{P\}$) [17]. Let T_{AB} be the 4×4 homogeneous rigid body transform between frames A and B. Then the closure constraints for the

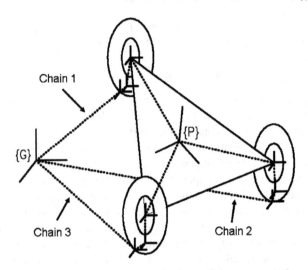

Fig. 16.7 Closure constraints: three kinematic chains in parallel must meet in frame $\{P\}$

robot are:

$$T_{GP,wheel1} = T_{GP,wheel2} = T_{GP,wheel3}. \qquad (16.11)$$

These can be interpreted as ensuring that the robot platform remains rigid. Equation (16.11) can be written as

$$
\begin{aligned}
T_{GP,wheel1} - T_{GP,wheel2} &= 0 \\
T_{GP,wheel1} - T_{GP,wheel3} &= 0,
\end{aligned}
\qquad (16.12)
$$

which are algebraic equations of the form $C(q) = 0$. To avoid having to solve a set of differential and algebraic equations (DAEs), $C(q)$ is differentiated to obtain:

$$\dot{C}(q) = \frac{\partial C}{\partial q}\,\dot{q} = J(q)\,\dot{q}. \qquad (16.13)$$

16.3.3.1 Constraint Stabilization

The Eq. (16.13) are velocity-level constraints, and during the simulation integration errors can accumulate leading to violation of the position-level constraints $C(q) = 0$. Many different algorithms have been proposed to deal with this well known issue in numerical integration of DAEs [22]. We choose a method based on the widely used Baumgarte stabilization method [23] used by Yun and Sarkar [24] because it is simple to implement, has a clear interpretation, and is effective for our simulation. In their approach, the authors suggests replacing (16.13) with:

$$J(q)\,\dot{q} + \sigma C(q) = 0, \quad \sigma > 0, \qquad (16.14)$$

which for any arbitrary initial condition C_0 has the solution of the form $C(t) = C_0 e^{-\sigma t}$. Since $\sigma > 0$, the solution converges exponentially to the desired constraints $C(q) = 0$ even if the constraints become violated at some point during the simulation. For our simulation, we found that values of σ between 1 and 10 produced good results ($\|C(q)\|_2 < 3 \times 10^{-4} \, \forall \, t$).

16.3.4 Definition of Platform Velocities

We are interested in the motion of the platform resulting from the input joint velocities $\dot{\theta}$. The time derivative of the coordinate transform relating the ground frame {G} and the platform frame {P} is

$$\dot{T}_{GP} = \begin{bmatrix} \dot{R}_{GP} & \dot{p}_{GP} \\ 0 & 0 \end{bmatrix}.$$

The linear velocity of the origin of the platform frame relative to the ground frame is $v_P = \dot{p}_{GP}$. The rotational velocity of the platform expressed in the platform frame is $\omega_P = \left(R_{GP}^T \dot{R}_{GP} \right)^{\vee}$, where the vee operator $\vee$ extracts the 3×1 vector components from the skew-symmetric matrix $R^T \dot{R}$ [17]. These output velocities are coupled in a 6×1 vector and are written as a linear combination of $\dot{\theta}$ and $\dot{\eta}$:

$$V_P = \begin{bmatrix} v_P \\ \omega_P \end{bmatrix} = \Phi_{V_P} \begin{bmatrix} \dot{\theta} \\ \dot{\eta} \end{bmatrix}. \tag{16.15}$$

These are the linear velocities of the platform frame in the global frame, and the angular velocities of the platform resolved in the local platform frame.

16.3.5 Forward Kinematics Equations

We now have all of the tools that we need to make a complete set of ordinary differential equations (ODEs) to model the robot/ground system. First, we collect the position and velocity variables of the system into vectors q and $\dot{q}$ as:

$$q = \begin{bmatrix} \theta \\ \eta \\ P_{PG} \\ \tilde{P}_c \end{bmatrix}, \quad \dot{q} = \begin{bmatrix} \dot{\theta} \\ \dot{\eta} \\ V_{PG} \\ \tilde{V}_c \end{bmatrix}, \tag{16.16}$$

where P_{PG} and $\tilde{P}_c$ are the position equivalents of V_{PG} and $\tilde{V}_c$, respectively.

Equation (16.10) relate the desired and actual joint velocities of the system. The rolling contact equation (16.4) are the non-holonomic system constraints. The stabilized holonomic constraints ensure that the wheels remain in the proper position and orientation relative to one another. The platform velocities are calculating according to (16.15). As all of these ODEs are linear in the velocity terms, they can be collectively written in the form:

$$M(q)\,\dot{q} = f(q) \qquad (16.17)$$

where

$$M(q)\,\dot{q} = \begin{bmatrix} I & 0 & 0 & 0 \\ -\Phi_{Vp1} & -\Phi_{Vp2} & I & 0 \\ J_1 & J_2 & J_3 & J_4 \\ 0 & I & 0 & -[CK] \end{bmatrix} \begin{bmatrix} \dot{\theta} \\ \dot{\eta} \\ V_{PG} \\ \tilde{V}_c \end{bmatrix},$$

$$f(q) = \begin{bmatrix} J_{in}\dot{\theta}_d \\ 0 \\ -\sigma\,C(q) \\ 0 \end{bmatrix},$$

where $\Phi_{Vp} = [\Phi_{Vp1}\ \Phi_{Vp2}]$ and $J(q) = [J_1\ J_2\ J_3\ J_4]$. The Eq. (16.17) completely describe the robot/ground system with inputs $\dot{\theta}_d$.

16.3.6 Adaptability of the Modeling Method

Our formulation is adaptable to other vehicle designs of arbitrary complexity: one simply has to create new coordinate transforms T_{GP} which reflect the geometry of the new system. All other equations will remain identical in structure to those presented here. This makes our modeling method versatile and powerful for realistic kinematic simulations of outdoor vehicles operating on rough terrains.

16.4 Results and Discussion

One of the advantages of our simulation is that it allows us to explore the motion of the wheeled mobile robot on uneven terrains of arbitrary shape. We ran the simulation on several different surfaces and for various inputs. MATLAB's ODE suite was used to solve Eq. (16.17) and the Spline Toolbox was used to generate the ground surfaces. We present results for two surfaces: a high plateau and a randomly-generated hilly terrain.

16.4.1 Descending a Steep Hill

Here we present a simulation of the three-wheeled mobile robot descending from a high plateau down a steep hill. To the authors' knowledge, this simulation, which precisely models the rolling motion of the wheels on a complex ground surface, is not possible with other existing methods. Figure 16.8 shows the three-wheeled robot on the ground surface.

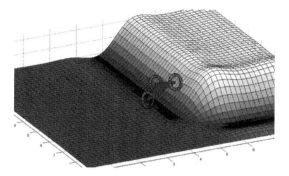

Fig. 16.8 The wheeled mobile robot on the plateau terrain

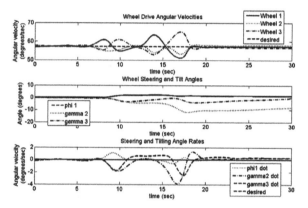

Fig. 16.9 Joint angles and rates: wheel drive rates, steering angle, and PVC angles

The simulation was run for 30 seconds with the following desired inputs:

- Steering rate $\dot{\phi}_1 = 0$
- Driving rates $\dot{\alpha}_{1,2,3} = 1\ rad/sec \approx 57.3\ deg/sec$
- PVC joint rates $\dot{\gamma}_{2,3} = 0$

Figure 16.9 plots the $\dot{\theta}$ inputs and the steering and PVC angles along with their desired values $\dot{\theta}_d$.

The platform velocities V_p are plotted in Fig. 16.10. Figure 16.11 plots the L_2 error in satisfaction of the holonomic constraint equations (16.12) and the rolling contact kinematic Eq. (16.4). Figure 16.11 shows that the constraint equations are well satisfied during the course of the simulation. This means that motion over the extreme terrain is possible with minimal wheel slip.

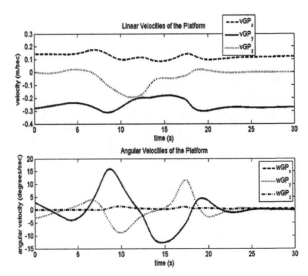

Fig. 16.10 The platform linear and angular velocities

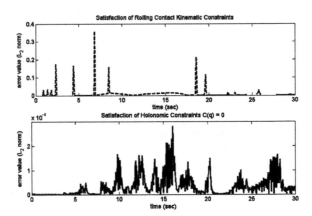

Fig. 16.11 The L_2 error in satisfaction of the holonomic and non-holonomic constraints

16.4.2 Random Terrain

Our simulation works for arbitrarily complex surfaces. Figure 16.12 shows the three-wheeled robot negotiating a randomly-generated ground surface. The inputs for this simulation were the same as for the plateau simulation in the previous section. Figure 16.13 plots the paths of the three wheel/ground contact points in the ground x-y plane. It also shows the projections of the wheel centers in that plane, to show that the wheels tilt as the robot traverses the uneven terrain. Figures 16.14 and 16.15 show the input joint velocities $\dot{\theta}$ and the output platform velocities V_p for the random terrain simulation.

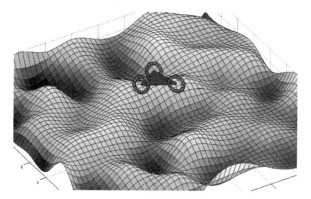

Fig. 16.12 The wheeled mobile robot on the random terrain

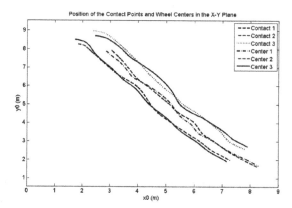

Fig. 16.13 The wheel/ground contact points and wheel centers in the xG-yG plane

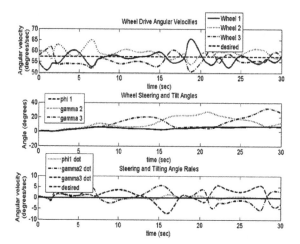

Fig. 16.14 Joint angles and rates: wheel drive rates, steering angle, and PVC angles

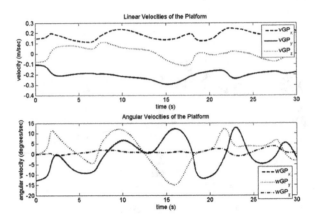

Fig. 16.15 The platform linear and angular velocities

16.5 Conclusion and Future Work

We have described a novel kinematic simulation of a three-wheeled mobile robot equipped with Passive Variable Camber (PVC) and moving on uneven terrain. PVC adds an extra degree of freedom at each wheel/platform joint, thereby allowing the wheel to tilt laterally. This extra motion allows the vehicle to better adapt to uneven terrain and reduce wheel slip, which is harmful to vehicle efficiency and performance.

Making use of concepts adapted from dextrous manipulator kinematics, our technique produces a model governed by a manageable set of linear ODEs. The resulting equations can tell us the instantaneous mobility (number of degrees of freedom) of the robot/ground system. We also showed a way of specifying joint velocity inputs which are compatible with system constraints. This could be useful on a real system in order to control the vehicle to minimize wheel slip. Our modeling technique is adaptable to vehicles of arbitrary number of wheels and joints.

Based on our simulation results, PVC has the potential to greatly improve the motion performance of wheeled mobile robots or any wheeled vehicle which moves outdoors on rough terrain by reducing wheel slip.

We are currently working on a number of extensions to this work. Under development is a way to do path planning for the PVC-equipped mobile robot. Our simulation will be used to verify that the robot can navigate from an initial position to any final configuration without wheel slip. Also, the kinematic model provides an excellent intermediate step to a full dynamic simulation. With a suitable friction model, the equations governing the wheel/ground contact kinematics can be easily extended to sliding contact. This could enhance the simulation by allowing a comparison between a vehicle with PVC and one without. PVC's effects on power consumption and localization ability will be explored in future versions of the simulation.

We are also designing and building a test set-up with a PVC-equipped wheel rolling on an uneven surface. The wheel will be instrumented to measure slip so that the efficacy of PVC for reducing wheel slip can be studied.

References

1. Alexander, J.C. and Maddocks, J.H. 1989. "On the Kinematics of Wheeled Mobile Robots". *J. Robot. Res.*, Vol. 8, No. 5, pp. 15–7.
2. Iagnemma, K., Golda, D., Spenko, M., and Dubowsky, S. 2003. "Experimental Study of High-speed Rough-terrain Mobile Robot Models for Reactive Behaviors". *Springer Tr. Adv. Robot.*, Vol. 5, pp. 654–663.
3. Lacaze, A., Murphy, K., and DelGiorno, M. 2002. "Autonomous Mobility for the DEMO III Experimental Unmanned Vehicles". *Proc. AUVSI 2002*, Orlando, FL.
4. Waldron, K. 1994. "Terrain Adaptive Vehicles". *ASME J. Mechanical Design*, Vol. 117, pp. 107–112.
5. Choi, B. and Sreenivasan, S. 1998. "Motion Planning of a Wheeled Mobile Robot with Slip-Free Motion Capability on a Smooth Uneven Surface". *Proc. 1998 IEEE ICRA*.
6. Divelbiss, A. and Wen, J. 1997. "A Path Space Approach to Nonholonomic Motion Planning in the Presence of Obstacles". *IEEE Trans. Robot. Automat.*, Vol. 13, No. 3, pp. 443–451.
7. Tarokh, M., McDermott, G., Hiyati, S., and Hung, J. 1999. "Kinematic Modeling of a High Mobility Mars Rover". *Proc. 1999 IEEE International Conference on Robotics and Automation*, May 10–15, 1999, Detroit, Michigan, USA.
8. Meghdari, A., Mahboobi, S., and Gaskarimahalle, A. 2004. "Dynamics Modeling of "Cedra" Rescue Robot on Uneven Terrains". *Proc. IMECE 2004*.
9. Tai, M. 2003. "Modeling of Wheeled Mobile Robot on Rough Terrain". *Proc. IMECE2003*.
10. Grand, C., BenAmar, F., Plumet, F., and Bidaud, P. 2004. "Decoupled Control of Posture and Trajectory of the Hybrid Wheel-Legged Robot Hylos". *Proc. 2004 IEEE International Conference on Robotics and Automation*, ICRA 2004, April 26–May 1, 2004, New Orleans, LA, USA.
11. Sreenivasan, S.V. and Nanua, P. 1999. "Kinematic Geometry of Wheeled Vehicle Systems". *Trans. ASME J. Mech. Des.*, Vol. 121, pp. 50–56.
12. Chakraborty, N. and Ghosal, A. 2004. "Kinematics of Wheeled Mobile Robots on Uneven Terrain". *Mech. Mach. Theor.* Vol. 39, No. 12, pp. 1273–1287.
13. Choi, B., Sreenivasan, S., and Davis, P. 1999. "Two Wheels Connected by an Unactuated Variable Length Axle on Uneven Ground: Kinematic Modeling and Experiments". *ASME J. of Mech. Des.*, Vol. 121, pp. 235–240.
14. Huntsberger, T., Aghazarian, H., Cheng, Y., Baumgartner, E., Tunstel, E., Leger, C., Trebi-Ollennu, A., and Schenker, P. 2002. "Rover Autonomy for Long Range Navigation and Science Data Acquisition on Planetary Surfaces". *Proc. 2002 IEEE ICRA*.
15. Lamon, P. and Siegwart, R. 2006. "3D Position Tracking in Challenging Terrain". *International Conference on Field and Service Robotics*, July 2007, Chamonix, France, STAR 25, pp. 529–540.
16. Montana, D. 1988. "The Kinematics of Contact and Grasp". *International Journal of Robotics Research*, Vol. 7, No. 3, pp. 17–32.
17. Murray, R., Li, Z., and Sastry, S. 1994. *A Mathematical Introduction to Robotic Manipulation*. CRC Press: Boca Raton, FL.
18. Montana, D. 1995. "The Kinematics of Multi-Fingered Manipulation". *IEEE Trans. Robot. Automat.*, Vol. 11, No. 4, pp. 491–503.
19. Han, L. and Trinkle, J.C. 1998. "The Instantaneous Kinematics of Manipulation". *Proc. 1998 IEEE ICRA*, pp. 1944–1949.

20. Kim, S.S. and Vanderploeg, M.J. 1986. "QR Decomposition for State Space Representation of Constrained Mechanical Dynamic Systems". *J. Mech. Trans. Automat. Des.*, Vol. 108, pp. 183–188.
21. Nash, J.C. 1990. *Compact Numerical Methods for Computers*. Second edition. Adam Hilger Publishing: Bristol.
22. Laulusa, A. and Bauchau, O.A. 2007. "Review of Classical Approaches for Constraint Enforcement in Multibody Systems". *Journal of Computational and Nonlinear Dynamics*, submitted for publication.
23. Baumgarte, J. 1983. "A New Method of Stabilization for Holonomic Constraints". *ASME Journal of Applied Mechanics*, Vol. 50, pp. 869–870.
24. Yun, X. and Sarkar, N. 1998. "Unified Formulation of Robotic Systems with Holonomic and Nonholonomic Constraints". *IEEE Trans. Robot. Automat.*, Vol. 14 No. 4, pp. 640–650.

Chapter 17
Behavior Emergence in Autonomous Robot Control by Means of Evolutionary Neural Networks*

Roman Neruda, Stanislav Slušný, and Petra Vidnerová

Abstract We study the emergence of intelligent behavior of a simple mobile robot. Robot control system is realized by mechanisms based on neural networks and evolutionary algorithms. The evolutionary algorithm is responsible for the adaptation of a neural network parameters based on the robot's performance in a simulated environment. In experiments, we demonstrate the performance of evolutionary algorithm on selected problems, namely maze exploration and discrimination of walls and cylinders. A comparison of different networks architectures is presented and discussed.

Keywords Robotics · Evolutionary algorithms · Neural networks · Behavior emergence

17.1 Introduction

One of the ultimate goals of the artificial intelligence is to develop mechanisms that would lead to autonomous intelligent agents. In the past, researchers often dealt with constrained environment (static, deterministic or fully observable world, instantaneous actions). In contrast to this traditional approach, behavior based robotics [1,2] relaxed some constraints, focusing on agents that work in dynamic, noisy and uncertain environments. However, their cognitive complexity is usually low. We believe, that evolutionary robotics can bring us more light into the problem of designing intelligent autonomous agents. It's main advantage is that it is an ideal framework for synthesizing agents whose behavior emerge from a large number of interactions

R. Neruda (✉), S. Slušný, and P. Vidnerová
Institute of Computer Science, Academy of Sciences of the Czech Republic, P.O. Box 5, 18207
Prague, Czech Republic
e-mail: roman.neruda@gmail.com

*This work was supported by the Grant Agency of the Charles University under the project number GA UK 7637/2007.

S.-I. Ao et al. (eds.), *Advances in Computational Algorithms and Data Analysis*,
Lecture Notes in Electrical Engineering 14,

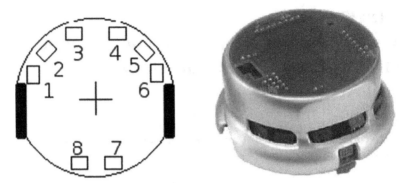

Fig. 17.1 Schema of the miniature Khepera robot

among their constituent parts [3]. It is the approach that connects robotics with two widely studied disciplines: evolutionary algorithms and neural networks.

Evolutionary robotics has been gaining increasing attention recently Fig. 17.1 shows schema of the miniature Khepera robot. The book [3] gives survey of the discipline. The straight navigation and obstacle avoidance task was solved in [4]. The example of neuro-ethological analysis can be found in works [5,6]. The authors resorted to a method traditionally employed by ethologists and neurophysiologists. The robot was put in a number of different situations while its internal variables were recorded. In our work, we present different approach, similar to extracting rules from neural network.

Till now, a lot of experiments have been done with various difficulty, ranging from box pushing robots [7] to predator and prey models [8, 9]. However, the detailed analysis of obtained results is often missing. In this work, we present some experiments with careful analysis.

In the following section we take a look at Khepera robots and related simulation software. Then, we introduce several neural network architectures, namely multilayer perceptron networks (MLP), Elman's networks (ELM), and radial basis function networks (RBF). Section 17.4 deals with evolutionary learning. Section 17.5 presents our experiments with Khepera robots. In both of them, the artificial evolution is guiding the self-organization process. In the first experiment we expect an emergence of behavior that guarantees full maze exploration. The second experiment shows the ability to train the robot to discriminate between walls and cylinders. In Section 17.6 we discuss and analyze the behaviors obtained in experiments. In the last section we draw some conclusions and present directions for our future work.

17.2 Khepera Robot

Khepera [10] is a miniature mobile robot with a diameter of 70 mm and a weight of 80 g. The robot is supported by two lateral wheels that can rotate in both directions and two rigid pivots in the front and in the back. The sensory system employs eight "active infrared light" sensors distributed around the body, six on one side and two on other side. In "active mode" these sensors emit a ray of infrared light and measure the amount of reflected light. The closer they are to a surface, the higher is the amount of infrared light measured. The Khepera sensors can detect a white paper at a maximum distance of approximately 5 cm.

In a typical setup, the controller mechanism of the robot is connected to the eight infrared sensors as input and its two outputs represent information about the left and right wheel power.

In our work, the controller mechanism is realized by a neural network. We typically consider architectures with eight input neurons, two output neurons and a single layer of neurons, mostly five or ten hidden neurons is considered in this paper.

17.3 Neural Networks

Neural networks are widely used in robotics for various reasons. They provide straightforward mapping from input signals to output signals, several levels of adaptation and they are robust to noise.

In this work we deal with three architectures of neural networks, multilayer perceptron networks, Elman's networks, and RBF networks.

A multilayer perceptron network (MLP) [11] is a one of the most widely used neural networks. It consists of several layers of *perceptrons* interconnected in a feedforward way (cf. Fig. 17.3).

The perceptron is a computational unit with n real inputs $\vec{x}$ and one real output y. It realizes the function (17.1).

$$y(\vec{x}) = \varphi \left(\sum_{i=1}^{n} w_i x_i \right) \tag{17.1}$$

$$\varphi(\xi) = 1/(1 + e^{-\xi t}) \tag{17.2}$$

$$y(\vec{x}) = \varphi \left(\sum_{i=1}^{n} w_i x_i \right) \tag{17.3}$$

$$s(\vec{x}^t) = \vec{x}^{t-1} \tag{17.4}$$

$$y(\vec{x}) = \varphi \left(\frac{\| \vec{x} - \vec{c} \|}{b} \right) \tag{17.5}$$

$$f_s(\vec{x}) = \sum_{j=1}^{h} w_{js} \varphi \left(\frac{\| \vec{x} - \vec{c}_j \|}{b_j} \right) \tag{17.6}$$

In contrast to MLP, the Elman's network [12] contains recurrent connections. For each hidden neuron there is a context neuron that holds a copy of the corresponding neuron activation at the previous time step. So the hidden neurons get inputs both from the input layer and the context layer that records the previous states (cf. Fig. 17.2).

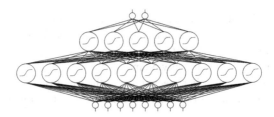

Fig. 17.2 Multilayer Perceptron Network (MLP). The neuron outputs are evaluated according to Eq. (17.1), x_i are the inputs of the neurons, w_i are synaptic weights, and φ is an activation function. One of the most common activation function is the logistic sigmoid function (17.2)

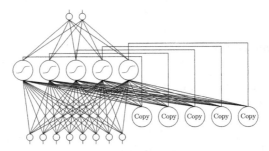

Fig. 17.3 Elman's Network. The neuron outputs are evaluated according to Eq. (17.3), the context neurons output their input values from the previous steps

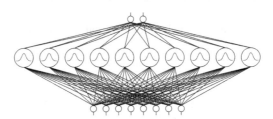

Fig. 17.4 RBF Network. Eq. (17.5) evaluates the output of a hidden RBF unit, f_s (17.6) is the output of the s-th output unit. φ is an activation function

An RBF neural network represents a relatively new neural network architecture [11]. It is a feed-forward neural network with one hidden layer of *RBF units* and linear output layer (cf. Figs. 17.3 and 17.4). In contrast to perceptrons the RBF unit realizes local radial function (17.5), such as Gaussian. The network output is given by Eq. (17.6).

17.4 Evolutionary Network Learning

The evolutionary algorithms (EA) [13, 14] represent a stochastic search technique used to find approximate solutions to optimization and search problems. They use techniques inspired by evolutionary biology such as mutation, selection, and crossover. The EA typically works with a population of *individuals* representing abstract encoding of feasible solutions. Each individual is assigned a *fitness* that is a measure of how good solution it represents. The better the solution is, the higher the fitness value it gets. The main advantage of EA is that – as a so-called weak method – it does not require auxiliary information or constrains in order to solve a given task (unlike e.g. gradient based neural learning methods that require collection of I/O examples and error function derivative). The only information needed is the performance evaluation of each individual.

In the course of EA run, the population evolves toward better solutions. The evolution starts from a population of random individuals, and it iterates in successive generations. In each generation, the fitness of each individual is evaluated. Individuals are stochastically selected from the current population based on their fitness, and modified by means of evolutionary operators. The most common recombination operators are *mutation* and *crossover*, they are usually encoding dependent, or even problem dependent. The new population is then used in the next iteration of the algorithm.

The above described neural networks used as robot controllers are encoded into individuals in order to use them in the evolutionary algorithm. While the architecture of the network (number of units and connections) is typically fixed, values of network weights are searched for by evolution.

In case of MLP and Elman's networks all weights undergo the adaptation. The RBF networks learning was motivated by the commonly used three-step learning [15]. Parameters of RBF network are divided into three groups: centers, widths of the hidden units, and output weights. Each group is then trained separately. The centers of hidden units are found by clustering (k-means algorithm) and the widths are fixed so as the areas of importance belonging to individual units cover the whole input space. Finally, the output weights are found by EA. The advantage of such approach is the lower number of parameters to be optimized by EA, i.e. smaller length of individual.

Unlike in more traditional genetic algorithm approaches, the individual is represented as a floating-point encoded vector of real parameters. Corresponding evolutionary operators for this case represent the uniform crossover the arithmetic crossover, and additive mutation. The rate of these operators is quite big, ensuring the exploration capabilities of the evolutionary learning. A standard roulette-wheel selection is used together with a small elitist rate parameter. Detailed discussion about the fitness function is presented in the next section with the respective experiments.

17.5 Experiments

Although evolution on real robots is feasible, serial evaluation of individuals on a single physical robot might require quite a long time. One of the widely used simulation software (for Khepera robots) is the Yaks simulator [16], which is freely

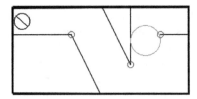

Fig. 17.5 In the maze exploration task, agent is rewarded for passing through the zone, which can not be sensed. The zone is drawn as the bigger circle, the smaller circle represents the Khepera robot. The training environment is of 60×30 cm

available. Simulation consists of predefined number of discrete steps, each single step corresponds to 100 ms.

To evaluate the individual, simulation is launched several times. Individual runs are called "trials". In each trial, neural network is constructed from the chromosome, environment is initialized and the robot is put into randomly chosen starting location. The inputs of neural networks are interconnected with robot's sensors and outputs with robot's motors. The robot is then left to "live" in the simulated environment for some (fixed) time period, fully controlled by neural network. As soon as the robot hits the wall or obstacle, simulation is stopped. Depending on how well the robot is performing, the individual is evaluated by value, which we call "trial score". The higher the trial score, the more successful robot in executing the task in a particular trial. The fitness value is then obtained by summing up all trial scores.

In this experiment, the agent is put in the 60×30 cm maze (cf. Fig. 17.5). The agent's task is to fully explore the maze. Fitness evaluation consists of four trials, individual trials differ by agent's starting location. Agent is left to live in the environment for 250 simulation steps.

The three-component $T_{k,j}$ motivates agent to learn to move and to avoid obstacles:

$$T_{k,j} = V_{k,j}(1 - \sqrt{\Delta V_{k,j}})(1 - i_{k,j}). \tag{17.7}$$

First component $V_{k,j}$ is computed by summing absolute values of motor speed in the k-th simulation step and j-th trial, generating value between 0 and 1. The second component $(1 - \sqrt{\Delta V_{k,j}})$ encourages the two wheels to rotate in the same direction. The last component $(1 - i_{k,j})$ encourage obstacle avoidance. The value $i_{k,j}$ of the most active sensor in k-th simulation step and j-th trial provides a conservative measure of how close the robot is to an object. The closer it is to an object, the higher is the measured value in range from 0.0 to 1.0. Thus, $T_{k,j}$ is in range from 0.0 to 1.0, too.

In the j-th trial, score S_j is computed by summing normalized trial gains $T_{k,j}$ in each simulation step:

$$S_j = \sum_{k=1}^{250} \frac{T_{k,j}}{250} \tag{17.8}$$

To stimulate maze exploration, agent is rewarded, when it passes through the zone. The zone is randomly located area, which can not be sensed by an agent. Therefore, Δ_j is 1, if agent passed through the zone in j-th trial and 0 otherwise. The fitness value is then computed as follows:

$$Fitness = \sum_{j=1}^{4} (S_j + \Delta_j) \tag{17.9}$$

Successful individuals, which pass through the zone in each trial, will have fitness value in range from 4.0 to 5.0. The fractional part of the fitness value reflects the speed of the agent and its ability to avoid obstacles.

All the networks included in the tests were able to learn the task of finding a random zone from all four positions. The resulting best fitness values (cf. Table 17.1) are all in the range of 4.3–4.4 and they differ only in the order of few percent. It can be seen that the MLP networks perform slightly better, RBF networks are in the middle, while recurrent networks are a bit worse in terms of the best fitness achieved. According to their general performance, which takes into account ten different EA runs, the situation changes slightly. In general, the networks can be divided into two categories. The first one represents networks that performed well in each experiment in a consistent manner, i.e. every run of the evolutionary algorithm out of the ten random populations ended in finding a successful network that was able to find the zone from each trial. MLP networks and recurrent networks with five units fall into this group. The second group has in fact a smaller trial rate because, typically, one out of ten runs of EA did not produced the optimal solution. The observance of average and standard deviation values in Table 17.1 shows this clearly. This might still be caused by the less-efficient EA performance for RBF and Elman networks.

The important thing is to test the quality of the obtained solution in a different arena, where a bigger maze is utilized (cf. Figs. 17.6, 17.7, and 17.8). Each of the architectures is capable of efficient space exploration behavior that has emerged during the learning to find random zone positions. The above mentioned figure shows that the robot trained in a quite simple arena and endowed by relatively small network of 5–10 units is capable to navigate in a very complex environment.

Table 17.1 Comparison of the fitness values achieved by different types of network in the experiments

Network type	Maze exploration				Wall and cylinder			
	Mean	Std	Min	Max	Mean	Std	Min	Max
MLP 5 units	4.29	0.08	4.20	4.44	2326.1	57.8	2185.5	2390.0
MLP 10 units	4.32	0.07	4.24	4.46	2331.4	86.6	2089.0	2391.5
ELM 5 units	4.24	0.06	4.14	4.33	2250.8	147.7	1954.5	2382.5
ELM 10 units	3.97	0.70	2.24	4.34	2027.8	204.3	1609.5	2301.5
RBF 5 units	3.98	0.90	1.42	4.36	1986.6	230.2	1604.0	2343.0
RBF 10 units	4.00	0.97	1.23	4.38	2079.4	94.5	2077.5	2359.5

Fig. 17.6 The agent is put in the bigger maze of 100 × 100 cm. Agent's strategy is to follow wall on it's left side

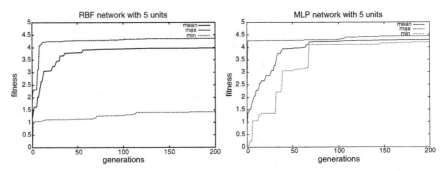

Fig. 17.7 Plot of fitness curves in consecutive populations (maximal, minimal, and average individual) for a typical EA run (one of ten) training the RBF network with five units (left) and MLP network with five hidden units (right)

Following experiment is based on the work [5, 17]. The task is to discriminate between the sensory patterns produced by the walls and small cylinders. As noted in [3], passive networks (i.e. networks which are passively exposed to a set of sensory patterns without being able to interact with the external environment through motor action), are mostly unable to discriminate between different objects. However, this problem can easily be solved by agents that are left free to move in the environment.

The agent is allowed to sense the world by only six frontal infrared sensors, which provide it with only limited information about environment. Fitness evaluation consists of five trials, individual trials differ by agent's starting location. Agent is left to live in the environment for 500 simulation steps. In each simulation step, trial score is increased by 1, if robot is near the cylinder, or 0.5, if robot is near the

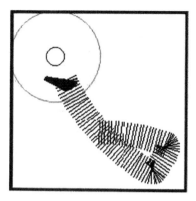

Fig. 17.8 Trajectory of an agent doing the walls and cylinders task. The small circle represents the searched target cylinder. The agent is rewarded in the zone represented by a bigger circle. It is able to discriminate between wall and cylinder, and after discovering the cylinder it stays in it's vicinity

wall. The fitness value is then obtained by summing up all trial scores. Environment is the arena of 40×40 cm surrounded by walls.

It may seem surprising that even this more complicated task was solved quite easily by relatively simple network architectures. The images of walls and cylinders are overlapping a lot in the input space determined by the sensors.

The results in terms of best individuals are again quite comparable for different architectures with reasonable network sizes. The differences are more pronounced than in the case of the previous task though. Again, the MLP is the overall winner mainly when considering the overall performance averaged over ten runs of EA. The behavior of EA for Elman and RBF networks was less consistent, there were again several runs that obviously got stuck in local extrema (cf. Table 17.1).

We should emphasize the difference between fitness functions in both experiment. The fitness function used in the first experiment rewards robot for single actions, whereas in the this experiment, we describe only desired behavior.

All network architectures produced similar behavior. Robot was exploring the environment by doing arc movements and after discovering target, it started to move there and back and remained in its vicinity.

17.6 Analysis of Successful Behaviors

Several behavioral patterns have been observed for the successful controllers. The most successful individuals exhibited a wall-following behavior which is under circumstances a successful strategy to explore a general maze. Depending on the initial position and orientation, either left wall or right wall following controllers have evolved.

In order to gain insight into the function of a controller, we have studied the partial I/O mappings from individual sensors to the motor control. We have chosen a typical left wall follower, a right wall follower, and an agent that exhibited general obstacle avoidance behavior without the maze exploration strategy.

For the wall following agents we expected to observe some kind of symmetry. First, note that for the left-wall-following strategy the most important sensors are sensors on the left side (i.e. 2 and 3), while for the right-wall-following strategy they are 4 and 5. If the maze has wide enough corridors, the sensor 6 (or 1, resp.) should not get many inputs at all. The back sensors 7–8 should reflect the situation that when they register a wall, it means that the left wall follower is turning right and the right wall follower is turning left.

Figures 17.9 and 17.10 show how the wheel speeds depend on left sensors for the left-wall follower and on the right sensors for the right-wall follower. The comparison of sensors 2 and 3 of left-wall follower with sensors 4 and 5 of right-wall follower shows that in case of an obstacle in front-left (front-right, resp.) each agent turns in the opposite direction, i.e. right (or left, resp.), thus following its strategy.

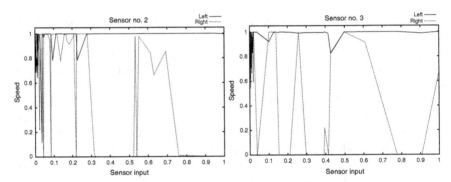

Fig. 17.9 Plot of I/O mappings from individual sensors to the motor control for the left-wall following agent

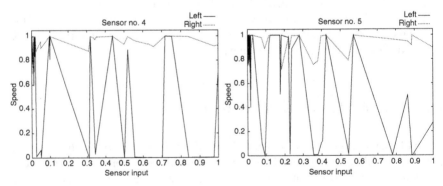

Fig. 17.10 Plot of I/O mappings from individual sensors to the motor control for the right-wall following agent

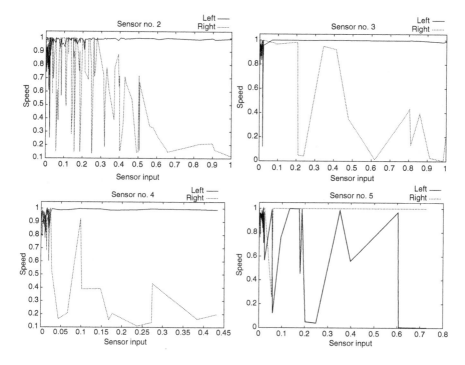

Fig. 17.11 Plot of sensor functions for the obstacle avoiding agent

Figure 17.11 shows the sensors 2–5 of obstacle avoiding agent. One can see a symmetric behavior, mainly comparing sensors 2 to 4 with sensor 5. When the obstacle is in front-left direction (sensor 2 is active), the agent has a tendency to power left engine more, i.e. it turns right from the obstacle. Inversely, the responses to sensor 5 mean powering the right wheel, i.e. turning left.

17.7 Conclusions

The main goal of this paper was to demonstrate the ability of neural networks trained by evolutionary algorithm to achieve non-trivial robotic tasks. There have been two experiments carried out with three types of neural networks and different number of units.

For the maze exploration experiment the results are encouraging, a neural network of any of the three types is able to develop the exploration behavior. The trained network is able to control the robot in the previously unseen environment. Typical behavioral patterns, like following the right wall have been developed, which in turn resulted in the very efficient exploration of an unknown maze. The best results achieved by any of the network architectures are quite comparable, with simpler perceptron networks (such as the five-hidden unit perceptron) marginally outperforming Elman and RBF networks.

In the second experiment it has been demonstrated that the above mentioned approach is able to take advantage of the embodied nature of agents in order to tell walls from cylindrical targets. Due to the sensor limitations of the agent, this task requires a synchronized use of a suitable position change and simple pattern recognition. This problem is obviously more difficult than the maze exploration, nevertheless, most of the neural architectures were able to locate and identify the round target regardless of its position.

The results reported above represent just a few steps in the journey toward more autonomous and adaptive robotic agents. The robots are able to learn simple behavior by evolutionary algorithm only by rewarding the good ones, and without explicitly specifying particular actions. The next step is to extend this approach for more complicated actions and compound behaviors. This can be probably realized by incremental learning one network a sequence of several tasks. Another – maybe a more promising approach – is to try to build a higher level architecture (like a type of the Brooks subsumption architecture [2]) which would have a control over switching simpler tasks realized by specialized networks. Ideally, this higher control structure is also evolved adaptively without the need to explicitly hardwire it in advance. The last direction of our future work is the extension of this methodology to the field of collective behavior [18–21].

References

1. Arking, R.C. (1998). *Behavior-Based Robotics*. Cambridge, MA:MIT Press.
2. Brooks, R.A. (1986). A robust layered control system for a mobile robot. *IEEE Journal of Robotics and Automation* 12(1), 14–23.
3. Nolfi, S. and Floreano, D. (2000). *Evolutionary Robotics – The Biology, Intelligence and Techology of Self-Organizing Machines.* The MIT Press.
4. Floreano, D. and Mondada, F. (1996). Automatic creation of an autonomous agent: Genetic evolution of neural-network drive robot. *From Animals to Animats 3: Proceedings of Third Conference on Simulation of Adaptive Behavior.*
5. Nolfi, S. (1996). Adaptation as a more powerful tool than decomposition and integration. *Proceedings of the Workshop on Evolutionary Computing and Machine Learning.*
6. Nolfi, S. and Tani, J. (1997). Learning to adapt to changing environments in evolving neural networks. *Adaptive Behavior* 5, 99–105.
7. Sprinkhuizen-Kuyper, I.G. (2001). Artificial evolution of box-pushing behaviour. Technical report, Universiteit Maastricht.
8. Cliff, D. and Miller, G.F. (1995). Tracking the red queen: Measurements of adaptive progress in co-evolutionary simulations. *Advances in Artificial Life: Proceedings of the Third European Conference on Artificial Life.*
9. Koza, J.R. (1991). Evolution and co-evolution of computer programs to control independently-acting agents. *From Animals to Animats: Proceedings of the First International Conference on Simulation of Adaptive Behavior.* February 1991, Paris, France.
10. K-Team corporation (2007). Khepera II documentation. http://k-team.com.
11. Haykin, S. (1998). *Neural Networks: A Comprehensive Foundation.* Englewood cliffs, NJ: Prentice Hall, 2nd edition. NJ, USA.

12. Elman, J. (1990). Finding structure in time. *Cognitive Science* 14, 179–214.
13. Fogel, D.B. (1998). *Evolutionary Computation: The Fossil Record.* MIT-IEEE Press.
14. Holland, J. (1992). *Adaptation in Natural and Artificial Systems.* MIT Press, reprinted edition.
15. Neruda, R. and Kudová, P. (2005). Learning methods for radial basis functions networks. *Future Generation Computer Systems*, 21, 1131–1142.
16. Carlsson, J. and Ziemke, T. (2001). Yaks – yet another khepera simulator. *Ruckert, S., Witkowski (Eds.), Autonomous Minirobots for Research and Entertainment Proceedings of the Fifth International Heinz Nixdorf Symposium.*
17. Nolfi, S. (1999). The power and limits of reactive agents. *Technical Report. Rome, Institute of Psychology, National Research Council.*
18. Baldassarre, G., Nolfi, S., and Parisi, D. (2002). Evolving mobile robots able to display collective behaviour. *Hemelrijk C. K (ed.), International Workshop on Self-Organisation and Evolution of Social Behaviour*, pp. 11–22. Zurich, Switzerland.
19. Baldassarre, G., Nolfi, S., and Parisi, D. (2006). Coordination of simulated robots based on self-organisation. *Artificial Life* 12(3), 289–311.
20. Martinoli, A. (1999). *Swarm Intelligence in Autonomous Collective Robotics: From Tools to the Analysis and Synthesis of Distributed Control Strategies.* Lausanne: Computer Science Department, EPFL.
21. Quinn, M., Smith, L., Mayley, G., and Husbands, P. (2002). Evolving teamwork and role-allocation with real robots. *Artificial Life* 8, 302–311.

Chapter 18
Swarm Economics

Sanza Kazadi and John Lee

Abstract The Hamiltonian Method of Swarm Design is applied to the design of an agent based economic system. The method allows the design of a system from the global behaviors to the agent behaviors, with a guarantee that once certain derived agent-level conditions are satisfied, the system behavior becomes the desired behavior. Conditions which must be satisfied by consumer agents in order to bring forth the "invisible hand of the market" are derived and demonstrated in simulation. A discussion of how this method might be extended to other economic systems and non-economic systems is presented.

Keywords Swarm engineering · Hamiltonian method of swarm design · Swarm economics

18.1 Introduction

Economic systems are inherently difficult to predict and direct due in large part to the nonlinear nature of the system. Like most complex systems, small variations in the activity of a single behavior or characteristic of any the parts of the system can have very large effects in the whole. As a result, much of economic theory intended to explain what people are doing at the micro and macro levels is incapable of explaining much of what happens in economic systems.

The fact that economic systems are hard to understand can easily be seen in the patterns of the economy, as of today, in 2008. The economy both in the United States and in the world is undergoing huge fluctuations with causes that have their basis

S. Kazadi (✉) and J. Lee
Jisan Research Institute, 515 S. Palm Ave., Unit 3, Alhambra, California 91803, USA
e-mail: skazadi@jisan.org

S.-I. Ao et al. (eds.), *Advances in Computational Algorithms and Data Analysis*,
Lecture Notes in Electrical Engineering 14,
© Springer Science+Business Media B.V. 2009

in human behavior. The current correction in the US housing market may be traced back to a great deal of enthusiasm in the development of the housing market during the early 2000s. The global system behavior is a result of millions of individual decisions by consumers, lenders, and governments. However, predicting that these decisions would cause such wide-ranging problems has been problematic at best, and disastrous at worst.

Economies are complex systems which encompass micro and macro behaviors, individual interaction, equilibriums, and, in most cases, some sense of self-regulation. Because of this overwhelming complexity, a quantitative form of economics has been difficult to observe. However, with more powerful computational power and the development of efficient control algorithms it is now possible to approach economics from a more quantitative, rather than theoretical, perspective. The use of these computational tools allows economists to examine many aspects of economic systems with remarkable flexibility and detail. Both local and global aspects of the economic system may be explored without the need for large data-collection enterprises or extensive human interaction. An accurate simulation can be run solely by itself and may be used to extensively examine basic economics laws and theories governing the physics of interaction of agents. An advantage of this method is the lack of the ceteris paribus[1] aspect of traditional economics. Observations qualified by ceteris paribus require that all other variables in a causal relationship are ruled out in order to simplify studies. Most economic systems are not simple enough to hold all other things constant, and examining them as though individual elements of the system can be isolated in order to be understood undermines an understanding of the true economic system. Computational studies do not require this kind of limitation, and the system may be examined in its full complexity.

Another salient advantage is an observer's ability to control the basic structure of interaction. Before a run, the simulation allows one to tinker with basic parameters of the system, such as sizes of budgets, rate of utility increase, and the magnitude of competition. By allowing such control, a user can predict results of economies in several types of real-life situations, which is key in understanding the scope of economic systems and the realistic range of our control.

Unlike traditional economic research tools, recent work in agent-based economic studies has begun to create mechanisms by which economic scenarios may be examined. These simulation tools allow millions or billions of interactions by simulated economic agents in relatively small amounts of time. By changing how these interactions work, the researcher can examine the short and long term effects of a multitude of interactions between agents and create an understanding of how these differing interactions cause varied global consequences.

Despite the power of the computational method of economic study, most computational studies have been limited in the sense that they've been more or less observational. That is, agents have been designed and the outcomes of repeated interactions have emerged over many iterations. The design has been changed and

---

[1] (Latin for "all other things unchanged").

new observations made. This tends to give an idea of how changes in agent behaviors cause global changes in the system. What it doesn't address directly is how one might force a global property to emerge from the interactions of agents. As a result, economists have tried to understand what occurs in economies in order to develop predictive models. Economists have not generally developed requirements for global effects to emerge, designed agents within the system based on these requirements, and validated these requirements using computational models.

Complex system design is a challenging field of science in which some to many independent interacting parts are combined so as to create a machine or system with a particular desired function or property set. A subset of the general field of complex systems is swarms, which are groups of bidirectionally communicating autonomous agents. Swarms are interesting for a number of reasons, the most important of which is the tendency of swarms to exhibit emergence, which allows them to undertake actions that are not explicitly part of the control algorithm. The most challenging thing in complex system design is ensuring that the different parts will interact with each other in a such a way as to generate a desired system behavior. This is particularly true for systems of autonomous agents. Since each agent is independent, the interactions can be very difficult to predict, a priori.

In parallel to developing computational economic models, a new field called swarm engineering has been emerging over the past decade. This field is a subfield of engineering in which swarms are designed around global goals which have been determined prior to the swarm's design phase. The individual agents' behaviors can be shown theoretically to lead to the swarm's global behaviors, giving the swarm's design a much more robust flavor.

In the swarm literature, there is little in the way of generally applicable principled approach to swarm design. Some researchers have built preliminary systems for monitoring or understanding the emergent behaviors of agents. However, these studies do not yet generalize to a methodology that works for a large number of swarm systems. As a result, no particular method exists for generating swarms of particular design.

In this chapter, we examine what we call the Hamiltonian Method of Swarm Design (HMOSD). This method is a principled approach to swarm design consisting of two main phases. In the first phase, the global goal(s) is(are) written in terms of properties that can be sensed and affected by the agents. The resulting equation(s) can then be used to develop requirements for the behaviors of the agents that lead to the global goal. The second phase consists of creating behaviors that satisfy these swarm requirements provably. Once these have been created, it can be asserted that the resulting swarm will have the desired global goals.

Though swarm engineering has typically been applied to robotic design and computation design, we broaden the scope here by applying it to an economic system. Real economic systems are systems of autonomous agents with bidirectional communication, satisfying a broad definition of a swarm. Thus, it stands to reason that swarm engineering techniques might be able to be applied to such a system so as to generate a predefined global behavior of the system. Many studies have been made which use agent-based simulations in which interactions between agents define what the economy will do. However, though these studies extracted global behaviors from their systems, they did not develop or apply a method of generating the global behavior, and then designing the system around that behavior. This study, which might be termed a study in swarm economics, is meant to examine the design phase of an economic system using the swarm engineering methodology.

The remainder of the paper is organized as follows. Section 18.2 examines the theoretical application of the HMOSD to a simple economic model. This section focuses on the properties of the agents that will give the economy a particular behavior. Section 18.3 presents the performance of the model under different expected agent behaviors. Section 18.4 offers some discussion and concluding remarks.

18.2 Swarm Engineering Basics

In this section, we give an overview of swarm engineering theory. We begin with a set of definitions that clarify and make rigorous some of the concepts behind swarm engineering. We continue with a theoretical description of the steps behind swarm design and proof of design efficacy.

18.2.1 Definitions

We assume that a system can be thought of as a closed set of objects together with a set of consistent dynamic properties. These properties need not have closed form expressions, but we assume that they are consistent in the sense that measurements or combinations of measurements cannot produce differing numerical values for any measurable quantity. Because the system is closed, the objects are not affected by anything outside of the system. As a result, in simulations involving an outside controller of an agent in the simulation, the controller and everything that affects it must be viewed as part of the system.

We define a property of the system to be a characteristic of the system that can be measured using a process that is independent of the characteristic. In what follows, we'll represent a system's property as P_i where the subscript i serves to identify the property. As an example, the temperature of a processor may be measured using the radiative emmissions of the processor, even though the measurement cannot affect the processor's temperature.[2]

We define an agent to be a situated subset of the system that exhibits autonomous control over at least one degree of freedom in the system. Autonomous control is control which does not exhibit a direct dependence on any part of the system other than the controlling element(s); the behavior of the agent also must not be attributable to the dynamic interactive equations that define the system.

---

[2] The radiative emmissions of the measuring device are likely to be much less important in determining the temperature of the device than internal processes. Thus, the effect of these emmissions is assumed to be negligible.

An autonomous agent is an agent that acts without the direct control of any outside influence. This means that outside of the things that it can sense, no part of the outside world affects any part of the agent's controller. While the agent can be affected by other things that it can sense, the effect of the senses is expected to be independent of the cause of the sensory input to its controller. Anything failing to meet this metric cannot be thought of as autonomous.

We may quantify this idea. Let the controller of an agent be defined by the way in which the agent responds to its memory state M and its sensory state S_s. Then, given any outside property of the system P_S, it is true that if the current state of the agent is given by S_a then

$$\frac{dS_a}{dt} = \frac{\partial S_a}{\partial M}\frac{dM}{dt} + \frac{\partial S_a}{\partial S_s}\frac{dS_s}{dt} + \frac{\partial S_a}{\partial P_S}\frac{dP_S}{dt}. \qquad (18.1)$$

That the final term is zero for all outside properties is a necessary and sufficient condition for autonomy. Now, this does not say that $\frac{dS_s}{dP_S}$ is zero. It simply means that the only way that this property may enter the controller is through the senses.

We define the behavior of a subset of the system to be the way in which properties of the subsystem change in time. I.e., if P_i is a property, then a behavior b_i is defined by

$$b_i = \frac{dP_i}{dt}. \qquad (18.2)$$

Behaviors often involve the interplay between more than one property. In this case, we require a formalism for describing such behaviors. Let us suppose that a system is made up of elements whose behavior is defined in terms of measurables $A = \{P_1,\ldots,P_n\}$. Then this system can have a behavior which is composed of all of the behaviors of the different measurables. That is,

$$\overrightarrow{b_A} = \left(\frac{dP_1}{dt},\ldots,\frac{dP_n}{dt}\right). \qquad (18.3)$$

These properties can be most easily thought of as composite properties of many agents or objects in the system. For instance, a star has a discernable size which is defined as a combination of the positional properties of the atoms making it up. Any single element of the system would be insufficient to describe the system. Thus, the property must be described in terms of the properties of all (or at least many) of the atoms in the system.

In many physical systems, there are properties that are derivations of other properties. These properties are not basic in the sense that they do not depend on dynamics of other properties. As an example, consider a point mass in our universe. We may define its position in terms of a variable $\overrightarrow{x}$. However, another property, the velocity $\overrightarrow{v}$, is a derivative property whose relationship with the basic positional property is given by

$$\overrightarrow{v} = \dot{\overrightarrow{x}}. \qquad (18.4)$$

It is possible to measure this property of the object, and so it is indeed a property of the system, as well as a behavior of the object. This duality of behavior and property can be resolved only by noting that behaviors are linked to properties, but the behaviors can only become properties if they, in fact, can be independently measured.

Emergence has been identified by many authors in the past in terms capturing the general idea that a system can have unintended global properties that are not explicitly built into its agents. The interest in swarm based systems seems to have come from this single observation. We now propose a rigorous definition of this property.

Suppose that we have a property P_j that is a function of another properties and behaviors of the system. That is, suppose that

$$P_j = f\left(b_1,\ldots,b_{n_b},P_1,\ldots,P_{j-1},P_{j+1},\ldots,P_{n_P}\right),\tag{18.5}$$

where n_b is the number of systems behaviors, and n_P is the number of systems properties. The number of behaviors is not necessarily equal to the number of properties of the system. The property P_j is an *emergent property* of the subsystem i if

$$\frac{\partial b_i}{\partial P_j} = 0.\tag{18.6}$$

That is, the property P_j is not a factor in the defining function of behavior b_i for any of the behaviors of the elements of the system. This means that the agent or agents in the system are acting independently of the property, and so the property is not a deliberate result of the design of the agent's behaviors. As a result, it cannot be viewed as part of the design of the agent(s), and so it satisfies the meaning of emergence.

Given the distinction between behaviors and properties above, we can also define emergent behaviors to be emergent properties that are themselves behaviors.

These definitions may be used to formally define various types of swarms of agents. Firstly, we define a swarm of agents to be a set of interacting agents within a system in which one agent's change in state can be perceived by at least one other agent and effects the state of the agent perceiving the change. Moreover, the subset must have the property that every agent has at least one state whose change will initiate a continual set of state changes that affects every other agent in the swarm.

Let us more rigorously define a swarm. Suppose that the state of the agents is specified by a set of varibles $\{S_i\}$. Then the set of agents is a swarm if

$$\frac{\partial S_i|_{(t>t_0)}}{\partial S_j|_{(t=t_0)}} \neq 0\tag{18.7}$$

[3] $\forall i \neq j$ for times t after some reference time t_0. That is, that the later states of agent i must depend on the current state of agent j.

---

[3] This can be rigorously defined as follows:

$$\frac{\partial S_i|_{(t>t_0)}}{\partial S_j|_{(t=t_0)}} = \lim_{\delta S_j \to 0} \frac{S_i(t,S_j+\delta S_j) - S_i(t,S_j)}{\delta S_j} \neq 0\tag{18.33}$$

for any time $t > t_0$.

Our definition of a swarm differs from others given in the literature in that it does not demand emergence from the system. However, emergent swarms are also interesting, and form the basis for most of the work in swarm engineering. Thus, we define a swarm of agents as an emergent swarm of agents with respect to property P_j if they exhibit an emergent behavior b_{P_j}. Note that this means that a swarm is defined only in terms of a specific property which yields the potential possibility that the group of agents is not a swarm with respect to another property P_k.

One of the unexpected results of this definition is that it does not exclude the potentiality of a centrally controlled swarm. The idea behind the swarm is that each element of the swarm is capable of initiating a cascade of state changes. How these are initiated is not important, and we can leave the possibility open that these go through a central controller, group of agents, or communication mechanism. Thus, we clarify these issues by defining a decentralized swarm to be a swarm that does not have a central communication or control mechanism. A centralized swarm is a swarm which is not decentralized.

The power of these definitions is that it is possible to test a set of agents in order to determine whether or not it is a swarm, if it is a centralized or decentralized swarm, and then whether or not it is an emergent swarm with respect to a specific property. For instance, it should be clear that a soccer team is a swarm, but it is not an emergent swarm with respect to, for instance, the team dispersion. Team members are very likely to use this information to affect their own behaviors. On the other hand, a swarm of ants is an emergent swarm with respect to food source exploitation, as it has the ability to exploit nearby food sources despite the absolute lack of knowledge on the part of the ants. This can be characterized by measuring the amount of exploitation of each food source when multiple food sources are available. Clearly this quantity is not part of the control algorithm of the agents.

18.2.2 Swarm Engineering

Swarms are difficult to engineer primarily because groups of independent interacting agents can exhibit very complex and unexpected behaviors for a very large number of different reasons [1–28]. Moreover, if the members of a group have specifications that are made independently, it is very difficult to guarantee that the specifications do not interact in an unexpected way. Moreover, proving that the interactions between the various agents have the desired outcome often requires the complete simulation of the group of agents. Finally, small perturbations to the system, which cause rather small changes in the behaviors of individual agents, can cause very large changes in the overall behavior of the system. This is, in fact, a foundational characteristic of the field of chaos.

It is important to create a new methodology for the generation of global behavior in a way that bypasses the difficulties presented here. We seek a method that is provable in the sense that the behaviors can be understood to generate the desired global behavior. The generated behaviors have well understood tolerances for perturbations within which the desired global behavior will still occur.

18.2.2.1 Swarm Engineering Equations

In this section, we explore differential equations which are relevant to swarm design. This method will assist in the determination of several things relevant to the overall global goal. The first thing needed is a set of different behaviors (also called castes) for the agents. The second thing is the set of sensors and actuators, with well-defined resolutions. Sensors with higher resolution (in the sense that they can measure the desired property with higher accuracy) can be used to affect the number of castes, as new ones can emerge at different times during the entire group action. This is the entire top-down portion of the design process.

Once the castes have been properly designed, the next step is to work out the specific sensory and actuation capabilities of the agents. This step consists of determining the actual hardware (either physical or virtual) that the agents will have, their computational capabilities, communication capabilities, etc. This hardware must make it possible for the agent to have the sensory and computational abilities determined in the top down part of the swarm design. Most notably, the agent must have the ability to determine what part of the phase space path it is on in order to properly determine its behavior (caste). The behaviors must also be developed at this step. If the behaviors have the ability to move the agent along the proper section of the path through phase space during the appropriate caste behavior, the global goal should be achieved.

18.2.2.2 Top Down

As our starting point we choose the global goal. It is described in terms of a set of properties of the swarm $G = \{P_1, \ldots, P_i, \ldots, P_{n_P}\}$ and their corresponding initial and final characteristics $G^0 = \{P_1^0, \ldots, P_i^0, \ldots, P_{n_P}^0\}$ and $G^F = \{P_1^F, \ldots, P_i^F, \ldots, P_{n_P}^F\}$. The initial and final characteristics may be numerical values as in a count-based characteristic or they may be functional, as in a trajectory. They may also be sets of potential initial or final states of the two forms.

Once these initial and final conditions have been determined, it is important to specify conditions under which the final characteristics become consequences of the initial conditions and the system dynamics.

Assume that function f from (18.5) is a differentiable function of the properties P_i, $i = 1, \ldots, n_p$ and the behaviors b_i, $i = 1, \ldots, n_b$. Then, in general case the following holds

$$b_j = \frac{dP_j}{dt} = \sum_{i=1}^{n_b} \frac{\partial P_j}{\partial b_i} \frac{db_i}{dt} + \sum_{i \neq j}^{n_p} \frac{\partial P_j}{\partial P_i} b_i. \tag{18.8}$$

For simplicity we assume that each property correspond with only one behavior, i.e., $n_P = n_b$, then

$$b_j = \frac{dP_j}{dt} = \sum_{\substack{i \neq j}}^{n_b} \left(\frac{\partial P_j}{\partial b_i} \frac{db_i}{dt} + \frac{\partial P_j}{\partial P_i} b_i \right) + \frac{\partial P_j}{\partial b_j} \frac{db_j}{dt}. \tag{18.9}$$

This expresses the idea that the change in the property is a function of the connectivity between other properties of the system and the behaviors which define this property. Thus, we wish to find a set of conditions such that

$$\lim_{\tau \to \infty} \int_0^\tau \sum_{\substack{i \neq j}}^{n_b} \left(\frac{\partial P_j}{\partial b_i} \frac{db_i}{dt} + \frac{\partial P_j}{\partial P_i} b_i \right) dt$$

$$+ \lim_{\tau \to \infty} \int_0^\tau \frac{\partial P_j}{\partial b_j} \frac{db_j}{dt} dt + P_j^0 = P_j^F. \tag{18.10}$$

This is the general swarm engineering condition, and must be fulfilled by the behavior and sensor sets. Behaviors of the system depend on behaviors of agents in that system. Those, in turn, depend on agents' sensors, memory state, behavioral strategy, and position.

Note that in Eq. (18.9), each of the entities b_i and $\frac{db_i}{dt}$ represent the behavior associated with P_i and its rate of change. These behaviors are changes in the property P_i, which can only happen through the action of the agents. This is a very powerful equation, as it indicates precisely which behaviors might be used to effect the global change. Clearly, Eq. (18.10) can be satisfied in a number of ways, with respect to the various individual properties. The stronger condition requires specific changes in all properties. I.e.

$$\lim_{\tau \to \infty} \int_0^\tau \frac{dP_i}{dt} dt = P_i^F - P_i^0. \tag{18.11}$$

In this case, each of the individual properties changes in a specific way, causing the overall change, assuming that P_j is single-valued.

We can imagine the change happening in a phase space, of sorts. In this phase space, each point represents a set of values for each of the properties. In order to ensure that the global goal is achieved, the system must follow a path through the state space. We can imagine a state space made up of n_b-dimensional vectors such that each point represents a different system state. The initial state would then be a point, and the motion of the system through state space would be achieved by the behaviors of the agents. That is, every action of the agents will move the system. The trick is to direct the agents' behaviors so that the path through state space will connect the initial point to the end point in a stable way. A system in which the final state is the outcome of any initial state is a system in which the final state is an attractor.

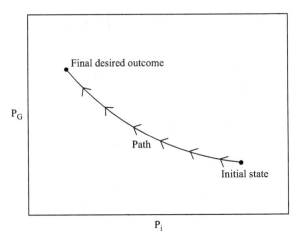

In general, not every configuration in phase space is allowed. If any group of properties is connected to one another, the connection may preclude certain areas of phase space. For instance, if one property is the distance between two objects, and a second is the distance between one of these objects and a third object, certain restrictions occur in the feasible points in phase space. In this case, the triangle inequality must apply, and this limits the range in phase space of the system.

Let us examine the possibility that the global property is constructed from other properties that are independent of one another. Then, the feasible region of the phase space is the cross product of the feasible regions of each of the individual properties. That is, if A_i is the set of all feasible choices for P_i, then

$$A = \bigotimes_{i=1}^{n_b} A_i \qquad (18.12)$$

is the set of all feasible points in the space. Suppose that each A_i is continuous. Then a piecewise linear path will connect the starting and ending points. Moreover, each of these linear segments may focus on a single property, indicating a specific task for the swarm. The situation is depicted in the next Figure in which a feasible region is clearly graphed, along with the engineered path through the phase space.

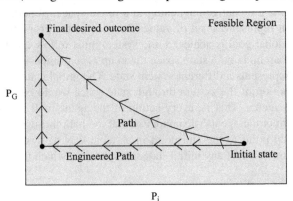

In this Figure, the path has two segments. The first segment illustrates the change of one of the properties, while the second illustrates the second. The agents that carry out this evolution of the system must therefore be able to sense the current state of both properties, determine when the system has reached the desired endpoint, and modify each of the system's components indicated by these properties.

As the system has two independent properties, our choice of how these actions can be achieved is quite open. We can have a single individual that works on one property or the other according to opportunity. On the other hand, one could build a system with two distinct castes of agents that each affect different properties simultaneously. Finally, we might have a single caste of agents that completes work on one property and then works on a second. This freedom is available to us once we realize that the properties are independent and therefore don't require synchrony. The engineer is free to choose how this is done.

Let us try to understand what this means. Each of the path segments is independent of one another, and so modification of this property only requires that some agent is capable of doing the modification. This, in turn, tells us a few practical things. First, it gives us an idea of the sensory capability of the agent. It must be able to discern under what conditions it should act with enough specificity to avoid changing any of the other properties and ending at the appropriate endpoint. Second, it must be able to carry out the behavior changing the system state. This gives us an idea of its physical requirements. Finally, the number of different behavioral states or different castes of agents is indicated by the number of different independent properties. These can work independently and in sequence or in tandem.

On the other hand, sometimes the properties are not independent of one another. In this case, the feasible region may not include the entire cross product space. This means that the path through phase space is much more constrained. The situation is depicted in the Figure below.

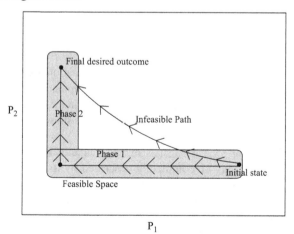

In this case, the path must necessarily result from the modification of property one first, and then property two. As a result, the swarm must consist of at least two disparate behavior sets. The first behavior set must move the system through phase space along property one, while the second behavior set must move the system through phase space along property two. This indicates that at different times, the swarm must behave in different ways. This is extremely important. It indicates a few things. First, it indicates two different behavioral castes. These castes are disparate in their behavior, and they act independently. This may be achieved practically by two different means. In the first, there are physically two different sets of agents which become active at the appropriate times. In the second, there is one set of agents, capable of discriminating between the two situations, and deciding how to behave based on the specific situation. Of course, in both situations, it is necessary that the agents carry sufficient computational and sensory machinery as to be able to discern which state the system is in, so that the correct behavior is achieved. This is identical to the previous situation, though in this case, the constraint on the behaviors is that they cannot happen in tandem. In this case the behaviors must occur independently and sequentially. Thus, the planning and possibly the behavior must be more precise.

If the feasible region does not include a complete path from the initial point to the desired final point, then the final completed task is impossible. In this case, the swarm may not be constructed using the sensors (properties) specified, though a path through a larger phase space enhanced by a new property might be able to connect disconnected phase spaces.

What these considerations allow us to do is to determine from the global property the agents' sensory requirements including resolution, the agents' physical capabilities, and the types of agents. This is a very useful set of information and can be used to determine how to build the set of agents for the task. This is the top-down portion of the swarm engineering methodology. Next, we consider the bottom up portion.

18.2.2.3 Bottom Up

Once we've worked out the top-down considerations, the remainder is relatively straightforward. The top-down considerations should clearly indicate how many behavioral castes there are, what sensory and/or computational capabilities are required, including the resolution of these sensory and computational capabilities, and how the different castes should be deployed. The remainder of the job consists of developing agents which meet these requirements. In this subsection, however, we'll examine their effect.

Let us assume that our swarm consists of $\{N_A\}$ agents. First, we may assume that the lth agent's state may be completely described by its memory state m_s^l, its internal state in_s^l, its sensor state s_s^l, its positional state p_s^l (which expresses its position and higher derivatives of position), and its behavioral strategy k^l. Note that k^l may be a function of time and it may be able to take on one of multiple states. Moreover, transitions may be triggered by sensor states. Then, we may express the global behavior b_j as a function of a number of things. First, the coupling between a global property and an agent behavior is defined, in part, by the positional state of the agent. We define the coupling between agent l and the global behavior

b_j by $C^l_{jk}(p^l_s, in^l_s)$. Secondly, we describe the individual behavior of the agent by $AB^l_{kl}(m^l_s, in^l_s, s^l_s)$. Then, the overall behavior may be expressed as

$$b_j = \sum_{l}^{N_A} C^l_{jk}(p^l_s, in^l_s) AB^l_{kl}(m^l_s, in^l_s, s^l_s).$$
(18.13)

[4] The trick, then, is to create behaviors that are dependant on realistic sensor states and internal states which provably satisfy Eq. (18.10). In many studies, Eqs. (18.13)–(18.35) are converted to average behavioral equations, greatly simplifying the required analysis.

Combining Eqs. (18.10) and (18.13), we obtain the general combined agent-swarm equations:

$$\lim_{\tau \to \infty} \int_0^\tau \sum_{i \neq j}^{n_b} \left(\frac{\partial P_j}{\partial b_i} \frac{db_i}{dt} + \frac{\partial P_j}{\partial P_i} b_i \right) dt$$

$$+ \lim_{\tau \to \infty} \int_0^\tau \frac{\partial P_j}{\partial b_j} \frac{db_j}{dt} dt + P^0_j$$

$$= \lim_{\tau \to \infty} \int_0^\tau \left[\sum_{l}^{N_A} C^l_{jk}(p^l_s, in^l_s) AB^l_{kl}(m^l_s, in^l_s, s^l_s) \right] dt$$

$$+ P^0_j = P^F_j.$$
(18.14)

Swarm engineering is concerned with balancing these equations linking the agent behaviors and the global desired behaviors.

18.3 Swarm Engineering Applied to Economic Systems

In this section, we will theoretically explore the application of the principles of swarm engineering to economic systems. In swarm engineering, we are primarily interested in generating group behaviors by utilizing careful examination of the desired global behavior and using this analysis to guide the design of agent-level behaviors capable of producing the desired global behavior. While this method still

---

[4] In the case that there is only one behavior, this is simplified to

$$b_j = \sum_{l}^{N_A} C^l_j(p^l_s, in^l_s) AB^l(m^l_s, in^l_s, s^l_s).$$
(18.34)

In the additional case that there is no memory, the expression simplifies further to

$$b_j = \sum_{l}^{N_A} C^l_j(p^l_s, in^l_s) AB^l(in^l_s, s^l_s).$$
(18.35)

This is the equation for uniform reactive agent swarms.

requires considerable input from the engineer, we have been able to use it to solve previously unsolved problems in deployment of swarms. In the present study, this means that we are interested in examining one or more global economic measurables and putting together a method of directly manipulating these by designing specific agent behaviors.

In economic systems, there are many global measurables. Each one is tied to local variables in a complicated and non-linear way. This makes the prediction of the global effect of a specific local behavior very difficult. As a result, it is often times simpler to utilize agent-based systems to get an idea of the effect of specific behaviors. The difficulty with utilizing agent-based systems in this way derives from the difficulty in creating a new system with specific desired qualities; the nonlinearity of complex systems makes this a very difficult thing to do. As a result, we utilize the swarm engineering methodology, which draws its initial motivation from the desired global outcome.

As our global property, we choose a truly dispersed property – that of the average cost of a commodity across vendors for sales of specific commodities. This property is interesting because it measures how much a consumer pays for goods and services that are worth a specific amount. If all vendors tend to end up with similar prices, this indicates that either the system is designed to enforce a specific price, or that there is some kind of communication between vendors that allows them to collude. We shall see that there are specific system designs that allow the former to occur without collusion or any communication between vendors.

We examine the design of consumer behavior as a method of controlling the average price. Vendors are modelled as profit maximizers who will increase their prices when all else is kept constant. The reaction of the consumers must be made in such a way that slow creeping price increase does not occur. We shall see that specific agent behaviors, designed properly, can limit the average prices to prices that very closely match the cost of vending the product.

18.3.1 Vendors and Consumers

We begin by modelling the main factors that affect the vendors in their decision to alter prices of commodities that they are selling. We begin with the assumption that all vendors will choose a price for a commodity that equals or exceeds his or her costs incurred during the sale of the commodity. The question then is what factors affect the change in the price?

We begin by assuming that the price function used by a vendor is a complicated function of several different values. That is, let the price be represented as

$$p_{v,c} = f(m_1, m_2, \ldots, m_n). \tag{18.15}$$

Then, each of these values m_i represents a factor in determining the price of the commodity.

There are many factors one might include in a decision about the cost of a commodity or in a decision about whether or not to increase the cost of a commodity. Among these factors are the demand for the commodity (D), the vendor's account balance (b), the total cost of the commodity to the vendor including the cost to put it on the shelf (space, cost outlays, and personnel) (c_c), any memorized or recorded data of the past l cycles ($\{m_j\}_{j=1}^{l}$), and the current income of the vendor (i). We assume, for the moment, that these are the main effectors of the cost of the commodity.

As we stated above, our goal is to examine the dynamics of the average price of specific commodities. This is the average price over all vendors of the commodity. I.e.,

$$P_{a,c} = \frac{1}{N_v} \sum_{v=1}^{N_v} p_{v,c} \qquad (18.16)$$

where $P_{a,c}$ is the average price for commodity c, N_v is the number of vendors, and $p_{v,c}$ is vendor v's price for the commodity c.

In real economic systems, the average price of a specific commodity typically remains stable or increases over time. However, theoretical prices should actually decrease or remain stable over time as the cost of production decreases. Moreover, the market is assumed to produce corrections to initially poorly priced items (i.e. items whose prices are much higher than the cost to produce it). We are interested in discovering what the minimal conditions are for consumers which will result in commodity prices that decrease or stabilize over time. This can be written mathematically as

$$\frac{dP_{a,c}}{dt} = \frac{1}{N_v} \sum_{i=1}^{N_v} \frac{dp_{v,c}}{dt} \leq 0. \qquad (18.17)$$

If a single vendor's prices start decreasing, then under competitive conditions, all vendors' prices should start decreasing. This being the case, we don't expect one vendor's price to increase while any of the other vendors' prices decrease. As a result, we can replace the requirement of (18.17) with

$$\frac{dp_{v,c}}{dt} \leq 0. \qquad (18.18)$$

If we begin by assuming that the vendors have a systematic method to their pricing choices, then we may write the prices faced by consumers as (18.15). Utilizing the various measurements indicated above, this means that

$$\frac{dp_{v,c}}{dt} = \frac{\partial f}{\partial D} \frac{dD}{dt} + \frac{\partial f}{\partial b} \frac{db}{dt} + \frac{\partial f}{\partial c_c} \frac{dc_c}{dt} + \sum_{j=1}^{l} \frac{\partial f}{\partial m_j} \frac{dm_j}{dt} + \frac{\partial f}{\partial i} \frac{di}{dt}. \qquad (18.19)$$

The term in (18.19) $\frac{\partial f}{\partial c_c} \frac{dc_c}{dt}$ would seem to have little to do with the consumers, and so cannot be directly affected by a behavioral change among consumers. We therefore ignore it as a potential design point. On the other hand, it is interesting to note that $\frac{db}{dt}$ is the rate at which the bank account changes. Thus, we identify this with the profit. If profit is Pr then,

$$Pr(t) = \frac{db}{dt} = D(t)(f(t) - c_c(t)). \qquad (18.20)$$

where $D(t)$ represents the number sold per time period. Moreover, this profit/loss may be memorized by the agent, affecting behavior. For each vending agent, the behavior can be different, but in general

$$m_k(t) = Pr(t - kt_p) = D(t - kt_p)(f(t - kt_p) - c_c(t - kt_p)) \qquad (18.21)$$

where t_p represents a time period and k represents the specific memory element being stored. k typically runs from 1 through N_m, the number of memory elements used in the function.

Since we are examining conditions that make $\frac{dp_{v,c}}{dt}$ a non-increasing function of time in the absence of inflation and supply variations, we want

$$0 \geq \frac{\partial f}{\partial D}\frac{dD}{dt} + \frac{\partial f}{\partial b}\frac{db}{dt} + \frac{\partial f}{\partial c_c}\frac{dc_c}{dt} + \frac{\partial f}{\partial m_{p/l}}\frac{dm_{p/l}}{dt} + \frac{\partial f}{\partial i}\frac{di}{dt}. \qquad (18.22)$$

As a result, we have that

$$\frac{\partial f}{\partial D}\frac{dD}{dt} \leq -\left(\frac{\partial f}{\partial b}\frac{db}{dt} + \frac{\partial f}{\partial c_c}\frac{dc_c}{dt} + \frac{\partial f}{\partial m_{p/l}}\frac{dm_{p/l}}{dt} + \frac{\partial f}{\partial i}\frac{di}{dt}\right) \qquad (18.23)$$

Inserting the results of (18.20) and (18.21) reveals that the actual form of this equation becomes

$$\frac{\partial f}{\partial D}\frac{dD}{dt} \leq -\left(\frac{\partial f}{\partial b}(D(t)(f(t) - c_c(t))) + \frac{\partial f}{\partial c_c}\frac{dc_c}{dt} + \frac{\partial f}{\partial i}\frac{di}{dt}\right)$$
$$-\left(\sum_k \left[\frac{\partial f}{\partial m_{p/l}}D(t - kt_p)(f(t - kt_p) - c_c(t - kt_p))\right]\right) \qquad (18.24)$$

In the case that the vendor simply reacts to current conditions, the relation takes the form

$$\frac{\partial f}{\partial D}\frac{dD}{dt} \leq -\left(\frac{\partial f}{\partial b}(D(t)(f(t) - c_c(t))) + \frac{\partial f}{\partial c_c}\frac{dc_c}{dt} + \frac{\partial f}{\partial i}\frac{di}{dt}\right). \qquad (18.25)$$

Now, we examine (18.24) to determine the form of f.

1. If the cost to the vendor increases, it is reasonable to expect the vendor to either increase or hold steady its prices. That is

$$\frac{dc_c}{dt} > 0 \Rightarrow \frac{\partial f}{\partial c_c} > 0. \qquad (18.26)$$

2. If the income increases, one can infer that the demand at a particular price has increased. Therefore, by increasing the price, the profit will increase. Thus, we expect that

$$\frac{\partial f}{\partial i} > 0. \tag{18.27}$$

3. If profit increases, one can infer that the demand at a particular price has increased. Therefore, by increasing the price, the profit will increase. Thus, we expect that

$$\frac{\partial f}{\partial b} > 0.$$

4. If the demand increases, typically the price increases. Therefore we expect that

$$\frac{\partial f}{\partial D} > 0.$$

These results together give us that

$$\frac{dD}{dt} \leq -\frac{1}{\frac{\partial f}{\partial D}} \left(\frac{\partial f}{\partial b} \left(D(t) \left(f(t) - c_c(t) \right) \right) + \frac{\partial f}{\partial c_c} \frac{dc_c}{dt} + \frac{\partial f}{\partial i} \frac{di}{dt} \right)$$

$$- \left(\sum_k \left[\frac{\partial f}{\partial m_{p/l}} D(t - kt_p) \left(f(t - kt_p) - c_c(t - kt_p) \right) \right] \right) \tag{18.28}$$

or in the case that the agents are purely reactive

$$\frac{dD}{dt} \leq -\frac{1}{\frac{\partial f}{\partial D}} \left(\frac{\partial f}{\partial b} \left(D(t) \left(f(t) - c_c(t) \right) \right) + \frac{\partial f}{\partial c_c} \frac{dc_c}{dt} + \frac{\partial f}{\partial i} \frac{di}{dt} \right). \tag{18.29}$$

We have just proved the following theorem.

Theorem 18.3.1 *If the condition in Eqs. (18.28) or (18.29) continually holds, then the price will be bounded above.*

These last two equations give the limits of the behavior of the consumer agents in a system composed of the vendor and consumer agents only. It indicates that the consumer agents must respond with a decrease in the demand for a commodity which is greater in magnitude than the magnitude of the right hand side of Eqs. (18.28) and (18.29). This is a severe design requirement on the consumer agents. However, as we will see in the next section, systems containing consumer agents which follow these restrictions do tend to have the desired global characteristics, while those that do not tend to have significantly higher to run-away prices.

18.3.2 Design of Consumer Agents

Our primary concern is that the consumer agents provably behave in such a way that the global average price remains bounded above. We have already seen in Section 18.2.1 that if the conditions in Eqs. (18.28) and (18.29) are obeyed, the goal will be achieved. That completes the top-down portion of the design problem. We now have an engineering requirement with which to work. We can now begin the bottom-up phase.

In this new phase, we must generate agents that satisfy this requirement. The general solution to the general equation given in (18.29) if $\frac{\partial f}{\partial i} = \frac{dc_c}{dt} = 0$, $\frac{\partial f}{\partial D} = \alpha$, and $\frac{\partial f}{\partial b} = \gamma$, the general solution is

$$D = e^{-\int_0^t -\frac{\gamma}{\alpha}(f(t') - c_c(t'))dt'}. \tag{18.30}$$

As a result of this general solution, it is clearly the case that, in order to react correctly in the next time frame, our agents must have the following capabilities.

1. The agents must be able to measure the price of the commodity.
2. The agents must be able to measure the demand for the commodity. In our simulations, it is a good estimate to know one's own probability of purchasing the commodity and multiplying by the population size.
3. The agents must be able to accurately estimate the cost to the vendor.

Thus, all agents must have this capability, and their behavior must be one of this family of behaviors. We can write this as an update rule. This becomes

$$D_{i+1} = D_i \left(1 - \frac{\gamma}{\alpha}(f_i - c_{c_i})\right). \tag{18.31}$$

This equation underscores the idea that the demand will remain constant when the price is near the cost. However, as the vendors will constantly be trying to increase the price, and the consumers will be reacting to increases, the actual average price will be greater than the cost to vendors. It is worth noting, of course, that in the real world, this cost is replaced by a very poorly defined notion of the "value" of an object. Since consumers have no idea, in general, how much a specific object actually costs in real terms, they must guess about it's value. However, despite this ignorance-driven inflation, the prices, once equilibrated, must respond to the same type of force.

In the next section, we describe our simulation and the behaviors of the agents carrying out repeated cycles of interactions between consumers and vendors. We generate a family of behaviors parametrized by a small number of parameters. Some values of the parameters generate behaviors that obey the requirements of (18.29) and some do not. We explore the effects of these parameters and demonstrate that they yield the expected global behaviors.

18.4 Simulation Design

We examine our theoretical results using a computer simulation that centers around the interactions between two types of agents: consumers and vendors. Our simulation functions by creating repeated interactions between the consumers and vendors as they learn and react to certain situations. Vendors have commodities to sell, and are designed to maximize profit. Consumers purchase commodities from vendors using money provided to them by jobs, and attempt to maximize consumption. The simulation proceeds by repeated "sessions" during which consumers visit vendors, evaluate what the vendors have to offer, and decide whether or not to buy. Vendors respond to changes in their products' marketability by changing prices in an attempt to increase their profit.

In our simulation, many details come into play. Both consumers and vendors have memory which help them decide on things such as which of the other class of agents to do business with, how to change prices, etc., and how to respond to current offerings. In the coming subsections, we explain these in detail, including motivating assumptions borrowed from economic theory. Our goal is to test our method of designing agents whose interactions produce a desired global goal, namely the control of the average price of a commodity. We describe, in addition to the agents' designs, the tools used to evaluate the function of the system.

18.4.1 Vendors

As soon as ABES is executed, the products are assigned a random cost. Each vendor sells a single commodity, and so must assign and manage the price of the single commodity. Each vendor calculates its own minimum price. Initially, the price is set at twice the cost to the vendor. All the profits made from a completed exchange is directly added to the vendor's bank account, the total amount of money that the vendor has. The vendor will restock its inventory when the number of products it holds reaches a user defined number if there is enough money in the bank to purchase more products. If the vendor fails to restock using the amount of money in the bank, then that vendor is considered bankrupt and is removed from the pool of vendors. As a result, the bankrupt vendor no longer participates in the interactions between vendors any consumer.

Each vendor's goal is to maximize its profit by any means. After a user-defined number of iterations, if the vendor has made more profit than it did in the previous period, the prices of the vendor's product are incremented by a constant, user defined percentage of the product's cost. This price update rule comes from the assumption that vendors will expect the same number (or nearly the same number) of products to sell the next period. A slight increase of price will increase the total profit. Conversely, if the vendor has made less profit, it reduces its prices by the same percentage. This behavior of decreasing the price derives from the assumption that the vendor will sell more the next period by slightly decreasing the price. This should increase the total profit.

18.4.2 Consumers

Each consumer interacts with its vendor in the same way: the consumer buys from the vendor if all of the conditions are met each time the consumer randomly chooses

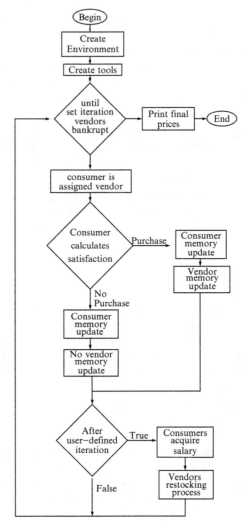

Fig. 18.1 This is a general flowchart of ABES

a vendor to buy the commodity from. Fig. 18.1 shows a general flowchart of ABES. We assume the commodity is something the consumer eventually must buy, like water. If the consumer waits long enough, it will be forced by necessity to purchase the commodity at any price. If the consumer has enough money, the item is in stock, the vendor is not bankrupt, and the consumer is "satisfied" with the product, the consumer will purchase the product. The consumer's satisfaction with the vendor's products is represented by a number from 0 to 100, and is affected by the length of time since the last purchase, the consumer's memory of the prices, and the vendor's profit margin. Along with the information in the consumer's memory, the consumer calculates its satisfaction toward the product. A random number from 0 to 100 is generated, and if the calculated satisfaction is higher than the generated number, then the consumer will be considered "satisfied" enough to buy the product. Thus, the higher the satisfaction of the consumer is, the more likely the consumer is to purchase an item from the vendor. Each consumer's cache of money is incremented by a user defined salary after some number of iterations, and decremented by the amount of each purchase.

The goal of the consumer in our simulation is maximize consumption at the lowest price and at the highest possible satisfaction. Our consumers are sensitive to the vendors' profit margin and will not purchase a product if the profit margin is too large. Whenever a vendor increases its price, consumer satisfaction decreases. As a result, consumers are less likely to purchase from the vendor. At some point in the simulation this will so aversely affect consumer satisfaction that very few of them will purchase the commodity. Once consumers cease purchasing, vendors react to a decrease in their income. Vendors, in turn, have no choice but to lower their price. Once the price has been lowered sufficiently, satisfaction returns to a high enough level for consumers to begin buying again. This consumer behavior keeps the vendors from constantly increasing their price and will result in a stablized price. However, as we will see in the next section, there are strict limits on even this behavior which yield control on global price levels.

In our simulation, we model the consumer satisfaction as

$$S = smax \left[1 - \left(\frac{1}{e^{t - [\alpha(profit) + (price - \gamma(pricemem))]}} \right) \right] \qquad (18.32)$$

Here S is the satisfaction, α is a constant that controls the consumer's aversion to profit margin, and γ is a constant that affects competition among vendors. Both of these variables can be initially assigned different values. Profit is the amount of money the vendors make after an exchange is complete. Price is the current price of the commodity and pricemem is the running average of the prices paid by the individual consumer during the last several interactions for the same commodity. The higher the exponent value, higher the satisfaction. Clearly, changing the value of α will alter the consumer's sensitivity toward the profit. Likewise, γ affects the consumer's sensitivity to prices much higher than those recently paid. This indirectly affects competition between vendors.

18.5 Simulation Data

In Section 18.2, we examined the theoretical basis for the design of consumer agents which, we expect, are capable of causing the "invisible hand of the market" to appear, limiting the prices of commodities. Section 18.3 described our simulation. This simulation consisted of two kinds of agents – consumers and vendors. The two types of agents interact with each other, and have conflicting goals. Moreover, the consumers have a limitation that they must have the commodity that is being sold, eventually. Such a commodity might be like water. The consumer agents have the limitation that the longer they go without the commodity in question, the more they're willing to tolerate to get it. As a result, there is potential for price gouging, leading to runaway prices.

In this section, we examine the behavior of the system under the action of the consumer agents. The agents' behaviors are controlled by the Eq. (18.32). In this equation, there are two main parameters, α and γ. By changing the values of these parameters, we can produce differing agents behaviors. Some of these behaviors satisfy Eq. (18.29) and some don't. We shall see that the desired outcome is achieved when Eq. (18.29) is satisfied.

18.5.1 The Effects of γ

In Eq. (18.32), we have two parameters, γ and α. γ primarily controls the effect of a high price with respect to previous experienced prices. A high value of γ indicates a high sensitivity to higher prices while a low γ value indicates little or no effect. The overall effect is akin to competition between individual vendors. With a high value of γ, the prices tend to stabilize near those of the agent with the lowest prices, while lower values do not tend to reinforce this.

We can understand this in terms of Eq. (18.29). The demand does not change on the left hand side if the prices are all the same. However, the first term on the right hand side is large enough that the equation does not hold. As a result, the price does not reduce, but rather stays constant once all vendors have synchronized their prices. The situation is depicted in Fig. 18.2.

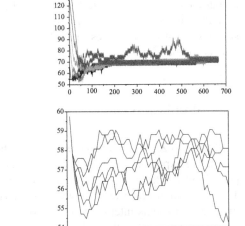

Fig. 18.2 With a high value of γ the prices are limited to the lowest price of all consumers. However, if this lowest price is itself high, the prices will not rebound, as can be seen in these figures

Fig. 18.3 Even with a high value of γ the prices can increase unboundedly if vendors continually increase their prices at similar rates

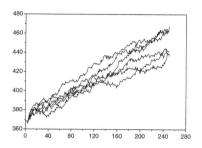

While γ tends to cause competition among vendors, it is not strong enough to cause the control of runaway prices. Consumers are generally stuck with the lowest of the vendor prices. We have seen that the failure of the system to satisfy the theoretical conditions translates to a failure of overall system to produce the desired property. If all of the vendors tend to increase their prices at the same rate (colluding or not), the effect on Eq. (18.32) is negligible, and so the condition is still not met. In this case, we can have runaway prices as well. This situation is depicted in Fig. 18.3.

18.5.2 Adding in α

It is clear that the competition between vendors is enough to hold most prices equal, but not strong enough to stabilize the cost of the commodities at prices that reflect their actual cost. This is interesting for a great many reasons, not the least of which is that this seems to contradict the "invisible hand of the market" that underlies much of economic theory. Clearly, more than simple competition is required to restore this property.

Satisfying Eq. (18.29) requires that another, stronger term become active. In Eq. (18.32), the parameter α controls the sensitivity of the consumer to the profit margin that the vendor is receiving. Very high values for α make the consumer intolerant of even small amounts of profit. On the other hand, small values for α make the consumer very tolerant of profits. We examine the effect of this.

The immediate effect is that the decrease in demand as a function of time becomes inextricably tied to the rate of increase of profit. If the profit increases, then the demand decreases. If α is high enough, the decrease exceeds any increase in overall profit associated with increasing the price. As a result, the condition in Eq. (18.29) is satisfied, and the price is controlled. The situation is depicted in Fig. 18.4.

Note that in Section 18.5.1, we kept α low, and the simulation had a global price increase over time. Only adding this very strong affector seems to hold prices low over time. The effect of this design element is so strong that it can take hold long after the price increase has begun, as illustrated in Fig. 18.5.

Fig. 18.4 With γ high or low, a high value of α is sufficient to control the prices of the commodity. This is expected due to Eq. (18.29), and confirmed in this simulation

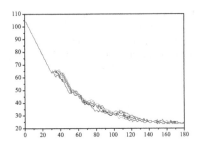

Fig. 18.5 If α is initially small, and γ high, the system exhibits slow price increase over time. However, if α is "turned on" at some later time, the system recovers its low-price configuration

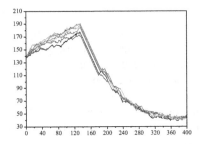

18.5.3 Examining (18.29)

One of the main guiding principles of this study has been the need to satisfy Eq. (18.29) in generating the consumer behavior. The reason is that we showed in Section 18.2 that if (18.29) is satisfied, then the behavior will lead to the desired global behavior. We now examine how closely our simulations adhere to this equation in generating the behaviors that limit commodity prices.

We can graph both sides of Eq. (18.29) as a function of the simulation iteration number. When we do this for both cases in which the price is controlled and cases in which the price is not controlled, we find that when a vast majority of the data follows Eq. (18.29), the prices are controlled. If this is not obeyed, even a bit more than intermittently, the prices are not controlled.

18.6 Perturbations of the Economic Swarm

In the previous sections, we saw that we are able to stabilize the prices using behaviors mediated by two behavioral parameters, alpha and gamma. Behaviors generated when alpha and gamma are relatively large tend to satisfy Eq. (18.29). Figure 18.6 illustrate the values of Eq. (18.29). We find that, as expected, the resultant behaviors cause the average prices to be limited. Other parameters, however, do not affect the agent behavior in a way that exerts much influence over the swarm's ability to achieve the desired global goal. As a result, these two parameters seem to be critical in determining whether or not the swarm will yield the global goal.

Fig. 18.6 These graphs illustrate the values of Eq. (18.29) as the simulation is run (top two) and histogram the number of times it is obeyed and not obeyed (bottom two). We find that when the equation is obeyed most of the time (first and third), the prices are controlled. However, when the equation is obeyed considerably less than all the time (second and fourth), the prices are not controlled. This supports our theoretical derivation of this condition

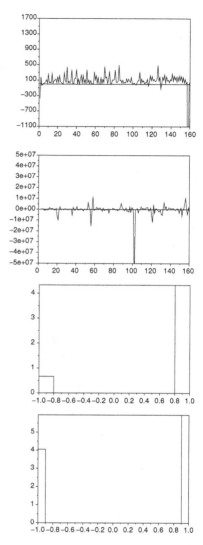

In this section, we examine a perturbation to the initial swarm developed in earlier sections. We examine the income effect in heterogeneous swarms. This refers to the effect of consumers having extra money to spend. Consumers tend to increase their consumption as their income increases. The question is whether or not this has an effect on the global behavior. If there is an effect on the global outcome, we examine how much of the swarm is required to see a change in the global outcome.

The income effect implies that people with the means will often purchase more items than those with lesser means. In previous simulations all consumers received identical wages and therefore exhibited equal purchasing power. In real systems, this is obviously not going to be the case because income differs depending on the consumers' professions. So we create a heterogeneous swarm by assigning different incomes to each agents. Moreover, we also examine the influence of the income effect applied to the swarm.

The income effect allows some consumers to purchase greater quantities of commodities than other consumers. The increase in purchasing power may create a leeway for vendors to subtly raise price. Since consumers are now capable of purchasing much larger quantities, the vendors might find it possible to increase their prices without losing profit, and therefore choose to leave the prices higher. Such an effect might destabilize the swarm, increasing prices over long period of time. We are interested in determining whether the swarm's design requirement from Eq. (18.29) is strong enough to offset the income effect in a heterogeneous swarm.

In order to examine the income effect in our swarm, we have extended ABES to execute the income effect with same parameters that produced the prior global goal. In the previous version of ABES, each consumer had only one opportunity to purchase a commodity per iteration regardless of his income. This limitation has been replaced with multiple purchase opportunities per consumers per iteration. Those consumers with higher incomes may purchase more per iteration than those with lower incomes. The amount that a consumer may purchase is proportional to his income. This simple modification emulates the income effect in the simulation leaving all other details unchanged.

As indicated above, the income effect may provide vendors a leeway to subtly inflate prices. As before, an increase in profits spurs vendors to raise prices. Wealthier consumer may still purchase at higher prices; the system might then produce various outcomes. Of interest to us is whether or not the effect of a greater income is overcome by the swarm's behavior as indicated by our prior analysis. Intuitively, one could make the argument that with more money available, prices might still be able to stabilize, but would tend to stabilize above the prior prices. Our analysis, however, indicates that as long as (18.29) is satisfied, the prices will not only remain stable, but they will stabilize at the same price.

In our simulations, we implement the income effect by giving consumers the opportunity to purchase multiple commodities in a single cycle. There are two types of interactions that cause the income effect. In the first interaction, consumers buy same commodities from multiple vendors; in the second interaction, consumers buy large numbers of the same commodity from a single vendor. The first type of interaction may occur, for example, when consumers purchase clothes. Many consumers purchase clothes from a wide range of clothing vendors, though they may be classified as interchangable products. The second interaction, however, is mostly likely to occur when consumers purchase consumables such as food or water. Consumers may purchase large quantities at once from a single vendor as there is no real reason to purchase it from many vendors at once.

As indicated in the Table 18.1, the income effect does result in an increase of purchasing power. The number of products purchased also increases. According to traditional economic analysis, this increase in demand should lead to an increase in the price of the commodity.

Table 18.1 As a result of the income effects, consumers purchase and consume significantly more of the commodity than simulations without the income effects

Effect	Mean
First income effect	$55,629.53 \pm 2,664.39$
Second income effect	$82,282.64 \pm 15,092.99$
No income effect	$35,649.37 \pm 25,956.33$

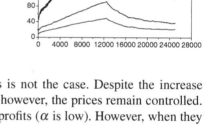

Fig. 18.7 Despite the income effect, with α high, the price stabilization takes place. In these graphs, the price of one commodity sharply drops when α is increased despite the strong income effect of the agents with greater salaries

However, as we can see in Fig 18.7 this is not the case. Despite the increase in the quantities of commodities purchased, however, the prices remain controlled. Initially, the agents are not cognizant of the profits (α is low). However, when they begin to take profit into account in their decisions, the prices drop dramatically. In fact, not only are the prices controlled, but they return to the same levels obtained when the income effect was not part of the simulation.

These data may be interpreted as indicating that the income effect, though changing the nature of the agents and increasing the consumption, does not affect the basic decision-making process that provides vendors with greater profits when prices are controlled. In terms of the consumer behavioral model, the parameter most responsible for this effect is alpha. Alpha mediates the sensitivity of consumers to unnecessary increases in price. When consumers are very sensitive to profit margin, vendors cannot overprice commodities. This local behavioral attribute is strong enough to stabilize prices independent of the volume of sales.

This data is significant because it indicates that, although one effect might be expected from a change in the conditions of the swarm's agents, the design of the swarm can be created so as to provide the opposite effect. This indicates a capability for the design of economic swarms and interactions that has heretofore been impossible.

18.7 Discussion and Conclusion

Designing swarms of agents is a very tricky business, owing to the nonlinear inter-actions of the various agents. As with all complex systems, swarms of a particular design might have a particular global behavior, but swarms with a very slight difference in behavior may have completely different global behaviors. As a result, predictive design has largely been avoided in the swarm literature.

In this chapter, we've explored a method of swarm design in which a specific global swarm behavior is developed prior to the design of the agents. The desired behavior, it has been shown, can be made to order once a set of requirements for agent design is worked out which will mathematically guarantee that the swarm accomplishes the task. Mathematical guarantee, which has eluded swarm researchers previously, is achieved by utilizing the global goal written in terms of the senses and actuators that the agents can be expected to have access to. Once the swarm condition has been met, the global goal may be achieved with agents meeting this condition.

It is interesting to note that this method of designing swarms is similar in form and function to the design of mechanical systems using the Lagrangian method. The power of this method lies in the ability of the engineer to create one or more properties whose numerical values are unique to the state that the system is in. The engineer, then, needs only chart a path through the allowed phase space of the system to the final desired value, hopefully utilizing behaviors which individual agents can accomplish on their own, with or without guidance from a central controller. The method can be applied to single properties or to vectors of properties, provided that the desired vector is well-defined in the same way a single property might be. We believe that the method is so powerful, in fact, that we now coin a term for this method: The Hamiltonian Method of Swarm Design.

This study, which examines the design of an agent based economic system, has demonstrated that in such systems, the achievement of global goals is possible when specific agent traits are required of the agents. It is interesting that such systems can exhibit control that typically comes from "the invisible hand of the market" or from a command economy. In fact, we have unmasked the "invisible hand of the market" in this study, revealing not only where it comes from but under what conditions it functions. It is interesting to ask, in light of the new method of controlling these swarms, what other economic indicators, trends, etc. can be commanded by the agents within the system.

Another interesting aspect of this study is just how fragile the system seems to be in terms of destabilizing under the improper behavior of one or a few agents. As we have seen in Section 18.4, when the inequality (18.29) is not obeyed, even a little, the prices become uncontrolled. It is interesting, then, to ask whether or not this system is stable in the sense that a few agents do not have the ability to drive the system into this uncontrolled region. This may give us insight into some of the interesting trends seen in recent years in economic systems including overvaluing of various commodities including .com stocks and housing prices. More research on this is clearly indicated.

One unexpected result of the current research is that we seem to have discovered that the income effect alone is not sufficient to raise prices of commodities. This unintutive result may be seen from the stability of prices in the simulations that utilized the income effect. This would seem to be in contradiction to traditional economic theory, which implies that prices would tend to rise. The stability may be seen to be caused by the engineered behaviors of the agents, rather than other market forces.

The current work also underscores the vast power that is enjoyed but not normally readily understood by consumers. It is clear that if consumers actually behaved in the way that our agents behave, the resulting economic system would be much more stable, in terms of commodity prices. The fact that this is now known might be used by policy makers in search of methods of stabilizing economies. With a properly implemented educational system, the public could bring its vast power to bear on economies without the need for governmental intervention or regulation. Whether or not people will actually adopt these behavioral norms is an entirely different question, though it is worth noting that the current method of swarm design brings this possibility to light.

In the future, we intend to apply this method to swarms of greater complexity than this one. We expect that this method of not only swarm design, but complex system design, may be applied to a large number of different systems including, but not limited to, systems of autonomous mechanical agents, computing systems, economic systems, and social systems. While some of this research is currently under way, we expect that the exploration of all fields to which this methodology might be applied will reveal an extraordinarily vast scope. Moreover, we expect that an extension to this work will be able to solve the problem originally posed ten years ago which led us to these results: "Is it possible that the global specification of a problem is enough to yield the basic requirements of the solution including all actuators, sensors, processing, and other capabilities of agents in the solution?" We believe the answer is yes.

References

1. E. Bonabeau, M. Dorigo, and G. Theraluz. Swarm Intelligence. New York: Oxford University Press, 1999.
2. M. Chang and J. Harrington, Jr. *Agent-Based Models of Organizations*. Handbook of Computational Economics II: Agent-Based Conputational Economics. March 24, 2005.
3. S. Das. *On Agent-Based Modeling of Complex Systems: Learning and Bounded Rationality*. Department of Computer Science and Engineering. La Jolla, CA, 92093–0404.
4. M. Fisher and C. Ghidini. *The ABC of Rational Agent Modeling*. Proceedings of the International Joint Conference on Autonomous Agents and Multiagent Systems: Part 2, Bologna, Italy, pp. 849–856, 2002.
5. C. Hommes. *Heterogeneous Agent Models in Economics and Finance*. Department of Quantitative Economics, University of Amsterdam, March 2005.
6. K. Judd, F. Kubler, and K. Schmudders. *Computational Methods for Dynamic Equilibria with Heterogeneous Agents*. World Congress, Cambridge University Press, pp. 243–290, 2003.
7. S. Kazadi. Swarm Engineering. Ph.D. thesis, California Institute of Technology, 2000.

8. S. Kazadi. *The Genesis of Swarm Engineering*. Proceedings of the SCI2003 Conference, Special Session on Swarm Engineering, Orlando, FL, 2003.
9. S. Kazadi. *On the Development of a Swarm Engineering Methodology*. Proceedings of IEEE Conference on Systems, Man, and Cybernetics, Waikoloa, Hawaii, pp. 1423–1428, October 2005.
10. S. Kazadi, P. Kim, J.S. Lee, and J. Lee. *Swarm Economics*. World Congress on Engineering and Computer Science 2007, San Francisco, CA, pp. 602–610, October 24-26, 2007.
11. S. Kazadi, J.R. Lee, and J. Lee. *Artificial Physics, Swarm Engineering, and the Hamiltonian Method*. World Congress on Engineering and Computer Science 2007, San Francisco, CA, pp. 623–632, October 24–26, 2007.
12. G. Klein, A Fallah-Seghrouchni, and P. Taillibert. *An Agent-Based Programming Method*. Proceedings of the International Joint Conference on Autonomous Agents and Multiagent Systems: Part 1, Bologna, Italy, pp. 4–7, 2002.
13. H. Knublauch. *Extreme Programming of Multi-Agent Systems*. Proceedings of the International Joint Conference on Autonomous Agents and Multiagent Systems: Part 2, Bologna, Italy, pp. 704–711, 2002.
14. B. LeBaron. *Agent Based Computational Finance*. Brandeis University, April 21, 2005.
15. P. Maes. *How to Do the Right Thing*. Connection Science Journal, 1(3), 3–24, 1989.
16. J. Mackie-Mason and M. Weldman. *Automated Markets and Trading Agents*. University of Michigan, Ann Arbor, MI 48109, USA.
17. R. Marks. *Market Design Using Agent-Based Models*. The Universities of Sydney and New South Wales Sydney. May 17, 2005.
18. P. Massonet, Y. Deville, and C. Neve. *From AOSE Methodology to Agent Implementation*. Proceedings of the International Joint Conference on Autonomous Agents and Multiagent Systems: Part 1, Bologna, Italy, pp. 27–34, 2002.
19. S. Mellouli, G. Mineau, and D. Pascot. *The Integrated Modeling of Multi-Agent Systems and Their Environment*. Proceedings of the International Joint Conference on Autonomous Agents and Multiagent Systems. Bologna, Italy, 2002.
20. P. Reitsma, P. Stone, J. Csirik, and M. Littman. *Randomized Strategic Demand Reduction — Getting More by Asking Less*. Proceedings of the International Joint Conference on Autonomous Agents and Multiagent Systems: Part 1, Bologna, Italy, pp. 162–163, 2002.
21. L. Said, T. Bouron, and A. Drogoul. *Agent-Based Interaction Analysis of Consumer Behavior*. Proceedings of the First International Joint Conference on Autonomous Agents and Multiagent Systems: Part 1, Bologna, Italy, pp. 184–190, 2002.
22. W.M. Spears and D.F. Gordon (1999). *Using Artificial Physics to Control Agents*. IEEE International Conference on Information, Intelligence, and Systems, November, 1999.
23. J. Stiglitz. Economics. W.W. Norton, New York, pp. 214–218, 2002.
24. L. Tesfatsion. *Agent-Based Computational Economics: A Constructive Approach to Economic Theory*. Economics Department, Iowa Sate University, Ames, IA 50011-1070.
25. K. Tumer, A. Ayogino, and D. Wolpert. *Learning Sequence of Actions in Collectives of Autonomous Agents*. Proceedings of the International Joint Conference on Autonomous Agents and Multiagent Systems: Part 1, Bologna, Italy, pp. 378–385, 2002.
26. A. Winfield, J. Sa, M.C. Gago, C. Dixon, and M. Fisher. *On Formal Specification of Emergent Behaviours in Swarm Robotics Systems*. International Journal of Advanced Robotic Systems, 2(4), 363–370, 2005.
27. S. Wolfram. A New Kind of Science. Champaign, IL, Wolfram Press, 2002.
28. G. Zhong, K. Takahashi, and S. Amamiya. *KODAMA Project: From Design to Implementation of a Distributed Multi-Agent System*. Proceedings of the International Joint Conference on Autonomous Agents and Multiagent Systems. Bologna, Italy, 2002.

Chapter 19
Machines Imitating Humans
Appearance and Behaviour in Robots

Qazi S. M. Zia-ul-Haque, Zhiliang Wang, and Xueyuan Zhang

Abstract The authors have synthesized the emotion in the speech of robot. The modeling of emotion in speech relies on a number of parameters among others, fundamental frequency (F0) level, voice quality, or articulation precision etc. As an initial work for synthesizing emotion in speech, we utilized the three voice features provided by the TTS engine of Microsoft Speech SDK i.e. pitch, rate and volume. Speech with these parameters controlled, was generated randomly with 20 sentences for each emotion and perception by human hearers were collected.

Keywords Machine imitating · Human appearance · Behavior · Robot

19.1 Modern Humanoid Robotics Research

In its early era, the field of robots focused to provide machines in industry to re- place man at work by performing laborious tasks for him with speed and accuracy. Robots now have started serving in the domestic environments, in hospitals, in mu- seum etc. where they have to face and deal with human directly. On the other hand man has desired to produce his mechanical replica for a long time. He has been sat- isfying his desire with dolls and puppets since ancient times. With the advancement of technology his efforts have turned him towards developing humanoid robots.

The first humanoid robot of modern era was presented at 1939 New York World Fair [1]. Since then efforts continued to produce more humanlike abstractions. Modern robotics research is also focusing to utilize these advancements to serve the humanity rather than producing robots just for fun or enjoyment. Robots are also expected to serve as companions/caregivers to elderly and children [2] and/or

Q.S.M. Zia-ul-Haque (✉), Z. Wang, and X. Zhang
University of Science and Technology Beijing, No.30 Xueyuan Road, Haidian District, Beijing 100083, China
e-mail: smzhaq@yahoo.com

S.-I. Ao et al. (eds.), *Advances in Computational Algorithms and Data Analysis,*
Lecture Notes in Electrical Engineering 14,

personal assistant, to prevent children accidents [3], for Robot Assisted Activities (RAA) and Robot Assisted Therapies (RAT) as a substitute for Animal Assisted Activities and Therapies (AAA/AAT) [4–7] etc. It will be useful for robots serving in these scenarios to utilize available human cooperation to perform the tasks more efficiently [8]. Human will feel more comfortable, pleasant and supporting with systems which (at least to some extent) possess ethical beliefs matching that of their own, do not make a decision or perform an action that is harmful to their moral values, and honour their basic social values and norms [9]. Thus where interaction with human is desired, the robots are desired to behave as social machines. Because of various level of social capabilities, social robots can be classified in four classes namely *socially evocative robots, socially communicative robots, socially responsive robots* and *sociable robots* [10].

19.2 Human-Robot Interaction (HRI)

Human Robot Interaction can be classified as active or passive interaction [11]. The most important factors to be considered to implement while designing an interactive robot are its capabilities to establish and maintain *Engagement, Trust* and *Motivation* with the user [12, 13]. Interactive robot design requires work from other research fields [14, 15] such as physiology, social psychology, artificial intelligence, and computer science in general and some other area specifically related to the application such as bariatrics, nutrition and behavior changes [12, 13]. The robots need further capabilities like initial contact, negotiating a collaboration, checking that other is still taking part in interaction, deciding to continue interaction or to end it etc. [16, 17].

19.2.1 Human Robot Collaboration

Human-Robot collaboration in joint activities to achieve common goals requires to maintain mutual beliefs, share relevant knowledge, coordinating actions, demonstrating commitment to do one's own part, helping the others to do their parts, avoiding from preventing others to complete their parts and completing the shared task, to communicate to establish and maintain a set of shared goals and beliefs and to coordinate their actions to execute shared plans [18].

It has been reported that human subjects although not too much, but comparatively rely more and feel less responsible while collaborating with a more humanlike robot than with a machinelike one. Also it is reported that people attribute less credit and more blame to robotic supervisors and subordinates as compared to robot peer. Finally it is claimed that the people feel more responsibility and attribute less credit or blame to robotic partners having machinelike appearance than those having humanlike appearance [19].

19.2.2 Nonverbal or Implicit Communication

Human often use nonverbal cues termed as *implicit communication* [20] to communicate to one another. Implicit non-verbal communication is helpful to understand the mental state, direction of attention on one another and to alter the behavior accordingly and to utilize the affective knowledge of one another. Thus the nonverbal information through social cues can improve the human-robot interaction and the efficiency to perform collaborative tasks [21]. Having an expressive face to provide non-verbal cues from expressions and indicating attention with movement both make a robot more compelling to interact with as a face to face interaction is the best model for interface in human-robot interaction [22].

An interesting demonstration of intuitive human-robot interaction was presented by Atienza and Zelinsky [23] where a robot through its active vision after detecting a human face follows the gaze of its human subject, picks up the object the human subject is looking at and hands it over to the subject. This way it fulfils the user's desire which is implicitly communicated to the robot through nonverbal communication by the gaze direction of user and not provided to the robot verbally.

19.2.3 Multi-person Interaction

Simultaneously interaction with multiple persons is also a challenging task for the interactive robots. While during the human-person interaction involving only one human faces the challenges of speech recognition, sound localization, tracking the human face, posture/gesture and expressive and cognitive capabilities, multi-person interaction puts further requirements of finding the current speaker and the addressee and to reply if the robot itself is the addressee, the information flow, appealing the intended interaction, the intended next speaker and focusing towards the speaker in time, attending interruption to its speech and to interrupt others smoothly [24, 25].

19.2.4 Issues in Human Robot Interaction

Various *social, moral, ethical* and *legal issues* are expected to arise with the increased sophistication of conscious machines [26]. A starting ground can be taken from animal rights as basis to build moral and social rules for such machines. Interaction can be enhanced by appropriate context suggesting more interacting activities but this can also increase the expectations of user thus a balanced context must be designed for interactive robot [27]. In addition to the conventional modalities of interaction (speech, gesture, haptics etc.) the physical activities and performance of robots should also be carefully designed to match the moral values (mutual distance for example) of the user interacting with it [28]. Breazeal and Kidd [29] have presented the issues such as *relationship issues, personality issues, cultural issues, quality issues, naturalness issues, user expectation issues* and*comparative media issues* desired to be addressed by HRI studies. According to Thomaz et al. [30] timing is also very important in human-robot interaction. Along with possessing high quality expressive behavior capabilities it is also important to express these behaviors at right time in a right manner [16]. Larger delays in a human beings and autistic children with only eye direction detection and intentionality detection systems. Kidd and Breazeal [31] have proposed measures to evaluate human robot interaction including *"Self-Report Measures"* using questionnaires, *"Physiological Measures"* such as galvanic skin and *"Behavioral Measures"* using data obtained from observations during HRI experiment. All three types have their own merits and demerits and to obtain reliable results a well balanced combination of three types of measures may provide the best evaluation measure. Burghart and Haeussling [21] have also suggested a network concept based sociological multilevel framework to evaluate the interaction at the levels of *Interaction Context, Interaction/Cooperation, Activity of Actors* and *Nonverbal Actions and Emotions*.

19.3 Humanlike Appearance

Role of appearance is claimed to be as important as its behavior for the robot to be recognized as a social identity [19, 32, 33]. A machine with a humanlike appearance is expected to elicit more natural response from human than one with a mechanical look. Having a humanlike face provides the benefit of universally recognized facial expressions, an understood focal point of interaction etc. Some researchers suggest that an iconic/minimal face to be sufficient as it provides a sense to project one's own emotions and expressions, and to apply their own identity to the robot whereas the completely realistic face may increase false expectation in users [34]. Experimental results from Hinds et al. also agree with these expectations [19]. Researchers thus are now focusing to develop human likeness in both appearances as well as in behavior of robots [35, 36]. Such machines are also expected to serve the studies in psychology and cognitive sciences to perform controlled experiments to understand the human-human interaction [15, 37, 38], which are rather not as easy with human beings. Results from research performed for human robot interaction have suggested hypothesis not only for the human-robot interaction but also for the human-human interaction [39, 40]. The work however is very preliminary and needs much more to be explored.

19.4 The Uncanny Valley

Most of the researchers have been using robots with mechanical appearance [41] and have rarely considered the human likeness of appearance in robots. The fear of falling into the uncanny valley hypothesized by Mori [42] in 1970 has restricted the developers of humanoid robots to avoid achieving this height of designs (Fig. 19.1).

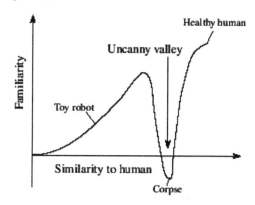

Fig. 19.1 The uncanny valley

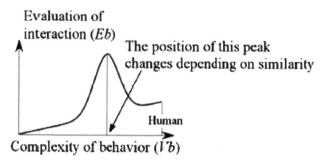

Fig. 19.2 Synergy effect

Japanese robotist Mori has hypothesized that as the designer tries to improve the human likeness, there comes the situation when the person interacting with the robot becomes conscious enough that little distances from human likeness give rise to eerie feelings. This negative behavior is named as the *Uncanny Valley* in the relation plotted between familiarity and human likeliness of the design as shown in Fig. 19.2. The eerie sensation generated in human by humanlike robots is supposed to be due to a reminder of mortality from the robots [37]. The effect has been suggested for both appearance and movement of the robots and that the overall response of human towards the humanlike entities can be obtained by combining their response to the movement (behavior) and that to the appearance [43]. Personal attributes such as age, gender, personality etc. of human user may also influence the depth and shape of the uncanny valley [15, 44, 45].

19.4.1 Effect of Behavior

Minato et al. [44] suggested that for different appearances the same behavior of the machine may elicit different response from human. Minato also presented a

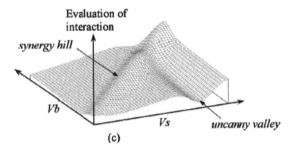

Fig. 19.3 Synthesizing the uncanny valley and synergic hill

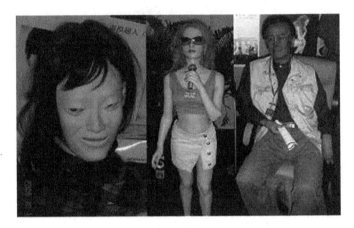

Fig. 19.4 Robots presented at China Robot Expo 2006

combined *"Synergy Effect"* at the point of matching of behavior and appearance (shown in Fig. 19.3) and suggested that by synthesizing the uncanny valley and synergic hill the uncanny effect caused by the appearance of robot can be reduced through its behavior (Fig. 19.4).

The uncanny valley however lacks sufficient experimental data in its support and is a question challenged nowadays. There are also researchers who do not believe in it at al claiming that humanlike machines can be appealing or disturbing at any level of similarity to real human. In contrast another theory called *"Path of Engagement"* has been proposed [46, 47].

The data both in support and opposition of uncanny valley is not sufficient enough to decide whether any such phenomenon exists or not. Another problem is that there is no universally defined method for quantitative evaluation of how much humanlike an imitation is so as to find its location on the similarity axis of Fig. 19.2. A possible solution to compute human-likeness of humanoid robot has been suggested in [48]. Once such a scale is defined and human-likeness is computed, the response the robots elicit in human can be used to examine the existence of uncanny valley.

19.5 Our Experiment

We are working towards the development of a humanlike robot. Since the role of humanlike appearance has been questioned among the researchers and developers and there are groups of people both in favor and against the extreme humanlike appearance, we wished to analyze human response towards such machines. We conducted a survey at the China Robot Expo 2006 where various robots with humanlike appearance were presented for the visitors to freely interact with. We wished to analyze human response towards these machines. We inquired the visitors through a questionnaire prepared with questions regarding their feelings and desires with these robots. We considered three of the robots with different levels of abstraction presented in the exhibition. There were also other humanoid robots presented with iconic appearance but since our interest was in interrogating the uncanny valley, we selected these three robots with different levels of humanlike appearance. Figure 19.4 shows the three robots we considered useful for our purpose. The first is our own head robot capable of generating few facial expressions and interaction with users through voice and facial expression but is not very fine in its appearance. The second is the female singing doll which presented different songs to the users with constant facial expressions and the third is the android designed to copy a real human appearance. This robot performs some humanlike movements in his hands and head/face and also interacts with people through voice.

The human subjects included people of both genders from almost all over the world almost including Americans, Asians, Europeans and others, and all were well grown adults of age group ranging from 20 to 40 years of age. However majority of them belonged to China. They included students as well as professionals from various backgrounds like technology, management, business and fine arts etc. We prepared a questionnaire to focus the response of people towards these robots. We have to select questions in such way to collect as much information as possible while keeping the number of questions to minimum so as to avoid the subjects from getting bored. We thus selected the ten questions inquiring:

1. Most liked robot
2. Most remarking feature
3. Features desired to be added or enhanced
4. Immediate feeling with first look
5. Feeling during interaction
6. Changes in feeling after interaction
7. Desired frequency of interactions
8. Expected length of interaction without getting bored
9. Suitable place for these robots in life
10. Gender desired of robot

19.6 Results

1. **Most Liked Robot:** 53% of the visitors liked the most humanlike robot (the third one) whereas almost 30% of them liked the singing doll. 11% visitors liked the least realistic but still much humanlike one (Fig. 19.1a) whereas 6% could not decide.
2. **Most Remarking Feature:** Nearly 45% of the visitors found the appearance of these robots to be the most attractive feature, almost 42% of them suggested the robots behavior during interaction while the rest 13% considered both appearance and behavior to be equally important.
3. **Features Desired to be Added or Enhanced:** Among the human subjects 12% suggested the appearance whereas 88% of subjects responded as the behavior to be the feature of these machines needing enhancement.
4. **Immediate Feeling with First Look:** 68% of the visitors were positively surprised with their first look and 26% said that they were attracted immediately, thus overall positive response was 94%. Only 3% responded to avoid and an equal amount of 3% claimed to fear i.e. only 6% of human subjects showed negative feeling from immediate look.
5. **Feeling During Interaction:** 88% of subjects said that they were attracted or amused with these robots whereas only 6% felt eerie. Remaining 6% were not clear about their feelings during interaction.
6. **Changes in Feeling After Interaction:** 56% of subjects said the attraction was increased. Initial fear in the 6% of subjects was reported to reduce with attraction. No changes in feelings were found in 29% of subjects and there were 9% of them who got bored.
7. **Desired Frequency of Interaction:** 18% of subjects showed interest in frequent interaction whereas 44% of them said that they would like to interact only occasionally, 23% preferred rare interaction and interestingly 15% of subjects disliked to interact with these machines ever.
8. **Expected Length of Interaction:** Answering about expected length of interaction without getting bored 38% responded it to be up to only few minutes, 32% expected it to span over hours, 15% said over a day and 15% of the people thought it to be any length of time.
9. **Possible Uses of These Robots:** About 48% of people considered best use of these robots as toys, 34% suppose them to be good companions, assistant or caregiver. Interestingly 3% of subjects said to find no place of these machines in their lives and 15% could not decide in what place these robots can be accepted.
10. **Gender:** About 68% of subjects considered the gender to be unimportant in these machines. 26% desired female robot and 2/3 of these were males whereas 6% all of whom were females desired these machines to look like a male human being.

19.7 Discussion

Responses from people showed that the human response is more positive with increase in the realism in imitation. Also the other hand, humanlike behaviour is found to be equally almost important as the humanlike appearance which is evident from the fact that enhancements in behaviour were desired by 88% of people, in robots whose appearances are already much humanlike and the most liked one is almost indistinguishable from natural human being (Fig 19.4c). The immediate feelings with these robots have been found positive and attractive where as negligibly small ratio of people expressed negative feelings. Further the negative feelings seem to be reduced after interaction. A small fraction of people got bored with this interaction but that was just loss of attraction and interest due to habituation and there was no sign of increase in fear or disturbance.

Very interestingly, in spite of finding these machines attractive, a very small number of people showed interest in frequent and long interactions with these imitations of human and most of the replies supported only occasional and short interactions with them. Almost only one-third of the visitors supposed these machines to have an active place of caregiver, assistant or companion in their lives. A large proportion just thought these not to be more than toys whereas there were also people who were not ready to give any place to these robots in their lives although they were attracted with them.

Personal features of these robots found to be almost unimportant as most of the people were happy with any gender present in robots. Only a small ratio was interested in robots with appearance of opposite gender and even smaller who were only females showed interest in same gender. Interestingly none of the male subjects desired the machines with masculine appearance.

19.8 Improving Behavior

Our findings encourage us to work towards the development of robots with humanlike appearance and behaviour. We have developed a robot with humanlike appearance. The robot is able to generate some facial expressions (shown in Fig. 19.5). Generation of six basic facial expressions follows the FACS model [49].

We have also been able to synthesize emotion in the speech of robot. The modeling of emotion in speech relies on a number of parameters among others, fundamental frequency (F0) level, voice quality, or articulation precision etc. A review of variations of speech features in emotional states can be found in [50]. Different synthesis techniques provide control over these parameters to very different degrees. As an initial work for synthesizing emotion in speech, we utilized the three voice features provided by the TTS engine of Microsoft Speech SDK i.e. pitch, rate and volume. Table 19.1 shows the selected values for pitch, volume and rate. Speech with these parameters controlled, was generated randomly with 20 sentences for each emotion and perception by human hearers were collected.

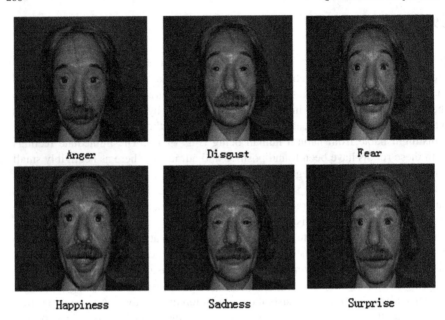

Fig. 19.5 Facial expressions of our humanoid robot

Table 19.1 Control values for pitch, volume and rate for emotion synthesis in speech

	Fear	Anger	Sadness	Happiness	Disgust	Neutral
Pitch	10	10	−3	8	−10	0
Volume	50	80	20	80	2	50
Rate	8	3	−3	5	−8	0

Table 19.2 Confusion matrix for perception of synthesized emotion in speech

	Fear	Anger	Sadness	Happiness	Disgust	Neutral
Fear	9	2	3	2	3	1
Anger	1	11	0	4	2	2
Sadness	4	1	10	0	4	1
Happiness	2	6	0	14	0	2
Disgust	3	0	3	0	9	1
Neutral	1	0	4	0	2	13

The confusion matrix for percentage of perceived emotions (horizontal) by human listeners for each generated emotion (vertical) is shown below (Table 19.2). Correct perception was only 55%. The speech was far from humanlike and is clearly recognized as mechanical voice. However we expect that the perception will be more appropriate when emotion is expressed during conversation (not randomly spoken sentences) because the context in which the speech is uttered may reduce the ambiguity of emotional contents.

19.9 Conclusion

The uncanny valley finds no proof for its existence. In addition to humanlike appearance, humanlike behavior is equally important to increase familiarity. Machines lagging in some features of appearance or behaviour may cause disappointment when not fulfilling expectations but there are almost negligible negative feelings (fear or danger) observed in humans interacting with these machines. Some people may avoid this technological achievement due to their social and cultural beliefs but it can not be considered as fear. Thus there is no support found for the concept of uncanny valley in the results of our study. However agreements to the hypothesis *"path of engagement"* is supported from people's desire of improved behavioural skills in humanlike machines.

We have been able to express emotions in speech and facial expressions of our humanlike robot. We are nowadays working on emotion recognition from speech and facial expressions of human subjects. Our future work surrounds to work out and implement emotion and behaviour models in the robot to incorporate humanlike behavior.

References

1. Ellen Thro. *"Robotics: Intelligent Machines for the New Century (New Edition)"*. Facts on File, New York, 2003.
2. Will Taggart and Sherry Turkle. *"Preliminary Remarks"*. A COGSCI 2005 Workshop and Corry D. Kidd. *"An Interactive Robot in Nursing Home; P"*, Stresa, Italy, July 25–26, 2005, pp. 56–61.
3. Altion Simo, Yoshifumi Nishida, and Koichi Nagashima. *"A Humanoid Robot to Prevent Children Accident"*. 12th International Conference on Virtual Systems and Multimedia 2006 (VSMM 2006), Xi' An, China, October 18–20, 2006.
4. Masahiro Fujita. *"On Activating Human Communication with Pet-Type Robot AIBO"*. Proceedings of the IEEE, Vol. 92, No. 11, November 2004.
5. Ronal C. Arkin, Masahiro Fujita, Tsuyoshi Takagi, and Rika Hasegawa. *"Ethological Modeling and Architecture for an Entertainment Robot"*. Proceedings of the 2001 IEEE International Conference on Robotics & Automation, Seoul, Korea, May 21–26, 2001.
6. Robins, B. et al. *"Sustaining Interaction Dynamics and Engagement in Dyadic Child-Robot Interaction Kinesics: Lessons Learnt from an Exploratory Study"*. 14th IEEE International Workshop on Robot and Human Interactive Communication (RO-MAN 2005), 2005, pp. 716–722.
7. Takanori Shibata, Kazuyoshi Wada, Tomoko Saito, and Kazuo Tanie. *"Human Interactive Robot for Psychological Enrichment and Therapy"*. Proceedings of the AISB Convention: Symposium on Robot Companions: Hard Problems and Open Challenges in Human-Robot Interaction, Hatfield, 2005, pp. 98–109.
8. Cynthia Breazeal. *"Social Interaction in HRI: The Robot View"*. IEEE Transactions on Cybernetics, Man and Systems – Part C, Application and Review, Vol. 34, No. 2, pp. 181–186, May 2004.
9. Wendell Wallach and Colin Allen. *"Android Ethics: Bottom-Up and Top-Down Approaches for Modeling Human Moral faculties"*. Towards Social Mechanism of Android Science. A COGSCI 2005 Workshop.
10. Cynthia Breazeal. *"Towards Sociable Robots"*. Robotics and Automation Systems, Vol. 42, pp. 167–175, 2003.

11. Cory D. Kidd and Cynthia Breazeal. *"Sociable Robot System for Real World Problems"*. 14th IEEE international Workshop on Robot and Human Interactive Communication (RO-MAN 2005), Nashville, TN, August 2005.
12. Gordon Cheng, Akihiko Nagakubo, and Yasuo Kuniyoshi. *"Continuous Humanoid Interaction: An Integrated Perspective – Gaining Adaptivity, Redundancy, Flexibility – In One"*. Robotics and Autonomous systems, Vol. 37, pp. 161–183, 2000.
13. Cory D. Kidd and Cynthia Breazeal. *"Designing a Sociable Robot System for Weight Maintenance"*. IEEE Consumer Communications and Networking Conference (IEEE CCNC 2006), Las Vegas, NV, January 7–10, 2006.
14. Julie Hillan. *"The Necessity of Enforcing Multidisciplinary Research and Development of Embodied Socially Intelligent Agents"*. Proceedings of the AISB Convention: Symposium on Robot Companions: Hard Problems and Open Challenges in Human-Robot Interaction, Hatfield, 2005, pp. 133–140.
15. Hiroshi Ishiguro. *"Android Science: Towards a New Cross Interdisciplinary Framework"*. Towards Social Mechanism of Android Science. A COGSCI 2005 Workshop, 27th Annual Conference of the Cognitive Science Society, Stresa, Piedmont, Italy, July 21–23, 2005, pp. 1–6.
16. Candace L. Sidner, Cory D. Kidd, Christopher Lee, and Neal Lesh. *"Where to Look: A Study of Human Robot Engagement"*. IUI'04, Maderia, Funchal, Portugal, January 13–16, 2004.
17. Candace Sidner, Christopher Lee, Cory Kidd, and Neal Lesh. *"Explorations in Engagement for Humans and Robots"*. Humanoids 2004, Santa Monica, CA, 2004.
18. Cynthia Breazeal, Andrew Brooks, David Chilongo, Jesse Gray, Guy Hoffman, Cory D. Kidd, Hans Lee, Jeff Lieberman, and Andrea Lockerd. *"Working Collaboratively with Humanoid Robots"*. Proceedings of IEEE-RAS/RSJ International Conference on Humanoid Robots 2004 (Humanoids 2004), Los Angeles, CA, November 2004.
19. Pamela J. Hinds, Teresa L. Roberts, and Hank Jones. *"Whose Job Is It Anyway? A Study of Human-Robot Interaction in Collaborative Task"*. Human-Computer Interaction, Vol. 19, pp. 151–181, 2004.
20. Cynthia Breazeal, Cory D. Kidd, Andrea Lockerd Thomaz, Guy Hoffman, and Matt Berlin. *"Effects of Nonverbal Communication on Efficiency and Robustness in Human-Robot Teamwork"*. IEEE/RSJ International Conference on Intelligent Robots and Systems (IROS) 2005, Edmonton, Alberta, Canada, August 2005.
21. Catherina Burghart and Roger Haeussling. *"Evaluation Criteria for Human Robot Interaction"*. Proceedings of the AISB Convention: Symposium on Robot Companions: Hard Problems and Open Challenges in Human-Robot Interaction, Hatfield, 2005, pp. 23–31.
22. Allison Bruce, Illah Nourbakhsh, and Reid Simmons. *"The Role of Expressiveness and Attention in Human-Robot Interaction"*. IEEE International Conference on Robotics and Automation, ICRA 2002, May 11–15, 2002, Washington, DC, USA.
23. Rowel Atienza and Alexandra Zelinsky. *"Intuitive Human-Robot Interaction Through Active 3D Gaze Tracking"*. Proceedings of 11th International Symposium of Robotics Research, Siena, Italy, October 2003.
24. Yosuke Matsusaka, Tsuyoshi Tojo, Sentaro Kubota, and Kenji Furukawa. *"Multi-Person Conversation via Multi-Modal Interface. A Robot Who Communicates with Multi-User"*. EUROSPEECH 99, Vol. 4, pp. 1723–1726, 1999.
25. Yosuke Matsusaka, Shinya Fujie, and Tetsunori Kobayashi. *"Modeling of Conversational Strategy for the Robot Participating in the Group Conversation"*. EUROSPEECH 2001, Scandinavia, 2001.
26. David J. Calverley. *"Android Science and the Animal Rights Movement: Are There Analogies"*. Towards Social Mechanism of Android Science. A COGSCI 2005 Workshop, Stresa, Italy, July 25–26, 2005, pp. 127–136.
27. Ben Robins, Kerstin Dautenhahn, Chrystopher L. Nehaniv, N. Assif Mirza, Dorothee Francois, and Lars Olsson. *"Sustaining Interaction Dynamics and Engagement in Dyadic Child-Robot Interaction Kinesics: Lessons Learnt from an Exploratory Study"*. 14th IEEE International Workshop on Robot and Human Interactive Communication (RO-MAN 2005), Nashville, TN, 2005, pp. 716–722.

28. Henrik I. Christensen and Elena Pacchierotti. *"Embodied Social Interaction for Robots"*. Proceedings of the AISB Convention: Symposium on Robot Companions: Hard Problems and Open Challenges in Human-Robot Interaction, Hatfield, 2005, pp. 40–45.
29. Cynthia Breazeal and Cory D. Kidd.*"Robots as an Interactive Media"*. Technical report, MIT Media Lab, 20.
30. Andrea Lockerd Thomaz, Matt Berlin, and Cynthia Breazeal. *"Robot Science Meets Social Science: An Embodied Computational Model of Social Referencing"*. Towards Social Mechanism of Android Science. A COGSCI 2005 Workshop, Stresa, Italy, July 25–26, 2005, pp. 7–17.
31. Cory D. Kidd and Cynthia Breazeal. *"Human-Robot Interaction Experiments: Lessons Learned"*. Kerstin Dautenhahn and Rene te Boekhorst (Eds.), Robot Companions: Hard Problems and Open Challenges in Robot-Human Interaction. Symposium at Social Intelligence and Interaction in Animals, Robots and Agents (AISB), University of Hertfordshire, Hatfield, England, April 2005, pp. 141–142.
32. Brain R. Duffy. *"Anthropomorphism and Robotics"*. Symposium on Animating Expressive Characters for Social Interactions, Imperial College, England, April 3–5, 2002.
33. Brain R. Duffy. *"Anthropomorphism and the Social Robot"*. Robotics and Autonomous Systems, Vol. 42, pp. 177–190, 2003.
34. Mike Blow, Kerstin Dautenhahn, Andrew Appleby, Crystopher L. Nehaniv, and David Lee. *"The Art of Designing Robot Faces – Dimensions for Human-Robot Interaction"*. HRI 2006, Salt Lake City, UT, March 2–3, 2006.
35. Minato, T. et al. *"Does Gaze Reveal the Human Likeness of an Android"*. Proceedings of 2005 4th IEEE International Conference on Development and Learning. Osaka, Japan, July 19–21, 2005, pp. 106–111.
36. MacDorman, K. et al. *"Assessing Human Likeness by Eye Contact in an Android Testbed"*. CogSci 2005 – 27th Annual Conference of the Cognitive Science Society, July 21–23, 2005, Stresa, Piedmont, Italy.
37. Karl F. MacDorman. *"Android as an Experimental Apparatus: Why Is There an Uncanny Alley and Can We Exploit It?"* Towards Social Mechanism of Android Science. A COGSCI 2005 Workshop.
38. Bryan Adams, Cynthia Breazeal, Rodney A. Brooks, and Brain Scassellati. *"Humanoid Robots: A New Kind of Tool"*. IEEE Intelligent Systems, Vol. 15, No. 4, pp. 25–31, July/August 2000.
39. Walters, M. et al. *"The Influence of Subject's Personality Traits on Predicting Comfortable Human-Robot Approach Distances"*. A COGSCI 2005 Workshop, Stresa, Italy, July 25–26, 2005, pp. 29–37.
40. Billy Lee. *"Interpersonal Perception and Android Design"*. A COGSCI 2005 Workshop, Stresa, Italy, July 25–26, 2005, pp. 50–55.
41. Stephen J. Cowley and Takayuki Kanda. *"Friendly Machines: Interaction Oriented Robots Today and Tomorrow"*. Proceedings of SACLA 2004, University of Kwa-Zulu Natal, Durban, South Africa, July 4–6, 2004.
42. Masahiro Mori. *"Bukimi no tani [The uncanny Valley]"* (Karl F. MacDorman and Takashi Minato, Trans.), Energy, Vol. 7, No. 4, pp. 33–35, 1970.
43. Dave Bryant. *"Why Are Monster-Movie Zombies so Horrifying and Talking Animals so Fascinating?"* Publisher unknown. Available at http://www.arclight.net/~pdb/nonfiction/uncanny-valley.html
44. Takashi Minato et al. *"Development of an Android Robot for Studying Human-Robot Interaction"*. Proceedings of IEA/AIE Conference, 17th International Conference on Innovations in Applied Artificial Intelligence, Ottawa, Canada, 2004, pp. 424–434.
45. Sarah Woods, Kerstin Dautenhahn, and Joerg Schulz. *"Child and Adults' Perspectives on Robot Appearance"*. AISB'05: Social Intelligence and Interaction in Animals, Robots and Agents Proceedings of the Symposium on Robot Companions: Hard Problems and Open Challenges in Robot-Human Interaction, Hatfield, UK, pp. 126–132.
46. David Hanson and Andrew Olney. *"We Can Build You: Scientific, Artistic and Ethical Implication or Robotically Emulating Humans"*. Available at http://hansonrobotics.com/press.php

47. David Hanson, Andrew Olney, Ismar A. Pereira, and Marge Zielke. *"Upending the Uncanny Valley"*. Proceedings of the 20th National Conference on Artificial Intelligence and the 17th Innovative Applications of Artificial Intelligence Conference, Pittsburgh, Pennsylvania, USA, 2005, pp. 1728–1729.

48. Zia-ul-Haque Qazi S.M., Zhiliang Wang, and Ihsan-ul-Haq. *"Human Likeness of Humanoid Robotics Exploring the Uncanny Valley"*. IEEE 2nd International Conference on Emerging Technologies, ICET 2006, Peshawar, Pakistan, November 13–14, 2006.

49. Paul Ekman and Wallace V. Friesen. *"Unmasking the Face."* Prentice Hall, Englewood Cliffs, NJ, 1975.

50. I. Murray and J. Arnott, *"Toward a Simulation of Emotion in Synthetic Speech: A Review of the Literature on Human Vocal Emotion."* Acoust. Soc. Ante., Vol. 93, No. 2, pp. 1097–1108, 1993.

Chapter 20
Reinforced ART (ReART) for Online Neural Control

Damjee D. Ediriweera and Ian W. Marshall

Abstract Fuzzy ART has been proposed for learning stable recognition categories for an arbitrary sequence of analogue input patterns. It uses a match based learning mechanism to categorise inputs based on similarities in their features. However, this approach does not work well for neural control, where inputs have to be categorised based on the classes which they represent, rather than by the features of the input. To address this we propose and investigate ReART, a novel extension to Fuzzy ART. ReART uses a feedback based categorisation mechanism supporting class based input categorisation, online learning, and immunity from the plasticity stability dilemma. ReART is used for online control by integrating it with a separate external function which maps each ReART category to a desired output action. We test the proposal in the context of a simulated wireless data reader intended to be carried by an autonomous mobile vehicle, and show that ReART training time and accuracy are significantly better than both Fuzzy ART and Back Propagation. ReART is also compared to a Naïve Bayesian Classifier. Naïve Bayesian Classification achieves faster learning, but is less accurate in testing compared to both ReART, and Bach Propagation.

Keywords Fuzzy ART · ReART · Back propagation · Naïve Bayesian classifier

20.1 Introduction

Fuzzy ART is an unsupervised Adaptive Resonance Theory (ART) network presented for classifying an arbitrary sequence of analogue input patters into stable recognition categories [1, 2]. In previous work the use of a standard Fuzzy ART

D.D. Ediriweera (✉) and I.W. Marshall
Computing Department, Lancaster University, InfoLab21, South Drive, Lancaster LA1 4WA, UK
e-mail: damjee@gmail.com

S.-I. Ao et al. (eds.), *Advances in Computational Algorithms and Data Analysis*,
Lecture Notes in Electrical Engineering 14,
© Springer Science+Business Media B.V. 2009

network for online neural control was investigated [3]. Further testing has revealed several weaknesses which limit its potential for online control applications. Therefore, ReART, a modified Fuzzy ART network designed to address these limitations, is presented here.

Fuzzy ART is made up of two neuron layers. The first layer represents its input neurons whereas the second layer represents its output categories. Fuzzy ART performs match based learning, and therefore the configuration of the output layer is dynamically determined based on the diversity of the presented inputs. The decision of creating a new output category depends on whether existing categories fail to match an input within a defined threshold. Resonance or learning occurs only when an input is successfully matched to an existing category, or when a new category is created to handle a distinctly new input. This approach allows Fuzzy ART to overcome the plasticity-stability dilemma [2], meaning, it is able to remain stable for known inputs while being plastic (adaptable) towards new ones.

The most challenging issue when applying Fuzzy ART for online control is its unsupervised classification nature. Based on previous experiments, it is revealed that Fuzzy ART often classifies similar input patterns together, with no regard to the class of input which they actually represent. This behaviour of unsupervised ART networks is verified by the work of Christopher and Daniel [4]. In the context of neural control this poses a problem since it means that a single Fuzzy ART category can no longer be mapped to a single output action because a single category might actually represent many input classes.

Further, since the classification process in Fuzzy ART is match based, it can only be controlled using the vigilance threshold. The vigilance value defines the level of similarity required for an input to be classified under an existing category. However, the vigilance value in Fuzzy ART is global to the entire network, and therefore it defines a single category size for the network. It is often found that inputs in the real world regularly represent input classes with varying sizes. Some input classes can be quite general and large in size, whereas others are quite specific and small. The mismatch between the category size of an ART network and the class sizes of inputs will often result in inefficient, or inaccurate categorising. Work presented here demonstrates that issues outlined above can emerge in the form of longer convergence times, unstable performance, and larger than optimal neural configurations, when Fuzzy ART is used for neural control.

This paper introduces ReART which uses a feedback based learning mechanism to overcome these limitations. The feedback mechanism drives the categorisation process by monitoring external feedback for each individual category and using this information to decide when new categories are required. Under this setup the vigilance parameter is typically set to a low value to encourage inputs to be classified into an existing category. Although this sometimes results in the network classifying different input classes in to a single category, such misclassifications are quickly detected when a category generates negative feedback. This feedback acts as a trigger for a new category to be created. This approach allows the network to quickly diagnose and correct misclassifications. Furthermore, since the vigilance value does not play a prominent role in the classification process the network is able to efficiently support input classes of different sizes. At the same time the network is still able to effectively select the best category for each input based on the direct access theorem which guarantees that if a matching output category, U, exists for an input, I, then ART would directly activate U when I is presented [5]. This is part of the Winner Take All (WTA) activation mechanism built in to Fuzzy ART. Work presented here demonstrates that the said modifications allow ReART to learn faster relative to Fuzzy ART, with greater accuracy, and less number of internal neurons.

20.2 Related Work

A review of existing work reveals several ART versions which have emerged since the original concept. Some of these include Fuzzy ART, ART-2, ART-3, DART, ARTMAP [6], ECART, Semi-supervised ART (SMART2) [4], Snap Drift learning (P-ART) [7], Flexible Adaptable-Size Topology (FAST) [8], Grow and Represent (GAR), and SF-ART. Although a majority of ART networks remain unsupervised, several attempts have been made at designing supervised and reinforced ART networks to cater for different requirements.

ARTMAP presents a supervised ART network capable of incremental learning of labelled input patterns. ARTMAP comprises two individual ART networks, ART1 and ART2, linked by an associative learning network. Input patterns are presented to ART1 and their labels are sent to ART2. When an input is presented, ART1 makes a prediction which is confirmed by associating it with the winning label of ART2. If a wrong prediction is made the network increases the vigilance of the winning neuron in ART1 which leads to a different candidate being chosen. The process occurs until the correct category is chosen. Resonance occurs only when the correct candidate is found.

Another approach to supervised ART is investigated in SMART2 [4]. SMART2 represents a modified ART2 network with a learning mechanism which allows learning only within the same class of inputs. This guarantees that similar input patterns from different input classes do not interfere in the learning processes of each other. To complement this, SMART2 also incorporates a mechanism of changing the learning rate depending on whether an input is classified correctly or not. The learning rate is high for inputs which are classified incorrectly. Based on numerical tests SMART2 is claimed to outperform ART2 for classification problems [4].

The Snap Drift algorithm presents a feedback based mechanism for improving the clustering process of ART [7]. This algorithm is designed for networks operating in non-stationary environments where new inputs are regularly received. Snap Drift works by altering the learning rate of individual ART categories depending on the feedback received by the system. This allows the system to snap away when performance is low, and drift when performance is high, hence the name Snap Drift. The literature indicates that the algorithm was successfully applied for generating automated service responses in a simulated active computer network [7].

A specific attempt to use an ART based neural network for neural control was made by Andres Perez [8]. This work investigates an approach of combining a

ART based Flexible Adaptable Size Topology (FAST) network with a reinforcement based action selector. Even though FAST does not employ a supervision mechanism, this work is significant here since it demonstrates the possibility of using ART for neural control. The literature indicates that the ART based neural controller was used for navigation control on a robot [8].

20.3 Reinforced ART (ReART)

In contrast to previous work, ReART uses a new feedback based mechanism to drive the entire categorisation process. ReART architecture is similar to that of Fuzzy ART. The network consists of two neuron fields F1 and F2 (see Fig. 20.1). F1 consists of the input neurons whereas F2 represents the dynamic category field.

The ReART learning algorithm can be summarised as follows:

$\rho \approx 0$ (set a low vigilance value)

while (i not empty)

{

present input ic (ic = complement coded i)

compute activation value Ti for all categories

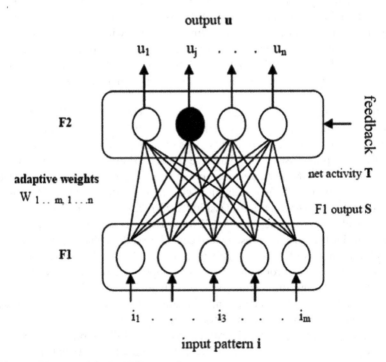

Fig. 20.1 Architecture of the ReART network

```
select category with max Ti as winner Uj
set Uj output to 1
receive feedback fi
if (fi is high)
adapt weights of Uj with new input
(same as Fuzzy ART, wj = U(ic,wj))
update performance of Uj, Ujp = fi
else
add 1 to Uj error count, Uje + +
if (Ujp is high and Uje is low)
(means poor current performance,
good previous performance, and
a low error count)
then high probability of a new category
endif
if (Ujp is high and Uje is high)
(means poor current performance,
good previous performance, and
a high error count)
then high probability of removing category Uj
endif
endif
}
```

Each individual class of input generally has its own desired output, and consequently a misclassified input rarely generates its desired output, and hence would not result in a positive feedback. Therefore, ReART learning is done only when an input classification receives positive feedback indicating that it was classified under the correct category. This speeds up the learning process by reducing the probability of inputs from different input classes interfering with each others learning.

The process of creating a new category is triggered when performance of an existing category is consistently low, or when a trend change is detected. Consistently low performance is a clear indicator of a single category attempting to classify inputs from two or more input classes. Similarly, a category with a history of positive feedback unexpectedly generating negative feedback normally indicates a new input class interfering with the learning of an existing category. ReART responds to misclassification by creating a new category tuned at classifying inputs which are causing the problem. This is the primary growth mechanism in ReART, and it is designed to create new output categories when new input classes are detected.

The decision of creating a new category is probabilistic. The probability of a new category being created is greatest as soon as a misclassification is detected, but decreases over time. Conversely the probability of removing a category is lowest when a misclassification is detected, but increases if performance does not improve over time. This approach offers a poorly performing category the chance to recover by separating its negative feedback generators into a separate category, but if performance does not improve the probability of it being removed approaches a maximum over time. This allows ReART to permanently remove categories which are struggling to improve. This is effective at removing poorly positioned categories which sometimes form between the boundaries of one or more input classes. The probabilistic approach for adding and removing categories is chosen to compensate for noise which might temporarily influence feedback.

20.4 Numerical Evaluation of ReART

The numerical evaluation of ReART was performed using the control architecture illustrated in Fig. 20.2. The figure outlines how ReART integrates with an external map function to achieve a functional neural controller. At a higher level the system works by ReART creating categories and the map function associating them with appropriate output actions. Both networks are driven by external feedback, and operate independently of each other. All outputs of the map function are either 1 or 0. The map function discovers desired output combinations for each ReART category based on a trial and error approach. Identical configurations of the map function were used for all experiments; hence its detailed workings are not discussed. Further information can be found here [3].

Several experiments were carried out to evaluate ReART. It was compared against three learning techniques: Fuzzy ART (using the same configuration as in Fig. 20.2); Back Propagation (BP) [9], and Naïve Bayesian Classification (NBC) [10].

The control problem simulated for the experiments is illustrated in Fig. 20.3. The selected control task is a neural based management of wireless communication on a mobile data reader. The objective of the controller is to optimize the power consumption of the wireless reader by managing communication distance, avoiding

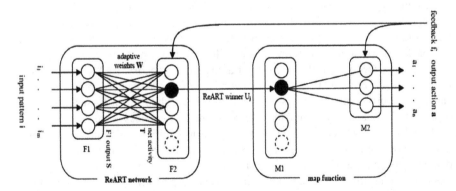

Fig. 20.2 Neural controller constructed using ReART

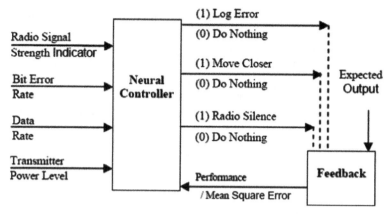

Fig. 20.3 Experiment setup

Table 20.1 Learning technique comparison

Network type	Max training accuracy achieved	When max training accuracy occurs (avg. epoch)	When ≈100 training accuracy occurs (avg. epoch)	Testing accuracy	Avg. no of categories	Avg. no of categories with multiple input classes
ReART	100%	71.30	11.86	98.42%	6.57	0.0
BP (4-5-3)	100%	1580.75	81.25	99.84%	5.00	0.0
Funzzy ART (0.2)	76.01%	27.88	Never	73.24%	2.97	2.3
Funzzy ART (0.4)	77.22%	41.61	Never	74.92%	5.75	2.9
Funzzy ART (0.6)	85.25%	64.28	Never	82.54%	12.42	3.7
Funzzy ART (0.8)	94.67%	129.59	Never	92.83%	31.92	4.6
NBC (50-50)	100%	8.69	5.13	95.51%	–	–
NBC (75-25)	100%	10.69	7.13	97.28%	–	–
NBC (90-10)	100%	11.84	8.27	98.02%	–	–

radio interference, adapting to host conditions, and by detecting errors. To achieve this, the network uses four inputs, and up to eight output combinations. The input set includes the Radio Signal Strength Indicator (RSSI), Bit Error Rate (BER), Data Rate (DR), and the Transmitter Power Level (TPL); all values monitored from the wireless reader. Possible output behaviours include, moving the reader closer to the transmitter, enforcing temporary radio silence, staying neutral (meaning continue current transmission), logging errors, and other possible combinations of the above.

A dataset of 1,200 input sets was recorded from a live environment. The dataset captured five distinct input classes. Each input class was assigned with a desired output action based on real world considerations. The desired output patterns were used to generate feedback for the ReART controller, and to calculate the Mean Square Error for BP. Feedback for ReART was binary, positive feedback was provided when an action was correct, and negative feedback otherwise. Experiments were run exhaustively, and all results were averaged over 500 independent runs. For all experiments excluding NBC, 75% of the dataset was used for training and 25% was reserved for testing. As indicated by Table 20.1, different proportions of training to testing data were used for NBC. In all cases, a 5% noise component was introduced to both training and test datasets. To simulate realistic conditions inputs were presented to the network in their natural order, each run starting at a random point in the training dataset.

20.5 Results and Analysis

Figure 20.4 compares the performance of ReART with BP and several Fuzzy ART configurations using different vigilance values. Results clearly illustrate ReART outperforming both BP and Fuzzy ART under tested conditions. ReART achieves an accuracy of over 90% within an average of 1 epoch, whereas to achieve the same BP requires an average 12.78 epochs, and the best Fuzzy ART configuration requires an average of 17.23 epochs. The BP configuration used for the experiment was selected after testing a range of configurations and therefore is believed to be optimal for the specified problem.

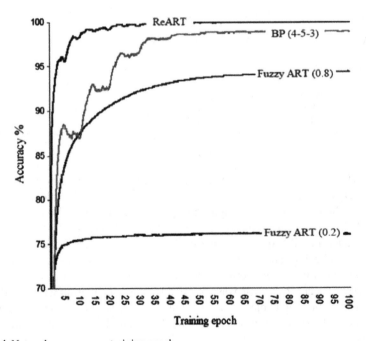

Fig. 20.4 Network accuracy vs. training epoch

The accuracy recorded in Fig. 20.4 indicates the number of correct actions observed for the most recent 100 inputs. An accuracy of 90 literally indicates 90 correct actions and 10 incorrect ones within the last 100 inputs. The axis indicating accuracy in Fig. 20.4 is scaled between 70 and 100 to improve clarity.

The notation BP (I, H, O) is used to identify a BP network with, I, input neurons, H, hidden neurons and, O, output neurons. Fuzzy ART (ρ) is used to identify a Fuzzy ART network with a vigilance value of ρ. NBC (x-y) is used to indicate x% of the dataset for training, and y% for testing.

ReART, compared with BP is able to learn faster. Table 20.1 reveals that both ReART and BP are able to reach a training accuracy near 100%, but ReART achieves this several magnitudes faster than BP. The extra training allows BP to generalise better as indicated by the higher BP testing accuracy of 99.84%, compared to the 98.42% of ReART. BP also uses relatively fewer internal neurons to achieve a similar level of accuracy; however, it is common knowledge that identifying the correct BP configuration is not straight forward.

A comprehensive comparison of ReART with Fuzzy ART is provided in Table 20.1. Figures here demonstrate the difficulties in selecting a global vigilance value to fit an entire dataset. Four separate Fuzzy ART configurations with different vigilance values were tested. No configurations were discovered which were able to efficiently classify the inputs correctly. Fuzzy ART (0.8) achieves the best accuracy of 92.83% but uses approximately 32 categories. In contrast Fuzzy ART (0.2) creates approximately 3 categories but fails to exceed an accuracy of 76.01%.

In Fuzzy ART the lower vigilance values struggle to precisely separate input patterns belonging to different input classes, whereas the higher vigilance values do a better job but results in multiple categories representing single input classes. This is clearly indicated by column nine of Table 20.1 which identifies the number of categories which were classifying multiple input classes at the end of the training. The percentage of such categories in the network has a direct relationship with the network vigilance and its performance. Fuzzy ART (0.2) had almost 77% of its clusters classifying multiple input classes with a maximum accuracy of 73.24%, whereas Fuzzy ART (0.8) had 14% of its clusters classifying multiple input classes with a maximum performance of 92.83%. Based on these results it is clear that none of the tested Fuzzy ART configurations were able to match the performance of ReART which achieved a training accuracy of 100% with only six to seven categories.

NBC (75-25) compared to ReART is able to learn faster. NBC (75-25) learns the dataset within an average of 7.13 epochs, while ReART requires an average of 11.86 epochs to achieve the same. Although NBC is able to learn faster, it performs less satisfactorily in testing. ReART demonstrates an average test accuracy of 98.42%, whereas NBC (75-25) is only able to reach 97.28%. This indicates that NBC is weaker at generalising compared to ReART. Accuracy of NBC (90-10) outperforms NBC (75-25) by approximately 1%, showing that NBC accuracy can be improved if a large percentage of the dataset is used for training.

In addition to its ability to learn quickly with high accuracy, the ReART based controller is also able to address the plasticity stability dilemma. The plasticity stability dilemma outlines a problem which prevents most neural networks from

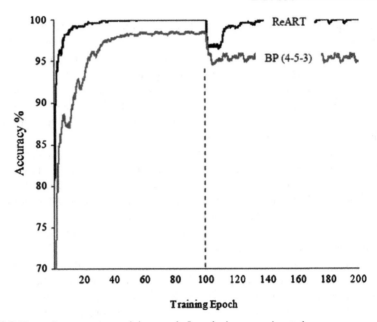

Fig. 20.5 Network accuracy vs. training epoch: Introducing a new input class

learning new inputs while preserving previously gathered knowledge. Even a popular and robust control mechanism such as BP is inherently troubled by this problem.
Figure 20.5 illustrates a comparison of BP and ReART when presented with the
plasticity stability dilemma. Here each network was initially allowed to train for 100
epochs. At the 100 epoch mark 50 new input patterns were introduced to the dataset.
New patterns represented a new input class with its own desired output action. Both
networks were then trained for an additional 100 epochs. The performance response
of both networks is presented in Fig. 20.5. The axis indicating network accuracy is
scaled between 70 and 100 to improve clarity.

Results indicate ReART to handle new inputs with greater effectiveness than
BP. The introduction of new inputs causes ReART accuracy to temporarily drop to
97%, but recovers quickly within few epochs to an approximate 100%. The ReART
network responded to the new input class by creating a new category to classify it.
This allows it to learn a new input with minimum impact on its existing knowledge
and accuracy. BP performance under identical conditions is less effective. BP is
able to reach an initial accuracy of 96% to 97% with 100 epochs of training. The
introduction of new inputs to BP causes its accuracy to drop to 94%, but unlike in
ReART the accuracy fails to recover beyond this during the rest of the 100 epochs
of training. The BP network recovers to its original accuracy approximately after
175 to 195 additional epochs of training. The result is as expected since BP learning
does not necessarily cater for network adaptability.

Compared to ReART, NBC is faster at learning new input to output mappings
if they are independent, and non-conflicting with existing knowledge. However,

ReART is faster when changing the mapping between an existing input category and its output. This is expected given that NBC stores input to output relationships in long term memory which it uses for calculating membership probabilities for new inputs. Therefore, it requires sufficient persistence before it can successfully change its long term memory to reflect new conditions. In contrary ReART is able to adapt faster since it does not employ the type of long term memory used by NBC.

20.6 Conclusion

Several limitations restricting the use of Fuzzy ART and other unsupervised ART networks in neural control were demonstrated. Fuzzy ART was modified to develop ReART, a feedback based ART network capable of addressing these limitations. ReART was utilized to construct a neural controller capable of online learning. The ReART based controller was compared through numerical testing with BP, NBC, and an identical Fuzzy ART based controller. Results indicate ReART outperform both BP and Fuzzy ART for the presented control problem. ReART learns several magnitudes faster than BP, and provides a similar level of training accuracy. Further, ReART learns faster, with greater accuracy and less internal categories than Fuzzy ART. It also avoids the Fuzzy ART problem of classifying multiple input classes under a single category. The ReART based controller also overcomes the plasticity stability dilemma, and is able to learn new inputs with a minimum impact on existing knowledge and accuracy. ReART compared to NBC is slower at learning, but provides a greater level of test accuracy. However, NBC is much simpler to design, and train compared to ReART.

Further work is planned on testing ReART and NBC for classifying more complex datasets to confirm whether these fast learning mechanisms are able to cope with more subtle categorisations. In addition, the performance of these techniques is to be tested using real feedback from a live environment with potentially greater noise and feedback errors. Work is currently ongoing on building several real world experiments for this purpose.

References

1. G. A. Carpenter, S. Grossberg. Fuzzy ART: Fast stable learning and categorization of analogue patterns by an adaptive resonance system. Neural Networks, Vol. 4, 1991, pp. 759–771.
2. G. A. Carpenter, S. Grossberg, Adaptive resonance theory. In M.A. Arbib (Ed.), The Handbook of Brain Theory and Neural Networks, Second Edition. Cambridge, MA: MIT, 2003, pp. 87–90.
3. D. D. Ediriweera, I. W. Marshall, Internally self organising neural network for online learning of perception to action mappings in sensor networks, Proceedings of London Communications Symposium, 2005, pp. 21–24.

4. C. J. Merz, D. C. St. Clair, W. E. Bond, SeMi-supervised adaptive resonance theory (SMART2). IJCNN, International Joint Conference on Neural Networks, Vol. 3, 7–11 June 1992, pp. 851–856.
5. A. Weitzenfeld, M. A. Arbib, A. Alexander, Chapter Eight in the Neural Simulation Language, Cambridge, MA: MIT, ISBN: 0-262-73149-5, July 2002, pp. 157–169.
6. G. A. Carpenter, S. Grossberg, J. H. Reynolds, ARTMAP: Supervised Real-Time Learning and Classification of Nonstationary Data by a Self-organizing Neural Network, Neural Networks (Publication), Vol. 4, 1991, pp. 565–588.
7. S. W. Lee. D. Palmer-Brown. C. M. Roadknight. Performance-guided neural network for rapidly self-organising active network management, Neurocomputing, special issue on Hybrid Neurocomputing, Dordrecht: Elsevier, 2004, 61C, pp. 520.
8. A. Pérez-Uribe. A non-computationally-intensive neurocontroller for autonomous mobile robot navigation. Chapter 8 in Biologically inspired robot behaviour engineering, Series: Studies in Fuzziness and Soft Computing, Vol. 109. In R. J. Duro, J. Santos, M. Grana (Eds.), Springer, Germany, 2002, pp. 215–238.
9. R. Beale, T. Jackson, Neural Computing: An Introduction, ISBN: 0852742622, Institute of Physics Publishing, London, UK, 1990, pp. 63–80.
10. Rish, Irina, An empirical study of the naive Bayes classifier, IJCAI 2001 Workshop on Empirical Methods in Artificial Intelligence, 2001.

Chapter 21
The Bump Hunting by the Decision Tree with the Genetic Algorithm

Hideo Hirose

Abstract In difficult classification problems of the z-dimensional points into two groups giving 0–1 responses due to the messy data structure, it is more favorable to search for the denser regions for the response 1 points than to find the boundaries to separate the two groups. For such problems which can often be seen in customer databases, we have developed a bump hunting method using probabilistic and statistical methods as shown in the previous study. By specifying a pureness rate in advance, a maximum capture rate will be obtained. In finding the maximum capture rate, we have used the decision tree method combined with the genetic algorithm. Then, a trade-off curve between the pureness rate and the capture rate can be constructed. However, such a trade-off curve could be optimistic if the training data set alone is used. Therefore, we should be careful in assessing the accuracy of the trade-off curve. Using the accuracy evaluation procedures such as the cross validation or the bootstrapped hold-out method combined with the training and test data sets, we have shown that the actually applicable trade-off curve can be obtained. We have also shown that an attainable upper bound trade-off curve can be estimated by using the extreme-value statistics because the genetic algorithm provides many local maxima of the capture rates with different initial values. We have constructed the three kinds of trade-off curves; the first is the curve obtained by using the training data; the second is the return capture rate curve obtained by using the extreme-value statistics; the last is the curve obtained by using the test data. These three are indispensable like the Trinity to comprehend the whole figure of the trade-off curve between the pureness rate and the capture rate. This paper deals with the behavior of the trade-off curve from a statistical viewpoint.

Keywords Bump hunting · Decision tree · Genetic algorithm

H. Hirose
Department of Systems Innovation and Informatics, Kyushu Institute of Technology,
Fukuoka, 820-8502 Japan
e-mail: hirose@ces.kyutech.ac.jp

S.-I. Ao et al. (eds.), *Advances in Computational Algorithms and Data Analysis*,
Lecture Notes in Electrical Engineering 14,

21.1 Introduction

The main objective of the article is to address the statistical aspects of the trade-off curve in the bump hunting method. However, we briefly review the study so far in this section before we discuss the main topic. We will explain, (1) what the bump hunting is, (2) the trade-off curve between the pureness rate and the capture rate, (3) the bump hunting using the decision tree with the genetic algorithm, (4) upper bound for the trade-off curve by using the extreme-value statistics, (5) actual lower bound for the trade-off curve, and (6) their summary.

21.1.1 What is the Bump Hunting?

Suppose that we are interested in classifying n points in a z-dimensional space into two groups according to their responses, where each point is assigned response 1 or response 0 as its target variable. For example, if a customer makes a decision to act a certain way, then we assign response 1 to this customer, and assign response 0 to the customer that does not. We want to know the customers' preferences presenting response 1. We assume that their personal features, such as gender, age, living district, education, family profile, etc., are already obtained.

Many classification problems have been dealt with elsewhere in rather simpler cases using the methods of the linear discrimination analysis, the nearest neighbor, logistic regression, decision tree, neural networks, support vector machine, boosting, etc. (see [1], e.g.) as fundamental classification problems. In some real data cases in customer classification, it is difficult to find the favorable customers, because many response 1 points and 0 points are closely located, resulting response 1 points are hardly separable from response 0 points [2, 3]. In such a case, to find the denser regions to the favorable customers is considered to be an alternative. Such regions are called the bumps, and finding them is called the bump hunting; see Fig. 21.1. The bump hunting has been studied in the fields of statistics, data mining, and machine learning [4–7].

21.1.2 Trade-Off Curve Between the Pureness Rate and the Capture Rate

By specifying a pureness rate p_0 in advance, where the pureness rate p is the ratio of the number of points of assigned response 1 to the total number of points assigned responses 0 and 1 in the target region, a maximum capture rate c_m will be obtained, where the capture rate c is the ratio of the number of points assigned response 1 to the number of points assigned responses 0 and 1 in the total regions. Then a trade-off curve between the pre-specified pureness rate p_0 and the maximum capture rate c_m can be constructed; see Fig. 21.2.

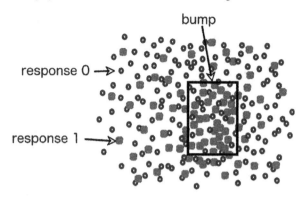

Fig. 21.1 The bump hunting for the denser regions to response 1 points which are hardly separable from response 0 points

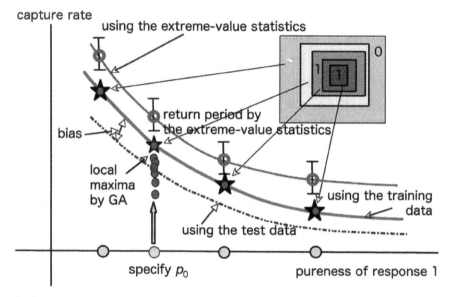

Fig. 21.2 Three trade-off curves between the pureness rate and the capture rate: (1) using the training data, (2) using the test data, (3) using the extreme-value statistics

Now, we let TP be true positive, TN be true negative, FP be false positive, and FN be false negative. Since a response 1 point in or outside the bump regions is considered to be TP or FN, respectively, and a response 0 in or outside the bumps is FP or TN, the pureness rate p can be defined by

$$p = \frac{\#TP}{\#TP + \#FP}$$

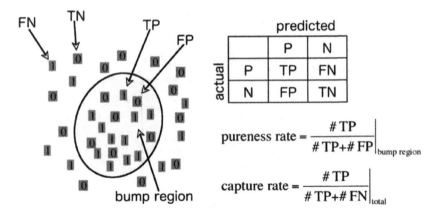

Fig. 21.3 The pureness rate and the capture rate

in the bump regions; the capture rate c can also be defined by

$$c = \frac{\#TP}{\#TP + \#FN}$$

in the total region [8], where "#" expresses the size of the samples; see Fig. 21.3.

You may remind similar curves, i.e., the ROC (Receiver Operator Characteristic) curve and the recall precision curve in medical and machine learning fields [9, 10]. Because, in a recall-precision curve, recall is defined by $\#TP/(\#TP + \#FN)$ which is identical to the capture rate, and precision is defined by $\#TP/(\#TP + \#FP)$ which is identical to the pureness rate; thus, a trade-off curve between the capture rate and the pureness rate seems to be equivalent to a recall-precision curve superficially [9, 10], e.g. However, we should note that these two are totally different from each other. As is seen in Fig. 21.4, it can be considered that our trade-off curve is constructed by collecting the skyline points consisting of many trade-off curves where each curve is corresponding to one classifier [11]. This trade-off curve is corresponding to the results obtained by a subset of the real customer data case as will b explained in Section 3.

21.1.3 Bump Hunting Using the Decision Tree with the Genetic Algorithm

In order to make future actions easier, we adopt simpler boundary shapes such as the union of z-dimensional boxes located parallel to some explanation variable axes for the bumps as shown in Fig. 21.1; that is, we use the binary decision tree. In decision trees, by selecting optimal explanation variables and splitting points to split the z-dimensional explanation variable subspaces into two regions from the top node

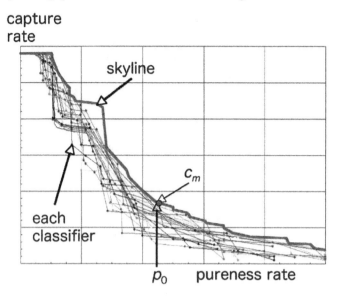

Fig. 21.4 Trade-off curve as a skyline curve consisting of many classifiers

to downward using the Gini's index as in the conventional method, we may obtain the number of response 1 points by collecting nodes where the pureness rates are satisfying if they turn out to be larger than the pre-specified pureness rate p_0. However, much response 1 points could be obtained if we locate adequate explanation variables to each branching knot. This is because the conventional algorithm has the property of local optimizer. Thus, we have developed a new decision tree method with the assistance of the random search methods such as the genetic algorithm (GA) specified to the tree structure, where the most adequate explanation variables are selected by the GA, but the best splitting points are obtained by using the Gini's index [12]; see Figs. 21.5 and 21.6. In the figure, a cross over method is shown in the GA procedure.

So far, we have been using the following evolution procedure in the GA:

1. The number of initial seeds is set to 30; here, the initial seeds mean the trees where the explanation variables to be allocated to each branch are randomly selected.
2. Obtain the capture rate to each seed, and select the top 20 best trees.
3. In the next generation, divide each tree to the left wing with or without the top node and the right wing with or without the top node, and combine the left wing and right wing trees of different parents to produce children trees; this is a cross over procedure; 40 children are then delivered, and select the top 20 best trees.
4. This evolution procedure is succeeded up to 20 generations.
5. At the final stage, select the best one rule (tree) to apply the future data.
6. The mutation rate is set to around 5%.

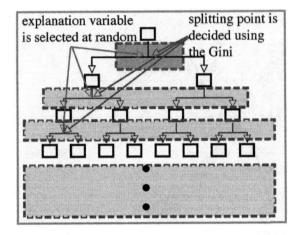

Fig. 21.5 The bump hunting procedure using the decision tree with the genetic algorithm

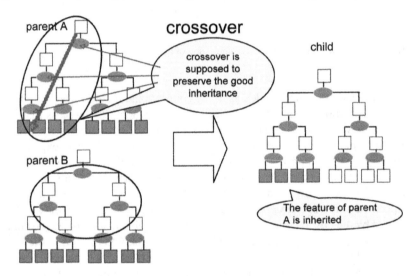

Fig. 21.6 The cross over procedure in the tree structure genetic algorithm

21.1.4 Upper Bound for the Trade-Off Curve by Using the Extreme-Value Statistics

As shown in [12], for the ten cases of the iteration procedures with ten different initial conditions in an simulated example, the converged solutions differ from each other when the initial value is set to different values. Our GA algorithm has a strong inclination of searching for the local maxima because we are using the tree structure in evolution procedure. Solutions obtained by the GA are primarily not global

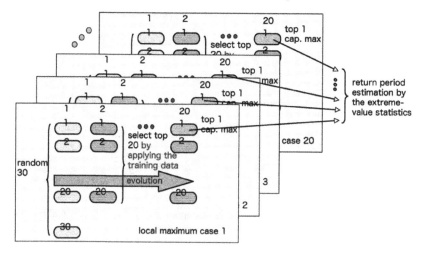

Fig. 21.7 The genetic algorithm procedure to provide the samples for the extreme-value statistics

optimal; this is a drawback of the algorithm. However, we have observed the existence of many local maxima with each different starting point in our GA procedure. This turns out to become an advantage; the use of the extreme-value statistics [13] can then be used to estimate the return period (expected global maximum capture rate), and the method did work successfully when the shape of the marginal density function of an explanation variable is simple, such as monotonic or unimodal [12, 14]. This property is also observed in a real customer database [11].

Thus, we add a function of

7. Estimating the upper bound capture rate by using the 20 trees where each tree is the best one tree at the final stage of the evolution procedure with each starting point in our GA; that is, we do procedures (1)–(6) for 20 cases, and we estimate the upper bound using these 20 local maxima. Figure 21.7 shows the genetic algorithm procedure to provide the samples for the extreme-value statistics. In Fig. 21.2, how we have obtained the trade-off curve for the upper bound is shown.

21.1.5 The Bias of the Capture Rates in the Trade-Off Curve

The solutions mentioned above are, however, the best fitted solutions [1]; that is, the rules are constructed by using the training data and the evaluations are performed also by using the same training data; so, the solutions could be optimistic [15, 16]. If we apply the rules obtained by the training data to a new test data case having the same data structure, we may no longer expect the same performance in the new data case. We have been aware that we should pay much attention to these kind of problems even though the size of the explanation variable is small [11].

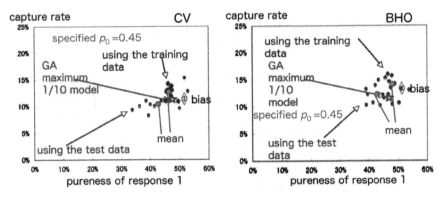

Fig. 21.8 The cross validation results and the bootstrapped hold-out results in a real data case. Pre-specified pureness rate is set to 0.45

The bootstrapped hold-out method (abridged by BHO) proposed [8, 17] or the cross validation method has been used to estimate the biases between the results using the training data and those using the test data [8, 11]. Here, the bootstrapped hold-out produces the tree by using the training data of randomly selected half size of the sample data without replacement, and we evaluate the performance of this rule by using the rest of the half size of the samples. The typical cross validation of ten-fold cross validation requires the ten performance evaluations. Thus, we also evaluate the ten cases of the GA procedure in the bootstrapped hold-out procedure [12]. In Fig. 21.8, we show a typical case for the cross validation results and the bootstrapped hold-out results in a real data case where p_0 is pre-specified to 0.45 [11]; we can guess the bias between the capture rate obtained by using the training data and that by using the test data. The real data case will be explained later.

21.2 Objective of the Research

Summarizing the above, the trade-off curves we are dealing with have three aspects. The first is the curve obtained by using the training data; we can apply the if-then-rules to the future data only to this curve. The second is the return curve obtained by using the extreme-value statistics; by estimating the ceiling for the capture rate, we can know where we are. The last is the curve obtained by using the test data; we can expect the actual capture rate for response 1. Figure 21.2 shows the relationships among these three curves. These three are indispensable like the Trinity to comprehend the whole figure of the trade-off curve between the pureness rate and the capture rate.

In our previous studies, however, we have not paid much attention to the statistical consideration of the GA outcomes. In this paper, we deal with this point.

21.3 Customer Data

A real customer data case we are dealing with is taken from a corresponding course in Japan [8, 11]. The number of customers is very large, say 160,000; thus, we will not use all these data because of the high computing cost. Therefore, we will treat 15,870 samples, randomly selected from the original database, where the number of response 1 (customers we are interested in) is 2,863; thus the mean pureness rate becomes 18.0%. The number of features of the customers is more than 60, but we will use 41 variables; the variables are continuous and discrete. We call this 1/10 model here. A much smaller case consisting of 1,635 samples was also investigated, where the number of response 1 is 290; the mean pureness rate is 17.7%. The number of variables is 44. We call this 1/100 model. Our primary objective here is how many response 1 samples can be captured if we require at least 40–50% pureness rate from a practical viewpoint using these two smaller models.

21.4 Extreme-Value Statistics Approach

As mentioned before, our genetic algorithm has a strong inclination of searching for the local maxima because we are using the tree structure in evolution procedure. This property turns out to be a merit to know the upper bound for the trade-off curve, although, in general, the genetic algorithm will not guarantee the global maximum. A set of samples collected from the local maxima could be samples for the extreme-value statistics for maxima. If the mother distribution function is a normal, exponential, log-normal, gamma, Gumbel, or Rayleigh type distribution, then the limiting distribution of the maximum values from the mother distribution follows the Gumbel distribution (see [13] e.g.). Thus, we apply the Gumbel distribution to the local maxima samples. In the following, we investigated the two cases, 1/100 model and 1/10 model, to assess the reliability for the trade-off curve due to the number of samples.

21.4.1 1/100 Model Real Data Case: When the Number of Samples is Small

For example in a data case where samples are drawn at random with 1/100 probability from a real customer data case, i.e., 1/100 model mentioned above, we have 20 local maxima of 48, 45, 48, 39, 56, 44, 32, 41, 56, 70, 40, 49, 42, 52, 38, 53, 47, 55, 34, 45, for the number of captures when we specify the pureness rate of 50%. If we fit the Gumbel distribution to the data, we can estimate the shape and scale parameters as 7.38 and 42.6. Then, the return capture rate (return period) for 500 trials is estimated to be 88.5. In Fig. 21.9, we can see that the histogram of the sampled data

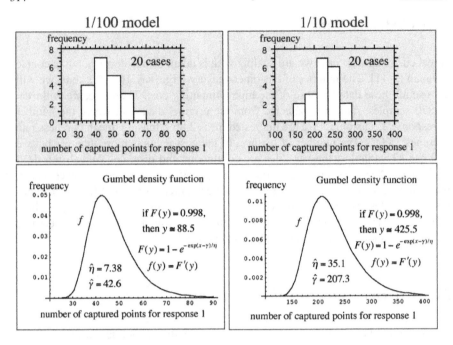

Fig. 21.9 Gumbel distribution fit to the 20 local maxima using the genetic algorithm

well expresses the fitted Gumbel distribution density function. Using this result, we can guess the number of return captures in the real full data case, which could be 8,661 and this corresponds to 30.5% capture rate. However, the results by applying the test data to the rules obtained by the training data was very pessimistic. The bias between the training data trade-off curve and test data one shown in Fig. 21.10 becomes very large because of large number of explanation variables, resulting that the rules obtained by the training data are not applicable to decide future action. In the figure, each point shows the mean value of the ten cases bootstrapped hold-out results.

21.4.2 1/10 Model Real Data Case: When the Number of Samples is Large

A much larger case of 1/10 model mentioned above is also investigated. The 20 local maxima are 207, 230, 251, 258, 255, 238, 170, 229, 204, 292, 247, 218, 281, 237, 230, 206, 195, 208, 193, 147, for the number of captures by the half training data, and the number of return capture is estimated as 425.5. Using this result, we can guess the number of return captures in the real full data case, which could be 4,290 and this corresponds to 14.9% capture rate. The results in this case are considered

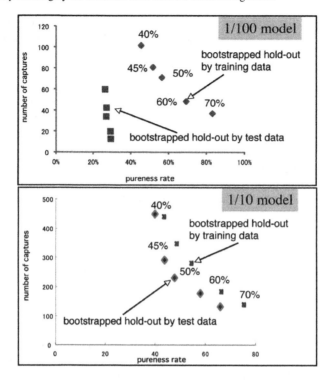

Fig. 21.10 Gumbel distribution fit to the 20 local maxima using the genetic algorithm

to be reliable because of small biases (see Figs. 21.9 and 21.10); however, the result obtained by using the 1/100 model is considerably optimistic. About two times of the response 1 points are obtained in the 1/100 model. We should, thus, pay much attention to the number of samples if the number of explanation variable is large.

21.4.3 How to Assess the Trade-Off Curves

As explained before, we can construct the three kinds of trade-off curves. First, we simply obtain the optimal tree with a pre-specified pureness rate using all the sampled data as the training data. By connecting these obtained points, we can construct the trade-off curve. We have to use these rules (trees) actually. This curve is, however, optimistic.

Therefore, to know the actual capture rate, we made the second trade-off curve. By dividing the sample data into two groups, training data and the test data, which is determined by the accuracy evaluation methods such as the cross validation and the bootstrapped hold-out, a number of points of pureness rate and capture rate are obtained for the training data set and the test data set. By averaging these points, we can obtain the bias between the training data results and the test data results. Then, we can apply this bias to the very original trade-off curve which is obtained by using all the sampled data. This estimated trade-off curve indicates the actual capture rate when we specify the pureness rate. In that sense, this curve can be a lower bound for the trade-off curve.

All the points in the trade-off curve are obtained by using the direct results from the genetic algorithm procedure. The maximum value obtained for the capture rate cannot necessarily be the real global maximum. However, using the extreme-value statistics, we can estimate the return period (return capture rate) as the global maximum value. Therefore, we are able to know where we are, i.e., we can estimate the upper bound ceiling for the trade-off curve.

These three kinds of the trade-off curves are, in a sense, like the Trinity to comprehend the whole figure of the trade-off curve. These three curves are typically illustrated in Fig. 21.2.

21.4.4 Reliability for the Return Period Due to the Number of Initial Seeds

All the initial seeds have been set to 30 so far. Considering that the property of the local convergence of the GA procedure, it would be better to provide much larger number of seeds to verify if the extreme-value statistics works well. Figure 21.11 shows the results of the number of captures in a data case resampled from the real customer data case when the number of seeds and the successive number of iterations are set to larger values. Here, the pre-specified pureness rate is 50% and the model is 1/100 scale. We can see that the extreme-value statistics work very well,

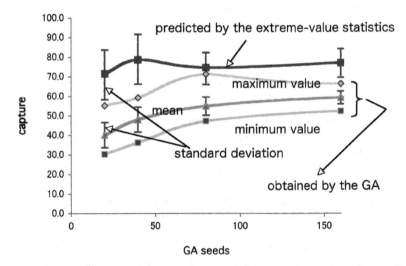

Fig. 21.11 Number of captures of response 1 versus number of the seeds in the genetic algorithm

and we can use this method even if the number of samples to the Gumbel distribution is small such as 20; we can see that the predicted number of captures by the extreme-value statistics preserve almost a constant value even though the converged local maxima are gradually becoming larger as the number of seeds becomes larger.

21.5 Concluding Remarks

In difficult classification problems, we have shown that the bump hunting method using the decision tree combined with the genetic algorithm is useful. In this paper, we have shown that the trade-off curve between the pureness rate and the capture rate can be characterized into three categories; (1) using all the sampled data as the training data, (2) using sampled data as the test data, and (3) using the extreme-value statistics. The trade-off curve obtained by using only the training data becomes optimistic. To know the actual capture rate, we should use the test data results; there is a bias between the result obtained by the training data and that by the test data. In addition, the trade-off curve by using the genetic algorithm cannot be the optimal. Then, we estimate the upper bound for the trade-off curve by using the exreme-value statistics. These three kinds of the trade-off curves are, in a sense, like the Trinity to comprehend the whole figure of the trade-off curves between the pureness rate and the capture rate; thus it is recommended to take into account these three categories altogether.

If the number of explanation variables is large to some extent and the number of samples is small, the bias between the result using the training data and that using the test data may become large even if the relaxation method is used, showing the pessimistic results. Reduction of the explanation variables and the increase of the number of samples will solve this difficulty.

The ROC curve and precision recall curve are compared to the proposed trade-off curve, showing that the trade-off curve is different from the precision recall curve even if they are superficially equivalent.

References

1. Hastie, T., Tibshirani, R., and Friedman, J.H.: Elements of Statistical Learning. New York: Springer (2001).
2. Hirose, H.: A method to discriminate the minor groups from the major groups. Hawaii Int. Conf. Stat. Math. Related Fields. (2005).
3. Hirose, H.: Optimal boundary finding method for the bumpy regions. IFORS2005.
4. Agarwal, D., Phillips, J.M., and Venkatasubramanian, S.: The hunting of the bump: On maximizing statistical discrepancy, SODA'06. (2006) 1137–1146.
5. Becker, U. and Fahrmeir, L.: Bump hunting for risk: A new data mining tool and its applications, Comput. Stat., 16 (2001) 373–386.
6. Friedman, J.H. and Fisher, N.I.: Bump hunting in high-dimensional data, Statistics and Computing. 9 (1999) 123–143.

7. Gray, J.B. and Fan, G.: Target: Tree analysis with randomly generated and evolved trees. Technical report. The University of Alabama (2003).
8. Hirose, H., Ohi, S., and Yukizane, T.: Assessment of the prediction accuracy in the bump hunting procedure. Hawaii Int. Conf. Stat. Math. Related Fields. (2007).
9. Davis, J. and Goadrich, M.: The relationship between precision-recall and ROC curves, Proc. 23rd Intl. Conf. Mach. Learn. (2006).
10. Fawcett, T.: An introduction to ROC analysis, Pattern Recog. Let. 27 (2006) 861–874.
11. Hirose, H., Yukizane, T., and Deguchi, T.: The bump hunting method and its accuracy using the genetic algorithm with application to real customer data. submitted.
12. Yukizane, T., Ohi, S., Miyano, E., and Hirose, H.: The bump hunting method using the genetic algorithm with the extreme-value statistics. IEICE Trans. Inf. Syst., E89-D (2006) 2332–2339.
13. Castillo, E.: Extreme Value Theory in Engineering. San Diego, CA, USA: Academic (1998).
14. Hirose, H., Yukizane, T., and Miyano, E.: Boundary detection for bumps using the Gini's index in messy classification problems. CITSA2006. (2006) 293–298.
15. Efron, B.: Estimating the error rate of a prediction rule: Improvements in cross-validation, JASA. 78 (1983) 316–331.
16. Kohavi, R.: A study of cross-validation and bootstrap for accuracy estimation and model selection. IJCAI. (1995).
17. Yukizane T., Hirose, H., Ohi, S., and Miyano, E.: Accuracy of the Solution in the Bump Hunting. IPSJ MPS SIG report, MPS06-62-04 (2006) 13–16.

Chapter 22
Machine Learning Approaches for the Inversion of the Radiative Transfer Equation

Esteban Garcia-Cuesta, Fernando de la Torre, and Antonio J. de Castro

Abstract Estimation of the constituents of a gas (e.g. temperature, concentration) from high resolution spectroscopic measurements is a fundamental step to control and improve the efficiency of combustion processes governed by the Radiative Transfer Equation (RTE). Typically such estimation is performed using thermocouples; however, these sensors are intrusive and must undergo the harsh furnace environment. In this paper, we follow a machine learning approach to learn the relation between the spectroscopic measurements and gas constituents such as temperature, concentration and length. This is a challenging problem due to the non-linear behavior of the RTE and the high dimensional data obtained from sensor measurements. We perform a comparative study of linear and neural network regression models, using canonical correlation analysis (CCA), principal component analysis (PCA), reduced rank regression (RRR), and kernel canonical correlation (KCCA) to reduce the dimensionality.

Keywords Feature extraction · Neural networks · Dimensionality reduction · Component analysis · Radiative transfer equation

22.1 Introduction

The regulation of harmful substances is getting tighter in several commercial plants using combustion processes (e.g. gas turbines, boilers, incinerators). The control and retrieval of temperature is an important factor to understand the mechanism

E. Garcia-Cuesta (✉) and A.J. de Castro
Physics Department, Carlos III University Avenida de la Universidad 30, 28911 Leganes (Madrid), Spain
e-mail: egc@fis.uc3m.es

F. de la Torre
Robotics Institute, Carnegie Mellon University Pittsburgh PA 15213, USA

S.-I. Ao et al. (eds.), *Advances in Computational Algorithms and Data Analysis,*
Lecture Notes in Electrical Engineering 14,
© Springer Science+Business Media B.V. 2009

of the combustion, and thus minimize environmental disruption and improve the efficiency of combustors [1–4]. Usually, thermocouples have been used to estimate temperature, however they have several drawbacks: they are intrusive and disturb the measurement, they must undergo the harsh furnace environment, and these measurements are taken at single points.

Several efforts have been made in the literature to reconstruct the flame temperature. Some studies make use of different spectral regions (ultraviolet, visible, infrared) [1]. The infrared sensing appears to be very promising due to the fact that hot gases in a flame, mainly carbon dioxide CO_2 and H_2O, exhibit important emission bands in it. Others are based on spectroscopic data and use tunable infrared laser and optical fibre in order to measure the difference between the energy emitted and received [2]. This is considered an active technique because it uses an infrared source in addition to the sensor system. This method is high sensitive, but its high complexity and high cost make it not suitable for routine operations in industrial furnaces.

To overcome some of the limitations of previous approaches, we use remote sensing infrared measurements (i.e. spectrometer) to infer gas constituents at a distance. The remote sensing approach is a passive technique and it does not disturb the measurement. Figure 22.1 illustrates this approach. The spectral characteristics of the flame are recorded with a spectrometer at a given distance. The forward model is described with a set of well known integral equations (see Section 22.2). The interaction of electromagnetic radiation with matter modifies to some extent the incident wave. The spectrometer measures the amount of energy emitted by the target over a range of wave numbers, and these measurements are related with the different properties of the target.

The challenge that we face in this paper is to invert the forward model. Analytical or numerical inversion of forward model is a challenging mathematical problem. In this paper, we make use of machine learning techniques to learn the relation between the flame energy spectrum and the temperature, concentration, and length.

The rest of the paper is organized as follows: in Section 22.2 we review previous work in the physics of RTE model and related work in machine learning approaches. In Section 22.3, we introduce the inverse problem as a supervised learning problem. Several techniques such as principal component analysis, reduced rank regression, canonical correlation, and kernel canonical correlation are summarized.

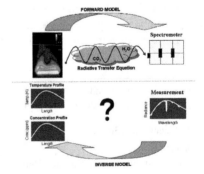

Fig. 22.1 (Top) RTE measurement corresponding to the forward model. (Down) Retrieval of combustion constituents formulated as inverse problem

Section 22.4 shows the results obtained applying these techniques. Conclusions and future work is given in Section 22.5.

22.2 Previous Work

The RTE explains how the propagation of radiation through a medium is affected by absorption, emission and scattering processes. RTE is common in different fields such as astrophysics, atmosphere, remote sensing, or geophysics, and has been studied in different ways over the last 40 years. In the next subsection, we make a physical review of the RTE and to previous machine learning approaches for dimensionality reduction and regression.

22.2.1 Physical Model

The most important by-products of combustion are carbon dioxide CO_2 and water vapor H_2O. These gases exhibit important emission bands in the infrared spectral region, and these types of emissions are governed by the RTE [5]. The RTE gives the spectral radiance L_i emitted by an inhomogeneous gas cloud at a given wave number i. The expression of the radiance L_i is given by:

$$L_i = \int_{z_0}^{z_1} B_i\{T(z), C(z)\}K_i(z)dz \tag{22.1}$$

$$K_i = \frac{d\tau_i}{dz}$$

$$\tau_i = \prod_{\Delta}\prod_{g}\tau_{ig}$$

$$\tau_{ig} = exp^{-(\alpha_{ig}\cdot c\cdot z\cdot f_{i\Delta})}$$

where B_i is Planck's law (standard black body emission), τ_i is the transmittance, $T(z)$ is the temperature profile, $C(z)$ is the concentration profile, and K_i is the so-called temperature weighting function that gives the relative contribution of the radiance coming from each region dz. In the transmittance formula, α_{ig} is the absorption coefficient of the gas g, c is its concentration, z is the path length and $f_{i\Delta}$ is a Lorentzian function that takes into account the spectral shape of the transmittance profile and takes values over the spectral range Δ.

During the last decade the progress in optoelectronic technologies has led to the fabrication of new sensors to measure radiance with unprecedented high resolution in spatial, spectral and temporal dimensions. For instance the spectrometer we used has a spectral range of 400–4,500 cm^{-1} and a spectral resolution of 32 cm^{-1} – 0.5 cm^{-1}. This allows new possibilities but also new challenges.

Work has been done in previous years to invert the RTE. In atmospheric sounding the objective is to retrieve the state of the atmosphere (temperature) and its constituents (e.g. water). In this context there are two main approaches to retrieval. The variational approach uses the forward model to calculate the radiance emitted by a specific atmospheric state. The measured radiance is compared with the estimated radiance and the state vector is adjusted to minimize the error [6]. Due to the fact that the forward problem complexity increases exponentially with the number of wave numbers, this approach can only be used if either the number of wave numbers or the number of iterations to converge is small. The second approach is based on the probability density of the pairwise [p(X), P(Y), P(Y, X)] which in practice is difficult to obtain.

22.2.2 Learning RTE Inversion

There have been some approaches that use supervised learning techniques to learn the inverse of RTE [7–9]. Due to the high resolution of the spectral measurements, one of the challenges of this type of problem is to develop supervised learning algorithms that can efficiently (memory and time) learn from very high dimensional data. To solve the course of dimensionality, a common approach has been to apply dimensionality reduction techniques. There are three main approaches to perform dimensionality reduction: feature extraction (linear or nonlinear), transformation of data, and feature selection (searching for subsets of original variables) [10].

[7, 8] make use of PCA to reduce the dimensionality, and linear and Neural Networks, respectively, to learn a mapping. [11] uses feature selection methods to choose a few wave numbers range, and show its effectiveness with synthetic data. The feature subset selection approach is interesting, because it allows the reduction of dimensionality and it also retains the semantic interpretation. Thus, a better understanding of the combustion properties is reached. Moreover, it allows for the design of specific sensors.

This paper differs from previous work in several aspects: firstly, we make a comparative study of several linear and non-linear dimensionality reduction and regression techniques. Secondly, we propose the use of KCCA and show how it outperforms previous methods. Thirdly, we apply these techniques in a gas retrieval ground remote sensing context where the length is unknown.

22.3 Supervised Learning for Temperature Retrieval

Equation 22.1 expresses the amount of energy emitted by each wave number i, and it is referred to as the forward problem. Typically, we would like to invert this equation, that is: from the radiance measure L_i provided by the spectrometer, we would like to know the temperature and concentration profiles. Given the measurements of energy represented by $\mathbf{x}_i \in \Re^{p \times 1}$ (see notation[1]), we would like to predict the flame constituents such as spatial temperature, length and concentration, represented by $\mathbf{y}_i \in \Re^{q \times 1}$. We would like to learn a mapping f such that $\mathbf{Y} = f(\mathbf{X})$, where $\mathbf{X} \in \Re^{p \times n}$ represents the energy emitted at different wave numbers and $\mathbf{Y} \in \Re^{q \times n}$ indicates the temperature profiles and concentration. Recall that each column corresponds to a different observation. For instance, each sample $\mathbf{x}_i$ is a spectrum of radiance in the infrared range of $2{,}110 \text{ cm}^{-1}$–$2{,}410 \text{ cm}^{-1}$ with $p = 6{,}000$. Likewise, each sample $\mathbf{y}_i$ is the corresponding temperature, length, and concentration with $q = 153$.

The inversion of Eq. 22.1 is non-linear and ill-posed since we are trying to retrieve $T(z)$, a continuous function, from finite measurements [12]. The non-linearity is due to the dependency of L with C and T. Also, recall that the mapping might not be unique; for instance, there could be more than one temperature/concentration profile that raises a particular radiance measurement.

In this section, a machine learning approach is adopted to learn the relation between the spectroscopist measurement and the gas constituents. Several techniques for a first dimensionality reduction have been compared: PCA, RRC, CCA and KCCA.

22.3.1 Principal Component Analysis (PCA)

PCA [13] is a standard linear dimensionality reduction technique, and it is optimal for Gaussian distributed classes. It is an energy-preserving transformation that decorrelates the data by projecting it into the first principal components. Following the notation of Section 22.3, let $\mathbf{B} \in \Re^{p \times k}$ be the first k principal components which contains the directions of maximum variation of the data. The k principal components $\mathbf{B}$ maximize $\max_{\mathbf{B}} \sum_{i=1}^{n} \|\mathbf{B}^T \mathbf{x}_i\|_2^2 = \max_{\mathbf{B}} \|\mathbf{B}^T \mathbf{\Gamma} \mathbf{B}\|_F$ under the constraint $\mathbf{B}^T \mathbf{B} = \mathbf{I}$, where $\mathbf{\Gamma} = \mathbf{X} \mathbf{X}^T = \sum_i \mathbf{x}_i \mathbf{x}_i^T$ is the covariance matrix (zero mean data). The columns of $\mathbf{B}$ form orthonormal bases that spans the principal subspace. If the effective rank of $\mathbf{X}$ is much less than d, we can approximate the column space of $\mathbf{X}$ with $k \ll p$ principal components. The sample $\mathbf{x}_i$ can be approximated as a linear combination of the principal components as $\mathbf{x}_i \approx \mathbf{B} \mathbf{c}_i$ where $\mathbf{c}_i = \mathbf{B}^T \mathbf{x}_i$. PCA bases are energy preserving and do not necessarily provide a meaningful representation of the signal; that is, it does not correspond to physical quantities. To avoid this problem, [11] proposes to select a subset of features on the original raw data based in the coefficients of the first orthonormal basis.

---

[1] Bold capital letters denote a matrix $\mathbf{D}$, bold lower-case letters a column vector $\mathbf{d}$. $\mathbf{d}_j$ represents the jth column of the matrix $\mathbf{D}$. d_{ij} denotes the scalar in the row i and column j of the matrix $\mathbf{D}$ and the scalar i-th element of a column vector $\mathbf{d}_j$. All non-bold letters represent scalar variables. $\|\mathbf{x}\|_2 = \sqrt{\mathbf{x}^T \mathbf{x}}$ designates Euclidean norm of $\mathbf{x}$. $\|\mathbf{A}\|_F = tr(\mathbf{A}^T \mathbf{A}) = tr(\mathbf{A} \mathbf{A}^T)$ designates the Frobenious norm of a matrix.

22.3.2 Reduced Rank Regression (RRR)

One of the drawbacks of PCA for supervised learning is the lack of dependency between the principal components of $\mathbf{X}$ and $\mathbf{Y}$. That is, a small signal common to both sets (relevant for regression) will be lost if performing independent PCA in each set [14]. PCA reduces the dimensionality optimally in the sense of reconstruction error, but it does not assure a better coupling between the new features c_i and the data to estimate y_i. That is, one could apply PCA separately to each set and then learn the mapping between them. However, this solution is suboptimal and assures the maximum variance within-set but not between-set.

A more direct approach of finding the direct mapping might be beneficial. A standard approach would be to perform direct regression between the variables $\mathbf{Y}$ and $\mathbf{X}$. For instance, finding the regression matrix $\mathbf{M}$ that minimizes $||\mathbf{Y} - \mathbf{MX}||_F$. The optimal $\mathbf{M}$ is given by $\mathbf{YX}^T(\mathbf{XX}^T)^{-1}$. For very high dimensional $\mathbf{X}$, it is likely that $(\mathbf{XX}^T)^{-1}$ is rank deficient. A common approach to solve this problem is to use reduced-rank regression (RRR). RRR minimizes $||\mathbf{Y} - \mathbf{MX}||_F$ subject to the constraint that $rank(\mathbf{M}) = k$.

A closed form solution for RRR [15] is accomplished by finding the principal directions of the augmented matrix $[\begin{smallmatrix}\mathbf{X}\\\mathbf{Y}\end{smallmatrix}]$, and use the QR-decompositions to extract subspaces for the rank-r (truncated). The optimal $\mathbf{M}$ is given by:

$$\mathbf{M} = \mathbf{Q}_Y \mathbf{F} \mathbf{Q}_X^T \qquad (22.2)$$

where $[\begin{smallmatrix}\mathbf{X}\\\mathbf{Y}\end{smallmatrix}] = [\begin{smallmatrix}\mathbf{U}_X\\\mathbf{U}_Y\end{smallmatrix}]\mathbf{SV}^T$ is the singular value decomposition over the augmented matrix, $\mathbf{U}_X = \mathbf{Q}_X\mathbf{R}_X$ and $\mathbf{U}_Y = \mathbf{Q}_Y\mathbf{R}_Y$, are the QR-decomposition, and $\mathbf{F} = \mathbf{R}_Y\mathbf{R}_X^{-1}$ maps from $\mathbf{Q}_X$ projections onto $\mathbf{Q}_Y$, or vice versa. See [15] for more details.

22.3.3 Canonical Correlation Analysis (CCA)

Another common approach used to reduce the dimensionality between two or more datasets that preserve discriminative information is CCA. CCA finds directions of maximum correlation between two datasets. In particular, CCA finds a set of bases vectors for two sets of variables such that the correlation between the projections of the variables onto these basis vectors are mutually maximized [16]. Consider the linear combinations $x = \mathbf{w}^T\mathbf{x}$ and $y = \mathbf{v}^T\mathbf{y}$ (x and y are called *canonical variates*). CCA finds the direction of $\mathbf{W}$ and $\mathbf{V}$ that maximizes:

$$\rho = \frac{E[xy]}{\sqrt{E[x^2]E[y^2]}} = \frac{E[\mathbf{w}^T\mathbf{x}\mathbf{y}^T\mathbf{v}]}{\sqrt{E[\mathbf{w}^T\mathbf{x}\mathbf{x}^T\mathbf{w}]E[\mathbf{v}^T\mathbf{y}\mathbf{y}^T\mathbf{v}]}} \qquad (22.3)$$

$$= \frac{\mathbf{w}^T\mathbf{C}_{xy}\mathbf{v}}{\sqrt{\mathbf{w}^T\mathbf{C}_{xx}\mathbf{w}\mathbf{v}^T\mathbf{C}_{yy}\mathbf{v}}}$$

where $\mathbf{C}_{xx}$ and $\mathbf{C}_{yy}$ are the within-sets covariance matrices of $\mathbf{X}$ and $\mathbf{Y}$ respectively and $\mathbf{C}_{xy} = \mathbf{C}_{yx}^T$ is the between-sets covariance matrix. The pairwise $\mathbf{W}$ and $\mathbf{V}$ can be found solving a generalized eigenproblem, $(C_{xx})^{-1} C_{xy} (C_{yy})^{-1} C_{yx} W = \lambda^2 W$ $(C_{yy})^{-1} C_{yx} (C_{xx})^{-1} C_{xy} V = \lambda^2 V$ (see [17]).

22.3.4 Kernel Canonical Correlation Analysis (KCCA)

In the previous section, we have reviewed linear dimensionality reduction methods. However, Eq. 22.1 describes a non-linear physical phenomenon. In this section, we introduce kernel canonical correlation analysis (KCCA), a non-linear extension of CCA. KCCA maps the data to a high dimensional space and performs linear CCA in this space. The mapping is implicitly defined with the kernel, and there is no need to explicitly compute the features in the high dimensional space (kernel trick). Therefore, let $\Phi_x : X \rightarrow F_x$ and $\Phi_y : Y \rightarrow F_y$ denote feature space mappings corresponding to possibly different kernel functions. The covariances in the feature space are represented by kernel matrices $K_x = \Phi_x \Phi_x^T$ and $K_y = \Phi_y \Phi_y^T$, and the spanned space is $\Im\{\Phi_x\}$ and $\Im\{\Phi_y\}$.

Since the canonical vectors $v_j \in \Im\{\Phi_x^T\}$ and $w_j \in \Im\{\Phi_y^T\}$ lie in the spaces spanned by the feature space mapped, we can represent them as linear combinations $v_j = \Phi_x^T \alpha_j$ and $w_j = \Phi_y^T \beta_j$ using $\alpha_j, \beta_j \in \Re^n$ as expansion coefficients. Therefore, the canonical variates are $a_j = \Phi_x v_j = K_x \alpha_j$ and likewise $b_j = \Phi_y w_j = K_y \beta_j$. As in the linear case, we have to find the canonical vectors in terms of expansion coefficients $\alpha_j, \beta_j \in \Re^n$. The solution can be reduced to an eigenproblem, where the objective is to find the canonical correlations between kernel feature spaces reducing the solution to linear CCA between kernel principal component scores

$$(C_x^T C_x)^{-1} C_x^T C_y (C_y^T C_y)^{-1} C_y^T C_x \psi_j = \lambda_j^2 \Psi_j \qquad (22.4)$$

$$(C_y^T C_y)^{-1} C_y^T C_x (C_x^T C_x)^{-1} C_x^T C_y \psi_j = \lambda_j^2 \xi_j \qquad (22.5)$$

where $C_x = \Phi_x U_x = K_x A_x$ and $U_x = \Phi_x^T A_x$ being U_x the principal components of Φ_x, and $a_j = K_x A_x \psi_j$ and $b_j = K_y A_y \xi_j$ are the kernel canonical variates. Likewise, we can obtain C_y (for a detailed explanation see [18]).

22.4 Experiments

In this section, we describe the experimental design and the results for synthetic data.

22.4.1 Experimental Design

The synthetic dataset used in this study has been generated with a simulator called CASIMIR [19], based on the well known experimental database HITRAN [20].

Fig. 22.2 Temperature profiles included in the dataset
used in the analysis

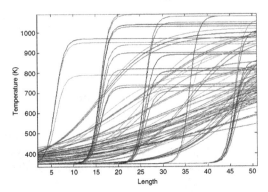

The parameter ranges used to generate this dataset are based on typical combustion environment conditions. The temperature ranges from 296 to 1100 K with several different profiles (see Fig. 22.2), the length range covers from 0.05 to 0.85 m, and the concentration values for CO_2 and H_2O have been selected as typical values from the combustion of fossil fuels at different temperatures.

22.4.2 Error description

To make a fair comparison between the different techniques, we need an error measurement for a new estimation $\hat{\mathbf{y}}_i \in \mathfrak{R}^{q \times 1}$ where the temperature (T) and length (L) are the parameters to estimate. Figure 22.4 shows the synthetic function that we want to retrieve and the retrieved function obtained with our method. Here it is worth highlighting two points. Firstly, the point to point measurement error has a major problem, because the error is presented in both axes (see the difference between the red point and the black point), a very small error in one of the axes (Δx_m) could lead to a big one in the other (Δy_m, see Fig. 22.3a), or the opposite (see Fig. 22.3b)). This phenomena can produce undesirable error measurements in the presence of sharp slopes. Figure 22.4 illustrates this phenomena, although the function 1 is more accurate in shape and surface, its point to point error is larger. Secondly, because of the data discretization, the minimum expected error is limited by the synthetic data resolution. Therefore, a higher discretization resolution in the estimated data would be useless.

Let ΔT be the resolution of the temperature discretization and ΔL be the length discretization. Then, we call $\varepsilon_0 = \Delta T * \Delta L$ the minimum error supported by the model that depends on the synthetic data discretization.

To solve previous problems, a new error measurement is proposed (see Fig. 22.5). P_{s1} and P_{s2} are the points that belong to the synthetic function and P_m is the estimated point. The new points P_{i1} and P_{i2} are calculated using a linear interpolation. Let Δx_m and Δy_m be the maximum error over each one of the axes when the error over the

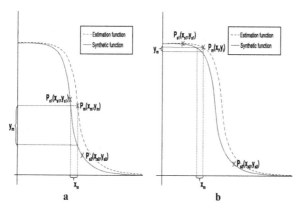

Fig. 22.3 Figure **a** the error is over the y dimension, and figure **b** the error is over the x dimension

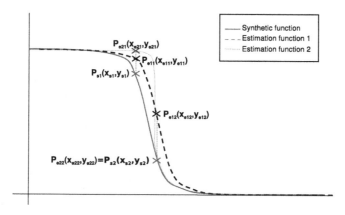

Fig. 22.4 Main problem of point to point error measurement approach

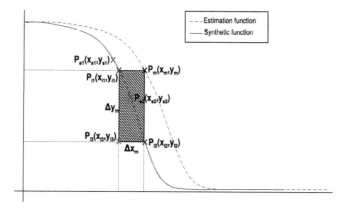

Fig. 22.5 Experimental error calculus

other is considered 0. Then, the surface error measurement is the area defined by the points $P_m, P_{i1}, P_{i2}, P_{i3}$. This area is the maximum squared error due to the fact that Δx_m and Δy_m are maximums. Then, the error is formulated as $\Delta \varepsilon_{mi} = \Delta x_{mi} * \Delta y_{mi}$, where m is the estimated point and i is the sample. Due to the minimum error supported, if $\Delta \varepsilon_{mi} < \varepsilon_0$ the error will be considered 0.

Because the surface error (see filled area in the Fig. 22.5) is the maximum squared error surface associated to the estimated point P_m, the final error and its mean error per point estimated are $\mathbf{E}_{mi} = \sqrt{\Delta \varepsilon_{mi}}$, and $ME_m = \frac{1}{n} \sum_{i=1}^{n} E_{mi}$ where n is the number of samples, and m the number of estimated points. Furthermore, in order to measure the overall performance of the techniques we define the total mean error as $TME = \frac{1}{m} \sum_{j=1}^{m} ME_m$. With this approach we are weighting the error and avoiding the problem that small differences on the shape of the function lead to larger unjustified error measurements.

22.4.3 Results

In this section, we review the results from applying the techniques described above for dimensionality reduction and regression to the combustion retrieval problem.

Let $\mathbf{Y} = f(\mathbf{C})$ be the mapping between the projected coefficients of the spectrum $\mathbf{C} \in \Re^{r \times n}$ and the temperature and concentration profiles, represented by $\mathbf{Y}$. $\mathbf{C}$ is obtained by any of the dimensionality reduction techniques explained in Section 22.3 (e.g. PCA, RRR, CCA, KCCA). To make a fair comparison in terms of parameters of the models, we have chosen $r = 80$ for each of the projections. We also fix the multilayer perceptron (MLP) architecture to a unique hidden layer with 80 hidden neurons. We split the dataset in 70% training and 30% testing.

Figure 22.6 shows the mean error computed as explained above. Recall that linear refers to a standard linear regression. We can observe in Fig. 22.6 and Table 22.1,

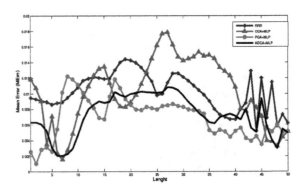

Fig. 22.6 ME_m error of the estimated temperature function. X axis is the length discretized in arbitrary units. Best performance is for KCCA+MLP model in solid line

Table 22.1 TME error of temperature estimation, and its standard deviation (SD)

Method	TME	Standard deviation
PCA+MLP	0.0071	0.0121
CCA+MLP	0.0104	0.0154
KCCA+MLP	0.0066	0.0110
RRC	0.0070	0.0114
KCCA+linear	0.2236	0.2415
CCA+linear	0.2251	0.2433

Table 22.2 Point to point mean absolute error per sample(MAEs) of temperature and concentration, and its standard deviation (SD)

Method	Temperature (MAEs/SD)K	Concentration (MAEs/SD)ppm
PCA+MLP	2.38/17.33	1.6E-4/1.2E-3
KCCA+MLP	2.27/ 4.65	1.5E-4/3.1E-4
CCA+MLP	8.81/25.18	6.0E-4/1.7E-3
RRC	2.30/ 9.57	1.2E-4/6.0E-4
KCCA+linear	35.80/54.86	2.0E-3/3.8E-3
CCA+linear	32.09/75.84	2.0E-3/5.0E-3

that KCCA + MLP (solid line) achieves the best results in terms of ME. This is not surprising, since PCA does not guarantee that the projection $\mathbf{C}$ has correlation with the predicted variable $\mathbf{Y}$. Surprisingly, CCA + MLP performance is inferior to that of PCA + MLP, this is due to several factors such as local minimum in MLP or CCA + MLP not getting the optimal number of components for CCA.

However, KCCA achieves better performance for the non-linear component. Table 22.2 shows the Mean Absolute Error profile per sample (MAEs). The MAEs for temperature and concentration are computed by $MAEs = \frac{1}{z}\frac{1}{n}\sum_{k=1}^{z}\sum_{j=1}^{n}|\mathbf{y}_{kj} - \hat{\mathbf{y}}_{kj}|$ where z is the discretized length, and n the number of samples. The MAEs gives an idea of the physical error although it is not a good measurement for comparing the different techniques for reasons already explained. In synthetic data experiments, the MAEs are below 1% relative error (2.27 K) which is an acceptable level of accuracy for most practical applications [2].

We also compare the MLP technique with linear regression techniques such as RRR and linear mapping. As can be observed in Fig. 22.6 and Table 22.2, RRR has a slight improvement over PCA + MLP or CCA + MLP. This indicates that the mapping, although non-linear, is not far from being linear. Furthermore, RRR has advantages with respect to the other techniques: firstly, it is extremely efficient (in space and time) to compute; secondly it is not prone to local minimum problems and the model for different dimensions can be computed recursively.

22.5 Conclusions

In this paper, we have presented a comparative study of different dimensionality reduction and regression techniques to invert the radiative transfer equation. KCCA+MLP has slightly outperformed other techniques in terms of ME and MAEs, and thus confirms the non-linearity of the RTE. Not surprisingly, KCCA overcomes the limitations of CCA and PCA to model non-linear structure between the spectrum values and the temperature profiles. The relative error in temperature achieved by KCCA+MLP is around 1%, which in the literature is considered to be "extremely accurate" [2]. Furthermore, using MLP for regression allows the the modelling of complicated dependencies.

On the other hand, RRR has worked surprisingly well and is computationally a very efficient method. RRR works especially well when the length is known a priori. This suggests a method of splitting the input data into several lengths and computing a local model for each of them. We are currently investigating extensions of this idea.

Although we have reported promising results in synthetic data, there are still several issues to solve in the future. We plan to gather real data and test the robustness of the algorithm and the validity of the RTE inversion in real environments. Moreover, we have fixed the number of projected components and the number of hidden units in the MLP, to make a fair comparison in terms of the parameters of the model. Each model may have its optimal number of projections and hidden units. More research needs to be done to address this problem. On the other hand, the performance of KCCA greatly depends on the choice of the kernels and their parameters and we are currently working on automatic methods to learn the kernels following prior work in computer vision [21].

Acknowledgements The author E. Garcia-Cuesta wishes to acknowledge the Spanish Ministry of Education and Science for financial support under the project TRA2005-08892-C02-01.

References

1. Romero, C., Xianchang, L., Shahla, K., and Rodney, R., "Spectrometer-based combustion monitoring for flame stoichiometry and temperature control", *Appl. Therm. Eng.*, 25, 659–676, 2005.
2. Lu, G., Yan, Y., and Colechin, M., "A digital imaging based multifuncional flame monitoring system", *IEEE trans. Instrum. Meas.*, 53, 1152–1158, 2004.
3. Deguchi, Y., Noda1, M., Fukuda, Y., Ichinose, Y., Endo, Y., Inada, M., Abe, Y., and Iwasaki, S., "Industrial applications of temperature and species concentration monitoring using laser diagnostics", *Meas. Sci. Technol.*, 13, R103–R115, 2002.
4. Liu, L.H., and Jiang, J., "Inverse radiation problem for reconstruction of temperature profile in axisymmetric free flames", *J Quant. Spectrosc. Radiat. Trans.*, 70, 207–215, 2001.
5. Goody, R.M. and Yung, Y.L., *Atmospheric Radiation. Theoretical Basis (Chap.2)*, Oxford University Press, New York, 1989.
6. Rodgers, C.D., "Retrieval of atmospheric temperature and composition from remote measurements of thermal radiation", *J. Geophys. Res.*, 14 (7), 609–624, 1976.

7. Aires, F., Chedin, A. Scott, N. A., and Rossow, W.B., "A regularized neural net approach for retrieval of atmospheric and surface temperatures with the IASI instrument", *J Appl. Meteorol.*, 41, 144–159, 2001.
8. Blackwell, W.J., "A neural-network technique for retrieval of atmospheric temperature and moisture profiles from high spectral resolution sounding data", *IEEE Trans. Geosci. Remote Sens.*, 43(11), 2535–2546, 2005.
9. Huang, H.L. and Antonelli, P., "Application of principal component analysis to high-resolution infrared measurement, compression and retrieval", *J. Appl. Meteorol.*, 40(3), 365–388, 2001.
10. Bishop, C.M., *Neural Networks for Pattern Recognition*, Oxford University Press, Oxford, UK, 1999.
11. Garcia-Cuesta, E., de Castro, A. and Galvan, I M., "Spectral high resolution feature selection for retrieval of combustion temperature profiles", *Lect. Notes Comput. SC*, 4224, 754–762, 2006.
12. McCornick, N.J., "Inverse radiative transfer problems: a review, *Nucl. Sci. Eng.*, 112(3), 185–198, 1992.
13. Jollife, I.T., "Principal Component Analysis (2nd Edition)", Springer: New York, 2002.
14. de la Torre Frade, F. and Black, M.J. "Dynamic Coupled Component Analysis" *IEEE Conference on Computer Vision and Pattern Recognition*, pp. 643–650, June 2001.
15. Brand, M.E., "Subspace mappings for image sequences", *Statistical Methods in Video Processing*, June 2002.
16. Shawe-Taylor, J. and Cristianini, N., *Kernel Methods for Pattern Analysis*, Cambridge University Press: Cambridge, UK, 2004.
17. Melzer, T., Reiter, M., Beschof, H., "Appearance models based on kernel canonical correlation analysis", *Pattern Recognition*, 36(9), 1961–1973, 2003.
18. Kuss, M. and Graepel, T., "The geometry of kernel canonical correlation analysis" *Technical Report No. 108 May 2003*. Available in: http://www.kyb.tuebingen.mpg.de/techreports.html
19. Garcia-Cuesta, E., "CASIMIR: Calculos Atmosfericos y Simulacion de la Transmitancia en el Infrarrojo", Technical Project Dept. of Physics University Carlos III, Madrid, 2003.
20. Rothman, LS. et al., "The HITRAN molecular spectroscopic database: edition of 2000 including updates through 2001", *J. Quant. Spectrosc. Radiat.*, tranf., 82, 5–44, 2003.
21. de la Torre Frade, F. and Vinyals, O., "Learning kernel expansions for image classification" *IEEE Conference on Computer Vision and Pattern Recognition*, June 2007.

Chapter 23
Enhancing the Performance of Entropy Algorithm using Minimum Tree in Decision Tree Classifier

Khalaf Khatatneh and Ibrahiem M.M. El Emary

Abstract In this chapter, we proposed a new algorithm called enhanced entropy algorithm has a major advantage of reducing the complexity and execution time w.r.t the original entropy algorithm and at the same time yields the same sequence that can be found by applying entropy algorithm. The presented Enhance Entropy Algorithm (EEP) finds the sequence of the minimum tree in a simple manner compared to entropy algorithm (which uses complex mathematical operations to find the sequence of the minimum tree) using simple logical operations.

Keywords Entropy Algorithm (EA) · Enhanced Entropy Algorithm (EEA) · Data Mining (DM) · Gain information measure (GINI)

23.1 Introduction

Classification builds a model based on historical data (training data set). Once the model is built, it is used to predict the class for a new instance. Many methods have been proposed to solve the classification problem (referred to as classifiers). One of the most popular and best classifier proposed so far is the decision tree classifier which mainly builds a model from the dataset in a form of tree. Multiple trees can be generated from the same dataset where all trees yield the same outcome for a given new instance to be classified. The possible trees for a dataset vary in their size

K. Khatatneh
Assist. Prof. of Computer Sciences, Information Technology Dept., Prince Abdullah Ben Ghazi College for Science and Information Technology, Al Balqa Applied University, Amman, Jordan

I.M.M.E. Emary (✉)
Assoc. Prof. of Computer Engineering, Computer Engineering Dept., Faculty of Engineering, Al Ahliyya Amman University, Amman, Jordan
e-mail: doctor_ebrahim@yahoo.com

S.-I. Ao et al. (eds.), *Advances in Computational Algorithms and Data Analysis*,
Lecture Notes in Electrical Engineering 14,

where the size of the tree depends on the sequence in which the dataset attribute is used to build the tree. However, the minimum tree is preferred because the minimum tree needs the shortest time to figure out the outcome of the model. One of the best algorithms that have been proposed to find the sequence that yields the minimum tree if used is the entropy algorithm. However, the major drawback of entropy algorithm is that it needs complex computations to find the sequence of minimum tree and here the complexity of computations grows exponentially when the numbers of attributes in a given dataset increase. So, in this chapter, we proposed a new algorithm called enhanced entropy algorithm has a major advantage of reducing the complexity and execution time w.r.t the original entropy algorithm and at the same time yields the same sequence that can be found by applying entropy algorithm. The presented Enhance Entropy Algorithm (EEP) finds the sequence of the minimum tree in a simple manner compared to entropy algorithm (which uses complex mathematical operations to find the sequence of the minimum tree) using simple logical operations. The main keywords used in this chapter are: Entropy Algorithm (EA), Enhanced Entropy Algorithm (EEA), Data Mining (DM) and Gain information measure (GINI).

23.1.1 Related Works

Data mining (DM) is defined as the knowledge discovery in database [1–3] where it attempts to apply machine learning, statistics, artificial intelligent concepts and techniques on databases in order to discover useful patterns and trends hidden in vast amount of data stored in form of files and database records. Classification [4] is a supervised data mining technique [5, 6] in which we can predict categorical class values for new instances based on the training data collected over time. One of the most popular classifiers proposed so far is the decision tree classifier where decision tree classifier builds a model from the training data in a form of tree. Once the model is built and verified, it can be used to predict the class for new given instances. Decision tree classifier is one of the best classifiers proposed for many reasons given by:

1. Can handle numeric and categorical data.
2. Needs reasonable time to construct the model.
3. It is an eager classifier.

Multiple trees [7] can be generated from the same dataset where all trees yield the same outcome (class value) for a given new instance to be classified. Possible trees for a dataset vary in their size. The size of the tree depends on the sequence in which the dataset attributes are used to build the tree. However, we prefer the tree with the smallest size because such tree needs the shortest time to figure out the outcome of the model. There are two algorithms proposed to find the sequence that yield the smallest tree if used; the first one is called entropy algorithm and the second one is called GINI algorithm. Entropy algorithm is usually used because it is suitable for almost all types of datasets while GINI algorithm is suitable for a small portion of datasets types. This chapter is organized from five sections. In Section 23.2, we explain the entropy algorithm. The proposed enhanced entropy algorithm is presented in details in Section 23.3. Section 23.4 deals with comparative study between the results of entropy algorithm performance and the corresponding one enhanced entropy algorithm performance. Finally Section 23.5 covers the main conclusion of this chapter and some of future work recommended to be done by others.

23.2 Entropy Algorithm Description

Entropy Algorithm computes the information gain value for each attribute in the dataset [8] excluding the class attribute. After that, the attributes are sorted in descending order based on their information gain values, the sorted attributes compromise the sequence that if used by the decision tree classifier construction process, the smallest tree will be generated. The information gain measure (abbreviated as iGain) for attribute A in dataset S is measured by the following function:

$$iGain(S, A) = Entropy(S) - \sum \frac{|S_v|}{|S|} * Entropy(S_v)... \tag{23.1}$$

Where:
  S_v is the subset of instances of S which have "V" as attribute "A" value,
  $|S_v|$ is the number of instances in the set S_v
  $|S|$ is the number of instances in the dataset S.
  The entropy of a set S of (positive class, negative class) training instances is:

$$Entropy(S) - plog_2(p) - nlog_2(n)... \tag{23.2}$$

Where:
  P is the number of instances with a positive class divided by number of instances in S.
  n is the number of instances with a negative class divided by number of instances in S.

23.3 The Proposed Enhanced Entropy Algorithm

The Enhanced Entropy Algorithm (EEA) that is proposed in this chapter try to find the sequence of the minimum tree faster than the original entropy algorithm. EEA finds the sequence of the minimum tree with the best possible way through the following steps:

1. For each attribute "A" in the dataset determine the classes distribution to find the class distribution, you should:

- Determine the attribute distance values.
- For each distance value V and class value C, determine the number of instances that have A = V and class value = C.

2. Sort the attributes in ascending order according to the number of zeros in the class distribution of every attribute.
3. In case of two attributes with the same number of zeros, find the difference between the maximum value and the minimum value in each attribute class distribution and sort them in ascending order according to the difference found.
4. In case of two attributes with the same number of zeros with the same difference, compute the information gain measure for both attributes using entropy algorithm and sort them in descending order according to the information gain.

23.3.1 The Tree Construction Process

The tree construction process with EEA passes through the following phases:

1. If all the training instances belong to the same class Ci, add one leaf with value Ci and stop; else go to step 2.
2. (a) For each attribute "A", determine the classes distribution and to find the classes distribution you should:
   - Determine the attribute distinct values.
   - For each distinct value V and class value C, determine the number of instances that have A = V and class value = C.
   (b) Sort the attribute in ascending order according to the number of zeros in the class distribution of every attribute.
   (c) In case of two attributes with the same number of zeros, find the difference between the maximum value and the minimum value in each attribute class distribution and sort them in ascending order according to the difference found.
   (d) In case of two attributes with same number of zeros with the same difference, compute the information gain measure for both attributes using entropy algorithm and sort them in ascending order according to the information gain.
3. Create a node labeled by the first attribute A in the sequence with branches holding the attribute distinct values.
4. For each branch that holds the distinct value V in the A, execute the following: if all training instances that satisfy V = A belong to the same class Ci, then create a leaf with the value Ci. Else, Go to step (3) with A as the next attribute in the sequence.

23.3.2 The Decision Tree Classifier (DTC)

The decision tree classifier (abbreviated as DTC) is a widely used technique for classification. As the name suggests, it builds a tree called decision tree where each leaf has an associated class and each internal node has a relational statement associated with it [9]. The tree construction process corresponds to the learning process and the model constructed is the decision tree. To construct the tree, we need to perform the following steps [10]:

1. If all the training instances belong to the same class C_i, add one leaf with value C_i and stop; else go to step (2).
2. Select the sequence on which the attributes will be used in the tree construction process arbitrary excluding the class attribute.
3. Create a node labeled by the first attribute A in the sequence with branches holding the attribute distinct values.
4. For each branch, holds the distinct value V in the A; execute the following: If all training instances that satisfy V = A belong to the same class Ci, then create a leaf with the value Ci. Else, go to step (3) with A as the next attribute in the sequence.

To classify a new instance, we start at the root and traverse the tree to reach a leaf. At the internal node, we evaluate the relational statement on the data instance to find which child to go to and the process continues till we reach a leaf node.

23.3.2.1 Case Study of DTC Classifier

In the following case, we want to train the following data set using DTC (data shown in Table 23.1).

Solution of the above task passes through the following phases:-

1. All instances are not of the same class, so we go to step (2).
2. We will select the following sequence: hair, height, eyes.
3. Starting from hair as the root, we create three branches one for blond and one for red and one for dark (as shown in Fig. 23.1a).
4. All instances that satisfy (hair = red) belong to class A, so we create a leaf under the branch that holds the value red with the value A. All instances that satisfy (hair = dark) belong to class B, so we create a leaf under the branch that holds

Table 23.1 The training dataset – DTC

Height	Hair	Eyes	Class
Short	Blond	Blue	A
Tall	Blond	Brown	B
Tall	Red	Blue	A
Short	Dark	Blue	B
Tall	Dark	Blue	A
Tall	Blond	Blue	A
Tall	Dark	Brown	B
Short	Blond	Brown	A

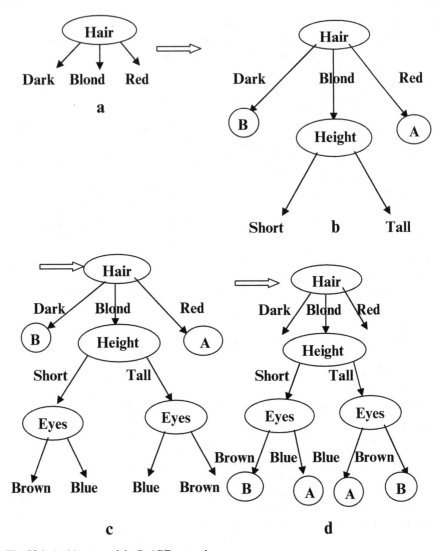

Fig. 23.1 Architecture of the ReART network

the value dark with the value B. Finally, instances that satisfy (hair = blond) don't belong to the same class, we create a node labeled by the next attribute in the sequence which is height with two branches one for tall and one for short (as shown in Fig. 23.1b).

5. The instances that satisfy(hair = blond and height = short) don't belong to the same class under the branch that holds the value short, we create a node labeled by the next attribute in the sequence which is eyes with two branches one for brown and one for blue. All instances that satisfy(hair = blond and height = tall) don't belong to the same class, so under the branch that holds the value short, we

create a node labeled by the next attribute in the sequence which is eyes with two branches one for brown and one for blue (as shown in Fig. 23.1c).

6. All the instances that satisfy(hair = blond and height = short and eyes = blue) belong to the same class A, so under the branch that holds the value blue we create a leaf labeled with the value A. All the instances that satisfy (hair = blond and height = short and eyes = brown) belong to the same class B, so under the branch that holds the value brown, we create a leaf labeled with the value B. All the instances that satisfy (hair = blond and height = tall and eyes = blue) belong to the same class A, so under the branch that holds the value blue, we create a leaf labeled with the value A. All the instances that satisfy (hair = blond and height = tall and eyes = brown) belong to the same class B, so under the branch that holds the value brown, we create a leaf labeled with the value B (as shown in Fig. 23.1d).

23.3.2.2 Case Study on EEA Classifier

In this subsection, we will execute the same case mentioned in Section 23.3.2.1 once again, however this time using the tree construction process with EEA that is described in Section 23.3. The steps in EEA are as follows:

1. All instances are not of the same class, so we go to step (2).
2. This step is composed of two important measures described as follows:

   • Find classes distribution for each attribute
   • For the height attribute, execute the following:

   (a) Height distinct values are: short and tall.
   (b) For the value "short", there are three instances, one of class A and two of class B, for the value "tall" there are five instances, two of class A and three of class B.

Conclusion: The class distribution for the height attribute is: **1 2 2 3**

• **For the hair attribute:**

   (i) Hair distinct values are: blond, red, dark.
   (ii) For the value "blond", there are four instances, two of class A and two of class B, for the value "red" there is one instance of the class A and no instances of class B. for the value "dark", there are three instances all of class B and no instances of class A.

Conclusion: The class distribution for hair attributes is: **2 2 1 0 0 3**

1. **For the eyes attribute:**

   (a) Eyes distinct values are: blue and brown.
   (b) For the value "brown", there are three instances, three of class B and no instances of class A. For the value "blue", there are five instances, three of class A and two of class B.

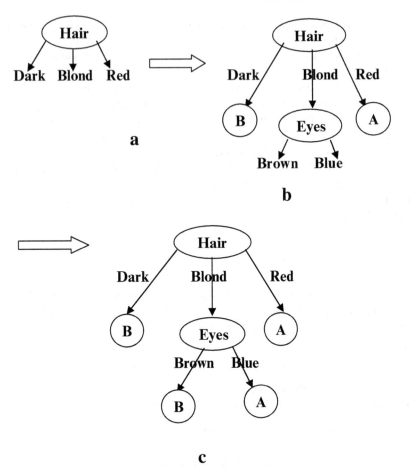

Fig. 23.2 Neural controller constructed using ReART

Conclusion: The class distribution for the eyes attribute is: **0 3 3 2**

- Height classes distribution contains no zeros, whereas hair classes distribution contains two zeros, and eyes classes distribution contains one zero, so the attributes after the sort is: hair, eyes, height.

There is no need to go through cases 3 and 4 as both cases didn't occur in this case; the remaining steps are the same as those mentioned in the case of the previous section. At the end, we reach to the tree illustrated in Fig. 23.2.

23.3.3 Native Bayesian Classifier

Native Bayesian classifier was built based the probability theory. When given a new instance D, it uses the distribution information to estimate for each class Cj, the probability that instance D does belong to class Cj (denoted by P(Cj | d)), the class with the largest P(Cj | d) is the class value for instance D [11]. Native Bayesian classifies a new instance using the following steps:

1. Find the probability for each attribute in the new instance with each class value.
2. Multiply the probabilities found in step (1) with the probability of each class value.
3. The class with the largest probability is the class value of the new instance.

23.3.3.1 Case Study of Native Bayesian Classifier

A tennis player wants to predict the probability of holding tennis match on a given day with relation to the weather condition <sunny, cool, high, and strong> using the following training data (Table 23.2):

The solution of this case passes through the following steps:-

1. The probability of the new instance with the class value "yes" is given by:

$P(sunny|y)*P(cool|y)*P(high|y)*P(strong|y) = (2/9)*(2/9)*(3/9)*(3/9) = 0.005$

The probability of the new instance with the class value "No" is given by:
$P(sunny|y)*P(cool|y)*P(high|y)*P(strong|y) = (3/5)*(1/5)*(4/5)*(3/5) = 0.0576$

2. The probability for (play tennis = Yes) is given by:

$P(y)*0.008 = (9/14)*0.005 = 0.003$

Table 23.2 Play tennis using Native Bayesian classifier

Day	Outlook	Temperature	Humidity	Wind	Play tennis
1	Sunny	Hot	High	Weak	No
2	Sunny	Hot	High	Strong	No
3	Overcast	Hot	High	Weak	Yes
4	Rain	Mild	High	Weak	Yes
5	Rain	Mild	Normal	Weak	Yes
6	Rain	Cool	Normal	Strong	No
7	Overcast	Cool	Normal	Strong	Yes
8	Sunny	Mild	High	Weak	No
9	Sunny	Cool	Normal	Weak	Yes
10	Rain	Mild	Normal	Weak	Yes
11	Sunny	Mild	Normal	Strong	Yes
12	Overcast	Mild	High	Strong	Yes
13	Overcast	Hot	Normal	Weak	Yes
14	Rain	Mild	High	Strong	No

The probability for (play tennis = No) is given by:
P(n)*0.0576 = (5/14)*0.0570 = 0.021

3. The "No" class have higher probability, so the class value for the new instance is No.

23.3.3.2 Case Study of the Tree Construction with EEA

Let us do the same case mentioned in Section 23.3.3.1 using the tree construction with EEA. Solution here passes through the following steps:-

1. All instances are not of the same class, so we go to step (2).
2. This step is composed of four important measures detailed as follows.

Step (1): Find classes distribution for each attribute

- **For the outlook attribute:**

    A. Outlook distinct values are: sunny, rain, overcast
    B. For the value "sunny", there are five instances, two of class Yes and three of class No, for the value rain, there are five instances, three of class Yes and two of class No. For the value overcast there are four instances, four of class yes and no instances of class no.

Conclusion: The class distribution for the outlook attribute is: **2 3 3 2 4 0**

- **For the temperature attribute:**

    A. Temperature distinct values are: hot, mild, cool
    B. For the value "hot", there are four instances, two of class Yes and two of class no. For the value "mild" there are seven instances, five of class yes and two of class no. For the value "cool", there are three instances, two of class Yes and one of class no.

Conclusion: The class distribution for the temperature attribute is: **2 2 5 2 2 1**

- **For the humidity attribute:**

    A. Humidity distinct values are: high and normal
    B. For the value "high", there are seven instances, three of class yes and four of class no. For the value "normal", there are seven instances, six of class yes and one of class no.

Conclusion: The class distribution for the humidity attribute is: **3 4 6 1**

- **For the windy attribute:**

    A. Windy distinct values are: false and true
    B. For the value "false", there are eight instances, six of class yes and two of class no. For the value "true", there are six instances, three of class yes and three of class no.

Conclusion: The class distribution for the windy attribute is: **6 2 3 3**

Step (2): Outlook class distribution contains one zero, temperature, humidity and windy classes distribution contains no zero, so the only fact we know about the sequence after this step is that outlook is the first attribute in the sequence.

Step (3): Temperature, humidity, and windy have the same number of zeros, the differences for temperature, humidity, and windy are 4(5–1), 5(6–1), and 4(6–2) respectively, humidity is the attribute with the largest difference while the difference for temperature and windy is the same, so we add humidity as the second attribute in the sequence.

Step (4): Temperature and windy have the same number of zeros and their differences are equal, so we compute the information gain measure for both using entropy, the information gain value for temperature is 0.029 while it is 0.048 for windy, so windy will precede temperature in the sequence.

Conclusion: Based on steps 2, 3 and 4, the sequence is: Outlook, Humidity, Windy, Temperature.

On the other hand, if entropy algorithm is used to find the sequence, the information gain for the attributes is:

- Information gain value for outlook is: 0.247.
- Information gain value for temperature is: 0.029.
- Information gain value for humidity is: 0.152.
- Information gain value for windy is: 0.048.

Therefore, the sequence would be: Outlook, Humidity, Windy, Temperature.

23.4 Simulated Results

The EEA algorithm gives the same results as entropy algorithm in a simpler manner. Our results prove that EEA is more efficient than the existing entropy algorithm in terms of complexity. Both algorithms (EEA and Entropy) contain common beginning section in which the dataset is loaded from a file for processing. In the following subsection, we will analyze a comparative study between the obtained performance of Entropy algorithm and EEA.

23.4.1 Performance Analysis of Entropy Algorithm

Entropy algorithm idea for finding the sequence that yield the minimum tree is simply to compute the information gain measure for the entire dataset attributes then sort them according to the information gain value. The sorted attribute list is the final sequence. However, when we think about implementing the idea using a programming language, we found that it is complicated and requires a number of nested loops to get the information gain value for each attribute. The complexity of any code written to implement the entropy algorithm will never be less than $O(n^3)$, this complexity was found by Quinlan in 1993 [12].

The idea behind the n3 complexity of entropy algorithm is that in order to find the information gain measure for an attribute, we have to scan all attributes in the dataset and for each attribute "A" we have to scan each distinct value "V" within the attribute "A" and for each distinct value, we to scan all distinct class values "C" in order to find the number of instances that satisfy "A"="V" and class value = "C". From the above facts, we conclude that we need three nested loops, one of the basic rules in algorithms science "three loops always means complexity of (n^3) [13].

One of the major disadvantages of entropy algorithm is that there is no best case or worst case when it is given a dataset to find the sequence of the minimum tree from the information gain measure for all attributes must be computed and we always have to iterate the three loops to find the sequence of the minimum tree.

23.4.2 Performance Analysis of EEA

EEA gives the same results as entropy algorithm in a simpler manner. However, this is not adequate to say that we came with a new algorithm, so the challenge was to prove that it is really more efficient than entropy algorithm in terms of complexity. EEA idea for finding the sequence that yields the minimum tree is to sort the attributes in the dataset according to the number of zeros in their class distribution. If two or more attributes having the same number of zeros, we find the difference between the maximum value and the minimum value in each attribute classes distribution and sort them ascending according to the difference. In case of two or more attributes with the same number of zeros and the same difference, we compute the information gain measure for the tied attributes using entropy, and sort them descending according to the information gain value.

From the above, we can conclude that EEA has best and worst case unlike the entropy algorithm that always have a constant case. In general, the best case happens when all attributes have different number of zeros, the attribute will be sorted according to the number of zeros and the sorted attributes list compromise the final sequence or when there are some attributes that have the same number of zeros but there is difference, so that the attributes with the same number of zeros can be sorted according to the difference and the final sequence is reached.

To see how the best case happen in EEA code, our designed software program computes the number of zeros as well as the difference between the maximum value and the minimum value in each attribute class distribution and then sorts the attributes in ascending order according to the number of zeros. In the case of two or more attributes with the same number of zeros, we sort them according to the difference value. If the attributes are sorted based on these concepts, the Boolean (sorted) will be evaluated to true and the software program will finish with the sorted attribute list is the final sequence.

In the best case, EEA needs two nested loops to find the number of zeros for each attribute, the same number of loops is required to find the difference between the maximum value and the minimum value in each attribute class distribution, the nested loops indicate that the complexity of EEA in the best case is $O(n^2)$.

In contrast, the worst case happen when all attributes have the same number of zeros and the difference between the maximum and minimum value in all attributes classes distribution is equivalent. In this case, EEA becomes the same as entropy algorithm. The information gain measure for all the dataset attributes must be measured, then the attributes is sorted according to the information gain value.

To see how the worst case happen in EEA, our designed software program computes the number of zeros as well as the difference between the maximum and the minimum value in each attribute class distribution. When the program tries to sort the first two attributes, it will find both attributes that have the same number of zeros and that difference between the maximum and minimum value in each attribute class distribution is equivalent, so the program will compute the information gain measure for both attributes and will sort them according to the information gain value. When the program reaches the third attribute, it will find that the third attribute number of zeros is the same as the number of zeros of the first two attributes, so the program finds the information gain measure for the third attribute and resort the first three attributes according to the information gain value. In general, for a dataset with n attributes, the program has to find the information gain measure for $(n-1)$ attributes in the worst case (we find the information gain measure for $(n-1)$ attributes not n because the final attribute is the class attribute which is excluded from the sequence determination process).

In the worst case, EEA is equivalent to entropy algorithm, the complexity in the worst case is the same as the complexity of entropy algorithm which is $O(n^3)$. Finally, we illustrate by examples how the best case and worst case happens in EEA, in the first case we prefer to a dataset in which the sequence of the minimum tree is found in best case complexity. In the second case, we refer to a dataset in which the sequence of the minimum tree is found in worst case complexity.

If EEA algorithm was used to solve the case mentioned in Section 23.3.2, the best case will happen because the attribute height contains no zero in its class distribution, the attribute eyes contains one zero in its class distribution, and finally the attribute hair contains two zeros in its class distribution, so to find the sequence of the minimum tree, we sort the attributes according to the number of zeros. The sorting yields the sequence; hair, eyes, height. In the mentioned case, there is no need to compute the information gain value for any attribute. If EEA algorithm was used to solve the best case mentioned in Section 23.4.2, the worst case will happen; all attributes have the same number of zeros in their class distribution. The difference between the maximum value and the minimum value in all attributes classes distribution is also equivalent. So, to find the sequence of the minimum tree, we have to compute the information gain value for all attributes.

23.5 Conclusion and Future Works

Data mining is defined as the knowledge discovery in database where it attempts to apply machine learning, statistics, and artificial intelligent concepts and techniques on database in order to discover useful patterns and trends hidden in vast amount of data stored in form of files and database records. Classification is one of the data mining techniques that aims to predict the class attribute value for a new instance using a model built based on historical data in which instances with known class values is stored.

Decision tree classifier is one of the most successful classifiers known so far. In decision tree classifier, the model is built in the form of a tree. The major drawback of decision tree classifier is that for a given dataset, we can find more than one possible valid tree for the same dataset. Trees found vary in size (number of branches and number of levels), the size of any possible tree depends on the sequence of the dataset attributes we choose to build the tree well. It is obvious that we always look for the tree with the minimum number of branches and levels (the minimum tree) because the minimum tree makes looking for a class value (trace starting from root to reach a leaf) easier and faster.

There are many algorithms have been proposed to find the sequence that yields the minimum tree including GINI algorithm and entropy algorithm. Entropy algorithm is usually used because it is suitable for almost all types of dataset types while GINI algorithm is suitable for a small portion of datasets types. The major drawback of entropy algorithm is that it needs to perform complex computations to find out the sequence of the minimum tree, here the complexity of computations required grows exponentially when the number of attributes in a given dataset increase. So, the proposed algorithm in this chapter (EEA) tries to find the sequence of the minimum tree in a dataset using few logical and mathematical operations that are simple compared to the complex operations needed by entropy algorithm to find the same sequence for the same dataset.

Comparing EEA to entropy algorithm in terms of algorithms complexity yield that entropy algorithm always need $O(n^3)$ to find the sequence of the minimum tree. There is no worst case nor best case, while EEA complexity in the worst case is $O(n^3)$ which is the same as entropy algorithm complexity, but in the best case it need a complexity of $O(n^2)$.

As a future work, we suggest to enhance EEA, so that there would be no worst case by eliminating the need to compute the information gain measure for two or more attributes if they have the same number of zeros and their difference is equivalent. Also, we recommend developing a software package with graphical user interface (GUI) to apply it on EEA for a given dataset.

References

1. Mendelson, S. and Smola A. (2003), Advances Lecture on Machine Learning, Spriger, Germany.
2. Michie, D., Spiegelhalter, D.J., and Taylor, C.C. (1994), Machine Learning, Neural and Statistical Classification, Ellis Horwood, USA.
3. Mitchell, T. (1997), Machine Learning, McGraw Hill, USA.
4. Duda, R.O., Hart, P.E., and Stock, D.G. (2001), Pattern Classification, Wiley-Interscience, USA.
5. Adrians, P. and Zantinge, D. (1996), Data Mining, Addison Wesley, USA.

6. Almasri, R. and Navathi, S.B. (2004), Fundamentals of Database Systems, 13:2-3, 197–210, Addison Wesley USA.
7. Liu, W. and White, A. (1997), The Importance of Attribute Selection Measures in Decision Tree Induction, Machine Learning, 15:1, 25–41.
8. Mingers, J. (1989), Am Emprical Comparison of Selection Measures for Decision Tree Induction, Machine Learning, 3:4, 319–342.
9. Apte, C. and Weiss, S. (1997), Data Mining with Decision Tree and Decision Rules, Future Generation Computer Systems, 13:2-3, 197–210.
10. Murthy, S. (1998), Automatic Construction of Decision Tree from data: A Multi Disciplinary Survey, Data Mining and Knowledge Discovery, 2:4, 345–389.
11. Wiss, S.M. and Kulikowsk, C.A. (1991), Computer Systems that Learn: Classification and Prediction Methods from Statistics, Neural Networks, Machine Learning and Expert Systems, Morgan Kaufmann, San Mateo.
12. Quinlan, J.R. (1986), Introduction of Decision Trees, Machine Learning 1(1): 81–106.
13. Naps, T. and Pothering, G. (1986), Introduction to Data Structure and Algorithm Analysis with Pascal, second edition 1986, West Publishing Company, USA.

Chapter 24
Numerical Analysis of Large Diameter Butterfly Valve

Park Youngchul and Song Xueguan

Abstract In this paper, a butterfly valve with the diameter of 1,800 mm was studied. Three-dimensional numerical technique by using commercial code CFX were conducted to observe the flow patterns and to measure flow coefficient, hydrodynamic torque coefficient and so on, when the large butterfly valve operated with various angles and uniform incoming velocity.

Keywords Computational Fluid Dynamics · Large diameter · Butterfly valve · Computer visualization · Flow patterns

24.1 Introduction

A large diameter butterfly valve shown in Fig. 24.1 is a type of flow control device that controls the flow of gas or liquid in a variety of process. It consists of a metal circular disc with its pivot axes at right angles to the direction of flow in the pipe, which when rotated on a shaft, seals against seats in the valve body. This valve offers a rotary stem movement of 90° or less in a compact design.

The importance of butterfly valves has been more and more increasing in the large pipe system. And there are so many reports on the characteristics, i.e. the flow coefficient, the torque coefficient, the pressure recovery factor and so on. Kerh et al. [1] performed an analysis of the butterfly valve on the basis of the experimental results. Sarpkara [2] theoretically treated the characteristics of a flat butterfly valve. Kimura and Tanaka [3] studied the pressure loss characteristics theoretically for a practical butterfly valve and so on.

P. Youngchul and S. Xueguan (✉)
CAE Lab, Department of Mechanical Engineering, Dong-A University, 840 Hadan-dong, Saha-gu, Busan 604-714, South Korea
e-mail: parkyc67@dau.ac.kr, songxguan@yahoo.com.cn

S.-I. Ao et al. (eds.), *Advances in Computational Algorithms and Data Analysis,*
Lecture Notes in Electrical Engineering 14,
349

Fig. 24.1 The large butterfly valve ($D = 1.8$ m)

With the development of the Computational Fluid Dynamics (CFD), the approach of using the technique of computational fluid dynamics has been substantially appreciated in mainstream scientific research and in industrial engineering communities. By now, the CFD simulation by commercial code has been proved its feasibility to predict the flow characteristic. There have been also many reports on valve using Computational Fluid Dynamics analysis. Huang and Kim [4] performed a three-dimensional numerical flow visualization of incompressible flows around the butterfly valve which revealed velocity field, pressure distributions by using commercial programs. Lin and Schohl [5] performed an analysis about the application of CFD commercial package in the butterfly valve field. Chern and Wang [6] employed a commercial package, STAR-CDTM, to investigate fluid flows through a ball valve and to estimate relevant coefficient of a ball valve.

Valve analysis in the past was performed using experiment methods, which required a number of equipments, times and funds. Especially for the large dimension valve, the only way is to reduce the prototype at an ideal scale, then do the analysis in a laboratory. Sometimes, this method can give a reasonable result by magnifying some parameters following some equations, however in most cases; it's not a good way to investigate the characteristic of butterfly valve.

Nowadays, due to the fast progress of the computer visualization and numerical technique, it becomes possible to do this by using numerical method. In this paper, a butterfly valve with the diameter of 1,800 mm was studied. Three-dimensional numerical technique by using commercial code CFX were conducted to observe the flow patterns and to measure flow coefficient, hydrodynamic torque coefficient and so on, when the large butterfly valve operated with various angles and uniform incoming velocity.

24.2 Important Aspects of Large Diameter Butterfly Valve

Before doing analysis of this large butterfly valve, the important parameters of this valve must be understood first. Generally speaking, for large butterfly valve, three parameters need to be considered if they will be investigated under different given circumstance.

- Flow coefficient (C_V)
- Cavitation parameters (σ)
- Hydrodynamic torque coefficient (C_T)

24.2.1 Flow Coefficients (C_V)

The flow coefficient is used to relate the pressure loss of a valve to the discharge of the valve at a giving valve opening angle. By using the C_V, a proper valve size can be accurately determined for most applications.

The most common formula used to relate the pressure differential to the flow is the Darcy-Wiesbach equation (Eq. 24.1).

$$\Delta H = K \frac{V^2}{2g} \tag{24.1}$$

where K is referred to as the resistance coefficient, ΔH is the net pressure differential or head loss in feet (meters), V is the flow velocity at the inlet in fps (m/s), and g is the gravitational constant 32.2 fps^2 (9.806 m/s^2). The parameter $V^2/2g$ is also referred to as the velocity head. As simple as the Darcy-Wiesbach equation is, the coefficient K has been used to denote a number of flow phenomena such as the cavitation parameter and numerous flow coefficients and constants.

However, a flow coefficient equation widely used by the valve industry (per ISA S75.01) is Eq. (24.2), it's also used in this paper.

$$C_{VISA} = \frac{Q_{gpm}}{\sqrt{\Delta P_{ISA}/S_g}} \tag{24.2}$$

Where ΔP is the pressure drop in units of psi, Qgpm is in units of gpm, and Sg is the specific gravity of the fluid.

It is very important to note that the ΔP used in the ISA C_{VISA} equation is not the net pressure drop ΔP_{net}. The pressure drop used in the ISA C_{VISA} equation is determined from testing requirements that include an additional 8 diameters in length of pipe and friction loss in the measured pressure drop. ISA S75.01 does not allow for the subtraction of the manifold loss between pipe taps and test valve from ΔP_{ISA}. However, the difference between ΔP_{net} and ΔP_{ISA} can be negligible when a flow coefficient Cv/d [2] is less than 20 (where d is in units of inches).

Note: Cv is not dimensionless. The units of Cv are a function of gpm and psi.

24.2.2 Cavitation Parameters

The cavitation parameter or index is a dimensionless ratio used to relate the conditions which inhibit cavitation $(P_2 - P_V)$ to the conditions which cause cavitation

(ΔP). There are numerous forms of the dimensionless number or parameter which have been used to mathematically describe cavitation. The most fundamental form is σ_2 of Eq. (24.3) which uses the downstream valve pressure (P_2), the vapor pressure of the liquid (P_V), and the pressure drop of the valve (ΔP). It is necessary to use σ_2 for determining specific cavitation effects such as size and scale effects. However, σ_2 can be difficult to use for scaling or sizing a control valve because of the lack or uncertainty of the value of the downstream valve pressure P_2. Equation (24.4) is an alternate form of the cavitation parameter σ that allows the use of the upstream pressure P_1 instead of the downstream pressure. This form of σ is much better suited for scaling and sizing valves. It is also important to notes that σ_2 can be converted to σ by just adding the value of 1.

$$\sigma_2 = \frac{P_2 - P_V}{\Delta P} \tag{24.3}$$

$$\sigma = \frac{P_1 - P_V}{\Delta P} = \sigma_2 + 1 \tag{24.4}$$

The cavitation parameter can be used to predict the pressure drop or discharge at which a control valve or orifice will begin to experience a given level of cavitation. If the σ calculated for the actual operating pressures of a valve is less than the value of σ for a cavitation limit, the valve will experience a level of cavitation more severe than that associated with the limit.

24.2.3 Hydrodynamic Torque Coefficient (C_T)

Dynamic flow torque is the flow torque produced by the flow forces on the butterfly valve disc about the valve shaft. Most butterfly valves produce a hydrodynamic flow torque that has the rotational direction that will open the valve disc at low opening degree and close the valve at high opening degree. The dimensionless flow torque coefficient that has been used by many valve vendors and designers is C_T of Eq. (24.5). It must be cautioned that there are other definitions and applications of C_T that are not dimensionless. For example, it is common to derive C_T from units of $ft \cdot lbs$ of torque, inches of disc diameter, and psi of pressure drop. This application is not dimensionless and produces coefficient values that are 1/12th that of a dimensionless coefficient.

$$C_T = \frac{Torque_{DynamicFlow}}{\Delta P d^3} = \frac{T}{\Delta P d^3} \tag{24.5}$$

24.3 Numerical Analysis

The main object of this part is to develop a model by using the commercial code ANSYS CFX 10.0, which accurately represents the flow behaviors and provide a three-dimensional numerical simulation of water around the large butterfly valve and estimate the pressure drop, flow coefficient and hydrodynamic torque coefficient and so on.

24.3.1 Numerical Method

The fluid flows large butterfly valve is simulated numerically. The incompressible and viscous fluid, which was modeled as water at 25°, is given a uniform velocity of 3 m/s at the inlet.

For our problem the hydraulic diameter is used because the water flows through a pipe from inlet to outlet. The inlet given velocity 3 m/s, and the kinematic viscosity of liquid water $8.9 \times 10^{-4} Pa \cdot s$ can be used to find Reynolds number (Eq. 24.6).

$$Re = \frac{\rho \cdot v \cdot D_{inlet}}{v} = \frac{997.0479 \text{kg/m}^3 \times 3m/s \times 1.8m}{8.90 \times 10^{-4} \text{Pa} \cdot \text{s}} = 6.05 \times 10^6 \qquad (24.6)$$

The Reynolds number is much greater than 10^5, which show that the flow through the valve is turbulent and the effect of the Reynolds Number is so small that it can be neglected [7].

To deal with the turbulence model, the Reynolds-averaged Navier-Stokes Equations (*RANS*) is utilized [8]. With RANS, a turbulence model is needed to close the momentum equations. Two-equation turbulence models such as the $k - \varepsilon$ model, $k - \omega$ model and Reynolds Shear Stress model have been primarily used in industry because of their robustness. After comparing the three models for valve opening of 55°, as a result, the $k - \varepsilon$ model is chosen because the $k - \varepsilon$ model does not involve the complex non-linear damping functions required for the other models and is therefore more accurate and more robust [9].

24.3.2 Grid Generation

Besides the turbulence model, another important factor affecting the accuracy of the simulation is the quality of "meshing". Theoretically, the more elements in the geometry, the higher mesh quality, and the better the accuracy of the results is. Simultaneity, a longer computer calculation time will be taken. Table 24.1 shows the accuracy and timing cost for different element numbers. For the sake of both precision and solving rate, as a result the middle level with about 948,721 elements is used. Correspondingly, the "Globe mesh size" is set to 150 and "Natural size" is set to 0.001.

In addition, the water flows through the butterfly valve, which causes a boundary layer close to the walls where there are large velocity and pressure gradients. Therefore, it is necessary to have a high concentration of nodes located near the walls to

Table 24.1 Various mesh size and calculation information

Total elements	142,536	948,721	1,275,822
Total nodes	34,715	231,054	327,351
Total CPU time (s)	1.456E4	5.826E4	1.672E5

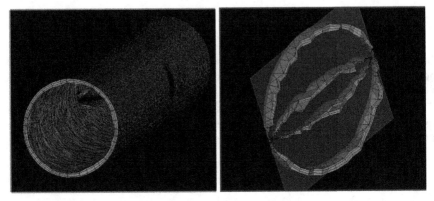

Fig. 24.2 Prism mesh around inner surfaces of pipe and at middle cut plane

fully resolve the velocity and pressure gradients. There are three layers prism mesh which is concentrated near the walls at a ratio of 1.2 from the inlet to the outlet. This wall-near domain is meshed to form 670,056 tetrahedron elements (see Fig. 24.2). One point should be noticed here, prism mesh can be difficult to smooth, so it is best to start with good quality tetra or tri surface mesh.

24.3.3 Boundary Condition

The prototype size is 1.8 m in diameter and it is manufactured from cast steel with machined inside surfaces. For a better result of CFD simulation, The CFX model of large butterfly valve is created at a 1:1 scale with no rough wall inside. The valve disc has a shape similar to "flying disk" and its maximum thickness in the middle of the valve is 360 mm and the minimum along the flange is 20 mm. some remote round corner or small character are removed. Some important boundary conditions are set as bellow.

24.3.3.1 Inlet Condition

Generally speaking, when we want to calculate the flow coefficient or torque coefficient, it doesn't matter about the inlet velocity or pressure, but when we mainly focus on the cavitation phenomenon, the pressure loss produced by butterfly valve becomes so important. So in this paper, two different patterns are applied. One is inlet velocity pattern, the fluid which is modeled as water, is given a uniform velocity of 3 m/s at the inlet and zero reference pressure at the outlet. Correspondingly, another model is the inlet given pressure model. Usually, a butterfly valve of 1.80 m in diameter is applied on a condition with a nominal head of 150 m, through calculation; a nominal head of 150 m can produce a static pressure of 15.7232 bar. In terms of this simplification without the influence of valve diameter, we can set the inlet pressure as 15.7232 bar.

The range of Reynolds Number of flow in this study has been identified larger than 10^5, hence the effect of the Reynolds Number is so small that it can be neglected.

24.3.3.2 Pipe Length

In terms of the research of Huang and Kim (1996), the upstream length (L1) and the downstream length (L2) should be at least two times and eight times of the diameter respectively. In terms of ISA standard, P_2 is measured at the point of 6 diameters downstream. So the downstream must be longer that 6D at least. Figure 24.3 shows the various velocity profile along the pipe at length of n*D(n $= -2 \sim 10$), Fig. 24.4 shows the various velocity profiles in the median vertical line from 2D upstream to 8D downstream.

Calculation shows the difference of the average velocity between 8 times diameters downstream and 10 times diameters downstream is less than 0.01%, which indicates that the length of the additional pipe (8D upstream and 12D downstream) can satisfy the accuracy requirement for the simulation.

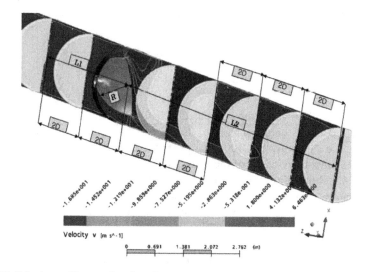

Fig. 24.3 Velocity profile at various lengths

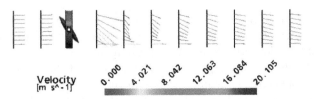

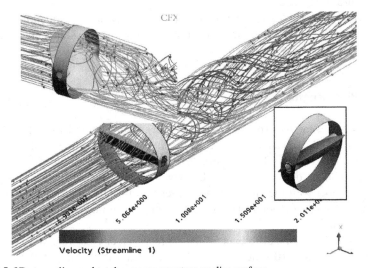

Fig. 24.4 Velocity vector at various median vertical lines

Fig. 24.5 3D streamline and total pressure counters on disc surface

In addition, no energy transfer and buoyancy is considered. Uniform temperature through the fluid flow is postulated. The convergence target in solve control is set to 0.0001.

24.4 Result and Discussion

Figures 24.5 and 24.6 show the computed velocity streamlines in 3D space and total pressure counters on the surface of disc at two opening angles. These figures demonstrate the capability of CFX to simulate the complicated flow in 3D space, also provides obvious evidence of the prediction's validity.

Figure 24.7 shows the vortices at the opening angle of 15°. In general, there are more than two vortices which can be observed at the downstream, more small vortices must exist, but not so obvious. The velocity direction also can be found in this figure.

Total pressure at XY plane and various opening angle is shown in Fig. 24.8. They are noted that the vortices and the maximum total pressure grow as the opening degree, α, decreases. The highest pressure and the maximum vortex zone occur at 15° opening around disc edge.

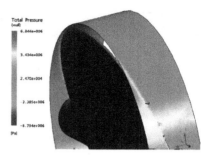

Fig. 24.6 Total pressure on disc surface, α = 15

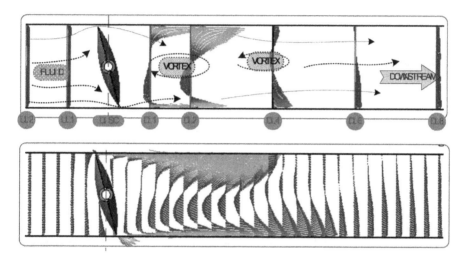

Fig. 24.7 Schematic flow pattern in the valve flow

This is because that the faster velocity zone becomes reducing to the smaller zone, but the magnitude of the velocity becomes larger; the boundary layer separation point becomes closer to the valve disc, and caused the vortex. As the opening degree decreases, the fluid flow passing the valve disc area becomes narrow. Then the fluids with the smaller angle are accelerated than the fluids with the larger angle. Induced pressure drop increases. The negative pressure gradient zone also increases. Finally the boundary layer separates to form the bigger and more complicated vortex.

Figure 24.9 shows the velocity profile at various central vertical lines from 4 diameters upstream ("UL4" for short) to 12 diameters downstream ("DL12" for short), for different opening degree. It's clear that along the line in front of 4 diameters of valve, the velocity profile is symmetrical and the values is close to the given condition (3 m/s), except for the area near the wall, but other velocity profiles are affected by the opening degree observably. Particularly, at the approximate height of disc edge's lowest point, each line has a severe velocity jump. Comparison reveals that

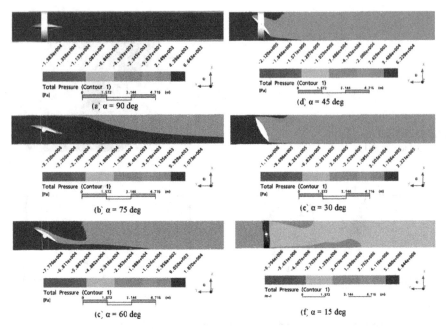

Fig. 24.8 Pressure counter at various opening degree, XY plane

the velocity profiles become more and more symmetrical and smooth, and the velocity values approximately decrease to original value as the opening degree increases.

The pressure drop and Torque, which are normalized by the maximum value of the pressure drop and Torque respectively is shown in Fig. 24.10. This figure can give more understanding about the relation between pressure drop and torque. It is mentioned that the trends of the torque and pressure drop are as similar as the valve opening degree increases, however the slope at each particular position is different. The reason is that, the dynamic torque is induced by the fluid dynamic pressure on the two sides of the disc, and the magnitude of the pressure is affected by the pressure drop at first hand. Hence, this result that these two lines have the similar trend in Fig. 24.10 can be comprehended.

Figure 24.11 compares the C_V by CFX simulation with that calculated by the experimental equation. As the valve opening degree increases, the value of C_V varies from zero to 3.7E5. Actually in terms of Eq. (24.3), C_V is just dependent on the root of pressure drop for a given velocity, so the trend of C_V is opposite. It's got to be verified here. However, it must be noticed that at valve opening less than $20°$s, the minimum error between CFX simulation and experimental data reach to 49.27958%. The reason why there is a so big error will be discussed later.

For getting the best shape of valve disc, this result and a chart from butterfly design criteria (Fig. 24.12) are merged together to check whether this design of valve disc measure up. Comparison show that even this disc shape is not the best design, its parameter is still in the range of general values, also can satisfy common

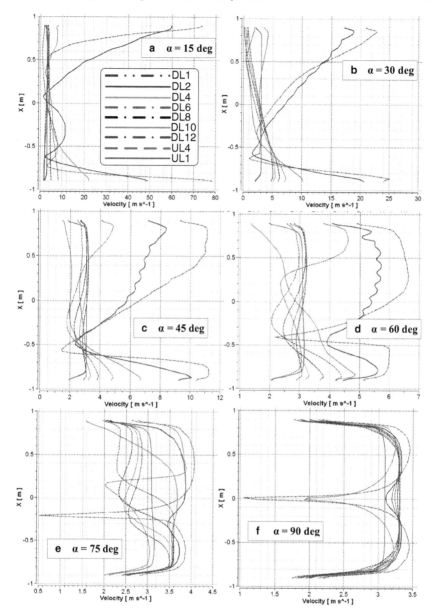

Fig. 24.9 Velocity profiles at various central vertical lines

application very well. It must be noticed that Fig. 24.11 shows the tendency of flow coefficient, but Fig. 24.12 represents the values of discharge coefficient. However, through reasonable change, including changing pressure drop to energy head and unite conversion, it can be compared at the same level.

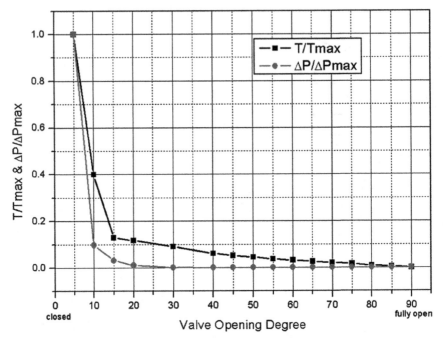

Fig. 24.10 Pressure drop ratio and Torque ratio at various opening angle, α

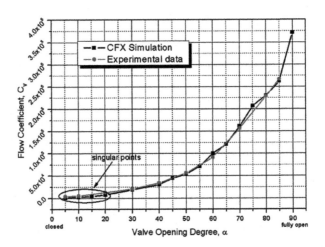

Fig. 24.11 CV at various opening degree, α

Figure 24.13 shows the experimental results and the simulation result of the C_T. It is displayed that the simulation data obtained by CFX agrees well with the experimental data get from a valve company. At an opening angle less than 50°, C_T creases slowly, but this does not mean the hydrodynamic torque creases very slowly

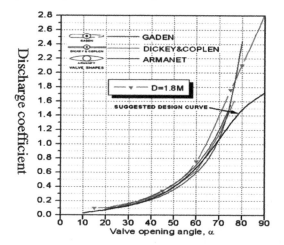

Fig. 24.12 Comparison of discharge coefficient at various opening degree, α

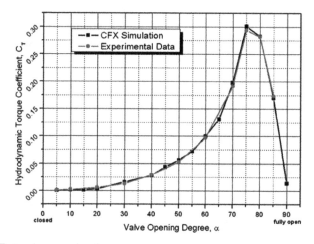

Fig. 24.13 CT at various opening degree, α

too, because the value of toque is the product of pressure drop and C_T, and product of pressure varies sharply at this area (see Fig. 23). The peak of the torque coefficient occurs at valve opening between 70° and 80°. Larger than 80°, the value of C_T drops to almost zero rapidly.

Figure 24.14 shows the cavatition index at different valve opening angle for the inlet given velocity of 3 m/s. Because of their interrelated calculation relation, the trends of these five parameters are very alike. It also implies that cavitation more possibly appears in the case with small valve opening angle. Mechanical vibrations induced by the cavitation could produce a significant random load on the butterfly valve components. It could result in the valve stem or disc failure by mechanical stress or in mechanical resonance of the system.

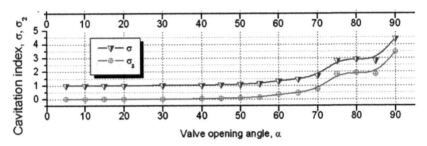

Fig. 24.14 Cavitation parameters at various opening angle, α

24.5 Conclusion

A computational model of a butterfly valve on demand has been developed. The result shows that it is possible to effectively use this type valve in common application. This type butterfly valve will produce little pressure drop and hydrodynamic torque at a given inlet velocity of 3 m/s. However, in some cases of small opening, it's suggested that a by-pass equipment to equalize the pressure on both sides of the disc should be employed because of the severe pressure drop, or a kind of cushion equipment should be designed and installed to lessen the impact when the valve is closed.

The turbulence and dynamic torque generated by the large butterfly valve become much stronger as the valve opening angle decreases. It's suggested to avoid fixing this type valve at a low opening angle.

The results of CFX simulation generated trend which agreed with the experimental data very well. It's proved that this commercial software can be used for the analysis of large butterfly valve. However, at some peculiar position, especially at the valve opening degree smaller than $20°$, it didn't agree well. This may be due to the disadvantage of the k–ε turbulent model of its own. It's suggested to use another turbulent model which is good at treatment of near-wall such as k–ω model and SST turbulent model.

In general, the result obtained by using commercial code ANSYS CFX 10.0 agrees with the experimental result very well. And this large diameter butterfly valve has much good characteristics.

References

1. T. Kerh, J. J. Lee and L. C. Wellford, Transient fluid-structure interaction in a control valve, Journal of Fluid Engineering 119, 1996, 354–359.
2. T. Sarpkara, Torque and cavitation characteristics of butterfly valve, ASME Journal of Applied Mechanics 28, 1961, 511–518.
3. T. Kimura and T. Tanaka, Hydrodynamic characteristics of a butterfly valve-prediction of pressure loss characteristic, ISA Transactions 34, 1995, 319–326.
4. C.D. Huang and R.H. Kim, Three-dimensional analysis of partially open butterfly valve flows, Transactions of the ASME 118, 1996, 562–568.

5. F. Lin and G. A. Schohl, CFD prediction and validation of butterfly valve hydrodynamic force, World Water Congress, 2004, pp. 1–8.
6. M. J. Chern and C. C. Wang, Control of volumetric flow-Rate of ball valve using V-port, Journal of Fluid Engineering 126, 2004, 471–481.
7. K. Ogawa and T. Kimura, Hydrodynamic characteristics of a butterfly valve-prediction of torque characteristic, ISA Transactions 34, 1995, 327–333.
8. ANSYS CFX 10.0 User's Manual, ANSYS, Inc.
9. B. Mohammadi and O. Pironneau, Analysis of the K-Epsilon Turbulence Model (Research in Applied Mathematics), Wiley (Import), Masson, Paris, August 1994.

Chapter 25
Axial Crushing of Thin-Walled Columns with Octagonal Section: Modeling and Design

Yucheng Liu and Michael L. Day

Abstract This chapter focus on numerical crashworthiness analysis of straight thin-walled columns with octagonal cross sections. Two important issues in this analysis are demonstrated here: computer modeling and crashworthiness design. In the first part, this chapter introduces a method of developing simplified finite element (FE) models for the straight thin-walled octagonal columns, which can be used for the numerical crashworthiness analysis. Next, this chapter performs a crashworthiness design for such thin-walled columns in order to maximize their energy absorption capability. Specific energy absorption (SEA) is set as the design objective, side length of the octagonal cross section and wall thickness are selected as design variables, and maximum crushing force (P_m) occurs during crashes is set as design constraint. Response surface method (RSM) is employed to formulate functions for both SEA and P_m.

Keywords Axial crushing · Thin-walled column · Octagonal section · Simplified model · Crashworthiness analysis · Design optimization · Response surface method (RSM) · Finite element analysis (FEA)

25.1 Simplified Modeling

25.1.1 Introduction

In engineering research, crash behaviors of the thin-walled structures during axial collapse and bending collapse have aroused a lot of interesting which include the axial and bending resistances, respectively. Previous researchers have thoroughly

---

Y. Liu (✉) and M.L. Day
Department of Mechanical Engineering, University of Louisville, Louisville, KY 40292, USA
e-mail: y0liu002@louisville.edu

S.-I. Ao et al. (eds.), *Advances in Computational Algorithms and Data Analysis,*
Lecture Notes in Electrical Engineering 14,
© Springer Science+Business Media B.V. 2009

investigated the crash behaviors of different types of the thin-walled structures and developed a series of mathematical equations to describe or predict the resistances. T. Wierzbicki, W. Abramowicz, and D. Kecman [1–6] applied different modes of deriving the mathematical equations that accurately predict the axial and bending resistances of the rectangular and square section columns. W. Abramowicz, N. Jones, A. Mamalis and other co-researchers [7, 8] studied the crushing behavior of the circular thin-walled columns when subjected to an axial impact or a bending moment during the crash. Y-C Liu and M. Day [9] studied the bending collapse behavior or the thin-walled columns with channel cross section and derived their bending resistances. However, besides above thin-walled structures, the thin-walled octagonal columns are also an important crashworthy structure due to its better characteristics in energy absorption and crushing [10]. Therefore it deserves special considerations and appropriate numerical models have to be developed to predict their crash behavior.

A. Mamalis, D. Manolakos and other co-researchers [11] simulated the crash behavior and energy absorption characteristics of steel thin-walled columns of octagonal cross section subjected to axial loading. In this article, based on a developed method for predicting crush behavior of multi-corner thin-walled columns subjected to axial compression [6], a general equation is derived to estimate the axial resistance of the thin-walled octagonal columns during the axial crushing. The derived axial resistance then can correctly predict the relationship between the axial load and the axial deformation of the octagonal columns.

An important application of the derived thin-walled columns' axial resistance is simplified modeling. Simplified modeling is a critical modeling technique which has been extensively applied in early design stage for product evaluation, crashworthiness analyses, and computer simulation. Simplified computer model is a finite element model that composed of beam and spring elements. Compared to detailed model, it requires less modeling work and consumes less computer resources during the modeling and simulation. In developing simplified thin-walled column models, the derived axial and bending resistances have been used to define nonlinear spring elements, which simulated the crash behavior of the models. Y-C Liu and M. Day [9, 14, 16], H-S Kim [12], and P. Drazetic [13] have created qualified simplified models for the thin-walled columns as well as an entire vehicle assembly and summarized general simplified modeling methodologies for such column members. In this section, the derived octagonal column's axial resistance is applied to develop the simplified octagonal column models. The developed simplified model is then used for crashworthiness analyses and the results are compared to those from the detailed model and from the published literature [11] for validation. Relatively good agreement is achieved through these analyses therefore both the derived axial resistance and the developed simplified model are validated. The explicit FE code LS-DYNA is used to simulate the axial compression of the steel thin-walled octagonal columns [15].

25.1.2 Axial Resistance

The method developed by W. Abramowicz and T. Wierzbicki [6] is used to predict the axial resistance of the thin-walled octagonal columns. According to the published method, the axial crushing resistance for a multi-corner thin-walled column equals to the appropriate crushing force per one element times the number of corner elements, n. The mean crushing force per one element is

$$\frac{P_m}{M_0} = \left\{ A_1 \frac{r}{t} + A_2 \frac{C}{H} + A_3 \frac{H}{r} \right\} \frac{2H}{\delta_{ef}} \tag{25.1}$$

and can be rewritten in the form:

$$\frac{P_m}{M_0} = 3(A_1 A_2 A_3)^{1/3} (C/t)^{1/3} \frac{2H}{\delta_{ef}} \tag{25.2}$$

where

$$M_0 = (\sigma_0 t^2)/4 \tag{25.3}$$

in (25.2): P_m is the mean crushing force; M_0 is fully plastic moment per unit length of section wall; A_1 through A_3 are coefficients and according to [6] $A_1 = 4.44$, $A_2 = \pi$ and $A_3 = 2.30$, respectively; C is edge length of octagonal cross section; t is wall thickness; 2H is length of one plastic fold generated when the column buckled (see Fig. 25.1), δ_{ef} is called "effective crushing distance" and $\delta_{ef}/2H$ was measured as 0.73 from previous experimental observations [6]. In (3), σ_0 is the material's energy equivalent flow stress that approximately equals to: $\sigma_0 = 0.92\sigma_u$.

Substitutes all the values into (25.2) and multiplies (25.1) by eight for the octagonal section, the axial resistance of the steel thin-walled octagonal columns can be approximated by the function:

$$\frac{P_m}{M_0} = 97.77(C/t)^{0.4} \tag{25.4}$$

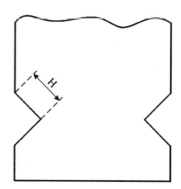

Fig. 25.1 Length of half plastic folding wave, H

Substitutes (25.3) to (25.4), the mean crushing force can be directly calculated based on the given octagonal column's dimensions and material properties as:

$$P_m = 24.44\sigma_0 t^{1.6} C^{0.4} \qquad (25.5)$$

To find the length of plastic fold 2H, minimizing Eq. (25.1) with respect to H by using $\partial P_m / \partial H = 0$ yields

$$H = 0.97 \sqrt[3]{t C^2} \qquad (25.6)$$

The calculated P_m is the axial resistance of the octagonal columns and will be used to determine the nonlinear spring element for developing the simplified model together with the calculated H.

25.1.3 Simplified Modeling of Thin-Walled Octagonal Column

In this part, a detailed thin-walled octagonal column model is firstly created using shell elements. Afterward, a simplified model is then developed using Liu and Day's method [9, 14]. The simplified model is composed of beam and nonlinear spring elements. The axial resistance demonstrated in above section is applied in determining the characteristics of the nonlinear spring elements. Both models are used for crashworthiness analyses in order to validate the developed simplified model as well as the presented axial resistance of the thin-walled octagonal column. Relatively good correlations are achieved through comparing the analysis results from the simplified model to those from the detailed model and from the published literature [11].

25.1.3.1 Detailed Model

The detailed model for the thin-walled octagonal column is shown in Fig. 25.2, which will then be used for the crashworthiness analysis. The detailed model uses the full integration shell element: 4-node Belytschko-Tsay shell element with five integration points through the thickness. Also, as shown in the figure, during the crash analysis, the beam model is fully constrained at one end. An initial velocity 1 m/s is applied on the other end to make the model move along the Z direction. Table 25.1 lists all the related conditions and information of the thin-walled straight beam. In the crashworthiness analysis, all the geometries and material properties of this model and the impact conditions come from the published literature [11], therefore the obtained analysis results are comparable. After the analysis, the dynamic results are documented and compared with those from the simplified model, which will be presented in the latter section.

Fig. 25.2 Detailed thin-walled octagonal column model

Table 25.1 Properties and impact conditions

Material properties	
Young's modulus	207 GPa
Density	7,830 kg/m³
Yield stress	250 MPa
Ultimate stress	448 MPa
Hardening modulus	630 MPa
Poisson's ratio	0.3
Geometries	
Total length	127 mm
Cross section	Edge 31.8 mm
Wall thickness	1.52 mm
Impact conditions	
Added mass	280 kg
Initial velocity	1 m/s
Crash time	10 s

25.1.3.2 Simplified Model

In order to develop the simplified model, the force – shortening characteristics of the nonlinear spring elements are determined. Applying Eq. (25.5) and (25.6) and substituting the material properties listed in Table 25.1, the calculated length of the plastic folding wave is 2H = 22.39 mm, and the mean crushing force is P_m = 78.5 kN. Thus, the nonlinear spring element for the straight hexagonal column then can be determined and its $F - \delta$ relationship is plotted in Fig. 25.3. With this property, the spring begins to deform when the crushing force reaches 78.5 kN. After its deformation reaches 22.39 mm, the spring fails and stops deforming.

Since LS-DYNA doesn't include the octagonal section as its standard section type, therefore the octagonal cross section has to be developed by the users. J. Hallquist [15] has provided a very good illustration and explanation of beam el-

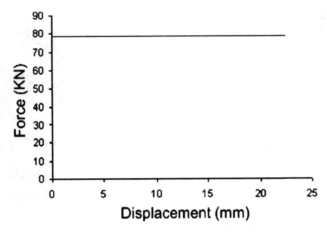

Fig. 25.3 F – δ curve of nonlinear spring element for simplified model

Fig. 25.4 Definition of integration points for defined
octagonal cross section

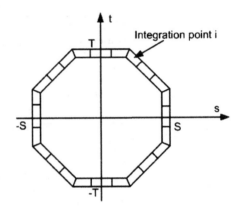

ements and also guided users how to develop arbitrary cross sections for the beam elements. According to that literature, in LS-DYNA cards, *SECTION_BEAM and *INTEGRATION_BEAM were modified to define the octagonal cross section. The Hughes-Liu beam with cross section integration is still used to define the beam elements and 24 integration points are defined along the octagonal cross section (Fig. 25.4). For each integration point, its normalized s and t coordinates $(-1 \leq s \leq 1, -1 \leq t \leq 1)$ and weighting factor are calculated and input to the *INTEGRATION_BEAM card to fully define the octagonal cross section. As shown in Fig. 25.4, the s or t coordinates equal to the integration point's coordinates (s_i, t_i) divided by the maximum coordinate of the cross section (S and T), and the weighting factor equals the area associated with the integration point divided by actual cross sectional area, which is 1/24.

The simplified finite element model (Fig. 25.5) for the thin-walled octagonal column is then created applying Liu and Day's method [9, 14, 16]. In the devel-

Fig. 25.5 Simplified thin-walled octagonal column model

oped simplified model, the Hughes-Liu beam elements are used for building a pure beam-element model and the developed cross sectional information is assigned. Afterwards, the entire beam is divided into several equal segments with the length 22.39 mm, which is the length of one plastic fold. Developed nonlinear spring elements connect these segments together, whose force – displacement relationship is displayed in Fig. 25.3. To ensure that the simplified model only deforms along its length, appropriate boundary conditions are applied.

25.1.4 Analyses and Comparison

Same crash analyses are performed on both detailed and simplified models. The results come from both models are compared and shown in Figs. 25.6 and 25.7 and Table 25.2.

From the results of the comparisons, it is verified that the developed simplified octagonal tube model is qualified for replacing the detailed model to be used for crash analyses. Figure 25.5 verifies that during the analyses, the detailed and simplified models showed similar axial shortening and underwent similar crushing force history. Table 25.2 shows that the numerical results yielded from the simplified model correlated to those from the detailed model very well. The errors between the peak and mean crushing force are below 10%, and the simplified model contains only about 1/8 of the number of elements of the detailed model. Also, in comparing the obtained force values to the published literature (from [11] the peak force

Fig. 25.6 Comparisons of crash results – thin-walled octagonal tube model: deformed configurations

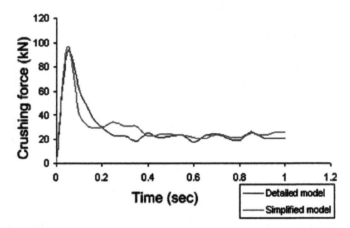

Fig. 25.7 Comparisons of crash results – thin-walled octagonal tube model: crushing forces

Table 25.2 Comparisons of crash results from detailed and simplified octagonal section tube models

	Detailed model	Simplified model	Difference (%)
Peak crushing force (kN)	91.8	96.3	4.9
Mean crushing force (kN)	28.5	29.6	3.9
Computer time	2 h 32 min	2 min 32 s	–
Elements	844	108	–

is 92.9 kN), it is seen that the detailed model created here yields same results as the model presented in [11], the tiny difference (about 1%) may be caused by the different techniques applied for creating the base plate and the impact interface.

In order to validate the simplified modeling method, two more simplified octagonal column models are created for the crash analyses. Compare to the model shown in Fig. 25.5, the first model has different geometries: whose sectional edge is 19.1 mm and thickness is 0.81 mm. The second model has the same geometries but different material properties: whose yield stress is 350 Mpa instead of 250 Mpa. Unlike generating detailed models, it is not necessary to re-create the simplified models. The presented simplified model still can be used and the users only have to change its cross sectional information or its material properties to obtain the new

Table 25.3 Comparisons of results from new created simplified models to published ones

		Simplified model results	Published results	Difference (%)
Model 1	Peak crushing force (kN)	96.3	92.9	3.7
	Mean crushing force (kN)	29.6	–	–
Model 2	Peak crushing force (kN)	12.6	11.1	13.5
	Mean crushing force (kN)	7.3	8.4	−13.0
Model 3	Peak crushing force (kN)	89.0	91.7	−2.9
	Mean crushing force (kN)	30.0	33.1	−9.4

simplified models. For the model which has different cross sections and wall thickness, Eqs. (25.5) and (25.6) are applied to determine its force-displacement characteristic, which is very similar to the $F - \delta$ relationship shown in Fig. 25.3.

All simplified models are used for the same crash analysis performed before and the analyses results together with the results come from the first model are compared to the published ones (in [11]) to verify the efficiency and accuracy of the developed simplified models and the presented modeling method (as shown in Table 25.3).

The results of the comparisons demonstrate that the results come from the developed simplified models correlate very well to the ones from the published literature. From Table 25.3, it is found that all of the errors are less than 15%. Meanwhile, the simplified models require much less modeling work and save much more computer resources than the detailed ones. Similarly, more simplified models can be developed for the thin-walled octagonal columns with different geometries and cross sections. These simplified models then can be used for crash analyses and are developed following the same steps as illustrated before.

25.2 Crashworthiness Design

25.2.1 Introduction

Practically, thin-walled columns most appear in truss and frame structures as major energy-absorbing components and absorb substantial amount of crash energy when the impact occurs. Therefore, such columns receive a lot of research interests [6, 10, 11]. In designing such columns, maximizing their energy absorption capability should always be a major objective. Hou et. al. [17] presented the optimal designs of straight hexagonal thin-walled columns with singly celled and triply celled configurations which provided the maximum energy absorption capability during the crashworthiness analyses. However, little effort has been spent on the optimization of the cross-sectional dimensions of the octagonal thin-walled columns. Thus, in this section, an optimum design is first performed for the cross-sectional profiles of such columns to maximize their capability of energy absorption.

During the optimum design, an advanced technique, response surface method (RSM) [17] is applied to approximately formulate the columns' energy absorption capabilities. This method has been employed to optimize several other thin-walled structures with crashworthiness criterion [17]. In those studies, the polynomial basis functions were used to model the energy absorption. On the basis of the previous researches, the RSM with the polynomial basis functions are used here to obtain the optimum design for the thin-walled octagonal section columns. Moreover, because of the high accuracy provided by quartic polynomial function [17], the quartic polynomial is employed in this design to formulate the response functions.

To seek for the optimal crashworthiness design, a set of designs are sampled from the design space using the factorial design, which have different cross-sectional dimensions. Finite element (FE) models are created for those designs and used for computer crashworthiness analyses to provide crash responses of these design samples, on which the response surface methods are constructed based.

25.2.2 Problem Description

In crashworthiness design, the energy absorption capacity of a thin-walled column is measured by the specific energy absorption (SEA), which is the crash energy absorbed per unit weight of the thin-walled column. Therefore, the major objective of this optimum design is to maximize the SEA for the octagonal thin-walled columns. The SEA is defined as:

$$\text{SEA} = \text{Total absorbed energy } E_{total}/\text{Total structural weight} \qquad (25.7)$$

Two factors have to be considered during this design. At first, based on the human safety issues, the maximum crushing force P_m that occurs during the crash should not exceed a certain criteria, which is an important issue in the vehicle design and manufacturing. Also, two design variables of the optimized octagonal cross section, its side length a and wall thickness t (as shown in Fig. 25.4), only vary between their upper and lower bounds. Thus, this optimization problem is formulated as

$$\text{Maximize: SEA(a, t)}$$
$$\text{Constraints: } P_m \leq \text{Criteria}, a^L \leq a \leq a^U \text{ and } t^L \leq t \leq t^U \qquad (25.8)$$

Where a^L, a^U, t^L and t^U are the lower and upper bounds of the design variables a and t, respectively.

25.2.3 Response Surface Method (RSM) [18]

RSM is extensively applied in modern industry for developing, improving, and optimizing processes. During a design process, such method is used to determine several input variables (independent variables) which potentially influenced the performance or quality of the system in order to achieve optimized responses from that
system. In this study, RSM is employed to determine the a and t of the octagonal thin-walled columns so as to maximize the SEA when impact occurs on such
columns.

In our problem, the response of the thin-walled box beam is SEA(a, t), which is
approximated using a series of the basic functions in a form of

$$\widehat{y}(x) = SEA(a,t) = \sum_{i=1}^{n} \beta_i \varphi_i(a,t) \tag{25.9}$$

where n represents the number of basic functions φ_i (a, t). Polynomials are used
here to construct these basic functions because the polynomial is generally used to
generate response surface models with a high accuracy.

In Eq. (25.9), the β_i, known as the regression coefficients, and the coefficient
vector $B = (\beta_1, \beta_2, \ldots, \beta_n)$ can be determined as

$$B = (\Phi^T \Phi)^{-1} \Phi^T y \tag{25.10}$$

where Φ denotes the matrix consisting of basic functions evaluated at the m sampling points, which is

$$\Phi = \begin{bmatrix} \varphi_1(a,t)_1 & \cdots & \varphi_n(a,t)_1 \\ \vdots & \ddots & \vdots \\ \varphi_1(a,t)_m & \cdots & \varphi_n(a,t)_m \end{bmatrix} \tag{25.11}$$

By substituting Eq. (25.11) into (25.9), the response surface model is created and
the response function SEA(a, t) can be fully determined.

25.2.4 Design Optimization – Straight Octagonal Columns

25.2.4.1 Sampling Design Points

A series of design sample points, (a, t), are selected from the design domain to
construct design samples. After that, FEA analyses are performed on these design
samples to provide crash response observations for octagonal column samples. The
response functions of SEA and P_m are then derived following the approach presented in above section. The design ranges of the side length a and wall thickness t

for the octagonal section are determined based on a previous literatures [11, 17] and the optimization problem is defined as

Maximize: SEA(a, t)

Constraints:$P_m \leq 70\,kN, 20\,mm \leq a \leq 44\,mm$ and $0.8\,mm \leq t \leq 1.6\,mm$

$$\text{(25.12)}$$

Within the design space of a and t, five-level full factorial design is used for sampling the design points. Consequently, 5×5 design sample points are selected, which evenly distributed along their design ranges. These 25 design sample points with different a and t values are selected from their design domains and 25 octagonal thin-walled columns are acquired based on these a and t values. Next, detailed FE models are first created for these columns and used for the computer crashworthiness analyses. The response functions of SEA and P_m are then derived from the FEA results.

25.2.4.2 FE Models and Crashworthiness Analysis

After collecting all the octagonal column samples, FE models are created for these columns and the FE models are used for the crashworthiness analyses. FEA results of the specific energy absorption SEA and the maximum crushing force P_m are obtained from the analyses and will later be used for constructing corresponding response surface models. The FE model information and analysis conditions are displayed in Fig. 25.2 and Table 25.1 except that here the total length of the columns is 300 m, the initial velocity is 15 m/s, and the crash time is 0.01 s. After the analyses, important FEA results are recorded and will be used for generating the response functions.

25.2.4.3 Response Surface Models

As illustrated in above section, the FEA analyses yielded crash responses for the 25 column design samples. In this section, the response surface (RS) models are constructed based on the FEA results. In generating these RS models, the quartic polynomial is selected as the basic functions. The regression coefficients β_i of these polynomials in the response functions are determined using Eq. (25.11) and (25.12). Since a and t are the only two variables involved in this design problem, the full quartic form of this problem is $1, a, t, a^2, at, t^2, a^3, a^2t, at^2, t^3, a^4, a^3t, a^2t^2, at^3, t^4$. Substituting the selected a and t values into Eq. (25.12), the matrix Φ is then fully determined.

With these regression coefficients the polynomial response functions are therefore derived based on the matrix Φ and the FEA results obtained from the previous crashworthiness analyses using Eq. (25.11). The response quartic polynomial functions for SEA and P_m are

Table 25.4 Optimal multicorner cross sectional designs

Models	Optimal design (mm)	SEA by RSM (kJ/kg)	SEA by FE (kJ/kg)	P_m(kN)
Octagonal straight column	a = 20 mm t = 1.6 mm	5.7208	5.7232	54.1
Hexagonal straight column [6]	a = 30 mm t = 2.29 mm	12.0472	12.0162	69.9
Hexagonal curved column	a = 60 mm t = 3 mm	6.4724	6.4616	36.6

$$SEA(a,t) = -0.9619 - 0.2127a + 20.5732t + 0.0357a^2 - 1.3231at - 5.7217t^2$$
$$- 0.0014a^3 + 0.0489a^2t - 0.2938at^2 + 6.1188t^3 + 1.4630 \times 10^{-5}a^4$$
$$- 3.6899 \times 10^{-4}a^3t - 4.7984 \times 10^{-3}a^2t^2 + 0.1716at^3 - 2.4645t^4 \qquad (25.13)$$

$$P_m(a,t) = -460.06 - 37.18a + 2518.21t + 1.456a^2 + 26.2493at - 3513.13t^2$$
$$- 0.0257a^3 - 0.5733a^2t - 8.6288at^2 + 2033.4t^3 + 1.4146 \times 10^{-4}a^4$$
$$+ 8.2755 \times 10^{-3}a^3t - 0.0914a^2t^2 + 4.4965at^3 - 445.31t^4 \qquad (25.14)$$

Afterwards, the optimal design can be obtained using constrained nonlinear multivariable optimization algorithm (fmincon), which is provided by MATLAB. In the optimal octagonal section, the side length a = 20 mm, t = 1.60 mm. According to Eq. (25.13) and (25.14), the corresponding SEA provided by the optimal design is 5.7232 kJ/kg and P_m is 54.1 kN. FEA results are also found for the optimal octagonal section column and all the optimized crash responses are listed in Table 25.4.

25.3 Conclusions

This chapter presents a modeling method to develop simplified models for the thin-walled octagonal column, which can be used for crashworthiness analysis. After that, an optimum design is performed to obtain optimal octagonal cross section for that column, which provides the best energy absorption capacity during the crashes.

In order to develop the simplified models, the axial collapse behavior of the thin-walled octagonal column was analyzed and mathematical equations were derived to predict its axial resistance. Such axial resistance is then used to develop the simplified thin-walled octagonal column models. Compares to the detailed model, the simplified model requires less modeling efforts and has smaller size, and save much more computer time and resources during dynamic analysis. The simplified model developed here has been validated through crashworthiness analyses and comparisons.

During the design optimization, SEA is set as the design objective, which represents the column's capacity of absorbing the crash energy. The cross sectional width a and the wall thickness t are selected as two design variables, and the highest crushing force occurs during the analyses is set as a design constraint. FEA, five-level full factorial design and RSM are employed in this study to formulate the optimum design problem and the optimal design is finally solved from the derived response surface functions (Eqs. (25.13, 25.14)).

The design ideas and methods introduced in this chapter can be applied for other applications, including the design and modeling of general thin-walled members.

References

1. T. Wierzbicki, W. Abramowicz, "On the Crushing Mechanisms of Thin-walled Structures", *Journal of Applied Mechanisms*, Vol. 50 (1983) pp. 727–734
2. W. Abramowicz, "Simplified Crushing Analysis of Thin-walled Columns and Beams", Rozprawy Inzynierskie, *Engineering Transactions*, Vol. 29, No. 1 (1981) pp. 5–26
3. D. Kecman, "Bending Collapse of Rectangular and Square Section Tubes", *International Journal of Mechanical Science*, Vol. 25, No. 9–10 (1983) pp. 623–636
4. T. Wierzbicki, L. Recke, W. Abramowicz, and T. Gholami "Stress Profiles in Thin-walled Prismatic Columns Subjected to Crush Loading – I. Compression", *Computers & Structure*, Vol. 51, No. 6 (1994) pp. 611–623
5. T. Wierzbicki, L. Recke, W. Abramowicz, T. Gholami, and J. Huang. "Stress Profiles in Thin-walled Prismatic Columns Subjected to Crush Loading – II. Bending", *Computers & Structure*, Vol. 51, No. 6 (1994) pp. 623–641
6. W. Abramowicz, T. Wierzbicki, "Axial Crushing of Multicorner Sheet Metal Columns", *Journal of Applied Mechanisms*, Vol. 53 (1989) pp. 113–120
7. W. Abramowicz and N. Jones, "Dynamic Progressive Buckling of Circular and Square Tubes", *International Journal of Impact Engineering,* Vol. 4, No. 4 (1986) pp. 243–270
8. A. Mamalis, D. Manolakos, M. Loannidis, and P. Kostazos, "Bending of Cylindrical Steel Tubes: Numerical Modelling", *International Journal of Crashworthiness*, Vol. 11, No. 1 (2006) pp. 37–47
9. Y-C. Liu, M. L. Day, "Bending Collapse of Thin-Walled Beams with Channel Cross-Section", *International Journal of Crashworthiness*, Vol. 11, No. 3 (2006) pp. 251–262
10. A. Mamalis, D. Manolakos, and A. Baldoukas, "Energy Dissipation and Associated Failure Modes When Axially Loading Polygonal Thin-Walled Cylinder", *Thin-Walled Structures*, Vol. 12 (1991) pp. 17–34
11. A. Mamalis, D. Manolakos, M. Loannidis, P. Kostazos, and C. Dimitriou, "Finite Element Simulation of the Axial Collapse of Metallic Thin-Walled Tubes with Octagonal Cross-Section", *Thin-Walled Structures*, Vol. 41 (2003) pp. 891–900
12. H. S. Kim, S. Y. Kang, I. H. Lee, S. H. Park, and D. C. Han. "Vehicle Frontal Crashworthiness Analysis by Simplified Structure Modeling using Nonlinear Spring and Beam Elements", *International Journal of Crashworthiness*, Vol. 2, No. 1 (1997) pp. 107–117
13. P. Drazetic, E. Markiewicz, and Y. Ravalard, "Application of Kinematic Models to Compression and Bending in Simplified Crash Calculation", *International Journal of Mechanical Science*, Vol. 35, No. 3/4 (1993) pp. 179–191
14. Y. C. Liu, M. L. Day, "Simplified Modeling of Thin-Walled Box Section Beam", *International Journal of Crashworthiness*, Vol. 11, No. 3 (2006) 263–272
15. J. Hallquist, "LS-DYNA 3D: Theoretical Manual", Livermore Software Technology Corporation, Livermore, CA, 1993

16. Y. C. Liu, M. L. Day, "Development of Simplified Thin-Walled Beam Models for Crashworthiness Analyses", *International Journal of Crashworthiness*, Vol. 12, No. 6 (2007) pp. 597–608
17. S. J. Hou, Q. Li, S. Y. Long, X. J. Yang, and W. Li, "Design Optimization of Regular Hexagonal Thin-Walled Columns with Crashworthiness Criteria", *Finite Elements in Analysis and Design*, Vol. 43 (2007) pp. 555–565
18. R. H. Myers, D. C. Montgomery, "Response Surface Methodology", Wiley, New York, 2002

Chapter 26
A Fast State Estimation Method for DC Motors

**Gabriela Mamani, Jonathan Becedas, Vicente Feliu,
and Hebertt Sira-Ramírez**

Abstract In this article a fast, non-asymptotic algebraic method is used for the estimation of state variables. The estimation is used to implement a position control scheme for DC motor. In addition the estimation of the Coulomb's friction coefficient of the servo motor model has been investigated. The approach is based on elementary algebraic manipulation that lead to specific formulae for the unmeasured states. The state estimation algorithm have been verified by simulation

Keywords State estimation · Algebraic identification · DC motors · State observer

26.1 Introduction

Control design method such as state feedback controls, which use the states in their control laws are designed under the assumption that all state variables are accessible for measurement. However, in many practical application it may not be economical or convenient to measure all of the state variables. An alternative approach is to use an estimation technique to provide estimates of the state variables which are not measured, for use in implementation of a feedback control law.

The foundation of linear state estimation was laid by [1], who developed the Kalman filter. Later [2] introduced a deterministic version of the Kalman filter, known as Luenberger observer. The theorical properties of the Kalman filter and the Luenberger observer are well understood and can be found in estimation and/or system theory textbook [3].

G. Mamani, J. Becedas (✉), and V. Feliu
Universidad de Castilla-La Mancha, Escuela Técnica Superior de Ingenieros Industriales,
Ciudad Real 13071, Spain
e-mail: jonathanbecedas@hotmail.com

H. Sira-Ramírez
CINVESTAV-IPN, Sección de Mecatrónica, Dept. Ingeniería Eléctrica, Avenida IPN, #2508,
Col. San Pedro Zacatenco, A.P. 14740, 07300 México, D.F., México

S.-I. Ao et al. (eds.), *Advances in Computational Algorithms and Data Analysis,*
Lecture Notes in Electrical Engineering 14,
© Springer Science+Business Media B.V. 2009

It is clear that for an observable system, represented in state space, the state estimation problem is intimately related to the computation of time derivates of the output signals, in a sufficient number. The main contribution of this chapter is that, we attempt an algebraic method, of non-asymptotic nature, for the estimation of states computing a finite number of time derivates of the output. The method is based on elementary algebraic manipulations that lead to specific formulae for the unmeasured states. Our approach uses the model of the system which is known almost most of times. In this work, we use an fast state estimation of continuous-time nature for the estimation of the state of a DC servomotor model in order to implement an PD control scheme. After the estimates of the state variables are obtained by algebraic method the Coulomb's friction coefficient is instantaneously estimated. The importance of this coefficient estimation is explained in [4] in order to appropriately control the system by compensating this non-linearity. The estimation method is based on elementary algebraic manipulations of the following mathematical tools: module theory, differential algebra and operational calculus. They were developed in [5].

26.2 Motor Model and Estimation Procedure

This section is devoted to explain the linear model of the DC motor and the algebraic estimation method. We assume that the linear model is affected by unknown perturbation due to the Coulomb's friction effects.

26.2.1 DC Motor Model

A common electromechanical actuator in many control systems is constituted by the DC motor, [6]. The DC motor used is fed by a servo-amplifier with a current inner loop control. We can write the dynamic equation of the system by using Newton's Second Law:

$$kV = J\ddot{\hat{\theta}}_m + v\dot{\hat{\theta}}_m + \hat{\Gamma}_c(\dot{\hat{\theta}}_m) \tag{26.1}$$

where J is the inertia of the motor $[\text{kg} \cdot \text{m}^2]$, v is the viscous friction coefficient $[\text{N} \cdot \text{m} \cdot \text{s}]$, $\hat{\Gamma}_c$ is the unknown Coulomb friction torque which affects the motor dynamics $[\text{N} \cdot \text{m}]$. This nonlinear friction term is considered as a perturbation, depending only on the sign of the angular velocity of the motor of the form $\mu \, sign(\dot{\hat{\theta}}_m)$ with μ constant. The parameter k is the electromechanical constant of the motor servo-amplifier system $[\text{Nm/V}]$. $\ddot{\hat{\theta}}_m$ and $\dot{\hat{\theta}}_m$ are the angular acceleration of the motor $[\text{rad/s}^2]$ and the angular velocity of the motor $[\text{rad/s}]$ respectively. The constant factor n is the reduction ratio of the motor gear; thus $\theta_m = \hat{\theta}_m/n$, where θ_m stands for the position of the motor gear and $\hat{\theta}_m$ for the position of the motor shaft. $\Gamma_c = \hat{\Gamma}_c n$, where Γ_c is the Coulomb friction torque in the motor gear. V is the motor input voltage $[\text{V}]$ acting as the control variable for the system. This is the input

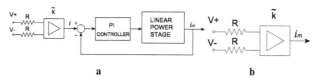

Fig. 26.1 Amplificador scheme of the Dc Motor. **a** Complete amplifier scheme. **b** Equivalent amplifier scheme

to a servo-amplifier which controls the input current to the motor by means of an internally PI current controller (see Fig. 26.1a). The electrical dynamics can be neglected because it is much faster than the mechanical dynamics of the motor. Thus, the servo-amplifier can be considered as a constant relation between the voltage and the current to the motor: $i_m = k_e V$ (see Fig. 26.1b), where i_m is the armature circuit current and k_e includes the gain of the amplifier, $\tilde{k}$, and the input resistance of the amplifier circuit R. The total torque delivered to the motor Γ_T is directly proportional to the armature circuit in the form $\Gamma_T = k_m i_m$, where k_m is the electromechanical constant of the motor. Thus, the electromechanical constant of the motor servo-amplifier system is $k = k_m k_e$.

In order to obtain the transfer function of the system the following perturbation-free system is considered:

$$KV = J\ddot{\theta}_m + v\dot{\theta}_m \tag{26.2}$$

where $K = k/n$. To simplify the developments, let $A = K/J, B = v/J$. The DC motor transfer function is then written as:

$$G(s) = \frac{\theta_m(s)}{V(s)} = \frac{\frac{K}{J}}{s^2 + \frac{v}{J}s} = \frac{A}{s(s+B)} \tag{26.3}$$

26.2.2 The Procedure of Fast State Estimation

Consider the second order perturbed system given in (26.1). Taking it into account and also the fact that $K = k/n$ and after some rearrangements, we have,

$$\ddot{\theta}_m + B\dot{\theta}_m + \Gamma^* = AV \tag{26.4}$$

where $\Gamma^* = \frac{\hat{\Gamma}_c}{nJ}$. We consider this parameter as a constant perturbation input and it will be identified in the next stage.

We proceed to compute the unmeasured states, the motor velocity, $\frac{d\theta_m}{dt}$, and the motor acceleration, $\frac{d^2\theta_m}{dt^2}$, as follows:

Taking Laplace transforms, of (26.4) yields,

$$(s^2\theta_m(s) - s\theta_m(0) - \dot{\theta}_m(0)) + B(s\theta_m(s) - \theta_m(0)) + \frac{\Gamma^*}{s} = AV(s) \tag{26.5}$$

we obtain multiplying out by s,

$$(s^3\theta_m(s) - s^2\theta_m(0) - s\dot{\theta}_m(0)) + B(s^2\theta_m(s) - s\theta_m(0)) + \Gamma^* = AsV(s) \tag{26.6}$$

Taking the third derivative with respect to the complex variable s, we obtain independence of initial conditions. Then (26.6) results in an expression free of the initial conditions $\dot{\theta}_m(0)$, $\theta_m(0)$ and the Coulomb's friction coefficient Γ^*:

$$\frac{d^3}{ds^3}\left[s^3\theta_m(s)\right] + B\frac{d^3}{ds^3}\left[s^2\theta_m(s)\right] = A\frac{d^3}{ds^3}[sV(s)] \tag{26.7}$$

Recall that multiplication by s in the operational domain corresponds to derivation in the time domain. After rearrangements we multiply both sides of the resulting expression by s^{-2}. We obtain

$$s\frac{d^3\theta_m(s)}{ds^3} + 9\frac{d^2\theta_m(s)}{ds^2} + 18s^{-1}\frac{d\theta_m(s)}{ds} + 6s^{-2}\theta_m(s) + B\left(\frac{d^3\theta_m(s)}{ds^3}\right) \tag{26.8}$$

$$+6s^{-1}\frac{d^2\theta_m(s)}{ds^2} + 6s^{-2}\frac{d\theta_m(s)}{ds}\right) = A\left(s^{-1}\frac{d^3V(s)}{ds^3} + 3s^{-2}\frac{d^2V(s)}{ds^2}\right)$$

In the time domain, we have:

$$-\frac{d}{dt}(t^3\theta_m) + 9t^2\theta_m - 18\int_0^t \sigma\theta_m(\sigma)d\sigma + 6\int_0^t\int_0^\sigma \theta_m(\lambda)d\lambda d\sigma - Bt^3\theta_m$$

$$+6B\int_0^t \sigma^2\theta_m(\sigma)d\sigma - 6B\int_0^t\int_0^\sigma \lambda\theta_m(\lambda)d\lambda d\sigma = A\left(-\int_0^t \sigma^3V(\sigma)d\sigma\right.$$

$$\left.+3\int_0^t\int_0^\sigma \lambda^2\theta_m(\lambda)d\lambda d\sigma\right) \tag{26.9}$$

From here we obtain the estimation of the motor velocity

$$\frac{d\theta_m}{dt} = \frac{1}{t^3}(6t^2\theta_m - 18\int_0^t \sigma\theta_m(\sigma)d\sigma + 6\int_0^t\int_0^\sigma \theta_m(\lambda)d\lambda d\sigma) + \frac{1}{t^3}(-Bt^3\theta_m$$

$$+6B\int_0^t \sigma^2\theta_m(\sigma)d\sigma - 6B\int_0^t\int_0^\sigma \lambda\theta_m(\lambda)d\lambda d\sigma) + \frac{1}{t^3}(A\int_0^t \sigma^3V(\sigma)d\sigma$$

$$-3A\int_0^t\int_0^\sigma \lambda^2V(\lambda)d\lambda d\sigma) \tag{26.10}$$

by multiply both sides of the resulting expression in (26.8) by s, we have:

$$s^2\frac{d^3\theta_m(s)}{ds^3} + 9s\frac{d^2\theta_m(s)}{ds^2} + 18\frac{d\theta_m(s)}{ds} + 6s^{-1}\theta_m(s) + B\left(\frac{d^3\theta_m(s)}{ds^3}\right.$$

$$+6\frac{d^2\theta(s)}{ds^2} + 6s^{-1}\frac{d\theta_m(s)}{ds}\right) = A\left(\frac{d^3V(s)}{ds^3} + 3s^{-1}\frac{d^2V(s)}{ds^2}\right) \tag{26.11}$$

which may be written in the time domain as:

$$
-\frac{d^2}{dt^2}(t^3\theta_m) + 9\frac{d}{dt}(t^2\theta) - 18t\theta_m + 6\int_0^t \theta_m(\sigma)d\sigma + B\left(\frac{d}{dt}(-t^3\theta_m)\right)
$$

$$
+6t^2\theta_m - 6\int_0^t \sigma\theta_m(\sigma)d\sigma\Bigg) = -At^3V + 3A\int_0^t \sigma^2V(\lambda)d\sigma \tag{26.12}
$$

We obtain the following expression for the motor acceleration, $\frac{d^2\theta_m}{dt^2}$:

$$
\frac{d^2\theta_m}{dt^2} = \frac{1}{t^3}\left(3t^2\frac{d\theta_m}{dt} - 6t\theta_m + 6\int_0^t \theta_m(\sigma)d\sigma + 3Bt^2\theta_m\right.
$$

$$
\left. -Bt^3\frac{d\theta_m}{dt} - 6B\int_0^t \sigma\theta_m(\sigma)d\sigma + At^3V - 3A\int_0^t \sigma^2V(\lambda)d\lambda\right) \tag{26.13}
$$

This expression may now be evaluated with the help of the already computed estimate of $\frac{d\theta_m}{dt}$.

26.3 Simulation

This section is devoted to show the good performance of the proposed state estimation method. The values of the motor parameters used in simulations are $A = 61.13\ (N/(V\cdot Kg\cdot s)$, $B = 15.15\ (N\cdot s/(Kg\cdot m)$, $k = 0.21\ (N\cdot m/V)$, $n = 50$, $\mu = 34.74\ (N\cdot m)/(kg\cdot m^2)$. The differential equation of the closed loop system is solved by using a $1\cdot 10^{-3}\ [s]$ fixed step fifth order Dormand-Prince method. We consider that there exists a servo amplifier used to supply voltage to the DC motor; this amplifier accepts control inputs from the computer in the range of $[-10, 10]\ [V]$. The signal used in the on-line estimation of the motor velocity are the input voltage to the DC motor and the motor position as a result of that input. In this case we have chosen the input to be a Bezier's eighth order polynomial with an offset of $0.8\ (V)$. We have considered the following initial conditions for the motor in order to show the robustness of the method to every initial conditions: $\theta_m(0) = 100$, $\dot\theta_m(0) = 0$. Both signals are depicted in Figs. 26.2a and b respectively.

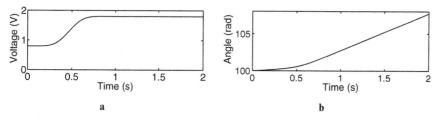

Fig. 26.2 Input and response of the motor. **a** Input to the DC motor. **b** Response of the DC motor

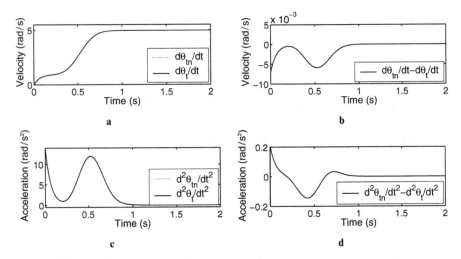

Fig. 26.3 Comparison between the estimations of the velocity and acceleration of the DC motor with the numerical method and with the algebraic state estimation. **a** Velocities estimation. **b** Difference between the two velocity estimations $\dot{\theta}_{tn} - \dot{\theta}_t$. **c** Accelerations estimation. **d** Difference between the two acceleration estimations $\ddot{\theta}_{tn} - \ddot{\theta}_t$

The results are compared with fifth order error numerical derivative. Figure 26.3a depicts the output of the velocity motor observer represented by $\frac{d\theta_t}{dt}$. We can compare it with that of the numerical derivative here represented by $\frac{d\theta_{tn}}{dt}$. Note that the two signals are superimposed. The difference between them is depicted in Fig. 26.3b and this is with 10^{-3} order. In Fig. 26.3c the estimation of the motor acceleration $\frac{d^2\theta_t}{dt^2}$ is depicted in addition to the numerical estimation $\frac{d^2\theta_{tn}}{dt^2}$. The difference $\frac{d^2\theta_t}{dt^2} - \frac{d^2\theta_{tn}}{dt^2}$ between them is depicted in Fig. 26.3d. Now, the difference is more noticeable because is required the first derivative of the signal to obtain the second one and in the case of the numerical estimation not knowledge of the system is used, this is the reason of the increasingly difference. On the other hand, the observer proposed take all the information of the system as possible providing more exacts estimations. This premise will be demonstrated in the application of Coulomb's friction estimation where more accuracy state estimation provides better parameter estimation.

In the new simulations robustness with respect noise of the algebraic state estimation is demonstrated. We consider noise in the measure (i.e in the motor position measure) with zero mean and 10^{-3} standard deviation. When noise appears in a measure, numerical estimation of derivatives of a signal is very imprecise and the estimation of bounded derivatives amplify the noise level. These signals are, customarily, quite noisy and the use of low pass filters become necessary to smooth them causing the well known dynamic delays that affect the performance of the obtained signals as a result. A solution to this problems may be the use of algebraic state estimator, which present robustness to the noise. In Fig. 26.4a the numerical

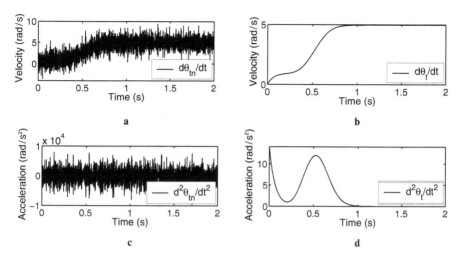

Fig. 26.4 Velocity estimations. **a** Numerical velocity estimation. **b** Algebraic velocity estimation. **c** Numerical acceleration estimation. **d** Algebraic acceleration estimation

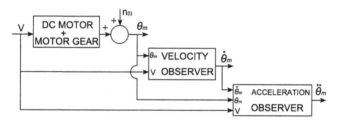

Fig. 26.5 Scheme of the algebraic observers implementation

estimation of the motor velocity is depicted. It is obvious the effect that the noise produces in the estimation. Figure 26.4b depicts the velocity estimation with the algebraic state estimator. In Fig. 26.4c the second derivative of the motor position is represented. Note that in this signal the noise level has been increased. At last, Fig. 26.4d depicts the second derivative of the motor estimated with the algebraic state estimator. Let us recall that in no one of the estimations filters have been used. An scheme of the observers implementation is depicted in Fig. 26.5.

This technique may be used in many applications such as estimation of parameters in which estimation of states are required and control of a feedback system because the estimators can be used in both open and closed loop due to the method does not require dependence between the system input and output.

26.3.1 Estimation Application

From the 1990s decade interest in controlling systems with gear reduction coupled in the motor shaft has increased. Researchers had to deal with non linearities which strongly affected the motor dynamics and which were produced by the friction torque [4]. In order solve that problem researchers used many techniques such as robust control schemes with high gain that minimized this effect [8] or the most modern techniques such as neural networks [9] which delay the obtaining of the non linearity parameters. We here propose a new and precise technique to obtain such parameters by using algebraic state estimators, which can be used in real time and in continuous time without the used of any sort of filter. It is well established that for a system operating at relatively high speed, the Coulomb's friction torque is a function of the angular velocity. For those systems, the Coulomb's friction is often expressed as a signum function dependent on the rotational speed. Consider system (26.4) with $\Gamma^* = \mu sign(\dot{\theta}_m)$.[1] From this equation, and due to the fact that the angular velocity and acceleration of the motor are obtained with the fast state estimation method, A and B are known, and we have

$$\mu sign(\dot{\theta}_m) = AV - (\ddot{\theta}_m)_e - B(\dot{\theta}_m)_e \qquad (26.14)$$

The term: $\mu sign(\dot{\theta}_m)$ is a perturbation produced by the Coulomb's friction torque, where μ is the scaled Coulomb's friction amplitude, or coefficient.[2] With the motor spinning only in one direction, Coulomb's friction coefficient will not change its sign, and can be considered as a constant. When the motor angular velocity is close to zero, the Coulomb's friction effect is that of a chattering high frequency signal.

$$\Gamma^* = \mu sign(V) = \left\{ \begin{array}{l} \mu \ (V > 0), \\ -\mu(V < 0) \end{array} \right\} \qquad (26.15)$$

Then, if the motor spins always in the same direction, in the identification time interval we have that $\Gamma^* = \mu$ and

$$\mu = AV - (\ddot{\theta}_m)_e - B(\dot{\theta}_m)_e \qquad (26.16)$$

Figure 26.6a depicts the estimation of the Coulomb's friction coefficient by using numerical state estimation (μ_n signal) and by using algebraic state estimation (μ_s signal). Note that the estimation μ_s is obtained from the beginning at time $t \approx 0$ and this value is maintained while the estimator works. Nevertheless, the estimation μ_n which uses numerical state estimations introduce an error until $t = 1$ (s), time at which the Bezier's trajectory finishes. The error of the two estimates with respect the real value of the Coulomb's parameter μ is depicted in Fig. 26.6b. $\varepsilon_{\mu_n} = \mu_n - \mu$ represents the error in the estimation with the numerical method and $\varepsilon_{\mu_s} = \mu_s - \mu$ represents the error in the estimation with the algebraic method. Note the error in the estimation with the algebraic state estimators has null error. This is because the state estimation with the proposed method provides an exact estimation of the bounded derivatives of the motor position due to the estimator uses all the information as posible from the system to estimate. Figure 26.7 depicts the results obtained in the

---

[1] The model $sign(\dot{\theta}_m)$ is defined as 1 if $\dot{\theta}_m > 0$ and as -1 if $\dot{\theta}_m < 0$.

[2] Note that $\Gamma^* = \frac{\hat{f}_c}{nJ} = \mu sign(\dot{\theta}_m)$ then, the Coulomb's friction coefficient is $\xi = Jn\mu$.

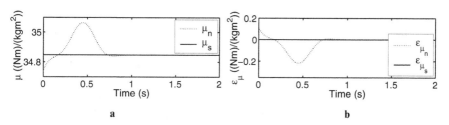

Fig. 26.6 Coulomb estimation with both numerical and algebraic methods, μ_n, μ_s numerical and algebraic estimates respectively. **a** Estimations of coulomb. **b** Estimation errors

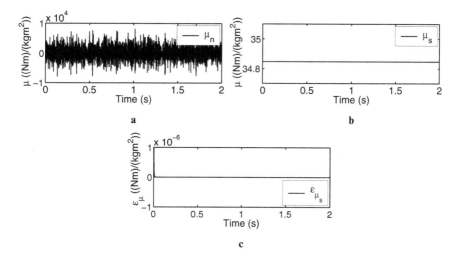

Fig. 26.7 Coulomb estimation. **a.** Numerical. **b** Algebraic **c** Algebraic error

Coulomb's parameter estimation with a noise in the measure of the motor position with zero mean and 10^{-3} standard deviation as done in the previous simulations. In Fig. 26.7a the estimation with numerical estimations of the states is depicted. Note that is impossible to identify any value in such a figure. On the other hand, Fig. 26.7b depicts an accurate estimation of the parameter estimated by using the algebraic method. The error of this last estimation is shown in Fig. 26.7c.

26.3.2 Control Application

DC motors is a topic of interest since is used as actuator in an extensive variety of robotics systems and one of the most used control methods is that based on proportional derivative *PD* controller. In [9] is an example of this. The inconvenient of this sort of control is the computation of the derivative action which always introduce noises in the control voltage input due to the on-line estimation of the input derivative to the controller. Sometimes filters are required to smooth that signal. In this subsection a control application for DC motors is proposed based on the on-line algebraic estimation of the motor velocity. A PD controller is proposed, $C_{pd}(s) = k_p + k_v s$, whose gains $\{k_p, k_v\}$ can be designed by locating all the poles in closed loop of the complete system (See Fig. 26.8) in the same location of the negative real axis. The stability condition on the closed loop expression $(1 + G_{m_0}(s)C_{pd}(s))$ leads to the following characteristic polynomial,

$$s^2 + (kvA + B)s + k_p A = 0 \qquad (26.17)$$

We can equate the corresponding coefficients of the closed loop characteristic polynomial (26.17) with those of a desired second order Hurwitz polynomial. Thus, we can choose to place all the closed loop poles at some real value using the following desired polynomial expression,

$$p(s) = (s+a)^2 = s^2 + 2as + a^2 \qquad (26.18)$$

where the parameter a, strictly positive, represents the common location of all the closed loop poles. Identifying the corresponding terms of the Eqs. (26.17) and (26.19) the parameters k_p and k_v may be uniquely obtained by computing the following equations,

$$k_p = \frac{a^2}{A}, \ k_v = \frac{2a - B}{A} \qquad (26.19)$$

With the previous estimation of the Coulomb's friction torque, Γ_c, a compensation term is introduced in the system in order to eliminate the effect of this perturbation. The compensation term is included in the control input voltage to the motor, and this is of the form:

$$\tilde{\Gamma} = \frac{\hat{\Gamma}_c}{k}(-sign(\dot{\theta}_m)) = \frac{\mu \cdot J \cdot n}{k}(-sign(\dot{\theta}_m)) \qquad (26.20)$$

when $\dot{\theta}_m \neq 0$. When $\dot{\theta}_m \approx 0$, the compensation term is included as done in the previous Eq. (26.20) but changing the function $sign(\dot{\theta}_m)$ by $sign(V)$. Figure 26.9 depicts the trajectory tracking of the motor with the PD controller with numerical computation of the motor velocity θ_{tn} and with the algebraic computation θ_t. The

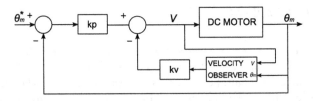

Fig. 26.8 Closed loop PD controller with algebraic observer implementation

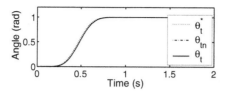

Fig. 26.9 Trajectory tracking of the closed loop system. θ_t^* Command trajectory. θ_{tn} system response with numerical PD. θ_t system response with algebraic PD

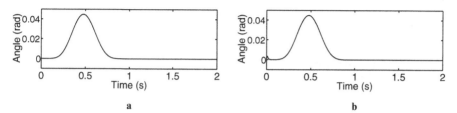

a   b

Fig. 26.10 Tracking errors of the closed loop systems. **a** Tracking error with numerical PD implementation. **b** Tracking error with algebraic PD implementation

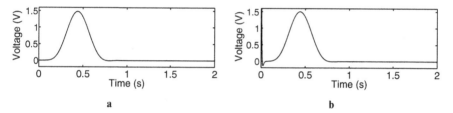

a   b

Fig. 26.11 Control input voltages to the DC motor. **a** With numerical PD implementation. **b** With algebraic PD implementation

two signals properly track the reference trajectory θ_t^* with good performance. In Fig. 26.10a the tracking error of the motor position when numerical PD is used is presented. By comparing such error with that of the Fig. 26.10b, in which the tracking error of the motor position with algebraic PD is depicted, we can observe that the two signals are the same in phase and magnitude. And it can be observe the same characteristic in Fig. 26.11a and Fig. 26.11b where the control input voltages to the DC motor are depicted. This accuracy tracking of both control schemes is due to the gains of the controllers which force the system to track the command trajectory by minimizing the error in the feedback. However, in real life we always find noises and errors which corrupt the measuring data, in this case, the encoder is not an infinite precisely measure system, therefore, noises are include the control system due to the limited precision of the apparatus. We consider a noise corrupting the data with zero mean and 10^{-3} standard deviation as considered in the previous simulations. The trajectory tracking of the motor position with numerical PD θ_{tn} and with algebraic PD θ_t with noisy measure is similar to those of the Fig. 26.9.

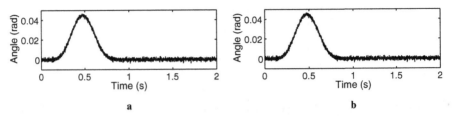

Fig. 26.12 Tracking errors of the closed loop systems with noise in the measure. **a** Tracking error with numerical PD. **b** Tracking error with algebraic PD

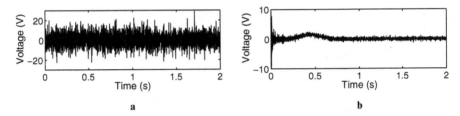

Fig. 26.13 Control input voltages to the DC motor in the system with noise. **a** With numerical PD. **b** With algebraic PD

Both trajectories follow properly the reference θ_t^*. Figures 26.12a and b depict the tracking errors of the previous signals respectively. Note the noise is introducing an aleatory component in the error. Although the two errors has the same amplitude the control input voltage to the DC motor of the numerical PD has not a smooth shape, on the contrary, such a voltage would saturate the amplifier which has the limits in ± 10 (V) (see Fig. 26.13a). In contrast, the control input voltage, when algebraic PD is used, has a smoother profile with much less control effort, therefore, such a signal would never saturate the amplifier. As a consequence, the amplifier would not suffer overheating.

26.4 Summary

A fast, non-asymptotic algebraic state estimation method has been successfully applied to estimates the Coulomb's friction coefficient. Performance studies show that the algebraic method provides satisfactory estimates even in the presence of significant noise levels. In addition, the state feedback controller is designed. Closed-loop simulation runs show that the state estimates calculated by algebraic method can be used efficiently.

References

1. Kalman, R. (1960). "A new approach to linear filtering and prediction problems", *Transaction of the ASME, Journal of Basic Engineering*, 83: 95–108.
2. Luenberger, D. G. (1971). "An introduction to observers", *IEEE Transaction on Automatic Control*, AC-16: 596–602.
3. Friedland, B. (1986). *Control System Design – An Introduction to State Space Methods*, New York: McGraw-Hill.
4. Olsson, H., Amstròm, K., and de Wit, C.C. (1998). "Friction models and Friction Compensation", *European Journal of Control*, Vol. 4, pp. 176–195.
5. Fliess, M. and Sira-Ramírez, H. (1998). "An algebraic framework for linear identification", *in ESAIM Control Optimization and Calculus of Variations*, Vol. 9, pp. 151–168.
6. Begamudre, R. D. (1998). "Flectro-Mechanical Energy Conversion with Dynamic of Machines," New York: Wiley.
7. Feliu V., Rattan, K. S., and Brown, H. B. (1993). "Control of flexible arms with friction in the joints", *IEEE Transactions on Robotics and Automation*, Vol. 9, N. 4, pp. 467–475.
8. Cicero, S., Santos, V., and de Carvahlo, B. (2006). "Active control to flexible manipulators", *IEEE/ASME Transactions on Mechatronics*, Vol. 11, N. 1, pp. 75–83.
9. Feliu, V. and Ramos, F. (2005). "Straing gauge based control of single-link flexible very light weight robots robust to payload changes", *Mechatronics*, Vol. 15, pp. 547–571.

Chapter 27
Flatness based GPI Control for Flexible Robots

Jonathan Becedas, Vicente Feliu, and Hebertt Sira-Ramírez

Abstract In this article, a new method to control a flexible robotic arm using a conventional direct current (DC) motor with a gear actuator strongly affected by non-linear friction torque is proposed. This control method does not require friction compensation and hence the estimation of this term because the control scheme is robust with respect to this effect. In addition, the only variables to measure are the motor shaft and tip angular positions. Velocity measurements, which always introduce errors and noises, are not required. The use of filters to estimate velocities and bounded derivatives are not needed. The Generalized Proportional Integral *GPI* controller is designed using a two-stage design procedure entitling an outer loop, designed under the assumption of no motor dynamics, and subsequently an inner loop which forces the motor response to track the control input position reference trajectory derived in the previous design stage. Velocity measurements, which always introduce errors and noises, are not required. Experimental results are presented.

Keywords Flexible robots · GPI control · Flat systems

27.1 Introduction

In the 1970s flexible robots arise as a new sort of robotic manipulators in engineering. With this new philosophy new applications appeared, most of all in the aerospace industry [1]. The reason of this interest is that flexible manipulators have

J. Becedas (✉)
Universidad de Castilla-La Mancha, Escuela Técnica Superior de Ingenieros Industriales, Edif. Politécnica, Ciudad Real 13071, Spain
e-mail: jonathanbecedas@hotmail.com

V. Feliu
Universidad de Castilla-La Mancha, Escuela Técnica Superior de Industriales, Edif. Politécnica, Ciudad Real 13071, Spain

H. Sira-Ramírez
CINVESTAV-IPN, Sección de Mecatrónica, Dept. Ingeniería Eléctrica, Avenida IPN, #2508, Col. San Pedro Zacatenco, A.P. 14740, 07300 México, D.F., México

S.-I. Ao et al. (eds.), *Advances in Computational Algorithms and Data Analysis,*
Lecture Notes in Electrical Engineering 14,
© Springer Science+Business Media B.V. 2009

many advantages with respect to conventional rigid robots. This mechanical structures require less material to be built, and moreover this materials are lighter in weight and cheaper. As a consequence, they require less power, require smaller actuators and are more manoeuvrable and transportable, have less overall costs, the payload to arm weight ratio is improved, the movements are faster and they are more safely operated due to the reduced inertia and compliant structure. Nevertheless, despite these favorable characteristics an important aspect must be considered when a flexible robot is used: the appearance of vibrations because of the high structural flexibility. Thus, a greater control effort is required to cancel structural vibrations and to achieve a good control response. Major research effort was made to flexible arms in the 1980s. Several papers appeared on this topic: Schmitz [2] and Cannon [3, 4] and Matsuno [5] are examples of controlling the endpoint position of a flexible robotic arm. In the years 1986 and 1987, Harahima and Ueshiba [6], Siciliano em et al. [7] and Rovner and Cannon [8] used different adaptive control schemes to account for changes in the loads. But in the 1990s the real problems appeared in controlling flexible manipulators with gear reductions coupled in the motor shaft. Researchers had to deal with non linearities such as the friction torque [9]. Robust control schemes with high gain minimized this effect [10]. Compensation of the Coulomb's friction torque accomplished by means of a feed-forward term in the control law was also used in [9]. The most modern technics have been applied to control flexible arms. Adaptive control [11], sliding control [12] and neural networks [13] are examples of these. But the problem with the friction torque goes on nowadays. In 2006, Cicero et al. in [13] used neural networks to compensate this friction effect and explains the necessity of an estimation of this friction model.

In this work we propose a control method that does not need any friction compensation model. The proposed output feedback control scheme is found to be robust with respect to the effects of the unknown friction torque and no estimation of this nonlinear phenomenon is therefore required. Our control scheme is truly an output feedback controller since it only uses the position of the motor and tip as given by an encoder and a pair of strain gauges respectively. Velocity measurements are not required in our control scheme. The estimation of the velocities is a widely recognized disadvantage in control schemes because always introduces errors in the signals and noises, and sometimes makes the use of suitable low pass filters necessary. This happens with classical PD and PID control schemes in which it is necessary to carry out the estimation of the velocities due to their *derivative* actions. Such control schemes can be extensively found in position control of rigid and flexible manipulators literature [14–17]. Becedas et al. in [18–20] in 2006–2007, introduced the proposed Generalized Proportional Integral control method for flexible manipulators. This work generalize the method and presents experimental results.

This chapter is organized as follows: In Section 27.2 the theoretical model of the flexible manipulator is described. Section 27.3 is devoted to explain the control method. Section 27.4 is devoted to show the experimental results obtained with the control scheme applied in a real platform. Finally, the main conclusions of this work are presented in Section 28.4.

27.2 Model Description

27.2.1 Flexible Beam Dynamics

A single-link flexible manipulator with tip mass is modeled that can rotate about de Z-axis perpendicular to the paper, as shown in Fig. 27.1. The beam is considered to be an Euler-Bernoulli beam and the axial deformation is neglected, so as to the gravitational effect because the mass of the flexible beam is floating over an air table which allows us to cancel the gravity effect and the friction with the surface of the table. Because structural damping always increases the stability margin of the system, a design without considering damping may provide a valid but conservative result. We study it under the hypothesis of small deformations with all its mass concentrated at the tip position because the mass of the load is bigger than that of the bar, furthermore the tip mass freely spins on its vertical axis, therefore there is no torque produced by the tip mass which influences in the beam dynamics. Thus the mass of the beam can be neglected. In other words, the flexible beam vibrates with the fundamental mode, therefore the rest of the modes are very far from the first one which can be neglected. Based on these considerations we propose the following model for the flexible link:

$$mL^2\ddot{\theta}_t = c\left(\theta_m - \theta_t\right) \qquad (27.1)$$

where m is the unknown mass in the tip position $[kg]$. L $[m]$ and $c = \frac{3EI}{L}$ $[N \cdot m]$ are the length of the flexible arm and the stiffness of the bar respectively. The stiffness depends on the flexural rigidity EI $[N \cdot m^2]$ and on the length of the bar L. θ_m $[rad]$ is the angular position of the motor gear.[1] θ_t and $\ddot{\theta}_t$ are the unmeasured angular position and angular acceleration $[rad/s^2]$ of the tip respectively.

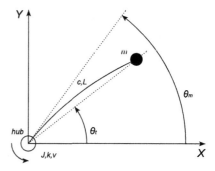

Fig. 27.1 Diagram of a single link flexible arm

---

[1] We denote by n the reduction ratio of the motor gear, thus $\theta_m = \hat{\theta}_m/n$ where $\hat{\theta}_m$ is motor shaft position.

27.2.2 DC Motor Dynamics

A common electromechanical actuator in many control systems is constituted by the DC motor. The DC motor used here is supplied by a servoamplifier with a current inner loop control. We can write the dynamic equation of the system by using Newton's Second law:

$$ku = J\ddot{\theta}_m + v\dot{\theta}_m + \hat{\Gamma}_c(\dot{\theta}_m) + \frac{\Gamma}{n} \qquad (27.2)$$

where J is the inertia of the motor $[kg \cdot m^2]$, v is the viscous friction coefficient $[N \cdot m \cdot s]$. $\ddot{\theta}_m$ and $\dot{\theta}_m$ are the angular acceleration of the motor $[rad/s^2]$ and the angular velocity of the motor $[rad/s]$ respectively. $\hat{\Gamma}_c$ is the unknown Coulomb friction torque which affects the motor dynamics $[N \cdot m]$. This nonlinear friction term is considered as a perturbation, depending only on the sign of the motor angular velocity. As a consequence, Coulomb's friction, when $\dot{\theta}_m \neq 0$, follows the model:

$$\hat{\Gamma}_c \cdot sign(\dot{\theta}_m) = \left\{ \begin{array}{c} \hat{\Gamma}_c(\dot{\theta}_m > 0) \\ -\hat{\Gamma}_c(\dot{\theta}_m < 0) \end{array} \right\} \qquad (27.3)$$

and, when $\dot{\theta}_m \approx 0$ the dynamics is similar to that on (27.3) but, in this case, depending on the sign of the voltage u. The parameter k is the known electromechanical constant of the motor servo-amplifier system $[Nm/V]$. Γ is the coupling torque measured in the hub $[N \cdot m]$ and n is the reduction ratio of the motor gear. u is the motor input voltage $[V]$. This variable is the control variable of the system. This is the input to a servo-amplifier which controls the input current to the motor by means of an internally PI current controller. This electrical dynamics can be rejected because this is faster than the mechanical dynamics of the motor. Thus, the servo-amplifier can be considered as a constant relationbetween the voltage and the current to the motor.

27.2.3 Complete System

The dynamics of the complete system, actuated by a DC motor is given by the following simplified model:

$$mL^2\ddot{\theta}_t = c(\theta_m - \theta_t) \qquad (27.4)$$

$$ku = J\ddot{\theta}_m + v\dot{\theta}_m + \hat{\Gamma}_c + \frac{\Gamma}{n} \qquad (27.5)$$

$$\Gamma = c(\theta_m - \theta_t) \qquad (27.6)$$

Equation (27.4) represents the dynamics of the flexible beam, Eq. (27.5) represents the dynamics of the DC motor and Eq. (27.6) represents the coupling torque measured in the hub and produced by the translation of the flexible beam, which is directly proportional to the stiffness of the beam and the difference between the positions of the motor and the tip respectively.

27.3 GPI Controller

In Laplace transforms notation, the flexible bar transfer function, obtained from (27.4), can be written as follows,

$$Gb(s) = \frac{\theta_t(s)}{\theta_m(s)} = \frac{\omega_0^2}{s^2 + \omega_0^2} \tag{27.7}$$

where $\omega_0 = \left(c/(mL^2)\right)^{1/2}$ is the unknown natural frequency of the bar due to the lack of precise knowledge of m. The coupling torque can be canceled in the motor by means of a compensation term. In this case the voltage applied to the motor is of the form,

$$u = u_c + \frac{\Gamma}{k \cdot n} \tag{27.8}$$

where u_c is the voltage applied before the compensation term. The system in (27.5) is then given by:

$$ku_c = J\ddot{\theta}_m + v\dot{\theta}_m + \hat{\Gamma}_c \tag{27.9}$$

The controller to be designed will be robust with respect to the unknown piecewise constant torque disturbances affecting the motor dynamics, $\hat{\Gamma}_c$. Then the perturbation free system to be considered is the following:

$$Ku_c = J\ddot{\theta}_m + v\dot{\theta}_m \tag{27.10}$$

Where $K = k/n$. To simplify the developments, let

$$A = K/J \tag{27.11}$$
$$B = v/J \tag{27.12}$$

The DC motor transfer function is then written as:

$$Gm(s) = \frac{\theta_m(s)}{u_c(s)} = \frac{A}{s(s+B)} \tag{27.13}$$

Figure 27.2 depicts the compensation scheme of the coupling torque measured in the hub.

The regulation of the load position $\theta_t(t)$ to track a given smooth reference trajectory, $\theta_t^*(t)$, is desired. For the synthesis of the feedback law we are using only the measured motor position θ_m and the measured coupling torque Γ.

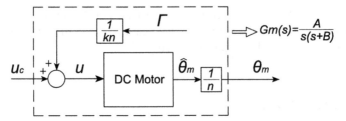

Fig. 27.2 Compensation of the coupling torque measured in the hub

One of the prevailing restrictions throughout our treatment of the problem is our desire of *not* to measure, or compute on the basis samplings, angular velocities of the motor shaft nor of the load.

27.3.1 Outer Loop Controller

Consider the model of the flexible link, given in (27.4). This subsystem is flat, with flat output given by θ_t. This means that all variables of the unperturbed system may be written in terms of the flat output and a finite number of its time derivatives (See [21]). The parametrization of θ_m in terms of θ_t is given, in reduction gear terms, by:

$$\theta_m = \frac{mL^2}{c}\ddot{\theta}_t + \theta_t = \frac{1}{\omega_0^2}\ddot{\theta}_t + \theta_t \tag{27.14}$$

System (27.14) is a second order system in which to regulate the tip position of the flexible bar, θ_t, towards a given smooth reference trajectory $\theta_t^*(t)$ is desired, with θ_m acting as an *auxiliary* control input. Clearly, if there exists an auxiliary open loop control input, $\theta_m^*(t)$, that ideally achieves the tracking of $\theta_t^*(t)$ for suitable initial conditions, it satisfies then the second order dynamics, in reduction gear terms (27.15).

$$\theta_m^*(t) = \frac{1}{\omega_0^2}\ddot{\theta_t}^*(t) + \theta_t^*(t) \tag{27.15}$$

Subtracting (27.15) from (27.14), an expression in terms of the angular tracking errors is obtained:

$$\ddot{e}_{\theta_t} = \omega_0^2\left(e_{\theta_m} - e_{\theta_t}\right) \tag{27.16}$$

where $e_{\theta_m} = \theta_m - \theta_m^*(t)$, $e_{\theta_t} = \theta_t - \theta_t^*(t)$. Suppose for a moment we are able to measure the angular position tracking error, e_{θ_t}, then the outer loop feedback incremental controller could be proposed to be the following PID controller,

$$e_{\theta_m} = e_{\theta_t} + \frac{1}{\omega_0^2}\left[-k_2\dot{e}_{\theta_t} - k_1 e_{\theta_t} - k_0\int_0^t e_{\theta_t}(\sigma)d\sigma\right] \tag{27.17}$$

In such a case, the closed loop tracking error e_{θ_t} evolves governed by,

$$e_{\theta_t}^{(3)} + k_2 \ddot{e}_{\theta_t} + k_1 \dot{e}_{\theta_t} + k_0 e_{\theta_t} = 0 \qquad (27.18)$$

The design parameters $\{k_2, k_1, k_0\}$, are then chosen so as to render the closed loop characteristic polynomial, into a Hurwitz polynomial with desirable roots. However, in order to avoid tracking error velocity measurements, we propose to obtain an *integral reconstructor* for the angular velocity error signal $\dot{e}_{\theta_t}$. We proceed by integrating the expression (27.16) once; and, later, by disregarding the constant error due to the tracking error velocity initial conditions. The estimated error velocity $[\dot{e}_{\theta_t}]_e$ can be computed in the following form:

$$[\dot{e}_{\theta_t}]_e = \dot{e}_{\theta_t}(t) - \dot{e}_{\theta_t}(0) = \omega_0^2 \int_0^t (e_{\theta_m}(\sigma) - e_{\theta_t}(\sigma)) d\sigma \qquad (27.19)$$

The integral reconstructor neglects the possibly nonzero initial condition $\dot{e}_{\theta_t}(0)$ and, hence, it exhibits a constant estimation error. When the reconstructor is used in the derivative part of the PID controller, the constant error is suitably compensated thanks to the integral control action of the PID controller. Substituting the integral reconstructor $[\dot{e}_{\theta_t}]_e$ (27.19) by $\dot{e}_{\theta_t}$ into the PID controller (27.17) and after some rearrangements we obtain:

$$(\theta_m - \theta_m^*) = \left[\frac{\gamma_1 s + \gamma_0}{s + \gamma_2} \right] (\theta_t^* - \theta_t) \qquad (27.20)$$

with the gains $\gamma_2 = k_2$, $\gamma_1 = (\omega_0^2 - k_1)/\omega_0^2$, and $\gamma_0 = (\omega_0^2 k_2 - k_0)/\omega_0^2$. The tip angular position can not be measured, but it certainly can be computed from the expression relating the tip position with the motor position and the coupling torque (Γ):

$$\Gamma = c(\theta_m - \theta_t) = mL^2 \ddot{\theta}_t \qquad (27.21)$$

Thus, the angular position θ_t is readily expressed as,

$$\theta_t = \theta_m - \frac{1}{c}\Gamma \qquad (27.22)$$

Figure 27.4 depicts the feedback outer loop control scheme. This is exponentially stable. To specify the parameters, $\{\gamma_2, \gamma_1, \gamma_0\}$, we can choose to locate the closed loop poles in the left half of the complex plane. All three poles can be located in the same point of the real line, $s = -a$, a being strictly positive, using the following polynomial equation,

$$(s+a)^3 = s^3 + 3as^2 + 3a^2 s + a^3 = 0 \qquad (27.23)$$

Where the parameter a represents the desired location of the poles. The characteristic equation of the closed loop system is,

$$s^3 + \gamma_2 s^2 + \omega_0^2(1+\gamma_1)s + \omega_0^2(\gamma_2 + \gamma_0) = 0 \qquad (27.24)$$

Identifying each term of the expression (27.23) with those of (27.24), the design parameters $\{\gamma_2, \gamma_1, \gamma_0\}$ can be uniquely specified if ω_0 is known.

27.3.2 Inner Loop Controller

Consider the model of the DC motor, given in (27.5). With some rearrangements and bearing in mind the compensation term of the coupling torque presented in Fig. 27.2, the perturbed system can be written as follows,

$$\ddot{\theta}_m = Au - B\dot{\theta}_m - \xi \tag{27.25}$$

where $\xi = \hat{\Gamma}_c/(n \cdot J)$ is an unknown piecewise constant perturbation produced by the Coulomb's friction torque. The unperturbed system is flat with flat output given by θ_m. Clearly, if an open loop control input $u^*(t)$ exists that *ideally* (i.e., under no perturbation inputs) achieves the tracking of $\theta_m(t)$ for suitable initial conditions, this must satisfy the second order dynamics of the unperturbed system:

$$\ddot{\theta}_m^*(t) = Au^*(t) - B\dot{\theta}_m^*(t) \tag{27.26}$$

So the nominal motor control input is computed from the flatness relation,

$$u^*(t) = \frac{1}{A}\ddot{\theta}_m^*(t) + \frac{B}{A}\dot{\theta}_m^*(t) \tag{27.27}$$

Given that the system is affected by a piece-wise constant torque input arising from the Coulomb friction term whenever the angular velocity is not settled to zero, the controller for the system should thus include a double integral compensation action which is capable of overcoming ramp tracking errors. The ramp error is mainly due to the integral angular velocity reconstructor, performed in the presence of constant, or piece-wise constant, torque perturbations characteristic of the Coulomb phenomenon before stabilization around zero velocity.

Subtracting (27.26) from (27.25) an expression in terms of the error in the system is obtained:

$$\ddot{e}_{\theta_m} = Ae_v - B\dot{e}_{\theta_m} - \xi \tag{27.28}$$

where $e_{\theta_m} = \theta_m - \theta_m^*(t)$ and $e_v = u - u^*(t)$.

We propose the following feedback controller, with a double integral control action:

$$Ae_v = B\dot{e}_{\theta_m} + \left\{ -k_3\dot{e}_{\theta_m} - k_2 e_{\theta_m} - k_1 \int_0^t e_{\theta_m}(\sigma)d\sigma \right.$$
$$\left. - k_0 \int_0^t \int_0^{\sigma_1} e_{\theta_m}(\sigma_2)d\sigma_2 d\sigma_1 \right\} \tag{27.29}$$

By substituting the previous Eq. (27.29) in (27.28) we obtain an expression for the closed loop system, which is evidently represented by an integro-differential equation for the output tracking error e_{θ_m} as an exponentially stable equilibrium point. The closed loop tracking error evolves governed by,

$$
\ddot{e}_{\theta_m} + k_3 \dot{e}_{\theta_m} + k_2 e_{\theta_m} + k_1 \int_0^t e_{\theta_m}(\sigma) d\sigma
$$
$$
+ k_0 \int_0^t \int_0^{\sigma_1} e_{\theta_m}(\sigma_2) d\sigma_2 d\sigma_1 = 0 \tag{27.30}
$$

The characteristic polynomial associated with this equation is easily shown to be

$$
p(s) = s^4 + k_3 s^3 + k_2 s^2 + k_1 s + k_0 = 0 \tag{27.31}
$$

Thus, the design problem is reduced to an appropriate choice of feedback controller gains so as to make the above polynomial Hurwitz.

We propose an *integral reconstructor* for the angular velocity error signal $\dot{e}_{\theta_m}$. When the Coulomb's friction does not change sign, we obtain:

$$
\dot{e}_{\theta_m}(t) - \dot{e}_{\theta_m}(0) = A \int_0^t e_v(\sigma) d\sigma - B[e_{\theta_m} - e_{\theta_m}(0)] - \xi\, t \tag{27.32}
$$

The integral based error velocity reconstructor $[\dot{e}_{\theta_m}]_e$ is proposed to be of the following form:

$$
[\dot{e}_{\theta_m}]_e = A \int_0^t e_{\theta_v}(\sigma) d\sigma - B e_{\theta_m}(t) \tag{27.33}
$$

The above proposed estimate of the angular velocity clearly exhibits a structural error of the *ramp* type.

The integral reconstructor neglects the effects of, possibly nonzero, initial conditions $\dot{e}_{\theta_m}(0)$ and $e_{\theta_m}(0)$ as well as the effects of a piecewise constant perturbation represented by ξ and resulting in a ramp signal error for the angular velocity estimate. This growing error is classically compensated by an iterated tracking error integral action. Thus, when the angular velocity integral reconstructor is used to replace the derivative part of the controller, the estimation error effects are suitably compensated thanks to the integral control action built into the controller. The integral reconstructor expression for $\dot{e}_{\theta_m}$ (27.33) is substituted into the proposed controller (27.28) and, after some rearrangements, the feedback control law is obtained as:

$$
u - u^*(t) = -\left[\frac{\alpha_2 s^2 + \alpha_1 s + \alpha_0}{s(s + \alpha_3)}\right](\theta_m^* - \theta_m) \tag{27.34}
$$

with the gains $\alpha_3 = k_3 - B$, $\alpha_2 = (B(B - k_3) + k_2)/A$, $\alpha_1 = k_1/A$, $\alpha_0 = k_0/A$.

The inner loop system in Fig. 27.3 is exponentially stable. As done with the outer loop, all poles can be located at the same real value, and k_3, k_2, k_1 and k_0 can be uniquely obtained equalizing the terms of the two following polynomials:

$$
(s + p)^4 = s^4 + 4ps^3 + 6p^2 s^2 + 4p^3 s + p^4 = 0 \tag{27.35}
$$
$$
s^4 + (\alpha_3 + B)s^3 + (\alpha_3 B + \alpha_2 A)s^2 + \alpha_1 A s + \alpha_0 A = 0 \tag{27.36}
$$

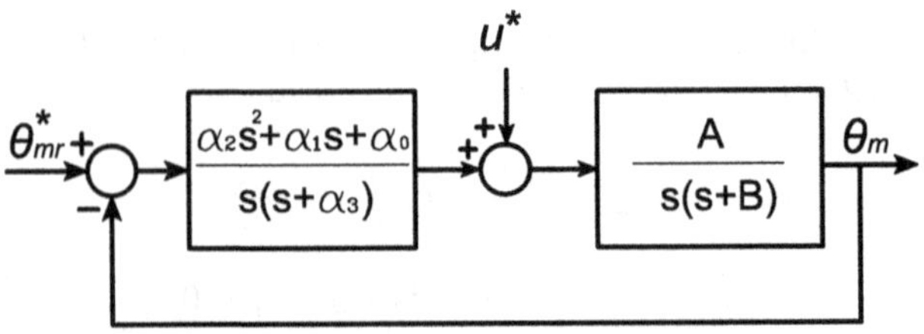

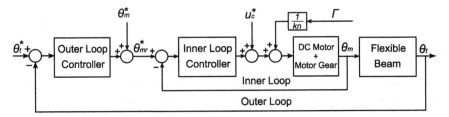

Fig. 27.3 Inner loop controller scheme

Fig. 27.4 Flexible link dc motor system controlled by a two stage GPI controller design

where the parameter p represents the common location of all the closed loop poles, this being strictly positive.

Figure 27.4 depicts the closed loop control system under which the outer and inner loop are implemented in practise.

27.4 Experimental Validation

This section is devoted to experimentally validate the previous algorithm.

27.4.1 Experimental Setup Description

Figure 27.5a depicts a picture of the used experimental platform constituted by a three legged metallic structure to support an Harmonic Drive mini servo DC motor RH-8D-6006-E036AL-SP(N) which has a reduction ratio characterized by $n = 50$. The frame makes possible the stably and free rotation of the motor in the horizontal plane around the vertical axis of the platform. The parameter values are: inertia $J = 6.87 \cdot 10^{-5}$ $[kg \cdot m^2]$, viscous friction $v = 1.041 \cdot 10^{-3}$ $[N \cdot m \cdot s]$ and electromechanical constant $k = 0.21$ $[N \cdot m/V]$. With these parameters, A and B of the transfer function of the DC motor in (27.13) can be computed as: $A = 61.14$ $[N/(V \cdot kg \cdot s)]$, $B = 15.15$ $[N \cdot s/(kg \cdot m)]$. The servoamplifier accepts control inputs from the computer in the range of $[-10, 10]$ $[V]$. The flexible bar is attached to the motor. The load floats over the surface of an air table, so the gravity effect and the friction of

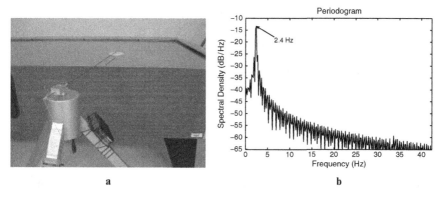

Fig. 27.5 **a** Photograph of the flexible arm platform. **b** Periodogram of the flexible beam oscillation

the load with the surface of the table are canceled. The values of the flexible beam parameters are: the length $L = 0.5$ $[m]$ and diameter $d = 3 \cdot 10^{-3}$ $[m]$. The flexural rigidity of the bar is $EI = 0.260$ $[N \cdot m^2]$, therefore the stiffness is $c = 1.585$ $[N \cdot m]$. The tip load parameters is a wood disc with mass $m = 3 \cdot 10^{-2}$ $[kg]$. With these parameters, the natural frequency of the bar characterized in transfer function (27.7) can be computed as: $\omega_0 = 14.54$ $[rad/s]$. The sensor system is integrated by an encoder embedded in the motor which allows us to know the motor position with a precision of $7 \cdot 10^{-5}$ $[rad]$. And a pair of strain gauges with gage factor 2.16 and resistance 120.2 $[\Omega]$. The sample time in the signals processing is 2 $[ms]$. In order to obtain the natural frequency of the system ω_0 to validate the one mode model proposed, a torque in the motor shaft was applied. Then the tip of the beam oscillated. The oscillation is translated as a peak in the periodogram[2] (see Fig. 27.5b). The estimation provided by the peak of the periodogram, observed at the abscissa axis is $f_0 \approx 2.4$ $[Hz]$, this is $\omega_0 \approx 2.4 \cdot 2\pi \approx 15.1$ $[rad/s]$. Note that in the periodogram only clearly appears one noticeable mode.

27.4.2 Outer Loop Design

We locate the poles of the outer loop at -12 in the real axis, thus we assure that the inner loop is faster than the outer one. The transfer function of the controller is given by the following expression:

$$\frac{\theta_m - \theta_m^*}{\theta_t^* - \theta_t} = \frac{1.044s - 27.82}{s + 36} \qquad (27.37)$$

---

[2] Recall that the periodogram of the signal $u(t)$, $t = 1, 2, ..., N$ is $|U_N(\omega)|^2$, where $U_N(\omega) = \frac{1}{\sqrt{N}} \sum_{t=1}^{N} U_N(2\pi k/N) e^{i2\pi kt/N}$, $k = 1, ..., N$ represents the discrete Fourier's transform (DFT) for $\omega = 2\pi k/N$.

The open loop reference control input $\theta_m^*(t)$ in (27.15) is given by,

$$\theta_m^*(t) = 4.7 \cdot 10^{-3} \ddot{\theta}_t^*(t) + \theta_t^*(t) \qquad (27.38)$$

$\theta_t^*(t)$ being the desired (rest-to-rest) reference trajectory of the angular displacement of the arm tip.

27.4.3 Inner Loop Design

Closed loop poles are placed at -110. Then the transfer function of the controller results:

$$\frac{u_c - u_c^*}{\theta_m^* - \theta_m} = \frac{1.08 \cdot 10^3 s^2 + 8.71 \cdot 10^4 s + 2.40 \cdot 10^6}{s(s + 424.85)} \qquad (27.39)$$

The feed-forward term in (27.27) is computed according to,

$$u_c^*(t) = 16.4 \cdot 10^{-3} \ddot{\theta}_m^*(t) + 0.25 \ddot{\theta}_m^*(t) \qquad (27.40)$$

27.5 Experimental Results

Figure 27.6a shows the commanded trajectory (θ_t^*) and the response of the closed loop system (θ_t) which is here compared with that of the simulations (such a response is here denoted by θ_{ts}). Note that the experimental response θ_t perfectly tracks the reference trajectory as done in simulations; all signals are superimposed. Figure 27.6b depicts a zoom of the tip position of the flexible arm at the beginning of the trajectory. Note that the tip position precisely follows the simulated response θ_{ts} and the error with respect the reference is almost insignificant. Figure 27.6c depicts a zoom of the system trajectory tracking at the end of the trajectory. There not appears overshoot in the experimental tracking so as to the simulated one. The error is minimized when the system reaches the steady state. Figure 27.7a depicts how the trajectory tracking error rapidly converges to zero, and thus a quite precise tracking of the desired trajectory is achieved. The experimental error can be compared with the simulated one Fig. 27.7b. The experimental error is corresponded with the previewed in simulations. Figure 27.8a depicts the experimental input voltage to the DC motor u. Note that this signal never reaches values which saturate the amplifier (i.e $+10V$, $-10V$). We can compare this signal with that of the simulations (see Fig. 27.8b) and observe that the two signals are similar in shape. Obviously the experimental signal is noisier because of the real behavior of the physical platform. In Fig. 27.9a the experimental trajectory tracking of the DC motor is presented. Although, the objective of the controller is to control the tip position of the flexible beam this picture shows that the inner loop has a good tracking performance.

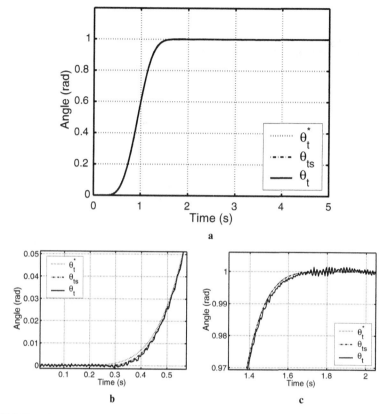

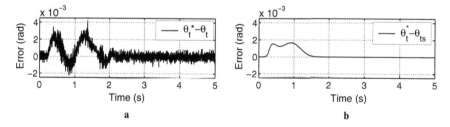

Fig. 27.6 Experimental tip position trajectory tracking. **a** Zoom at the beginning of the trajectory. **b** Zoom at the end of the trajectory

Fig. 27.7 Trajectory tracking error. **a** Experimental. **b** Simulated

Figure 27.9b depicts the coupling torque produced in the hub Γ. This signal is compared with the simulated coupling toque Γ_s. Although the experimental coupling torque is noisier than the simulated one, we can observe that the two signals are similar, they have the same phase and amplitude.

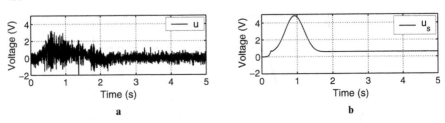

Fig. 27.8 Control voltage to the DC motor. **a** Experimental. **b** Simulated

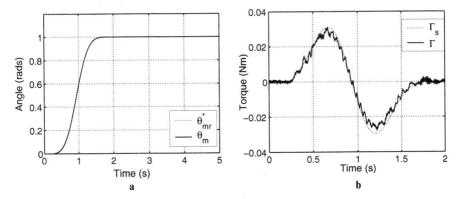

Fig. 27.9 a Experimental motor position tracking. **b** Coupling torque measured by the strain gauges

27.6 Conclusions

A two stage *GPI* controller design scheme has been proposed for the regulation and reference trajectory tracking of a single-link flexible arm. In many articles in the literature, the control of a flexible arm has been approached from classical control schemes, such as classical PD and PID. Generally speaking, those classical control schemes yield good results; nevertheless, they do require the estimation, or on-line computation, of angular velocities for the motor and the arm. These signals are, customarily, quite noisy and the use of low pass filters become necessary to smooth them causing the well known dynamic delays that affect the performance and accuracy of the tracking maneuver. The GPI control scheme here proposed only requires the measurement of the angular position of the motor and that of the tip velocity measurements which always introduce noises and errors are not required. In addition, the controller proposed is found to be robust with respect nonlinear friction effects. Therefore, estimation of these is not required.

Acknowledgements This research was supported by the Junta de Comunidades de Castilla-La Mancha via project ref.: PBI-05-057 and European Social Fund. and by the Spanish Ministerio de Educacin y Ciencia via project ref.: DPI2006-13834.

References

1. Maizza-Neto O.,"Modal analysis and control of flexible manipulator arms", *Ph.D. dissertation, Dept. Mechanical Engineering, MIT, Cambridge*, 1974.
2. Schmitz E., "Experiments on the end-point control of a very flexible one-link manipulator", *Ph.D. dissertation, Dept. Aeronaut. Astronaut., Standford University*, 1985.
3. Cannon R. H. and Schmitz E., "Precise control of flexible manipulators", *Robotic Research*, ed. Brady and Paul, MIT Press, 1984. (Proceedings of the First International Symposium on Robotics Research, Bretton Woods, USA August, 1983.), p.841–861.
4. Cannon R. H. and Rosenthal D. E.,"Experiments in control of flexible estructures with non-colocated sensors and actuators", *Journal of Guidance and Control*, V7, N5, pp. 546–553, 1984.
5. Matsuno F., Fukushima F. and coworkers., "Feedback control of a flexible manipulator with a parallel drive mechanism", *Int. Journal of Robotics Research*, V6, N4, 1987.
6. Harahima F. and Ueshiba T., "Adaptive control of flexible arm using end-point position sensing", *Japan-USA Symposium on Flexible Automation*, Osaka, Japan, 1986.
7. Siciliano B., Yuan B. S. and Book W. J., "Model reference adaptive control of a one link flexible arm", *IEEE Conference on Decision and Control*, Athens, Greece, 1986.
8. Rovner D. M. and Cannon R. H., "Experiments towards on-line identification and control of a very flexible one-link flexible manipulator", *International Journal of Robotics Research*, V6, N4, 1987.
9. Olsson H., Amström K. J. and Canudas de Wit C., "Friction models and friction Compensation", *European Journal of Control*, V4, pp. 176–195, 1998.
10. Feliu V., Rattan K.S. and Brown H.B., "Control of flexible arms with friction in the joints", *IEEE Transactions on Robotics and Automation*, V9, N4, pp. 467–475, 1993.
11. Yang T.C., Yang J.C.S. and Kudva P., "Load adaptive control of a single-link flexible manipulator", *IEEE Transactions on Systems, Man and Cybernetics*, V22, N1, pp. 85–91, 1992.
12. Chen Y. and Hsu H., "Regulation and vibration control of a FEM-based single-link flexible arm using sliding-mode theory", *Journal of Vibration and Control*, V7, N5, pp. 741–52, 2001.
13. Cicero S., Santos V. and de Carvahlo B., "Active control to flexible manipulators", *IEEE/ASME Transactions on Mechatronics*, V11, N1, pp. 75–83, 2006.
14. Rivetta C., Briegel C. and Czarapata P., "Motion control design of the SDSS 2.5 mts telescope", *Proceedings of SPIE*, Santa Clara, USA, pp. 212–221, 2000.
15. Fiene J. and Niemeyer G., "Switching motor control: an integrated amplifier design for improved velocity estimation and feedback", *IEEE International Conference on Robotics and Automation* V5, pp. 4504–4509, 2004.
16. Feliu V. and Ramos F., "Strain gauge based control of single-link flexible very lightweight robots robust to payload changes.", *Mechatronics*, V15, pp. 547–571, 2004.
17. Yuan K. and Liu L., "Achieving minimum phase transfer function for a noncollocated single-link flexible manipulator", *Asian Journal of Control*, V2, N3, pp. 179–191, 2000.
18. Becedas J. and coworkers., "A fast on-line algebraic estimation of a single-link flexible arm applied to GPI control", *The 32nd Annual Conference of the IEEE Industrial Electronics Society*, Paris, France, 2006.
19. Becedas J., Trapero J. R., Sira-Ramirez H. and Feliu V., "Fast identification method to control a flexible manipulator with parameters uncertainties", *IEEE International Conference on Robotics and Automation*, Roma, 2007.
20. Becedas J., Feliu V. and Sira-Ramirez H., "GPI control for a single-link flexible manipulator", *The International World Congress on Engineering and Computer Sciene. Intenational Conference on Modelling Simulation and Control*, San Francisco, 2007.
21. Sira-Ramirez H. and Agrawal S., *Differentially Flat Systems*, Marcel Dekker, New York, 2004.

Chapter 28
Estimation of Mass-Spring-Dumper Systems

Jonathan Becedas, Gabriela Mamani, Vicente Feliu,
and Hebertt Sira-Ramírez

Abstract In this chapter a procedure for parameters identification using an algebraic identification method for a continuous time constant linear system is described.
A specific application in the determination of the parameters mass-spring-damper
system is made. The method is suitable for simultaneously identifying, both, the
spring constant and the damping coefficient. It is found that the proposed method is
computationally fast and robust with respect to noises. The identification algorithm
is verified by simulation results. The estimations are carried out on-line.

Keywords Algebraic identification · System identification · Mass-spring-damper
system

28.1 Introduction

Parameter identification is used to obtain an accurate model of a real system, and
the completed model provides a suitable platform for further developments of design or control strategies investigation. On-line parameter identification schemes
are actually used to estimate system parameters, monitor changes in parameters and
characteristics of the system, and for diagnostic purposes related to a variety of technological areas. The identification schemes can be used to update the value of the
design parameters specified by manufacturers.

In this article, an on-line algebraic method of non-asymptotic nature for the estimation of the mass-spring-damper system is used. The spring constant and the

J. Becedas (✉), G. Mamani, and V. Feliu
Universidad de Castilla-La Mancha, Escuela Técnica Superior de Ingenieros Industriales, Ciudad
Real 13071, Spain
e-mail: jonathanbecedas@hotmail.com

H. Sira-Ramírez
CINVESTAV-IPN Sección de Mecatrónica, Dept. Ingeniería Eléctrica, Avenida IPN, #2508,
Col. San Pedro Zacatenco, A.P. 14740, 07300 México, D.F., México

S.-I. Ao et al. (eds.), *Advances in Computational Algorithms and Data Analysis,*   411
Lecture Notes in Electrical Engineering 14,
© Springer Science+Business Media B.V. 2009

damping coefficient are simultaneously estimated. The input variables to the estimator are the force input to the system and the displacement of the mass. The identification method is based on elementary algebraic manipulations. This method is based on the following mathematical tools: module theory, differential algebra and operational calculus. They were developed in [1]. A differential algebraic justification of this article follows similar lines to those encountered in [2–5]. Let us recall that those techniques are not asymptotic.

Parameter estimation has been an important topic in system identification literature. The traditional theory is well developed, see [6,7]. The most well known technique for parameter estimation is the recursive least square algorithm. This chapter basically focuses on an on-line identification method, of continuous-time nature, for mechanical systems. The approach uses the model of the system, which is known most of times. The advantages are: it does not need any statistical knowledge of the noises corrupting the data; the estimation does not require initial conditions or dependence between the system input and output; and the algorithm is computed on-line.

We mention that the algebraic method has also been applied in [8] in the area of signal processing applications and in [9] in flexible robots estimation with good results. In this last work the algebraic method independence to the input signal design is also demonstrated.

Finally, this estimation method can be applied in a wide range of applications in which appears the mass-spring-damper model, such as vibration control [10], impact dynamics [11], estimation of contact parameters [12] and control in robotics [13] among others.

This chapter is structured as follows: in Section 28.2, the mass-spring-damper model is introduced and the algebraic identification method is presented. After the identification method is outlined simulation results are presented to confirm the accuracy of the parameters estimation. This is accomplished in Section 28.3. Finally, Section 28.4 is devoted to concluding remarks.

28.2 Mass-Spring-Damper Model and Identification Procedure

This section is devoted to explain the linear model of the mass-spring-damper system and the algebraic identification method.

28.2.1 Mass-Spring-Damper Model

Although vibrational phenomena are complex, some basic principles can be recognized in a very simple linear model of a mass-spring-damper system. Such a system contains a mass m $[kg]$, a spring with spring constant k $[N/m]$, which serves to restore the mass to a neutral position, and a damping element which opposes the motion of the vibratory response with a force proportional to the velocity of the system, the constant of proportionality being the damping constant c $[Ns/m]$. An ideal mass-spring-damper system can be described with the following.

$$F_s = -kx \tag{28.1}$$

$$F_d = -cv = -c\dot{x} = -c\frac{dx}{dt} \tag{28.2}$$

This equations system is derived by the Newton's law of motion which is

$$\sum F = ma = m\ddot{x} = m\frac{d^2x}{dt^2} \tag{28.3}$$

where a is the acceleration $[m/s^2]$ of the mass and x $[m]$ is the displacement of the mass relative to a fixed point of reference. The above equation combine to form the equation of motion: a second-order differential equation for displacement x as a function of time t $[s]$:

$$m\ddot{x} + c\dot{x} + kx = F \tag{28.4}$$

With rearrangements it is obtained:

$$\ddot{x} + \frac{c}{m}\dot{x} + \frac{k}{m}x = \frac{F}{m} \tag{28.5}$$

An scheme of the system is depicted in Fig. 28.1.

Next, to simplify the equation, the following parameters are defined: $B = \frac{c}{m}$, $K = \frac{k}{m}$ and $f = \frac{F}{m}$, and the second order system is obtained,

$$\ddot{x} + B\dot{x} + Kx = f \tag{28.6}$$

The mass-spring-damper transfer function is then written as:

$$G(s) = \frac{x(s)}{f(s)} = \frac{1}{(s^2 + Bs + K)} \tag{28.7}$$

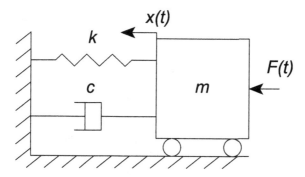

Fig. 28.1 Mass-spring-damper system scheme

In the parameter identification scheme we will compute, in an algebraic form: B and K from linear identifiability. From these relation, and due to the fact that m is known, we have

$$k = Km \tag{28.8}$$

$$c = Bm \tag{28.9}$$

28.2.2 The Procedure of Algebraic Identification

Consider the second order system given in (28.6). c and k are unknown parameters and they are not linearly identifiable. Nevertheless, the parameter $\frac{c}{m}$ denoted by B and the parameter $\frac{k}{m}$ denoted by K are linearly identifiable.

The unknown system parameters B and K can be computed as follows:

Taking Laplace transforms, of (28.6) yields,

$$s^2 x(s) - sx(0) - \dot{x}(0) + B(sx(s) - x(0)) + Kx(s) = f(s) \tag{28.10}$$

Taking derivative with respect to the complex variable s, twice, independence of initial conditions is obtained. Thus (28.10) results in an expression free of the initial conditions $\dot{x}(0)$ and $x(0)$.

$$\frac{d^2}{ds^2}\left[s^2 x(s)\right] + B\frac{d^2}{ds^2}\left[s^2 x(s)\right] = \frac{d^2}{ds^2}[f(s) - Kx(s)] \tag{28.11}$$

The terms of (28.11) can be developed as:

$$\frac{d^2}{ds^2}[s^2 x(s)] = 2x + 4s\frac{dx}{ds} + s^2\frac{d^2 x}{ds^2} \tag{28.12}$$

$$\frac{d^2}{ds^2}[sx(s)] = 2\frac{dx}{ds} + s\frac{d^2 x}{ds^2} \tag{28.13}$$

Recall that multiplication by s in the operational domain corresponds to derivation in the time domain. To avoid derivation, after replacing the expressions (28.12, 28.13), in Eq. (28.11), both sides of the resulting expression are multiplied by s^{-2} to obtain:

$$B(2s^{-2}\frac{dx}{ds} + s^{-1}\frac{d^2 x}{ds^2}) + Ks^{-2}\frac{d^2 x}{ds^2} = s^{-2}\frac{d^2 f(s)}{ds^2} - \frac{d^2 f}{ds^2} \tag{28.14}$$
$$-2s^{-2} - 4s^{-1}\frac{dx}{ds} - s^{-2}2x$$

In the time domain, one obtains the first equation for the unknown parameters B and K. This is a linear equation of the form

$$Bp_{11}(t) + Kp_{12}(t) = q_1(t) \tag{28.15}$$

where p_{11}, p_{12} and q_1 are

$$p_{11}(t) = -2\int^{(2)} tx + \int t^2 x \tag{28.16}$$

$$p_{12}(t) = \int^{(2)} t^2 x \tag{28.17}$$

$$q_1(t) = \int^{(2)} (t^2 f - 2x) + 4\int tx - t^2 x \tag{28.18}$$

Notation[1]

The expression (28.14) is multiplied both sides by s^{-1} once more, leads to a second linear equation for the estimates B, and K.

This linear system can be represented in matrix form as:

$$PX = Q \tag{28.19}$$

where P is a matrix whose coefficients are time dependant, X is the column vector of the unknown parameters, and Q is a column vector whose coefficients are also time dependant. This is, in general form,

$$\begin{bmatrix} p_{11}(t) & p_{12}(t) \\ p_{21}(t) & p_{22}(t) \end{bmatrix} \begin{bmatrix} B \\ K \end{bmatrix} = \begin{bmatrix} q_1(t) \\ q_2(t) \end{bmatrix} \tag{28.20}$$

$p_{21}(t) = \int p_{11}(t), p_{22}(t) = \int p_{12}(t), q_2(t) = \int q_1(t)$ being,

$$p_{21}(t) = -2\int^{(3)} tx + \int^{(2)} t^2 x \tag{28.21}$$

$$p_{12}(t) = \int^{(3)} t^2 x \tag{28.22}$$

$$q_2(t) = \int^{(3)} (t^2 f - 2x) + 4\int^{(2)} tx - \int t^2 x \tag{28.23}$$

The estimates of the parameters K and B can be readily obtained by solving the following linear equation

$$K = \frac{[-p_{21}(t)q_1(t) + p_{11}(t)q_2(t)]}{[p_{11}(t)p_{22}(t) - p_{12}(t)p_{21}(t)]} \tag{28.24}$$

$$B = \frac{[p_{22}(t)q_1(t) - p_{12}(t)q_2(t)]}{[p_{11}(t)p_{22}(t) - p_{12}(t)p_{21}(t)]} \tag{28.25}$$

The time realization of the elements of matrixes P and Q can be written in a State Space framework via time-variant linear (unstable) filters to make the physical implementation of the estimator easier in the real time platform:

---

[1] $\int^{(n)} \phi(t)$ representing the iterated integral $\int_0^t \int_0^{\sigma_1} \ldots \int_0^{\sigma_{n-1}} \phi(\sigma_n) d\sigma_n \ldots d\sigma_1$ with $(\int \phi_{(t)}) = (\int^{(1)} \phi_{(t)}) = \int_0^t \phi(\sigma) d\sigma$.

$$p_{11}(t) = x_1 \tag{28.26}$$
$$\dot{x}_1 = -t^3 \theta_m(t) + x_2$$
$$\dot{x}_2 = 6t^2 \theta_m(t) + x_3$$
$$\dot{x}_3 = -6t \theta_m(t)$$

$$p_{12}(t) = y_1 \tag{28.27}$$
$$\dot{y}_1 = y_2$$
$$\dot{y}_2 = t^3 V(t) + y_3$$
$$\dot{y}_3 = -3t^2 V(t)$$

$$q_{11}(t) = t^3 \theta_m(t) z_1 \tag{28.28}$$
$$\dot{z}_1 = -9t^2 \theta_m(t) + z_2$$
$$\dot{z}_2 = 18t \theta_m(t) + z_3$$
$$\dot{x}_3 = -6\theta_m(t)$$

$$p_{21}(t) = \xi_1 \tag{28.29}$$
$$\dot{\xi}_1 = \xi_2$$
$$\dot{\xi}_2 = -t^3 \theta_m(t) + \xi_3$$
$$\dot{\xi}_3 = 6t^2 \theta_m(t) + \xi_4$$
$$\dot{\xi}_4 = -6t \theta_m(t)$$

where $\xi_2 = p_{11}$

$$p_{22}(t) = \rho_1 \tag{28.30}$$
$$\dot{\rho}_1 = \rho_2$$
$$\dot{\rho}_2 = \rho_3$$
$$\dot{\rho}_3 = t^3 V(t) + \rho_4$$
$$\dot{\rho}_4 = -3t^2 V(t)$$

where $\rho_2 = p_{12}$

$$q_{21}(t) = \psi_1 \tag{28.31}$$
$$\dot{\psi}_1 = t^3 \theta_m(t) + \psi_2$$
$$\dot{\psi}_2 = -9t^2 \theta_m(t) + \psi_3$$
$$\dot{\psi}_3 = 18t \theta_m(t) + \psi_4$$
$$\dot{\psi}_4 = -6\theta_m(t)$$

where $\psi_2 = q_{11}$.

The matrix $P(t)$ is not invertible at time $t = 0$. This means that no estimation of the parameters is done at this time. But $P(t)$ is certainly invertible after an arbitrarily small time $t = \varepsilon > 0$, then accuracy estimation of the motor parameters is obtained in a very short period of time. In practice we initialize the estimator after the small arbitrary time interval $t = \varepsilon$ to assure that the estimator obtain good estimates. From $t = 0$ to $t = \varepsilon$ there exists many singularities because of the divisions by the zero value (see Eqs. (28.24) and (28.25), such singularities occur when $p_{11}(t)p_{22}(t) - p_{12}(t)p_{21}(t) = 0$ in the K estimator and $p_{11}(t)p_{22}(t) - p_{12}(t)p_{21}(t) = 0$ in the B estimator.

Since the available signals θ_m and V are noisy, the estimation precision yielded by the estimator in (28.24)–(28.25) will depend on the Signal to Noise Ratio (SNR). We assume that θ_m and V are perturbed by an added noise with unknown statistical properties. In order to enhance the SNR, we simultaneously filter the numerator and denominator by the same low-pass filter. Taking advantage of the estimator rational form, the quotient will not be affected by the filters. This invariance is emphasized with the use of different notations in frequency and time domain:

$$K = \frac{F(s)\left[-p_{21}(t)q_1(t) + p_{11}(t)q_2(t)\right]}{F(s)(p_{11}(t)p_{22}(t) - p_{12}(t)p_{21}(t))} \tag{28.32}$$

$$B = \frac{F(s)\left[p_{22}(t)q_1(t) - p_{12}(t)q_2(t)\right]}{F(s)(p_{11}(t)p_{22}(t) - p_{12}(t)p_{21}(t))} \tag{28.33}$$

Remark *Invariant low-pass filtering is based on pure integrations of the form $F(s) = 1/s^p$, $p \geq 1$. We assumed high frequency noises. This hypothesis were motivated by recent developments in Non-standard Analysis, towards a new non stochastic noise theory. More details in [14].*

28.3 Simulation Results

This section is devoted to demonstrate the good performance of the theoretical algorithm previously explained. On the one hand, signals without any sort of noise are used in the implementation to show the time in which an ideal estimation is obtained. On the other hand, the input signals to the estimator are corrupted with a stochastic noise $n(t)$ with zero mean and standard deviation 10^{-2} in the measure of the position. Figure 28.2 depicts the implementation scheme of the estimator. Note that in estimations without noise the input $n(t)$ is zero. Independence with respect an specific design of the input to the system is also demonstrated by using two different inputs to the system: step input and sinusoidal input.

The parameters used in the estimation are depicted in Table 28.1.

The parameters $B = \frac{c}{m}$ and $K = \frac{k}{m}$ which will be estimated by the estimator will have values of $2\ [Ns/(mkg)]$ and $3\ [N/(mkg)]$ respectively.

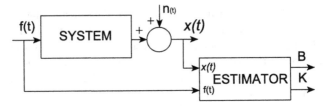

Fig. 28.2 Implementation scheme of the estimator

Table 28.1 System parameters used in simulations

Parameter	Value
m	1 [kg]
c	2 [Ns/m]
k	3 [N/m]

28.3.1 Estimation Without Noise in the Measure

In this subsection estimation of the system parameters are obtained by using ideal input signals to the estimator. In Fig. 28.3a the step input to the system is shown. The response of the system to this input is depicted in Fig. 28.3b. The estimator has such signals as input. The estimation of the parameters B and K are almost immediately obtained (see Fig. 28.4). At time $t = 0.04$ [s] we get good estimates of the parameters with null error with respect the ideal values $B = 2$ [Ns/(mkg)] and $K = 3$ [N/(mkg)]. Until time $t = 0.01$ [s] the estimator has been initialized to zero value. With this estimates and bearing in mind the mass value $m = 1$ [kg], from Eqs. (28.8) and (28.9), the real values of $k = 3$ [N/m] and $c = 2$ [Ns/m] can be obtained respectively.

In the simulation with sinusoidal signal as input to the system are obtained the same results. The sinusoidal signal has amplitude value of 1 [N] and frequency value 1 [rad/s]. Figure 28.5a depicts the sinusoidal input to the system, whereas Fig. 28.5b depicts the response of the system to such an input. The estimation of the values $B = 2$ [Ns/(mkg)] and $K = 3$ [N/(mkg)] is shown in Fig. 28.6. In this case, we have initialized the estimator to zero value within 0.03 [s]. The estimates are obtained at time $t = 0.06$ [s] and the values are maintained until the end of the experiment (see Fig. 28.6).

28.3.2 Estimation with Noise in the Measure

In this case, the response signals of the system $x(t)$ are corrupted by stochastic noise with zero mean and 10^{-2} standard deviation. We consider noise in this signal

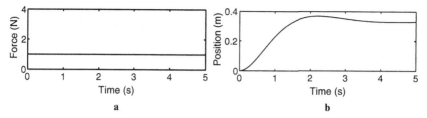

Fig. 28.3 a Step input to the system. **b** Response of the system to step input

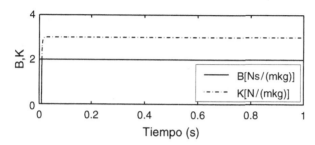

Fig. 28.4 Estimation of B and K with step input

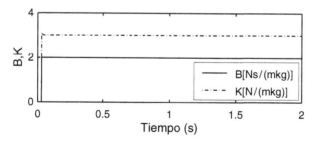

Fig. 28.5 a Sinusoidal input to the system. **b** Response of the system to sinusoidal input

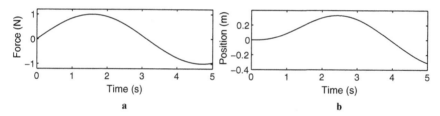

Fig. 28.6 Estimation of B and K with sinusoidal input

because is the only variable to measure in an experimental platform by means of sensors such as accelerometers, vision system,... The step and sinusoidal inputs to the system are the same that the used in Section 28.3 (see Fig. 28.3a and Fig. 28.5a).

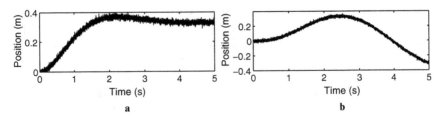

Fig. 28.7 a Response of the system to step input and noise in the measure of the mass position. **b** Response of the system to sinusoidal input and error in the measure of the mass position

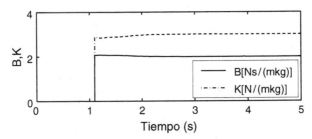

Fig. 28.8 Estimation of B and K with step input and noise in the measure of the mass position

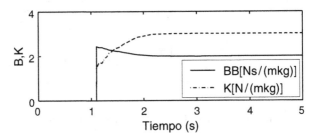

Fig. 28.9 Estimation of B and K with sinusoidal input and noise in the measure of the mass position

The input $x(t)$ corrupted by noise to the estimator is depicted in Fig. 28.7a in the case in which the input to the system is the step signal, and in Fig. 28.7b in the case in which the input to the system is the sinusoidal signal. Note that the noise strongly affects the signals.

The estimation of the parameters B and K when the input $f(t)$ is a step are depicted in Fig. 28.8. The estimator has been initialized to zero until time $t = 1.1$ $[s]$ and at time $t = 2.5$ $[s]$ the values of $B = 2$ $[Ns/(mkg)]$ and $K = 3$ $[N/(mkg)]$ are obtained. When the input $f(t)$ is sinusoidal the results are those depicted in Fig. 28.9 with the same results.

The difference between the ideal case and this case in which noise in the measure appears is that the time in the estimation is different. Whereas in the estimations without noise the values are obtained in 0.3 $[s]$ after the initialized zero value, the estimations with noise are carried out in 1.4 $[s]$, after such initialization. Nevertheless, no error appears in every estimation. Further, the different input signals to the system do not affect the estimated values.

28.4 Conclusions

Parameter identification using an on-line, non asymptotic, algebraic identification method for continuous-time constant linear systems has been proposed for the estimation of unknown parameters of a mass-spring-damper model.

In this research, an algebraic technic to estimate the physical values of the spring and damper of the system is presented. This is based on a bunch of (unstable) filters that vary on time and which are combined with classical low pass filters. The resultant expressions are obtained from derivative algebraic operations, including the unknown constants elimination through derivations with respect the complex variable s.

The only input variables to the estimator are the input force to the system and the displacement of the mass. Among the advantages of this approach we find: this method is independent of initial conditions; the methodology is also robust with respect to zero mean high frequency noises as seen from digital computer based simulations; the estimation is obtained in a very short period of time and good results are achieved; a direct estimation of the parameters is achieved without translation between discrete and continuous time domains; and the approach does not need a specific design of the inputs needed to estimate the parameters of the plant because exact formulas are proposed. Therefore its implementation in regulated closed loop systems is direct.

Acknowledgements This research was supported by the Junta de Comunidades de Castilla-La Mancha via project ref.: PBI-05-057 and European Social Fund. and by the Spanish Ministerio de Educacin y Ciencia via project ref.: DPI2006-13834.

References

1. Fliess, M., Sira-Ramirez, H.,"An algebraic framework for linear identification", *ESAIM Contr. Optim. Calc. Variat.*, V9, pp. 151–168, 2003.
2. Fliess, M., Sira-Ramirez, H., "On the non-calibrated visual based control of planar manipulators: An on-line algebraic identification approach", *IEEE-SMC-2002*, V5, 2002.
3. Sira-Ramirez, H., Fossas, E., Fliess, M.,"Output trajectory tracking in an uncertain double bridge buck dc to dc power converter: An algebraic on-line parameter identification approach", *in Proc. 41st IEEE Conf. Decision Control*, V3, pp. 2462–2467, 2002.
4. Sira-Ramirez, H., Fliess, M., "On the discrete time uncertain visual based control of planar manipulator: An on-line algebraic identification approach", *Proc. 41st IEEE Conf. Decision Control*, 2002.

5. Fliess, M., Mboup, M., Mounier, H., Sira-Ramírez, H., *Questioning Some Paradigms of Signal Processing Via Concrete Example*, Editorial Lagares, Mexico City, 2003.
6. Ljung, L., *System Identification, Theory for Users*, 2nd ed. Prentice-Hall PTR, USA, 1999.
7. Bar-Shalom, Y. , Rong Li, X., Thiagalingam Kirubarajan, *Estimation with Applications to Tracking and Navigation, Theory, Algorithms and Software*, Wiley, New York, 2001.
8. Sira-Ramirez, H., Trapero, J., Feliu, V., "Frequency identification in the noisy sum of two sinusoidal signal", *17th International symposium on Mathematical Theory of Network and Systems*, Kyoto, Japan, 2006.
9. Becedas, J., Trapero, J., Sira-Ramirez, H., Feliu, V., "Fast identification method to control a flexible manipulator with parameters uncertainties", *IEEE International Conference on Robotics and Automation*, Rome, 2007.
10. Yunfeng Li, Horowitz, R., "Active suspension vibration control with dual stage actuators in hard disk drives", *Proc. Am. Control Conf.*, V4, pp. 2786–2791, 2001.
11. Rivin, Eugene, I., *Stiffness and Damping in Mechanical Design*, Marcel Decker, New York, 1999.
12. Erickson, D., Weber, M., Sharf, I., "Contact stiffness and damping estimation for robotic systems", *Int. J. Robot. Res.*, V22, N1, pp. 41–57, 2003.
13. Faik, S., Witteman, H., "Modeling of impact dynamics: A literature survey", *International ADAMS User Conference*, Berchtesgaden, Germany, 2001.
14. Fliess, M., "Analyse non standard du bruit," *C.R. Acad. Sci. Paris*, Ser. I 342. 2006.

Chapter 29
MIMO PID Controller Synthesis with Closed-Loop Pole Assignment

Tsu-Shuan Chang and A. Nazli Gündeş

Abstract For certain classes of linear, time-invariant, multi-input multi-output plants, a systematic synthesis is developed for stabilization using Proportional + Integral + Derivative (PID) controllers, where the closed-loop poles can be assigned to the left of an axis shifted away from the origin. The real-parts of the closed-loop poles can be smaller than any given negative value for some of these classes. The classes that admit PID-controllers with this property of small negative real-part assignability of closed-loop poles include stable and some unstable plants.

Keywords Simultaneous stabilization and tracking · PID control · Integral action · Stability margin

29.1 Introduction

Proportional + Integral + Derivative (PID) controllers are preferred in many control designs since they are simple, have low-order, provide integral-action and hence, achieve asymptotic tracking of step-input references (e.g., [1]). Although the simplicity of PID-controllers is desirable due to easy implementation and tuning, the order constraint presents a major restriction that only certain classes of plants can be controlled by using PID-controllers. Rigorous PID synthesis methods based on modern control theory are explored recently in e.g., [2–4]. Sufficient conditions for PID stabilizability of linear, time-invariant (LTI), multi-input multi-output (MIMO) plants were given in [4] and several plant classes that admit PID-controllers were identified.

The systematic PID-controller design method given in [4] allows freedom in several of the design parameters. It may be possible to choose these parameters

T.-S. Chang (✉) and A.N. Gündeş
Department of Electrical and Computer Engineering, University of California, Davis, CA 95616
e-mail: chang@ece.ucdavis.edu, angundes@ucdavis.edu

S.-I. Ao et al. (eds.), *Advances in Computational Algorithms and Data Analysis,*
Lecture Notes in Electrical Engineering 14,

appropriately to achieve various performance goals. One important criterion for control design is to assign the closed-loop poles sufficiently far from the imaginary-axis of the complex plane in order to have small time-constants. Therefore, it is desirable for the closed-loop poles to have real-parts less than $-h$ for a pre-specified positive constant h. This design objective is achievable for certain plant classes as identified in this work.

All plant classes that admit PID-controllers are necessarily strongly stabilizable, although strong stabilizability is not sufficient for existence of PID-controllers [4]. The integral-constant of the PID-controller can be non-zero only if the plant has no zeros at the origin. Stable plants are obviously strongly stabilizable and they admit PID-controllers. The additional objective of assigning values less than $-h$ to the real-part of the closed-loop poles can be achieved only for certain values of h as shown in Proposition 29.2.1-(i). The restriction on h is removed for stable plants that have no finite zeros with real-parts larger than the given $-h$; the closed-loop poles can be assigned to the left of this $-h$ for any chosen value of h as shown in Proposition 29.2.1-(ii)-(iii). The unstable plant classes here have no finite zeros with real-parts larger than the given $-h$. Propositions 29.2.2 and 29.2.3 present systematic PID-controller synthesis methods for closed-loop pole assignment to the left of the finite zero with the largest negative real-part.

The main results presented in Section 29.2 start with the problem statement and basic definitions. The three plant classes under consideration are studied under three subsections. Several illustrative single-input single-output (SISO) and MIMO examples are given for the stable plant case; the choice of the free parameters can be optimized with a chosen cost function. Section 29.3 gives concluding remarks.

Although we discuss continuous-time systems here, all results apply also to discrete-time systems with appropriate modifications.

Notation: Let $\mathbb{C}$, $\mathbb{R}$, $\mathbb{R}_+$ denote complex, real, positive real numbers. For $h \in \mathbb{R}_+ \cup \{0\}$, let $\mathcal{U}_h := \{s \in \mathbb{C} \mid \mathcal{R}e(s) \geq -h\} \cup \{\infty\}$. If $h = 0$, $\mathcal{U}_h = \mathcal{U}_0 := \{s \in \mathbb{C} \mid \mathcal{R}e(s) \geq 0\} \cup \{\infty\}$ is the extended closed right-half complex plane. Let $\mathbf{R_p}$ denote real proper rational functions of s. For $h \geq 0$, $\mathbf{S}_h \subset \mathbf{R_p}$ is the subset with no poles in $\mathcal{U}_h$. The set of matrices with entries in $\mathbf{S}_h$ is denoted by $\mathcal{M}(\mathbf{S}_h)$; $\mathbf{S}_h^{m \times m}$ is used instead of $\mathcal{M}(\mathbf{S}_h)$ to indicate the matrix size explicitly. A matrix $M \in \mathcal{M}(\mathbf{S}_h)$ is called $\mathbf{S}_h$-stable; $M \in \mathcal{M}(\mathbf{S}_h)$ is called $\mathbf{S}_h$-unimodular iff $M^{-1} \in \mathcal{M}(\mathbf{S}_h)$. The H_∞-norm of $M(s) \in \mathcal{M}(\mathbf{S}_h)$ is $\|M\| := \sup_{s \in \partial \mathcal{U}_h} \bar{\sigma}(M(s))$, where $\bar{\sigma}$ is the maximum singular value and $\partial \mathcal{U}_h$ is the boundary of $\mathcal{U}_h$. We drop (s) in transfer-matrices such as $G(s)$ where this causes no confusion. We use coprime factorizations over $\mathbf{S}_h$; i.e., for $G \in \mathbf{R_p}^{m \times m}$, $G = Y^{-1}X$ denotes a left-coprime-factorization (LCF) and $G = N_g D_g^{-1}$ denotes a right-coprime-factorization (RCF) where $X, Y, N_g, D_g \in \mathcal{M}(\mathbf{S}_h)$, $\det Y(\infty) \neq 0$, $\det D_g(\infty) \neq 0$. For MIMO transfer-functions, we refer to transmission-zeros simply as zeros; blocking-zeros are a subset of transmission-zeros. If $G \in \mathbf{R_p}^{m \times m}$ is full (normal) rank, then $z_o \in \mathcal{U}_h$ is called a transmission-zero of $G = Y^{-1}X$ if $\mathrm{rank} X(z_o) < m$; $z_b \in \mathcal{U}_h$ is called a blocking-zero of $G = Y^{-1}X$ if $X(z_b) = 0$ and equivalently, $G(z_b) = 0$.

29.2 Main Results

Consider the LTI, MIMO unity-feedback system $Sys(G,C)$ shown in Fig. 29.1, where $G \in \mathbf{R_p}^{m \times m}$ and $C \in \mathbf{R_p}^{m \times m}$ are the plant and controller transfer-functions. Assume that $Sys(G,C)$ is well-posed, G and C have no unstable hidden-modes, and $G \in \mathbf{R_p}^{m \times m}$ is full rank.

We consider the realizable form of proper PID-controllers given by (29.1), where $K_p, K_i, K_d \in \mathrm{I\!R}^{m \times m}$ are the proportional, integral, derivative constants, respectively, and $\tau \in \mathrm{I\!R}_+$ (see [5]):

$$C_{pid}(s) = K_p + \frac{K_i}{s} + \frac{K_d\, s}{\tau s + 1}. \tag{29.1}$$

For implementation, a (typically fast) pole is added to the derivative term so that C_{pid} in (29.1) is proper. The integral-action in C_{pid} is present when $K_i \neq 0$. Subsets of PID-controllers are obtained by setting one or two of the three constants equal to zero: (29.1) becomes a PI-controller C_{pi} when $K_d = 0$, an ID-controller C_{id} when $K_p = 0$, a PD-controller C_{pd} when $K_i = 0$, a P-controller C_p when $K_d = K_i = 0$, an I-controller C_i when $K_p = K_d = 0$, a D-controller C_d when $K_p = K_i = 0$.

Definition 29.2.1 *(a)* $Sys(G,C)$ is said to be $\mathbf{S}_h$-stable iff the closed-loop transfer-function from (r,v) to (y,w) is in $\mathcal{M}(\mathbf{S}_h)$. *(b)* C is said to $\mathbf{S}_h$-stabilize G iff C is proper and $Sys(G,C)$ is $\mathbf{S}_h$-stable. *(c)* $G \in \mathbf{R_p}^{m \times m}$ is said to admit a PID-controller such that the closed-loop poles of $Sys(G,C)$ are in $\mathcal{U}_h$ iff there exists $C = C_{pid}$ as in (29.1) such that $Sys(G,C_{pid})$ is $\mathbf{S}_h$-stable. We say that G is $\mathbf{S}_h$-stabilizable by a PID-controller, and C_{pid} is an $\mathbf{S}_h$-stabilizing PID-controller. ∎

Let $G = Y^{-1}X$ be any LCF of G, $C = N_c D_c^{-1}$ be any RCF of C; for $G \in \mathbf{R_p}^{m \times m}$, $X, Y \in \mathcal{M}(\mathbf{S}_h)$, $\det Y(\infty) \neq 0$, and for $C \in \mathbf{R_p}^{n_u \times n_y}$, $N_c, D_c \in \mathcal{M}(\mathbf{S}_h)$, $\det D_c(\infty) \neq 0$. Then $C\, \mathbf{S}_h$-stabilizes G if and only if

$$M := YD_c + XN_c \tag{29.2}$$

is $\mathbf{S}_h$-unimodular (see [6,7]).

The problem addressed here is the following: Suppose that $h \in \mathrm{I\!R}_+$ is a given constant. Is there a PID-controller C_{pid} that stabilizes the system $Sys(G,C_{pid})$ with a guaranteed stability margin, i.e., with real parts of the closed-loop poles of the system $Sys(G,C_{pid})$ less than $-h$? It is clear that this goal is not achievable for some plants. Even when it is achievable, it may be possible to place the closed-loop poles to the left of a shifted-axis that goes through $-h$ only for certain $h \in \mathrm{I\!R}_+$.

Fig. 29.1 Unity-Feedback
System $Sys(G,C)$

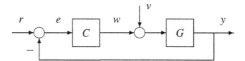

Let $\hat{s}$ and $\hat{G}$, $\hat{C}_{pid}$ be defined as

$$\hat{s} := s + h, \quad \text{equivalently}, \quad s =: \hat{s} - h \,; \tag{29.3}$$

$$\hat{G}(\hat{s}) := G(\hat{s} - h)\,; \tag{29.4}$$

$$\hat{C}_{pid}(\hat{s}) := C_{pid}(\hat{s} - h) := K_p + \frac{K_i}{\hat{s} - h} + \frac{K_d(\hat{s} - h)}{\tau(\hat{s} - h) + 1}. \tag{29.5}$$

Then $C_{pid}(s)$ $\mathbf{S}_h$-stabilizes $G(s)$ if and only if $\hat{C}_{pid}(\hat{s})$ $\mathbf{S}_0$-stabilizes $\hat{G}(\hat{s})$. For any $\alpha \in \mathbb{R}_+$, an RCF of $\hat{C}_{pid}(\hat{s})$ is given by

$$\hat{C}_{pid} = \left(\frac{(\hat{s} - h)}{\hat{s} + \alpha} \hat{C}_{pid} \right) \left(\frac{(\hat{s} - h)}{\hat{s} + \alpha} I \right)^{-1}. \tag{29.6}$$

We consider plant classes that admit PID-controllers and identify values of $h \in \mathbb{R}$ such that the closed-loop poles lie to the left of $-h$. A necessary condition for existence of PID-controllers with nonzero integral-constant K_i is that the plant $G(s)$ has no zeros (transmission-zeros or blocking-zeros) at $s = 0$ [4]. Therefore, all plants under consideration are assumed to be free of zeros at the origin (of the s-plane). The three specific classes are defined as follows:

1. The first class of plants $\mathcal{G}_{ph}$ is the set of $\mathbf{S}_h$-stable $m \times m$ plants that have no (transmission or blocking) zeros at $s = 0$; i.e., for a given $h \in \mathbb{R}_+ \cup \{0\}$, let $\mathcal{G}_{ph} \subset \mathbf{S}_h{}^{m \times m}$ be defined as

$$\mathcal{G}_{ph} := \{ G(s) \in \mathbf{S}_h{}^{m \times m} \mid \det G(0) \neq 0 \}. \tag{29.7}$$

For $G(s) \in \mathcal{G}_{ph}$, with $\hat{G}(\hat{s}) := G(\hat{s} - h)$, $\det G(0) \neq 0$ is equivalent to $\det \hat{G}(h) \neq 0$. Clearly, the plants $G \in \mathcal{G}_{ph}$ may have transmission-zeros or blocking-zeros anywhere in $\mathbb{C}$ other than $s = 0$.

2. The second class of plants $\mathcal{G}_{zh}$ is the set of $m \times m$ plants that have no (transmission or blocking) zeros in $\mathcal{U}_h$; i.e., for a given $h \in \mathbb{R}_+ \cup \{0\}$, let $\mathcal{G}_{zh} \subset \mathbf{R}_\mathbf{p}{}^{m \times m}$ be defined as

$$\mathcal{G}_{zh} := \{ G(s) \in \mathbf{R}_\mathbf{p}{}^{m \times m} \mid G^{-1}(s) \in \mathbf{S}_h{}^{m \times m} \}. \tag{29.8}$$

In the SISO case, this class represents plants without zeros in $\mathcal{U}_h$ that have zero relative degree. Some plants in the set $\mathcal{G}_{zh}$ are not $\mathbf{S}_h$-stable; therefore, these plants either have poles in $\mathcal{U}_0$, or they are $\mathbf{S}_0$-stable but some poles have negative real-parts larger than the specified $-h$. Obviously, the plants in $\mathcal{G}_{zh}$ satisfy the necessary condition for existence of PID-controllers with nonzero integral-constant K_i since the fact that they have no zeros in $\mathcal{U}_h$ implies that they have no zeros at $s = 0$.

3. The third class of plants $\mathcal{G}_\infty$ is the set of $m \times m$ strictly-proper plants that have no (transmission or blocking) zeros in $\mathcal{U}_h$ except at infinity. For a given $h \in \mathbb{R}_+ \cup \{0\}$, let $\mathcal{G}_\infty \subset \mathbf{R}_\mathbf{p}{}^{m \times m}$ be defined as

$$\mathcal{G}_\infty := \{ G(s) \in \mathbf{R}_\mathbf{p}{}^{m \times m} \mid \frac{1}{s + a} G^{-1}(s) \in \mathbf{S}_h{}^{m \times m} \text{ for any } a > h \}. \tag{29.9}$$

In the SISO case, this class represents plants without zeros in $\mathcal{U}_h$, that have relative degree one. Some plants in the set $\mathcal{G}_\infty$ are not $\mathbf{S}_h$-stable; therefore, these plants either have poles in $\mathcal{U}_0$, or they are $\mathbf{S}_0$-stable but some poles have negative real-parts larger than the specified $-h$. Obviously, the plants in $\mathcal{G}_\infty$ satisfy the necessary condition for existence of PID-controllers with nonzero integral-constant K_i since the fact that they have no zeros in $\mathcal{U}_h$ implies that they have no zeros at $s = 0$.

The set $\mathcal{G}_{ph} \cap \mathcal{G}_{zh}$ corresponds to $\mathbf{S}_h$-stable plants with no poles and no zeros in $\mathcal{U}_h$ (including infinity). The set $\mathcal{G}_{ph} \cap \mathcal{G}_\infty$ corresponds to $\mathbf{S}_h$-stable plants with no poles in $\mathcal{U}_h$ and no zeros in $\mathcal{U}_h$ except at infinity.

29.2.1 Plants with No Poles in $\mathcal{U}_h$

We start our investigation by considering the $\mathbf{S}_h$-stable plant class $\mathcal{G}_{ph}$ described in (29.7). In Proposition 29.2.1-(i), we obtain a sufficient condition on h for existence of PID-controllers that $\mathbf{S}_h$-stabilize the plant $G \in \mathcal{G}_{ph}$ such that none of the closed-loop poles are in $\mathcal{U}_h$. We propose a systematic PID-controller synthesis procedure with explicit definitions of the controller parameters. In Proposition 29.2.1-(ii), we consider the subclass $\mathcal{G}_{ph} \cap \mathcal{G}_{zh}$ of $\mathcal{G}_{ph}$, where the plants have no (transmission or blocking) zeros in $\mathcal{U}_h$, i.e., $G \in \mathcal{G}_{ph}$ such that $G^{-1} \in \mathcal{M}(\mathbf{S}_h)$. For these $\mathbf{S}_h$-unimodular plants, there exist stabilizing PID-controllers such that none of the closed-loop poles are in $\mathcal{U}_h$ for any $h \in \mathbb{R}_+$. In Proposition 29.2.1-(iii), we consider the subclass $\mathcal{G}_{ph} \cap \mathcal{G}_\infty$, where the plants are strictly-proper and have no finite (transmission or blocking) zeros in $\mathcal{U}_h$, i.e., $G \in \mathcal{G}_{ph}$ such that $\frac{1}{s+a}G^{-1} \in \mathcal{M}(\mathbf{S}_h)$ for any $a > h$. For these plants, there exist stabilizing PID-controllers such that none of the closed-loop poles are in $\mathcal{U}_h$ for any choice of $h \in \mathbb{R}_+$. Proposition 29.2.1-(ii) and (iii) indicate that PID-controllers can be designed so that the closed-loop poles have negative real-parts less than any $-h$ if the open-loop poles *and* (finite) zeros are not in $\mathcal{U}_h$. A methodology leading to explicit design parameter choices is proposed for each special case.

Proposition 29.2.1 (*Controller synthesis for $\mathbf{S}_h$-stable plants*):
Let $h \in \mathbb{R}_+$ and $G(s) \subset \mathcal{G}_{ph}$ be given.
(i) ($\mathbf{S}_h$-*stable plants with zeros in* $\mathcal{U}_h$): Define $\Theta(\hat{s})$ as

$$\Theta(\hat{s}) := \hat{G}(\hat{s})(\hat{K}_p + \frac{\hat{K}_d(\hat{s}-h)}{\tau(\hat{s}-h)+1}) + \frac{\hat{G}(\hat{s})G(0)^{-1} - I}{\hat{s}-h} . \qquad (29.10)$$

If the given $h \in \mathbb{R}_+$ satisfies (29.11) for some $\hat{K}_p \in \mathbb{R}^{m \times m}$, $\hat{K}_d \in \mathbb{R}^{m \times m}$ and $\tau < 1/h$, where

$$h < \frac{1}{2} \| \Theta(\hat{s}) \|^{-1}, \qquad (29.11)$$

then there exists an $\mathbf{S}_h$-stabilizing PID-controller, and C_{pid} can be designed as follows: Choose any $\hat{K}_p \in \mathbb{R}^{m \times m}$, $\hat{K}_d \in \mathbb{R}^{m \times m}$, $\tau \in \mathbb{R}_+$ satisfying $\tau < 1/h$. Let $K_p = (\alpha + h)\hat{K}_p$, $K_d = (\alpha + h)\hat{K}_d$, $K_i = (\alpha + h)G(0)^{-1} = (\alpha + h)\hat{G}(h)^{-1}$, where

$\alpha \in \mathbb{R}_+$ satisfies

$$h < \alpha < \| \Theta(\hat{s}) \|^{-1} - h \, . \tag{29.12}$$

Then an $\mathbf{S}_h$-stabilizing PID-controller C_{pid} is given by

$$C_{pid} = (\alpha + h)\hat{K}_p + \frac{(\alpha + h)G(0)^{-1}}{s} + \frac{(\alpha + h)\hat{K}_d \, s}{\tau s + 1} \, . \tag{29.13}$$

(ii) ($\mathbf{S}_h$-stable plants with no zeros in $\mathcal{U}_h$): Let $G(s) \in \mathcal{G}_{ph} \cap \mathcal{G}_{zh}$. Then there exists an $\mathbf{S}_h$-stabilizing PID-controller, and C_{pid} can be designed as follows: Choose any nonsingular $\hat{K}_p \in \mathbb{R}^{m \times m}$, any $K_d \in \mathbb{R}^{m \times m}$, and $\tau \in \mathbb{R}_+$ satisfying $\tau < 1/h$. Choose any $g \in \mathbb{R}_+$ satisfying

$$g > 2h \, . \tag{29.14}$$

Define $\widetilde{\Phi}(\hat{s})$ as

$$\widetilde{\Phi}(\hat{s}) := \hat{K}_p^{-1} \left[\, \hat{G}^{-1}(\hat{s}) + \frac{K_d(\hat{s} - h)}{\tau(\hat{s} - h) + 1} \, \right] . \tag{29.15}$$

Let $K_p = \tilde{\beta}\hat{K}_p$, $K_i = g\tilde{\beta}\hat{K}_p$, where $\tilde{\beta} \in \mathbb{R}_+$ satisfies

$$\tilde{\beta} > \| \, \widetilde{\Phi}(\hat{s}) \, \| \, . \tag{29.16}$$

Then an $\mathbf{S}_h$-stabilizing PID-controller C_{pid} is given by

$$C_{pid} = \tilde{\beta}\hat{K}_p + \frac{g\tilde{\beta}\hat{K}_p}{s} + \frac{K_d \, s}{\tau s + 1} \, . \tag{29.17}$$

(iii) ($\mathbf{S}_h$-stable strictly-proper plants): Let $G(s) \in \mathcal{G}_{ph} \cap \mathcal{G}_\infty$. Then there exists an $\mathbf{S}_h$-stabilizing PID-controller, and C_{pid} can be designed as follows: Let $Y(\infty)^{-1} := sG(s)|_{s \to \infty}$. Choose any $K_d \in \mathbb{R}^{m \times m}$, and $\tau \in \mathbb{R}_+$ satisfying $\tau < 1/h$. Choose any $g \in \mathbb{R}_+$ satisfying

$$g > h \, . \tag{29.18}$$

Define $\widecheck{\Psi}(\hat{s})$ as

$$\widecheck{\Psi}(\hat{s}) := \left[\, \hat{G}^{-1}(\hat{s}) + \frac{K_d(\hat{s} - h)}{\tau(\hat{s} - h) + 1} \, \right] \frac{(\hat{s} - h)}{(\hat{s} - h + g)} Y(\infty)^{-1} - \hat{s}I \, . \tag{29.19}$$

Let $K_p = \tilde{\delta}Y(\infty)$, $K_i = g\tilde{\delta}Y(\infty)$, where $\tilde{\delta} \in \mathbb{R}_+$ satisfies

$$\tilde{\delta} > \| \widecheck{\Psi}(\hat{s}) \| \, . \tag{29.20}$$

Then an $\mathbf{S}_h$-stabilizing PID-controller C_{pid} is given by

$$C_{pid} = \tilde{\delta}Y(\infty) + \frac{g\tilde{\delta}Y(\infty)}{s} + \frac{K_d \, s}{\tau s + 1} \, . \tag{29.21}$$

■

Remark: Condition (29.11) is obviously satisfied if $h = 0$, i.e., $\mathcal{U}_h = \mathcal{U}_0$; therefore, there exists a PID-controller C_{pid} of the form in (29.1) that stabilizes a given stable plant G, where the closed-loop poles of the system $Sys(G, C_{pid})$ may be anywhere in the open left-half complex plane. ∎

Proof of Proposition 29.2.1: **(i)** Substitute $\hat{s} = s + h$ as in (29.3–29.5). With $\hat{G}(\hat{s}) = I^{-1}\hat{G}$, write $\hat{C}_{pid}(\hat{s})$ as in (29.6). By (29.2), $\hat{C}_{pid}$ in (29.13) stabilizes $\hat{G}(\hat{s})$ if and only if $\hat{M}(\hat{s})$ is $\mathbf{S}_0$-unimodular:

$$\hat{M}(\hat{s}) = \frac{(\hat{s} - h)}{\hat{s} + \alpha}I + \hat{G}(\hat{s})\frac{(\hat{s} - h)}{\hat{s} + \alpha}\hat{C}_{pid}$$

$$= I - \frac{(\alpha + h)}{\hat{s} + \alpha}I + \frac{(\hat{s} - h)}{\hat{s} + \alpha}\hat{G}(\hat{s})\hat{C}_{pid} = I + \frac{(\alpha + h)(\hat{s} - h)}{\hat{s} + \alpha}\Theta(\hat{s}) . \qquad (29.22)$$

In (29.22), $\Theta(\hat{s}) \in \mathcal{M}(\mathbf{S}_0)$ since $\frac{\hat{G}(\hat{s})G(0)^{-1} - I}{\hat{s} - h} = \frac{\hat{G}(\hat{s})\hat{G}(h)^{-1} - I}{\hat{s} - h} \in \mathcal{M}(\mathbf{S}_0)$. If (29.11) and (29.12) hold, then $h < \alpha$ and $\alpha + h < \|\Theta(\hat{s})\|^{-1}$ imply

$$\left\| \frac{(\alpha + h)(\hat{s} - h)}{\hat{s} + \alpha}\Theta(\hat{s}) \right\| \le (\alpha + h)\left\| \frac{\hat{s} - h}{\hat{s} + \alpha} \right\| \|\Theta(\hat{s})\| = (\alpha + h)\|\Theta(\hat{s})\| < 1;$$

hence, $\hat{M}(\hat{s})$ in (29.22) is $\mathbf{S}_0$-unimodular by the "small-gain theorem" (e.g., [7]). Therefore, $\hat{C}_{pid}(\hat{s})$ stabilizes $\hat{G}(\hat{s})$; hence, C_{pid} is an $\mathbf{S}_h$-stabilizing controller for G. **(ii)** Write the controller $C_{pid}(s)$ given in (29.17) as

$$C_{pid}(s) = \left(\frac{s}{s + g}C_{pid}\right)\left(\frac{sI}{s + g}\right)^{-1} = \left(\tilde{\beta}\hat{K}_p + \frac{K_d s}{(\tau s + 1)}\frac{s}{(s + g)}\right)\left(\frac{sI}{s + g}\right)^{-1}. \qquad (29.23)$$

Substitute $\hat{s} = s + h$ into (29.23) to obtain an RCF of $\hat{C}_{pid}(\hat{s})$ as in (29.6), with $\alpha = g - h$. Then

$$\hat{C}_{pid}(\hat{s}) = \left(\tilde{\beta}\hat{K}_p + \frac{K_d(\hat{s} - h)}{(\tau(\hat{s} - h) + 1)}\frac{(\hat{s} - h)}{(\hat{s} - h + g)}\right)\left(\frac{(\hat{s} - h)}{\hat{s} - h + g}I\right)^{-1}, \qquad (29.24)$$

where $(1 - \tau h) \in \mathbb{R}_+$ and $(g - h) \in \mathbb{R}_+$ by assumption. Since $G(s) \in \mathcal{G}_{ph} \cap \mathcal{G}_{zh}$ implies $G^{-1}(s) \in \mathcal{M}(\mathbf{S}_h)$, we have $\hat{G}^{-1}(\hat{s}) \in \mathcal{M}(\mathbf{S}_0)$. By (29.2), $\hat{C}_{pid}(\hat{s})$ in (29.17) stabilizes $\hat{G}(\hat{s})$ if and only if $\tilde{M}_\beta(\hat{s})$ is $\mathbf{S}_0$-unimodular:

$$\tilde{M}_\beta(\hat{s}) = \frac{(\hat{s} - h)}{\hat{s} - h + g}I + \hat{G}(\hat{s})\frac{(\hat{s} - h)}{\hat{s} - h + g}\hat{C}_{pid} = \hat{G}(\hat{s})\tilde{\beta}\,\hat{K}_p$$

$$+ \left[I + \hat{G}(\hat{s})\frac{K_d(\hat{s} - h)}{(\tau(\hat{s} - h) + 1)}\right]\frac{(\hat{s} - h)}{(\hat{s} - h + g)} = \tilde{\beta}\,\hat{G}(\hat{s})\hat{K}_p\left(I + \frac{1}{\tilde{\beta}}\hat{K}_p^{-1}\right)$$

$$[\hat{G}^{-1}(\hat{s}) + \frac{K_d(\hat{s} - h)}{(\tau(\hat{s} - h) + 1)}]\frac{(\hat{s} - h)}{(\hat{s} - h + g)}) = \tilde{\beta}\hat{G}(\hat{s})\hat{K}_p(I + \frac{1}{\tilde{\beta}}\tilde{\Phi}(\hat{s})\frac{(\hat{s} - h)}{(\hat{s} - h + g)}),$$

$$\qquad (29.25)$$

where $\hat{K}_p$ is nonsingular and $G^{-1}(s) \in \mathcal{M}(\mathbf{S}_h)$ implies $\hat{G}^{-1}(\hat{s}) \in \mathcal{M}(\mathbf{S}_0)$ by assumption. If (29.14) holds, then $2h < g$ and $\tilde{\beta} > \|\tilde{\Phi}(\hat{s})\|$ imply

$$\|\frac{1}{\tilde{\beta}}\tilde{\Phi}(\hat{s})\frac{(\hat{s}-h)}{(\hat{s}-h+g)}\| \le \frac{1}{\tilde{\beta}}\|\tilde{\Phi}(\hat{s})\|\|\frac{(\hat{s}-h)}{(\hat{s}-h+g)}\| = \frac{1}{\tilde{\beta}}\|\tilde{\Phi}(\hat{s})\| < 1;$$

hence, $\tilde{M}_\beta(\hat{s})$ in (29.25) is $\mathbf{S}_0$-unimodular. Therefore, $\hat{C}_{pid}(\hat{s})$ an $\mathbf{S}_0$-stabilizing controller for $\hat{G}(\hat{s})$; hence, C_{pid} is an $\mathbf{S}_h$-stabilizing controller for G.

(iii) Substitute $\hat{s} = s + h$ as in (29.3–29.5). Then an LCF of $\hat{G}(\hat{s})$ is

$$\hat{G}(\hat{s}) = \hat{Y}^{-1}\hat{X} := I^{-1}\hat{G}(\hat{s}) .$$

Write the controller $C_{pid}(s)$ given in (29.21) as

$$C_{pid}(s) = (\frac{s}{s+g}C_{pid})(\frac{sI}{s+g})^{-1} = (\tilde{\delta}Y(\infty) + \frac{K_d s}{(\tau s+1)}\frac{s}{(s+g)})(\frac{sI}{s+g}I)^{-1}. \quad (29.26)$$

Substitute $\hat{s} = s + h$ into (29.26) to obtain an RCF of $\hat{C}_{pid}(\hat{s})$ as in (29.6), with $\alpha = g - h$. Then

$$\hat{C}_{pid}(\hat{s}) = (\tilde{\delta}Y(\infty) + \frac{K_d(\hat{s}-h)}{(\tau(\hat{s}-h)+1)}\frac{(\hat{s}-h)}{(\hat{s}-h+g)})(\frac{(\hat{s}-h)}{\hat{s}-h+g}I)^{-1}, \quad (29.27)$$

where $(1 - \tau h) \in \mathbb{R}_+$ and $(g - h) \in \mathbb{R}_+$ by assumption. Since $G(s) \in \mathcal{G}_{ph} \cap \mathcal{G}_\infty$ implies $(s+a)G(s) \in \mathcal{M}(\mathbf{S}_h)$ and $\frac{1}{(s+a)}G^{-1}(s) \in \mathcal{M}(\mathbf{S}_h)$ for $a > h$, we have $(\hat{s} - h+a)\hat{G}(\hat{s}) \in \mathcal{M}(\mathbf{S}_0)$ and $\frac{1}{(\hat{s}-h+a)}\hat{G}^{-1}(\hat{s}) \in \mathcal{M}(\mathbf{S}_0)$; similarly, $(\hat{s}+\tilde{\delta})\hat{G}(\hat{s}) \in \mathcal{M}(\mathbf{S}_0)$ and $\frac{1}{(\hat{s}+\tilde{\delta})}\hat{G}^{-1}(\hat{s}) \in \mathcal{M}(\mathbf{S}_0)$. By (29.2), $\hat{C}_{pid}(\hat{s})$ in (29.26) stabilizes $\hat{G}(\hat{s})$ if and only if $\tilde{M}_\delta(\hat{s})$ is $\mathbf{S}_0$-unimodular:

$$\tilde{M}_\delta(\hat{s}) = \frac{(\hat{s}-h)}{(\hat{s}-h+g)}I + \hat{G}(\hat{s})\frac{(\hat{s}-h)}{(\hat{s}-h+g)}\hat{C}_{pid}$$

$$= (\hat{s}+\tilde{\delta})\hat{G}(\hat{s})(\frac{1}{(\hat{s}+\tilde{\delta})}\hat{G}^{-1}(\hat{s})\frac{(\hat{s}-h)}{(\hat{s}-h+g)}I + \frac{1}{(\hat{s}+\tilde{\delta})}I\frac{(\hat{s}-h)}{(\hat{s}-h+g)}\hat{C}_{pid})$$

$$= (\hat{s}+\tilde{\delta})\hat{G}(\hat{s})(\frac{\tilde{\delta}Y(\infty)}{(\hat{s}+\tilde{\delta})} + \frac{1}{(\hat{s}+\tilde{\delta})}[\hat{G}^{-1}(\hat{s}) + \frac{K_d(\hat{s}-h)}{(\tau(\hat{s}-h)+1)}]\frac{(\hat{s}-h)}{(\hat{s}-h+g)})$$

$$= (\hat{s}+\tilde{\delta})\hat{G}(\hat{s}) (I + \frac{1}{(\hat{s}+\tilde{\delta})}[\frac{(\hat{s}-h)}{(\hat{s}-h+g)}\hat{G}^{-1}(\hat{s})Y(\infty)^{-1} - \hat{s}I$$

$$+ \frac{K_d(\hat{s}-h)}{(\tau(\hat{s}-h)+1)}\frac{(\hat{s}-h)}{(\hat{s}-h+g)}Y(\infty)^{-1}])Y(\infty)$$

$$= (\hat{s}+\tilde{\delta})\hat{G}(\hat{s}) (I + \frac{1}{(\hat{s}+\tilde{\delta})}\tilde{\Psi}(\hat{s}))Y(\infty) . \quad (29.28)$$

Then $\widetilde{\Psi}(\hat{s}) \in \mathcal{M}(\mathbf{S}_0)$ and therefore $M_\delta(\hat{s})$ in (29.28) is $\mathbf{S}_0$-unimodular since $\delta > \|\widetilde{\Psi}(\hat{s})\|$ implies

$$\left\| \frac{1}{(\hat{s}+\tilde{\delta})} \widetilde{\Psi}(\hat{s}) \right\| \leq \left\| \frac{1}{(\hat{s}+\tilde{\delta})} \right\| \left\| \widetilde{\Psi}(\hat{s}) \right\| = \frac{1}{\tilde{\delta}} \left\| \widetilde{\Psi}(\hat{s}) \right\| < 1.$$

Therefore, $\hat{C}_{pid}(\hat{s})$ an $\mathbf{S}_0$-stabilizing controller for $\hat{G}(\hat{s})$; hence, C_{pid} is an $\mathbf{S}_h$-stabilizing controller for G.   ■

The systematic PID-controller design method of Proposition 29.2.1 is illustrated by the following examples in [8]. Given $h \in \mathbb{R}_+$ and $G \in \mathcal{G}_h$, define

$$\rho := max\{x \,|\, p = x + jy, \ \text{where } p \text{ is a pole of } G(s)\}; \qquad (29.29)$$

then $-h > \rho$ since $G \in \mathcal{G}_h$. We also define

$$\gamma(\hat{K}_p, \hat{K}_d) := \| \Theta(\hat{s}) \|^{-1} . \qquad (29.30)$$

Example 29.2.1 Consider the plant transfer-function

$$G(s) = \frac{(s+5)(s^2+8s+32)}{(s+2)(s+8)(s^2+12s+40)} . \qquad (29.31)$$

By (29.29), $\rho = -2$. Suppose that $h = 1$ and we choose $\hat{K}_p = 2.5$, $\hat{K}_d = 0.2$, $\tau = 0.05$. We compute $\gamma = 4.7 > 2h = 2$, and set $\alpha = 0.5\gamma$. The closed-loop poles are -1.79, -2.66, $-4.93 \pm j2.53$, -6.87, -42.58, which all have real-parts less than $-h = -1$. For a given $(\hat{K}_p, \hat{K}_d)$, there may exist a maximum value h_{max} such that condition (29.11) is violated. In this case, we can obtain that h_{max} is about 1.81.   ■

Example 29.2.2 Consider the quadruple-tank apparatus in [9], which consists of four interconnected water tanks and two pumps. The output variables are the water levels of the two lower tanks, and they are controlled by the currents that are manipulating two pumps. The transfer-matrix of the linearized model at some operating point is

$$G = \begin{bmatrix} \frac{3.7b_1}{62s+1} & \frac{3.7(1-b_2)}{(23s+1)(62s+1)} \\ \frac{4.7(1-b_1)}{(30s+1)(90s+1)} & \frac{4.7b_2}{90s+1} \end{bmatrix} \in \mathbf{S}_0^{2\times2}. \qquad (29.32)$$

One of the two transmission-zeros of the linearized system dynamics can be moved between the positive and negative real-axis by changing a valve. The adjustable transmission-zeros depends on parameters b_1 and b_2 (the proportions of water flow into the tanks adjusted by two valves). For the values of b_1, b_2 chosen as $b_1 = 0.43$ and $b_2 = 0.34$, the plant G has transmission-zeros at $z_1 = 0.0229 > 0$ and $z_2 = -0.0997$. By (29.29) $\rho = -1/90 = -0.0111$. Suppose that $h = 0.004$. Choose $\tau = 0.05$, and

$$\hat{K}_p = \begin{bmatrix} -22.61 & 37.61 \\ 72.14 & -43.96 \end{bmatrix}, \quad \hat{K}_d = \begin{bmatrix} 5.28 & 6.21 \\ 6.53 & 7.84 \end{bmatrix} . \qquad (29.33)$$

Fig. 29.2 Step responses for Example 29.2.3

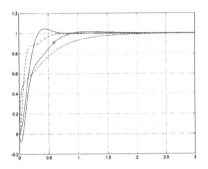

We can compute $\gamma = 0.0099 > 2h = 0.008$, and set $\alpha = 0.5\gamma$. The maximum of the real-parts of the closed poles can now be computed as -0.0059, which is less than $-h = -0.004$. Thus the requirement is fulfilled. In this example, h_{max} is very small due to the fact that ρ is very close to the imaginary-axis. ∎

Example 29.2.3 The PID-synthesis procedure based on Proposition 29.2.1 involves free parameter choices. Consider the same transfer-function as in (29.31) of Example 2.1. Let $h = 1$, choose $\tau = 0.05$, and set $\alpha = 0.5\gamma$ as before. If we choose $(\hat{K}_p = 2.5, \hat{K}_d = 0.2)$, then the dotted line in Fig. 29.2 shows the closed-loop step response. However, if we choose $(\hat{K}_p = 2, \hat{K}_d = -0.1)$, then we obtain a completely different step response as shown by the dash-dotted line. It is natural to ask then if the free parameters can be chosen optimally in some sense. Consider a prototype second order model plant, with $\zeta = 0.7$ and $\omega_n = 6$; i.e.,

$$T_m = \frac{\omega_n^2}{s^2 + 2\zeta\omega + \omega_n^2} \ . \tag{29.34}$$

We want the closed-loop step response $s_m(t)$ using the model plant T_m to be as close as possible to the actual step response $s_o(t)$. The solid line shows the step response using T_m. Consider the cost function

$$error = \frac{1}{3}\int_0^3 (s_m(t) - s_o(t))^2 dt, \tag{29.35}$$

where $s_o(t)$ is the step response for any choice of $(\hat{K}_p, \hat{K}_d)$. By plotting the contour of the error in terms of $(\hat{K}_p, \hat{K}_d)$, we find the global minimum of the error occurring at $(\hat{K}_p = 1.47, \hat{K}_d = -0.15)$. The step response for this choice of $(\hat{K}_p, \hat{K}_d)$ is shown by the solid line with a circle, which is closer to the model step response than the other two. ∎

29.2.2 Plants with No Zeros in $\mathcal{U}_h$

Consider the class $\mathcal{G}_{zh}$ of $m \times m$ plants with no (transmission or blocking) zeros in $\mathcal{U}_h$ as described in (29.8). The plants in $\mathcal{G}_{zh}$ obviously have no zeros at $s = 0$

since they have no zeros in $\mathcal{U}_h$. The plants $G \in \mathcal{G}_{zh}$ may not be $\mathbf{S}_h$-stable but $G^{-1} \in \mathcal{M}(\mathbf{S}_h)$; an LCF of $G(s)$ is

$$G = Y^{-1}X = (G^{-1})^{-1}I . \tag{29.36}$$

The plants in $\mathcal{G}_{zh}$ are strongly stabilizable, and they admit $\mathbf{S}_0$-stabilizing PID-controllers [4]. Proposition 29.2.2 shows that these plants also admit $\mathbf{S}_h$-stabilizing PID-controllers for any pre-specified $h \in \mathbb{R}_+$, and proposes a systematic PID-controller synthesis.

Proposition 29.2.2 (*Controller synthesis for plants with no zeros in $\mathcal{U}_h$*):
Let $G \in \mathcal{G}_{zh}$. Then there exists an $\mathbf{S}_h$-stabilizing PID-controller C_{pid}. Furthermore, C_{pid} can be designed as follows: Choose any nonsingular $\hat{K}_p \in \mathbb{R}^{m\times m}$. Choose any $K_d \in \mathbb{R}^{m\times m}$, and $\tau \in \mathbb{R}_+$ satisfying $\tau < 1/h$. Choose any $g \in \mathbb{R}_+$ satisfying

$$g > 2h . \tag{29.37}$$

Define $\Phi(\hat{s})$ as

$$\Phi(\hat{s}) := \hat{K}_p^{-1} \left[\hat{G}^{-1}(\hat{s}) + \frac{K_d(\hat{s}-h)}{\tau(\hat{s}-h)+1} \right] . \tag{29.38}$$

Let $K_p = \beta \hat{K}_p$, $K_i = g \beta \hat{K}_p$, where $\beta \in \mathbb{R}_+$ satisfies

$$\beta > \| \Phi(\hat{s}) \| . \tag{29.39}$$

Then an $\mathbf{S}_h$-stabilizing PID-controller C_{pid} is given by

$$C_{pid} = \beta \hat{K}_p + \frac{g \beta \hat{K}_p}{s} + \frac{K_d s}{\tau s + 1} . \qquad \blacksquare \tag{29.40}$$

Proof of Proposition 29.2.2: Substitute $\hat{s} = s + h$ as in (29.3–29.5). Then an LCF of $\hat{G}(\hat{s})$ is $\hat{G}(\hat{s}) = \hat{Y}^{-1}\hat{X} := (\hat{G}^{-1}(\hat{s}))^{-1}I$. Write the controller $C_{pid}(s)$ given in (29.40) as

$$C_{pid}(s) = \left(\frac{s}{s+g} C_{pid} \right) \left(\frac{sI}{s+g} \right)^{-1} = \left(\beta \dot{K}_p + \frac{K_d s}{(\tau s + 1)} \frac{s}{(s+g)} \right) \left(\frac{sI}{s+g} \right)^{-1}. \tag{29.41}$$

Substitute $\hat{s} = s + h$ into (29.41) to obtain an RCF of $\hat{C}_{pid}(\hat{s})$ as in (29.6), with $\alpha = g - h$. Then

$$\hat{C}_{pid}(\hat{s}) = \left(\beta \hat{K}_p + \frac{K_d(\hat{s}-h)}{(\tau(\hat{s}-h)+1)} \frac{(\hat{s}-h)}{(\hat{s}-h+g)} \right) \left(\frac{(\hat{s}-h)}{\hat{s}-h+g}I \right)^{-1} , \tag{29.42}$$

where $(1 - \tau h) \in \mathbb{R}_+$ and $(g - h) \in \mathbb{R}_+$ by assumption. By (29.2), $\hat{C}_{pid}(\hat{s})$ in (29.41) stabilizes $\hat{G}(\hat{s})$ if and only if $M_\beta(\hat{s})$ is $\mathbf{S}_0$-unimodular:

$$M_\beta(\hat{s}) = \hat{Y}(\hat{s}) \frac{(\hat{s}-h)}{\hat{s}-h+g} I + \hat{X}(\hat{s}) \frac{(\hat{s}-h)}{\hat{s}-h+g} \hat{C}_{pid} = \hat{G}^{-1}(\hat{s}) \frac{(\hat{s}-h)}{\hat{s}-h+g} I$$

$$+ \frac{(\hat{s}-h)}{\hat{s}-h+g} \hat{C}_{pid} = \beta \hat{K}_p + [\hat{G}^{-1}(\hat{s}) + \frac{K_d(\hat{s}-h)}{(\tau(\hat{s}-h)+1)}] \frac{(\hat{s}-h)}{(\hat{s}-h+g)}$$

$$= \beta \hat{K}_p (I + \frac{1}{\beta} \hat{K}_p^{-1} [\hat{G}^{-1}(\hat{s}) + \frac{K_d(\hat{s}-h)}{(\tau(\hat{s}-h)+1)}] \frac{(\hat{s}-h)}{(\hat{s}-h+g)})$$

$$= \beta \hat{K}_p (I + \frac{1}{\beta} \Phi(\hat{s}) \frac{(\hat{s}-h)}{(\hat{s}-h+g)}) , \qquad (29.43)$$

where $\hat{K}_p$ is unimodular and $G^{-1}(s) \in \mathcal{M}(\mathbf{S}_h)$ by assumption. If (29.37) holds, then $2h < g$ and $\beta > \|\Phi(\hat{s})\|$ imply

$$\|\frac{1}{\beta} \Phi(\hat{s}) \frac{(\hat{s}-h)}{(\hat{s}-h+g)}\| \leq \frac{1}{\beta} \|\Phi(\hat{s})\| \|\frac{(\hat{s}-h)}{(\hat{s}-h+g)}\| = \frac{1}{\beta} \|\Phi(\hat{s})\| < 1;$$

hence, $M_\beta(\hat{s})$ in (29.43) is $\mathbf{S}_0$-unimodular. Therefore, $\hat{C}_{pid}(\hat{s})$ an $\mathbf{S}_0$-stabilizing controller for $\hat{G}(\hat{s})$; hence, C_{pid} is an $\mathbf{S}_h$-stabilizing controller for G. ∎

29.2.3 Strictly-Proper Plants with No Other Zeros in $\mathcal{U}_h$

Consider the class $\mathcal{G}_\infty$ of $m \times m$ strictly-proper plants that have no other (transmission or blocking) zeros in $\mathcal{U}_h$ as described in (29.9). Since the plants in $\mathcal{G}_\infty$ have no zeros in $\mathcal{U}_h$ other than the one at $s = \infty$, they obviously have no zeros at $s = 0$. The plants $G \in \mathcal{G}_{zh}$ are not all $\mathbf{S}_h$-stable but $\frac{1}{s+a} G^{-1} \in \mathcal{M}(\mathbf{S}_h)$ for any $a > h$; an LCF of $G(s)$ is

$$G = Y^{-1}X = (\frac{1}{s+a}G^{-1})^{-1}(\frac{1}{s+a}I) ; \qquad (29.44)$$

in (29.44), $G(\infty) = 0$, and $Y(\infty)^{-1} = (s+a)G(s)|_{s \to \infty} = sG(s)|_{s \to \infty}$. The plants in $\mathcal{G}_\infty$ are strongly stabilizable, and they admit $\mathbf{S}_0$-stabilizing PID-controllers [4]). Proposition 29.2.3 shows that these plants also admit $\mathbf{S}_h$-stabilizing PID-controllers for any pre-specified $h \in \mathbb{R}_+$, and proposes a systematic PID-controller synthesis procedure.

Proposition 29.2.3 (Controller synthesis for strictly-proper plants):
Let $G \in \mathcal{G}_\infty$. Then there exists an $\mathbf{S}_h$-stabilizing PID-controller, and C_{pid} can be designed as follows: Let $Y(\infty)^{-1} := sG(s)|_{s \to \infty}$. Choose any $K_d \in \mathbb{R}^{m \times m}$, and $\tau \in \mathbb{R}_+$ satisfying $\tau < 1/h$. Choose any $g \in \mathbb{R}_+$ satisfying

$$g > h . \qquad (29.45)$$

Define $\Psi(\hat{s})$ as

$$\Psi(\hat{s}) := [\,\hat{G}^{-1}(\hat{s}) + \frac{K_d(\hat{s}-h)}{\tau(\hat{s}-h)+1}\,]\frac{(\hat{s}-h)}{(\hat{s}-h+g)}Y(\infty)^{-1} - \hat{s}I\,. \tag{29.46}$$

Let $K_p = \delta Y(\infty)$, $K_i = g\,\delta Y(\infty)$, where $\delta \in \mathbb{R}_+$ satisfies

$$\delta > \|\,\Psi(\hat{s})\,\|\,. \tag{29.47}$$

Then an $\mathbf{S}_h$-stabilizing PID-controller C_{pid} is given by

$$C_{pid} = \delta Y(\infty) + \frac{g\,\delta Y(\infty)}{s} + \frac{K_d\,s}{\tau s + 1}\,. \qquad\blacksquare \tag{29.48}$$

Proof of Proposition 29.2.3: Substitute $\hat{s} = s+h$ as in (29.3–29.5). Then an LCF of $\hat{G}(\hat{s})$ is

$$\hat{G}(\hat{s}) = \hat{Y}^{-1}\hat{X} := (\,\frac{1}{\hat{s}-h+a}\hat{G}^{-1}(\hat{s})\,)^{-1}(\,\frac{1}{\hat{s}-h+a}I\,).$$

Write the controller $C_{pid}(s)$ given in (29.48) as

$$C_{pid}(s) = (\frac{s}{s+g}C_{pid})(\frac{sI}{s+g})^{-1} = (\delta Y(\infty) + \frac{K_d\,s}{(\tau s+1)}\frac{s}{(s+g)})(\frac{sI}{s+g})^{-1}. \tag{29.49}$$

Substitute $\hat{s} = s+h$ into (29.49) to obtain an RCF of $\hat{C}_{pid}(\hat{s})$ as in (29.6), with $\alpha = g - h$. Then

$$\hat{C}_{pid}(\hat{s}) = (\delta Y(\infty) + \frac{K_d(\hat{s}-h)}{(\tau(\hat{s}-h)+1)}\frac{(\hat{s}-h)}{(\hat{s}-h+g)})(\frac{(\hat{s}-h)}{\hat{s}-h+g}I)^{-1}\,, \tag{29.50}$$

where $(1-\tau h)\in\mathbb{R}_+$ and $(g-h)\in\mathbb{R}_+$ by assumption. By (29.2), $\hat{C}_{pid}(\hat{s})$ in (29.49) stabilizes $\hat{G}(\hat{s})$ if and only if $M_\delta(\hat{s})$ is $\mathbf{S}_0$-unimodular:

$$M_\delta(\hat{s}) = \hat{Y}(\hat{s})\frac{(\hat{s}-h)}{(\hat{s}-h+g)}I + \hat{X}(\hat{s})\frac{(\hat{s}-h)}{(\hat{s}-h+g)}\hat{C}_{pid}$$

$$= \frac{1}{(\hat{s}-h+a)}\hat{G}^{-1}(\hat{s})\frac{(\hat{s}-h)}{(\hat{s}-h+g)}I + \frac{1}{(\hat{s}-h+a)}I\frac{(\hat{s}-h)}{(\hat{s}-h+g)}\hat{C}_{pid}$$

$$= \frac{1}{(\hat{s}-h+a)}\delta Y(\infty) + \frac{1}{(\hat{s}-h+a)}[\,\hat{G}^{-1}(\hat{s}) + \frac{K_d(\hat{s}-h)}{(\tau(\hat{s}-h)+1)}\,]\frac{(\hat{s}-h)}{(\hat{s}-h+g)}$$

$$= \frac{(\hat{s}+\delta)}{(\hat{s}-h+a)}(\frac{\delta I}{\hat{s}+\delta} + \frac{1}{(\hat{s}+\delta)}[\hat{G}^{-1}(\hat{s}) + \frac{K_d(\hat{s}-h)}{(\tau(\hat{s}-h)+1)}]\frac{(\hat{s}-h)Y(\infty)^{-1}}{(\hat{s}-h+g)})Y(\infty)$$

$$= \frac{(\hat{s}+\delta)}{(\hat{s}-h+a)}(I + \frac{1}{(\hat{s}+\delta)}\,\Psi(\hat{s})\,)Y(\infty). \tag{29.51}$$

Then $\Psi(\hat{s}) \in \mathcal{M}(\mathbf{S}_0)$ since

$$
\Psi(\hat{s}) = \hat{G}^{-1}(\hat{s}) \frac{(\hat{s}-h)}{(\hat{s}-h+g)} Y(\infty)^{-1} - \hat{s}I + \frac{K_d(\hat{s}-h)}{(\tau(\hat{s}-h)+1)} \frac{(\hat{s}-h)}{(\hat{s}-h+g)} Y(\infty)^{-1}
$$

$$
= (\hat{s}-h+a)\hat{Y}(\hat{s}) \frac{(\hat{s}-h)}{(\hat{s}-h+g)} Y(\infty)^{-1} - \hat{s}I + \frac{K_d(\hat{s}-h)}{(\tau(\hat{s}-h)+1)} \frac{(\hat{s}-h)}{(\hat{s}-h+g)} Y(\infty)^{-1}
$$

$$
= \hat{s}\left[\frac{(\hat{s}-h)}{(\hat{s}-h+g)} \hat{Y}(\hat{s}) Y(\infty)^{-1} - I \right] + \left[(a-h)\hat{Y}(\hat{s}) + \frac{K_d(\hat{s}-h)}{(\tau(\hat{s}-h)+1)} \right] \frac{(\hat{s}-h)Y(\infty)^{-1}}{(\hat{s}-h+g)},
$$

and $\hat{Y}(\infty) = Y(\infty)$ implies $\left[\frac{(\hat{s}-h)}{(\hat{s}-h+g)} \hat{Y}(\hat{s}) Y(\infty)^{-1} - I \right]$ is strictly-proper. Therefore $M_\delta(\hat{s})$ in (29.51) is $\mathbf{S}_0$-unimodular since $\delta > \|\Psi(\hat{s})\|$ implies

$$
\left\| \frac{1}{(\hat{s}+\delta)} \Psi(\hat{s}) \right\| \leq \left\| \frac{1}{(\hat{s}+\delta)} \right\| \|\Psi(\hat{s})\| = \frac{1}{\delta} \|\Psi(\hat{s})\| < 1.
$$

Therefore, $\hat{C}_{pid}(\hat{s})$ an $\mathbf{S}_0$-stabilizing controller for $\hat{G}(\hat{s})$; hence, C_{pid} is an $\mathbf{S}_h$-stabilizing controller for G.   ■

29.3 Conclusions

Systematic PID-controller designs were proposed in this work, where closed-loop poles are placed in the left-half complex plane to the left of the plant zero with the largest negative real-part. The plants under consideration are either stable, or unstable but their finite zeros are in the region of stability. The proposed synthesis method allowed freedom in the choice of parameters. Illustrative examples were given, including one that demonstrates how this freedom can be used to improve an SISO system's performance. Extending the optimal parameter selection to MIMO systems would be a challenging goal. Future directions will involve extension to more general classes of unstable MIMO plants. Optimal parameter selections for the MIMO case will also be explored.

References

1. Aström, K. J. and Hagglund, T. (1995). *PID Controllers: Theory, Design, and Tuning, Second Edition*. Instrument Society of America, Research Triangle Park, NC.
2. Ho, M.-T., Datta, A., and Bhattacharyya, S. P. (1998). An extension of the generalized Hermite-Biehler theorem: relaxation of earlier assumptions. *Proceedings of 1998 American Control Conference*, 3206–3209.
3. Silva, G. J., Datta, A., and Bhattacharyya, S. P. (2005). *PID Controllers for Time-Delay Systems*. Birkhäuser, Boston, MA.
4. Gündeş, A. N. and Ozguler, A. B. (2007). PID stabilization of MIMO plants. *IEEE Trans. Automatic Control*, 52:1502–1508.

5. Goodwin, G. C., Graebe, S. F., and Salgado, M. E. (2001). *Control System Design.* Prentice-Hall, Upper Saddle River, NJ.
6. Gündeş, A. N. and Desoer, C. A. (1990). *Algebraic Theory of Linear Feedback Systems with Full and Decentralized Compensators.* Lecture Notes in Control and Information Sciences, 142, Springer, Germany.
7. Vidyasagar, M. (1985). *Control System Synthesis: A Factorization Approach.* MIT Press, Cambridge, MA.
8. Chang, T. S. and Gündeş, A. N. (2007). PID controller design with guaranteede stability margin for MIMO systems. *Proc. World Congress Eng. Comput. Sci.*, San Francisco, USA, 52: 747–752.
9. Johansson, K. H. (2000). The quadruple-tank process: A multivariable laboratory process with an adjustable zero. *IEEE Trans. Contr. Sys. Technol.*, 8:456–465.

Chapter 30
Robust Design of Motor PWM Control using Modeling and Simulation

Wei Zhan

Abstract A robust design method is developed for Pulse Width Modulation (PWM) motor speed control. A first principle model for DC permanent magnetic motor is used to build a Simulink model for simulation and analysis. Based on the simulation result, the main factors that contributed to the average speed variation are identified using Design of Experiment (DOE). A robust solution is derived to reduce the average speed control variation using Response Surface Method (RSM). The robustness of the new design is verified using the simulation model.

Keywords Design of experiment · Monte Carlo analysis · Pulse width modulation · Response surface method

30.1 Introduction

Pulse Width Modulation [1, 2] (PWM) is commonly used in industry. For instance, in many automotive applications, a battery/alternator is often used as the power source. The battery/alternator has an output voltage of about 12 V with small variations. When a voltage different from this value is needed to control an actuator, one can either use a hardware voltage regulator or PWM control. The hardware solution is usually not desirable due to the high cost and packaging issues associated with it. The idea of using PWM control is simple: the power to the actuator is effectively reduced by an amount that can be adjusted by changing the PWM duty cycles. A 50% PWM control command is shown in Fig. 30.1, where the PWM frequency is defined to be 1/Ts, and the PWM duty cycle is defined to be On/Ts * 100%. When the command is "high", the constant voltage source is connected to the actuator. When the

W. Zhan
Department of Engineering Technology and Industrial Distribution, Texas A&M University, College Station, TX 77843-3367
e-mail: Zhan@entc.tamu.edu

S.-I. Ao et al. (eds.), *Advances in Computational Algorithms and Data Analysis,*
Lecture Notes in Electrical Engineering 14,

439

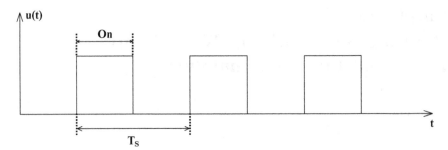

Fig. 30.1 A typical PMW control command

command is "low", the constant voltage source is disconnected from the actuator. A rule of thumb for PWM duty cycle selection is given by Barr [2] as

$$DutyCycle = \frac{DesiredVoltage}{NominalVoltage} \times 100\% \qquad (30.1)$$

This is a very attractive solution. However, it was found by Zhan [3] that the results are usually not accurate. Even if the PWM duty cycle is selected correctly, in real world applications, one gets large variation in the result. The causes for the variation can be from the nonlinearity in the system, the variation in the system parameters and the voltage source etc.

Using the Six Sigma [4] approach, the robustness of the PWM control of motor speed was discussed by Zhan [3]. Details and many references in controlling DC permanent magnetic motors can be found [5]. In this paper, we try to generalize the result by Zhan [3] by adding the temperature and load torque factors in the robustness analysis. In Section 30.2, the model for the motor developed by Zhan [6] is modified so that the temperature and its impact on the motor parameters are taken into consideration. The baseline performance of PWM control is established using Monte Carlo analysis [7, 8] in Section 30.3. In Section 30.4, DOE analysis [9, 10] is conducted to find the main factors contributing to the average speed variation. The Response Surface Method [11, 12] RSM) is used in Section 30.5 to find a robust design. The new design is also verified and compared to the baseline using Monte Carlo analysis. Conclusions are drawn in Section 30.6.

30.2 Modeling

A DC permanent magnetic motor can be modeled by the following equations when a voltage source with output $e_a(t)$ is applied to the motor armature:

$$\frac{di_a(t)}{dt} = \frac{1}{L_a}e_a(t) - \frac{R_a}{L_a}i_a(t) - \frac{1}{L_a}e_b(t)$$

$$T_m(t) = K_i i_a(t)$$

$$e_b(t) = K_b \frac{d\theta(t)}{dt} \tag{30.2}$$

$$\frac{d^2\theta(t)}{dt^2} = \frac{1}{J}T_m(t) - \frac{1}{J}T_L(t)$$

where

- K_i is the torque constant, in Nm/A.
- K_b is the back-emf constant, in V/(rad/s).
- $K_b = K_i$ (This can be easily derived based on the fact that energy goes in is equal to energy comes out).
- $i_a(t)$ is the armature current, in A.
- R_a is the armature resistance, in Ω.
- $e_b(t)$ is the back emf, in V.
- $T_L(t)$ is the load torque, in Nm.
- $T_m(t)$ is the motor torque, in Nm.
- $\theta(t)$ is the rotor displacement, in rad.
- L_a is the armature inductance, in H.
- $e_a(t)$ is the applied motor voltage, in V.
- J is the rotor inertia, in $kg - m^2$.

The armature resistance R_a and torque/back emf gain K_i are dependent on the temperature T:

$$R_a(T) = R[1 + 0.0039(T - 20)] \tag{30.3}$$

$$K_i(T) = K[1 - 0.0021(T - 20)] \tag{30.4}$$

where K is the nominal torque gain and R is the nominal resistance of the coil, both at 20°C.

The first equation in Eq. (30.2) is derived using the Kirchhoff Voltage Law: the sum of the voltage drops across the resistor, the inductor, and the back emf is equal to the applied voltage. The second equation in Eq. (30.2) is simply from the linear approximation of the torque-current curve. The third equation in Eq. (30.2) is from the fact that the back emf voltage is proportional to the angular velocity of the rotor. The fourth equation in Eq. (30.2) is derived using Newton's Second Law.

When the motor is disconnected from the voltage source, the current quickly goes to 0. As a result, the only torque applied to the motor is the load torque. Depending on the switching circuit, there may be a transient period, but the effect of the transient is negligible on the motor speed. Therefore, the motor dynamics is governed by the Newton's Second Law:

$$\frac{d^2\theta(t)}{dt^2} = -\frac{1}{J}T_L(t) \tag{30.5}$$

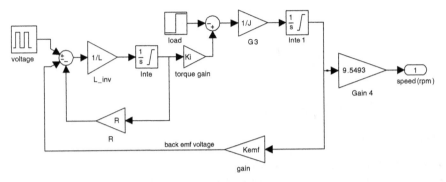

Fig. 30.2 Simulink model for a DC permanent magnetic motor

When the motor is switched from on to off or from off to on, the final rotational position of the rotor at one stage becomes the initial condition for the next stage.

Based on Eqs. (30.2)–(30.5), we build a Simulink model (Fig. 30.2). The values for the motor model parameters such as K_i, R_a, L_a, J, the temperature, and the control parameter Duty Cycle are assigned in a Matlab script file. Equations (30.3) and (30.4) are also evaluated in the same script file for simplicity. It is worth noting that system is a switched linear control system. The lower bound in the integrator "inte" is set to 0. Thus, the current transient does not create a negative torque. An interested reader can verify that the model shown in Fig. 30.2 represents the switched linear control system defined by Eqs. (30.2)–(30.5) when one is only interested in the motor speed. The model in Fig. 30.2 allows us to simulate thousands of different parameter values by using the Matlab script file for statistic analysis purpose.

It was discovered by Zhan [3] that peak-to-peak speed is mainly determined by the PWM frequency and the average speed is mainly determined by the PWM duty cycle. In this paper, the focus will be on the average speed. Thus, without loss of generality, we assume that the PWM frequency is 40 Hz.

Using the model, it can be easily shown that with the following motor/control parameter set:

- $K_i = 0.02\,\text{Nm}/\text{A}$
- $R_a = 0.1\,\Omega$
- $L_a = 1.0\text{e-}4\,\text{H}$
- $J = 9.0\text{e-}5\,\text{kg}\,\text{m}^2$
- $V = 12\,\text{V}$
- DC = 20%
- $T_{\text{load}} = 0.3\,\text{Nm}$
- $T = 20°\text{C}$

a 28.6% duty cycle gives an average speed of 3,000 rpm. The "rule of thumb" defined in Eq. (30.1) gives a duty cycle of 64.83% in order to achieve 3,000 rpm average speed. Simulation result shows that using the duty cycle suggested by the rule of thumb would result in a 4,570 rpm average speed. Clearly, the rule of thumb can have significant errors when applied to motor PWM control.

30.3 Baseline Result

The average error defined by the following formula

$$average\ error = average\ speed - target\ speed \qquad (30.6)$$

will be used as a metric for performance. To establish the baseline performance, we use the Simulink model developed in Fig. 30.2 with the following assumption on the model parameters:

- Resistance at 20°C: normal distribution, mean value $1.0e-1\,\Omega$, $\sigma = 5.0e\text{-}3\,\Omega$
- Inductance: normal distribution, mean value $1.0*e\text{-}4\,H$, $\sigma = 5.0e\text{-}6\,H$
- Inertia: normal distribution, mean value $9.0e\text{-}5\,kg\,m^2$, $\sigma = 4.5e\text{-}6\,kg\,m^2$
- Torque gain at 20°C: normal distribution, mean value $2.0e\text{-}2\,Nm/A$, $\sigma = 1.0e\text{-}3\,Nm/A$
- Voltage: normal distribution, mean value $12\,v$, $\sigma = 1.0\,V$
- Temperature: uniform distribution, range: $-10°C \sim 60°C$
- Load: $T_{load} = 0.3\,Nm$, $\sigma = 0.015\,Nm$

where σ is the standard deviation of the normal distribution.

The random values for model parameters were generated using Minitab. One thousand set of values were generated in a random manner with the specified probability distributions. A Matlab script was written to read each set of the values, run the model and record the output. The target average speed is set to be 3,000 rpm. From previous section, the duty cycle for PWM control should be set to 28.6% based on the nominal value of the motor model parameters.

The probability plot of the simulation result in Fig. 30.3 shows that the average motor speed error has a normal distribution. Using the simulation result, the baseline

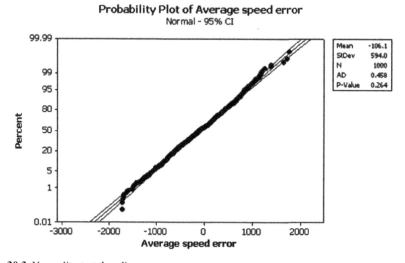

Fig. 30.3 Normality test: baseline

Summary for Average speed error: baseline

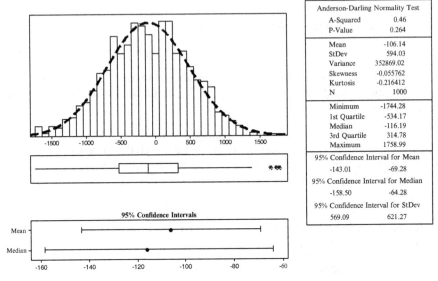

Anderson-Darling Normality Test	
A-Squared	0.46
P-Value	0.264
Mean	-106.14
StDev	594.03
Variance	352869.02
Skewness	-0.055762
Kurtosis	-0.216412
N	1000
Minimum	-1744.28
1st Quartile	-534.17
Median	-116.19
3rd Quartile	314.78
Maximum	1758.99

95% Confidence Interval for Mean
-143.01	-69.28

95% Confidence Interval for Median
-158.50	-64.28

95% Confidence Interval for StDev
569.09	621.27

Fig. 30.4 Baseline performance

performance is plotted in Fig. 30.4. The mean of the average motor speed error is equal to -106.1 rpm and standard deviation is equal to 594 rpm.

30.4 Determining the Main Factors

To understand how each parameter in the model affects the average motor speed, we conduct a DOE for PWM motor speed control. The 2-level full factorial DOE with the following variables is chosen:

- Resistance: $0.085\,\Omega, 0.115\,\Omega$
- Inductance: 8.5e-5 H, 1.15e-4 H
- Inertia: 7.65e-5 kg m^2, 1.04e-4 kg m^2
- Torque: 0.017 Nm/A, 0.023 Nm/A
- Voltage: 8 V, 16 V
- Temperature: $-10°C, 60°C$
- Duty cycle: 25%, 35%
- Load: 0.297 Nm, 0.33 Nm

The full factorial DOE is chosen because the simulation is reasonably fast. Unlike actual testing, time and cost are not a concern for the simulation. For the 2-level full factorial DOE, there are 2^8 (256) experiments to be conducted. If one uses the same approach by testing the actual hardware, other factorial design of DOE can be used to reduce the number of experiments. Replicates and randomization of runs are not

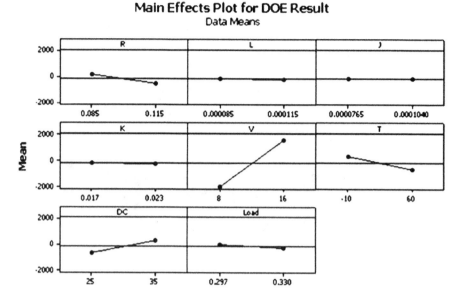

Fig. 30.5 Main effects for average motor speed variation

needed for simulation since the model gives the exactly same result for the same set
of parameters. An interested reader is referred to Taguchi [9, 10] for more details
in DOE.

Based on the simulation result from DOE, the main effects plot is shown in
Fig. 30.5. It can be seen that for the given range of the parameters, the applied voltage causes the mean to vary by the largest amount. Therefore, the applied voltage is
the top contributor to the average speed variation.

The temperature and PWM duty cycle also have significant impact on the average motor speed. The other parameters, R, L, J, K, and $Load$ have much less impact
on the average speed variation. One can also use the simulation result to find the interactions between different factors. The conclusion is similar: interactions between
voltage and other parameters are more significant than other ones.

Since temperature is usually not monitored, voltage is usually monitored, and
duty cycle is a control variable that can be adjusted if necessary, we focus on the
voltage and duty cycle. This will be justified in later sections. It is worth noting that
the top three main factors are not related to the hardware parameters. This means
tightening the tolerance bands of the design parameters for the motor is not a good
approach to solve the motor speed variation problem.

30.5 Response Surface Method

To further investigate the impact of PWM duty cycle and battery voltage on the
average motor speed, we carry out the following steps to find the response surface
(i.e., the average motor speed as a function of PWM duty cycle and battery voltage):

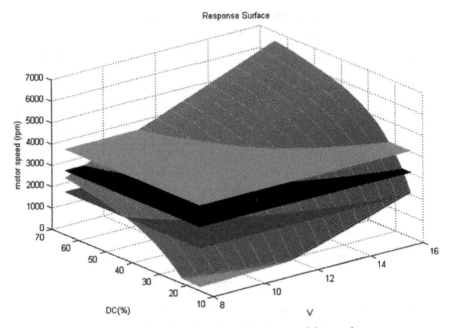

Fig. 30.6 Average motor speed as a function of applied voltage and duty cycle

- Assume the nominal value for all the motor parameters, temperature, and load torque.
- Vary the battery voltage from 8 to 16 V by incremental of 0.5 V.
- Vary the PWM duty cycle from 15% to 65% by incremental of 0.3%.
- Simulate the model and record the average speeds.

The simulation result is plotted in Fig. 30.6. Horizontal planes passing through target speed of 2,000, 3,000, and 4,000 rpm are also plotted in the same figure. It can be seen from the response surface that for any given PWM duty cycle there is a large variation to the average speed as the voltage varies.

The intersection of the response surface and one of the horizontal planes defines the relationship between Voltage and PWM duty cycle for a given target speed

$$DC = f(V) \tag{30.7}$$

Since the curve specified by Eq. (30.7) lies on a plane, we can plot it in a two dimensional plane. This process can be repeated for any target speed, the results for 2,000, 3,000, and 4,000 rpm are shown in Fig. 30.7.

The function in Eq. (30.7) can be estimated by using numerical curve fitting method. For example, if the target speed is 3,000 rpm, then Eq. (30.7) can be approximated by

$$DC = 0.0267V^4 - 1.4984V^3 + 31.734V^2 - 303.28V + 1134.6 \tag{30.8}$$

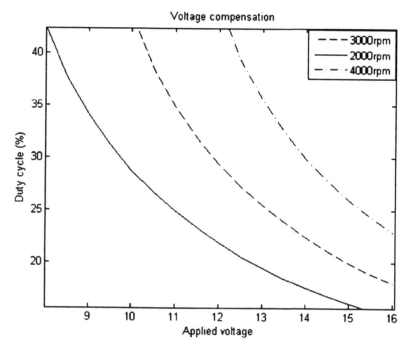

Fig. 30.7 Duty cycle as a function of voltage

If we adjust the PWM duty cycle using Eq. (30.8), then we will have a constant average motor speed of 3,000 rpm. This leads us to a new PWM control design: For any given target speed, the intersection of the constant target speed plane and the response surface is determined by off-line simulation. The result leads to a curve-fitting equation such as Eq. (30.8). These equations are coded in the controller. The duty cycle of PWM control is adjusted according to the value of the voltage being applied to the motor using these equations.

Of course the actual result will have variation since we still have the part-to-part variation and temperature, load torque variation. To compare the new design to the baseline design, we conduct the Monte Carlo analysis with the new design. The same set of randomly generated parameters are applied to the model with the PWM duty cycle replaced by the function defined in Eq. (30.8).

The small p-value in the probability plot of the simulation results in Fig. 30.8 tells us that the average motor speed error is not a normal distribution. Using the simulation result, the performance of the new design is plotted in Fig. 30.9. The mean value is reduced from −106.14 rpm for the baseline to −51.05 rpm for the new design. The standard deviation is reduced by 39% or 231 rpm. The reduction of the standard deviation is a significant improvement over the baseline design.

It is worth mentioning that the result shown in Fig. 30.9 is achieved without tightening the tolerance bands for the motor parameters. Also, the temperature and its impact on the motor parameters are taken into consideration in the Monte Carlo

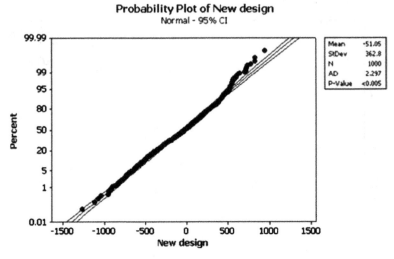

Fig. 30.8 Normality test: new design

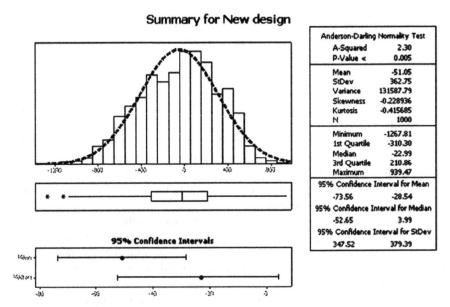

Fig. 30.9 Performance of new design

analysis, even though we decided in Section 30.4 to leave the temperature factor alone and focus on the voltage and duty cycle. The analysis shows that it was a reasonable assumption to ignore the temperature factor. If a temperature sensor is available, then hopefully we can further reduce the variation. This is still under investigation.

30.6 Conclusions

A first principle model, which takes temperature into consideration, is used to simulate the PWM control of DC permanent magnetic motors. Using DOE, the main contributing factors to the average speed variation are found. A new robust PWM control design is derived using RSM. The average speed error variation for the new design is reduced by 39% or 231 rpm compared to the baseline PWM motor speed control. This is achieved without tightening the tolerance bands for the motor parameters and monitoring of the temperature. In other words, the robustness of the PWM control is achieved without any additional cost. As one can easily see that the methodology used in this paper is not limited to motor speed control. The basic robustness design concept can be used in many other areas.

References

1. A.K. Gelig and A.N. Churilov, "Stability and Oscillations of Nonlinear Pulse-Modulated Systems", Birkhauser, Boston, MA, 1998.
2. M. Barr, "Pulse Width Modulation", Embedded Systems Programming, September 2001, 14, no. 10, 103–104.
3. W. Zhan, "Robust Design for Pulse Width Modulation Control Using Six Sigma Tools", The 12th Annual International Conference on Industry, Engineering & Management Systems, Florida, March 2007, pp. 445–452.
4. M. Harry and R. Schroeder, "Six Sigma: The Breakthrough Management Strategy Revolutionizing the World's Top Corporations", 1st edition, Doubleday, New York, 2000.
5. R. Valentine, "Motor Control Electronics Handbook", McGraw-Hill, New York, 1998.
6. W. Zhan, "Sensorless Speed Control for DC Permanent Magnetic Motors", The Ninth IASTED International Conference on Control and Applications, Montreal, May 2007, pp. 116–120.
7. G. Casella, "Monte Carlo Statistical Methods", Springer, New York, September 2004.
8. J.S. Liu, "Monte Carlo Strategies in Scientific Computing", Springer, New York, June 2001.
9. G. Taguchi, "Taguchi on Robust Technology Development", ASME Press, New York, 1993.
10. G. Taguchi, "System of Experimental Design", Unipub/Kraus/American Supplier Institute, Dearborn, MI, 1987.
11. R.H. Myers and D.C. Montgomery, "Response Surface Methodology", Wiley, New York, 1995.
12. E. del Castillo, "Process Optimization: A Statistical Approach", International Series in Operations Research & Management Science, Vol. 105, American Institute Physics, 2007.

Chapter 31
Modeling, Control and Simulation of a Novel Mobile Robotic System

Xiaoli Bai, Jeremy Davis, James Doebbler, James D. Turner, and John L. Junkins

Abstract We in the Department of Aerospace Engineering at Texas A&M University are developing an autonomous mobile robotic system to emulate six degree of freedom (DOF) relative spacecraft motion during proximity operations. The base uses an active split offset castor (ASOC) drive train to achieve omni-directional planar motion with desired tracking position errors in the ± 1 cm range and heading angle error in the $\pm 0.5°$ range. With six independently controlled wheels, we achieve a nominally uniform motor torque distribution and reduce the total disturbances with system control redundancy. A CAD (Computer-aided Design) sketch of our one-third scale model prototype is shown.

Keywords Modeling · Simulation · Mobile Robotic System · Active split offset castor · Controlled wheels

31.1 Introduction

We in the Department of Aerospace Engineering at Texas A&M University are developing an autonomous mobile robotic system to emulate six degree of freedom (DOF) relative spacecraft motion during proximity operations [1, 2]. The base uses

X. Bai (✉), J. Davis, and J. Doebbler
Graduate Research Assistant, Department of Aerospace Engineering, Texas A&M University, College Station, TX 77840, USA
e-mail: x0b9473@aeromail.tamu.edu

J.D. Turner
Research Professor, Department of Aerospace Engineering, Texas A&M University, College Station, TX 77840, USA

J.L. Junkins
Regents Professor, Distinguished Professor, Royce E. Wisenbaker '39 Chair in Engineering, Department of Aerospace Engineering, Texas A&M University, College Station, TX 77840, USA

S.-I. Ao et al. (eds.), *Advances in Computational Algorithms and Data Analysis*, Lecture Notes in Electrical Engineering 14,
© Springer Science+Business Media B.V. 2009

Fig. 31.1a Base robot
prototype

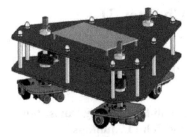

Fig. 31.1b Platform on the
base

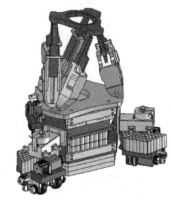

an active split offset castor (ASOC) drive train to achieve omni-directional planar motion with desired tracking position errors in the ± 1 cm range and heading angle error in the $\pm 0.5°$ range. With six independently controlled wheels, we achieve a nominally uniform motor torque distribution and reduce the total disturbances with system control redundancy. A CAD (Computer-aided Design) sketch of our one-third scale model prototype is shown in Fig. 31.1a.

A micro-positioning Stewart platform is mounted on the full-size moving base, as shown in Fig. 31.1b. For the Stewart platform, the base plate and top plate are connected by six extensible legs; the parallel nature provides higher stiffness, higher loading capacities, and higher frequency compared with typical serial positioning devices. A complete dynamic model and a robust adaptive controller for the Stewart platform have been developed using a novel automatic differentiation method in [3, 4].

The novel modeling approach makes it easy to modify the model assumptions and eliminate complicated derivative calculations. The robust adaptive control law guarantees that the tracking errors are asymptotically stable under even large parameter errors and slow-changing or constant external disturbances.

This chapter focuses on modeling and control issues for the mobile base as well as the overall system simulation. Section 31.2 presents a dynamic model for the mobile base using a Lagrangian approach. Two control laws for the base are compared. One uses an input-output feedback method to design a dynamic control law. The other uses a kinematic control method. In Section 31.3, the overall system simulation approach is described and the simulation results are shown. Conclusions are presented in Section 31.4.

31.2 Dynamics and Control of the Moving Base

Using ASOCs to provide true omni-directional mobility was introduced by H. Yu [5]. While kinematics and implementation have been discussed in [5], no dynamic models have been studied for mobile robots driven by ASOCs. In fact, there have been many papers written about solving the problem of dynamic model formulation under non-holonomic constraints using the kinematic model, but only a few papers addressing the integration of the non-holonomic kinematic controller with the dynamics of the mobile robot [6]. "Perfect velocity tracking" is always assumed in the available papers, which may not be the truth in reality [7]. This chapter formulates a rigorous approach to track the reference path while taking account of the system dynamics. The mobile base we propose in this chapter not only achieves true omni-directional motion without wheel reorientation, but also achieves higher precision control with higher loading capacity compared with the models discussed in [6–8]. Furthermore, our modeling and control approaches are very general and applicable to a significant class of similar systems.

31.2.1 Kinematic Equations

A top view representation of the entire base assembly with the frame definitions is shown in Fig. 31.2. The three vertex points of the triangular base are all pivot points, each of which is connected to a castor. The castor is connected to two wheels through shafts. Each castor is free to rotate about its pivot point and each wheel is independently driven by a mounted motor.

Kinematic equations are derived as follows. Knowing the velocity of the mass center of the triangular base together with the measured castor angles, and utilizing

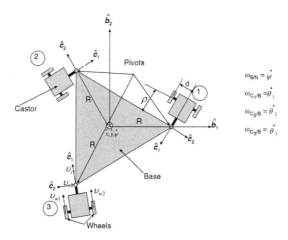

Fig. 31.2 Top view of the base

non-slippage constraints for the six wheels (the wheels roll without slip and also can not have side slip), the velocity of each wheel and shaft is uniquely defined. The velocity of the triangular base includes its translational velocity $\mathbf{v}_b = [\dot{x}, \dot{y}, 0]^\mathsf{T}$ and rotational velocity $\dot{\psi}$ about its center of mass, where x and y are the base positions in the inertial frame and ψ is the rotational angle between the base body frame and the inertial frame.

The velocity of the *ith* pivot point is computed by

$$\mathbf{v}_{p_i} = \mathbf{v}_b + \mathbf{BI}[-\dot{\psi}Rsin(\phi_i), \dot{\psi}Rcos(\phi_i), 0]^\mathsf{T} \tag{31.1}$$

where ϕ is the angle between $\hat{\mathbf{b}}_1$ and the line connecting base center of mass to the *ith* pivot point, and for the symmetric base, $\phi_1 = 0°$, $\phi_2 = 120°$, and $\phi_3 = 240°$; $\mathbf{BI}$ is the direction cosine matrix that transforms components of a vector in the $\mathbf{B}$ frame $\{\hat{\mathbf{b}}_1, \hat{\mathbf{b}}_2, \hat{\mathbf{b}}_3\}$ to the same vector with components in the inertial frame with the form

$$\mathbf{BI} = \begin{bmatrix} \cos(\psi) & -\sin(\psi) & 0 \\ \sin(\psi) & \cos(\psi) & 0 \\ 0 & 0 & 1 \end{bmatrix} \tag{31.2}$$

R is the distance from the base center of the mass to the pivot point. The inertial velocity of the pivot point is expressed in the castor $\mathbf{C}$ frame $\{\hat{\mathbf{c}}_1, \hat{\mathbf{c}}_2, \hat{\mathbf{c}}_3\}$ as $\mathbf{v}_{c_i} = [v_{c_1}, v_{c_2}, 0]$ using

$$\mathbf{v}_{c_i} = \mathbf{IC}\mathbf{v}_{p_i} \tag{31.3}$$

where

$$\mathbf{IC} = \begin{bmatrix} \cos(\psi + \theta_i) & \sin(\psi + \theta_i) & 0 \\ -\sin(\psi + \theta_i) & \cos(\psi + \theta_i) & 0 \\ 0 & 0 & 1 \end{bmatrix} \tag{31.4}$$

and θ_i is the rotational angle between the *ith* castor $\mathbf{C}$ frame and the base $\mathbf{B}$ frame. Angular velocity of the castor with respect to the inertial frame is

$$\omega_i = \dot{\theta}_i + \dot{\psi} \tag{31.5}$$

Using the no side slip constraints, the angular velocity can also be expressed as

$$\omega_i = v_{c_2}/\rho; \tag{31.6}$$

where ρ is the distance from the pivot point to the center of the wheels along $\hat{\mathbf{c}}_1$ direction.

The velocities of the two wheels are

$$v_{w_1} = v_{c_1} - d\omega_i \tag{31.7}$$
$$v_{w_2} = v_{c_1} + d\omega_i \tag{31.8}$$

where d is the distance from the pivot point to the center of the wheels along $\hat{\mathbf{c}}_2$ direction. The velocities of the shafts are solved for in the same way.

31.2.2 Dynamics and Control Law

Since we ignore the system small vertical motions in this chapter, no potential energy is involved. The energy of the mobile base is the summation of the kinetic energy of the triangular base, three castors, six shafts and six wheels. The same approach developed below will be used in the future to derive a three dimensional model, which includes the potential energy change because of the floor irregularities.

Twelve generalized coordinates

$$\mathbf{q} = [x, y, \psi, \theta_1, \theta_2, \theta_3, \Pi_1, \Pi_2, \Pi_3, \Pi_4, \Pi_5, \Pi_6]^T$$

are chosen for the base system, which includes two translational and one angular position of the triangular base, three angular positions of the castors, and six angular positions of the wheels, respectively. The corresponding twelve generalized velocities are not independent. Firstly, since Eqs. 31.5 and 31.6 describe the same quantity, we have the first set of constraints

$$w_i - v_{c_2}/\rho = 0 \tag{31.9}$$

Secondly, using the rolling without slip constraints for the wheels, we have

$$R_w \dot{\Pi}_1 - v_{w_1} = 0 \tag{31.10}$$
$$R_w \dot{\Pi}_2 - v_{w_2} = 0 \tag{31.11}$$

where R_w is the radius of the wheels. Equations 31.9–31.11 provide three constraints for each castor, leading to nine kinematic constraints in total. Since these non-slippage constraints are non-holonomic, we need to use Lagrangian multipliers to formulate the equations of motion (EOM) using 12 coordinates with nine constraints, while not trying to reduce the system to a minimal order representation by substituting the constraint relationships.

The final EOM is formulated as

$$M(\mathbf{q})\ddot{\mathbf{q}} + N(\mathbf{q}, \dot{\mathbf{q}})\dot{\mathbf{q}} + G(\mathbf{q}) = B(\mathbf{q})\mathbf{u} + C^T \lambda \tag{31.12}$$

The kinematic constraints are not dependent on time and can be expressed as

$$C\dot{\mathbf{q}} = 0 \tag{31.13}$$

where $\dot{\mathbf{q}}$ is the generalized velocity and $M(\mathbf{q})$ is the 12×12 symmetric, bounded, positive definite mass matrix; $N(\mathbf{q}, \dot{\mathbf{q}})\dot{\mathbf{q}}$ represents the centripetal and Coriolis torque, $G(\mathbf{q})$ is the gravity torque, $B(\mathbf{q})$ is the matrix that transforms the input $\mathbf{u}$ to the generalized force, C is the 9×12 constraint matrix, λ is the Lagrangian multiplier, and $C^T \lambda$ is the constraint force. It has been proven that these kind of non-holonomic dynamic systems can not achieve asymptotical stability using smooth time-invariant state feedback [9]. The stabilization methods proposed so far include discontinuous time-invariant stabilization, time-varying stabilization, and hybrid feedback [10].

To simplify the controller for real-time application, null space formulation is used to find a solution [8]. Choose $S(\mathbf{q})$ to be a 12×3 matrix, which is formed by three smooth and linearly independent vector fields spanning the null space of C, leading to

$$S^{\mathrm{T}} C^{\mathrm{T}} = 0 \tag{31.14}$$

According to Eqs. 31.13 and 31.14, since the constrained generalized velocity is always in the null space of C which is characterized by Eq. 31.13, a vector $\mathbf{v}(t) \in R^3$ can be constructed such that

$$\dot{\mathbf{q}} = S(\mathbf{q})\mathbf{v} \tag{31.15}$$

Notice that the choice of S and $\mathbf{v}$ is not unique. In general, $\mathbf{v}$ is an abstract variable and may not have any physical meaning.

Differentiating Eq. 31.15, substituting it into Eq. 31.12, and pre-multiplying the resulted equation by S^{T}, we get the transformed EOM

$$S^{\mathrm{T}} M S \dot{\mathbf{v}} + (S^{\mathrm{T}} M \dot{S} + S^{\mathrm{T}} N S)\mathbf{v} + S^{\mathrm{T}} G = S^{\mathrm{T}} B \mathbf{u} \tag{31.16}$$

The new system state space equation is

$$\dot{\mathbf{x}} = \begin{bmatrix} S\mathbf{v} \\ \mathbf{0} \end{bmatrix} + \begin{bmatrix} 0 \\ I \end{bmatrix} \tau \tag{31.17}$$

when the state-space vector is chosen as $\mathbf{x} = [\mathbf{q}^{\mathrm{T}}, \mathbf{v}^{\mathrm{T}}]^{\mathrm{T}}$, and the control input is

$$\tau = (S^{\mathrm{T}} M S)^{-1} (S^{\mathrm{T}} B \mathbf{u} - (S^{\mathrm{T}} M \dot{S} + S^{\mathrm{T}} N S)\mathbf{v} - S^{\mathrm{T}} G). \tag{31.18}$$

Since for our 6-DOF motion emulation, the requirement for the mobile base robot is to provide a satisfactory trajectory that can extend the planar workspace of the upper high precision Stewart platform, a Lyapunov method is used to design such an input-output feedback control law that guarantees the position and orientation errors of the mobile base are asymptotically stable and is simpler to apply than either discontinuous or time-varying control. The output equation is defined as

$$\mathbf{Y} = h(\mathbf{q}) = [x; y; \psi] \tag{31.19}$$

Using Eq. 31.15, the output velocity equation is expressed as

$$\dot{\mathbf{Y}} = \frac{\partial h}{\partial \mathbf{q}} S\mathbf{v} = JS\mathbf{v} = \phi\mathbf{v} \tag{31.20}$$

where $J = \frac{\partial h}{\partial \mathbf{q}}$, and ϕ is the decoupling matrix. The necessary and sufficient condition for the input-output linearization is that the decoupling matrix has full rank. For our case, the determinant of ϕ is 1, thus it never becomes singular. Using the Lyapunov method, we design a control law for τ to track reference position $\mathbf{Y}_r(t)$ and reference velocity $\dot{\mathbf{Y}}_r(t)$. The Lyapunov function is defined as

$$V = 1/2 (\mathbf{Y}_r - \mathbf{Y})^{\mathrm{T}} K_1 (\mathbf{Y}_r - \mathbf{Y}) + 1/2 (\dot{\mathbf{Y}}_r - \dot{\mathbf{Y}})^{\mathrm{T}} K_2 (\dot{\mathbf{Y}}_r - \dot{\mathbf{Y}}) \tag{31.21}$$

It is easy to prove that using the control law in Eq. 31.22, the output tracking errors are asymptotically stable

$$\tau = (JS)^{-1}(\ddot{\mathbf{Y}}_r + K_1 \mathbf{e} + K_2 \dot{\mathbf{e}} - J\dot{S}\mathbf{v} + K_3 \varepsilon) \tag{31.22}$$
$$\mathbf{e} = Y_r - Y$$
$$\dot{\mathbf{e}} = \dot{Y}_r - \dot{Y}$$
$$\dot{\varepsilon} = \mathbf{e}$$

The input torques from the six motors are solved for using Eq. 31.18. Notice that B is a 12×6 matrix for this redundantly controlled system. The pseudoinverse is used for the inverse of $S^T B$, yielding a minimal motor control effort.

31.2.3 Simulation Results

Firstly, as one of the criteria to validate the dynamic model, we numerically solve the EOM when there is no external input and then check whether the system kinematic energy remains constant or not. Theoretically, the energy should be constant. To avoid the constraint drift during integration, we utilize a constraint stabilization method which was first proposed in [11]. Dynamic response when the center of the base is doing a pure rotating motion has been validated through the simulation. The relative errors for the energy change δ_e, no side slip constraint δ_{pr} and rolling without slip constraint δ_{ns}, are calculated as follows and are shown in Figs. 31.3–31.5.

$$\delta_e = (E(t) - E(t_0))/E(t_0) \tag{31.23}$$
$$\delta_{pr} = (C_{pr}(t) - C_{pr}(t_0))/C_{pr}(t_0) \tag{31.24}$$
$$\delta_{ns} = (C_{ns}(t) - C_{ns}(t_0))/C_{ns}(t_0) \tag{31.25}$$

where $E(t)$ is the system energy at time t, $C_{pr}(t) = \dot{\theta}_1 + \dot{\psi} - v_{c_2}/\rho$ is the first castor no side slip constraint defined in Eq. 31.9, and $C_{ns}(t) = R_w \dot{\Pi}_1 - v_{w_1}$ is the first wheel rolling without slip constraint defined in Eq. 31.10.

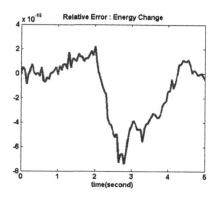

Fig. 31.3 Energy variation history

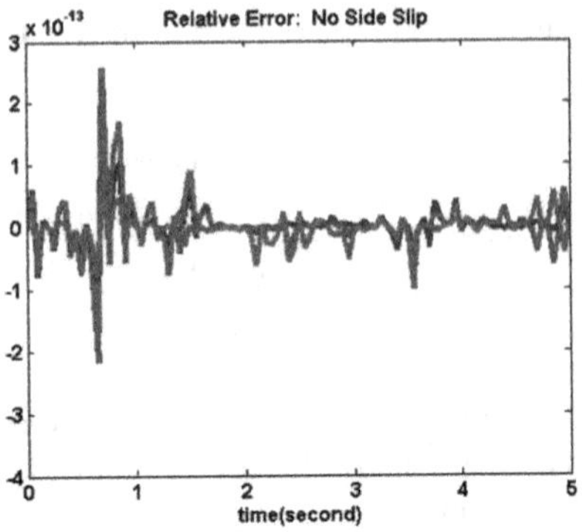

Fig. 31.4 No side slip
constraint

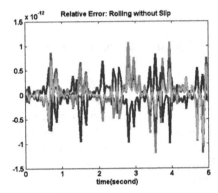

Fig. 31.5 Rolling without slip
constraint

Secondly, the mobile base is commanded to track a constant velocity reference
trajectory with initial position and velocity errors with all the wheels starting in
a pulling position. We compare the tracking results using the dynamic controller
proposed in Eq. 31.22 with a kinematic controller.

For the kinematic controller, knowing the reference trajectory of the mass center of the base, the top level controller generates a commanded base velocity
$\mathbf{V}_c = \dot{\mathbf{Y}}_r - K(\mathbf{Y} - \mathbf{Y}_r)$, where all the symbols are defined as in Eq. 31.21. Through
the unique mapping from the base center velocities to the wheel velocities, the kine-
matic controller commands the motors to track the commanded wheel velocities.

Parameters of Animatics SmartMotor 2315DT [12] are utilized in the simulation.
For the dynamic controller, we achieve the commanded torque through command-
ing a corresponding quantized motor input voltage. The kinematic controller runs
through a PID loop internal to each motor to achieve the commanded velocity.

Simulation results are shown in Figs. 31.6–31.8. We find that both controllers can
achieve satisfactory tracking while the dynamic response can be changed through
tuning gains. In addition, the dynamic controller is more efficient than the kinematic
controller in terms of the control efforts involved, although the dynamic controller
needs more computation time than the kinematic controller. For the kinematic controller, since each wheel does not have any information about other wheels, they
may fight each other during the motion. We point out that no friction models are
implemented in these simulations.

Fig. 31.6 Tracking errors comparison

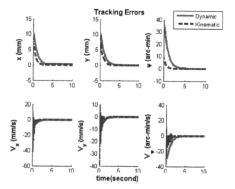

Fig. 31.7 Current comparison

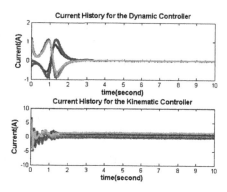

Fig. 31.8 Voltage comparison

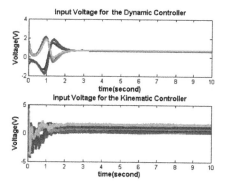

31.3 Overall System Simulation System Equations of Motion

Mobile manipulators have received significantly increased interest in the industrial, military, and public communities for their mobility combined with the manipulator's dexterous abilities. But most current work treats the system without considering dynamic interactions, and only copes with holonomic constraints, or just considers the kinematic interactions [13, 14]. Since our final aim is to do a high fidelity emulation, the previous simplifications are not acceptable to us at the dynamic modeling level. We formulate a complete model of the Stewart platform and a mobile base separately at first. For each subsystem, a robust controller is designed to account for the dynamic interaction forces. All the generalized coordinates with the dynamic interaction forces are solved simultaneously to generate the true system dynamic response.

For the 6-DOF Stewart platform system, we choose the mass center location $\mathbf{R}_c = [X_m, Y_m, Z_m]^T$, three Euler angles $\theta_m = [\theta_{1_m}, \theta_{2_m}, \theta_{3_m}]$ of the top plate, and the position and orientation of the base plate $\mathbf{q}_{b_m} = [x, y, \psi]^T$ as the generalized coordinates. Let $\mathbf{q}_m = [X_m, Y_m, Z_m, \theta_{1_m}, \theta_{2_m}, \theta_{3_m}]^T$. Note that all of these coordinates are expressed with respect to the inertial frame. The moving base is modeled using twelve generalized coordinates $\mathbf{q}_b = [x_b, y_b, \psi_b, \theta_1, \theta_2, \theta_3, \Pi_1, \Pi_2, \Pi_3, \Pi_4, \Pi_5, \Pi_6]^T$. Let $\mathbf{q}_{\bar{b}} = [x_b, y_b, \psi_b]^T$ and $\mathbf{q}_{\bar{o}} = [\theta_1, \theta_2, \theta_3, \Pi_1, \Pi_2, \Pi_3, \Pi_4, \Pi_5, \Pi_6]^T$. Since the base plate of the Stewart platform and the triangular base of the mobile robot are rigidly connected, we assume that $\mathbf{q}_{\bar{b}} = \mathbf{q}_{b_m}$. Thus for the Stewart platform, the dynamic equations are expressed as

$$\begin{bmatrix} M_{1_m} & M_{12_m} \\ M_{12_m}^T & M_{2_m} \end{bmatrix} \begin{pmatrix} \ddot{\mathbf{q}}_m \\ \ddot{\mathbf{q}}_{\bar{b}} \end{pmatrix} + \begin{bmatrix} N_1 \\ N_2 \end{bmatrix} = \begin{bmatrix} Q_{1_l} \\ Q_{2_l} \end{bmatrix} + \begin{bmatrix} 0_{6\times 6} \\ Q_d \end{bmatrix} \qquad (31.26)$$

where the mass matrix of the Stewart platform M_m is

$$M_m = \begin{bmatrix} M_{1_m} & M_{12_m} \\ M_{12_m}^T & M_{2_m} \end{bmatrix}$$

Q_{1_l} and Q_{2_l} are the generalized forces generated by the Stewart platform actuator motors and projected on the generalized coordinates $\mathbf{q}_m$ and $\mathbf{q}_{\bar{b}}$, Q_d are the interaction forces/torques projected on the $\mathbf{q}_{\bar{b}}$ coordinates, and N_1 and N_2 are other nonlinear terms. Details about solving for the compact form of the mass matrix M can be found in [3, 4, 15].

For the mobile moving base, the final organized dynamic equations, which have the Lagrangian multipliers eliminated by taking the derivatives of Eq. 31.13 and substituting Eq. 31.12 into them, are

$$\begin{bmatrix} M_{1_b} & M_{12_b} \\ M_{12_b}^T & M_{2_b} \end{bmatrix} \begin{pmatrix} \ddot{\mathbf{q}}_{\bar{b}} \\ \ddot{\mathbf{q}}_{\bar{o}} \end{pmatrix} + \begin{bmatrix} N_{1_b} \\ N_{2_b} \end{bmatrix} = \begin{bmatrix} Q_{1_b} \\ Q_{2_b} \end{bmatrix} + \begin{bmatrix} K_1 Q_d \\ K_2 Q_d \end{bmatrix} \qquad (31.27)$$

where the mass matrix of the mobile base M_b is

$$M_b = \begin{bmatrix} M_{1_b} & M_{12_b} \\ M_{12_b}^T & M_{2_b} \end{bmatrix}$$

In Eq. 31.27

$$K = C^{\mathrm{T}}(CM_b^{-1}C^{\mathrm{T}})^{-1}CM_b^{-1} - I \tag{31.28}$$

$$= \begin{bmatrix} K_1 & K_3 \\ K_2 & K_4 \end{bmatrix} \tag{31.29}$$

$$Q_{1b} = [K_1\,K_3]Q_t \tag{31.30}$$

$$Q_{2b} = [K_2\,K_4]Q_t \tag{31.31}$$

K_1 is a 9×3 matrix, K_2 is a 3×3 matrix, K_3 is a 9×9 matrix, and K_4 is a 3×9 matrix; Q_t are the generalized forces resulting from the base motors; C is the constraint matrix; I is the 12×12 identity matrix; N_{1_b} and N_{2_b} are the remaining nonlinear terms. Equations 31.26 and 31.27 can be reformulated as

$$M_t \begin{bmatrix} \ddot{q}_m \\ \ddot{q}_b \\ \ddot{q}_o \\ Q_d \end{bmatrix} + \begin{bmatrix} N_1 \\ N_2 \\ N_{1_b} \\ N_{2_b} \end{bmatrix} = \begin{bmatrix} Q_{1_t} \\ Q_{2_t} \\ Q_{1_b} \\ Q_{2_b} \end{bmatrix} \tag{31.32}$$

where

$$M_t = \begin{bmatrix} M_{1_m} & M_{12_m} & 0_{6\times9} & 0_{6\times3} \\ M_{12_m}^{\mathrm{T}} & M_{2_m} & 0_{3\times9} & -I_{3\times3} \\ 0_{9\times6} & M_{1_b} & M_{12_b} & -K_1 \\ 0_{3\times6} & M_{12_b}^{\mathrm{T}} & M_{2_b} & -K_2 \end{bmatrix}$$

Equation 31.32 is the final form to generate the overall system dynamic response.

31.3.1 System Simulation Results

A tracking scenario is designed as

$$x_b = 0.1t \tag{31.33}$$

$$x_s = 10^{-3}\sin\left(\frac{2\Pi}{0.1}t\right) + 0.1t \tag{31.34}$$

where x_b is the reference tracking position for the base robot along x direction; x_s is the reference tracking position for the Stewart platform along x direction; and the commanded motions along all the other directions for both the base robot and the Stewart platform are zero.

In the simulation, the bearing frictions and scrubbing torques are included in the dynamic model as the disturbances. The bearing frictions M_b are assumed to have the form

$$M_b = -c_f\dot{q} \tag{31.35}$$

where c_f is the bearing friction coefficient obtained from experimental data and $\dot{q}$ is the wheel velocity. Usually the scrubbing torque is defined as the torque needed to twist a single wheel around its vertical axis [5]. We use a simple way to calculate this torque M_s according to a linear form

$$M_s = kr + b \tag{31.36}$$

where rotation radius r is the distance from the center of the rotation of the two wheels to the individual wheel, k and b are the coefficients obtained by fitting experimental data using the linear equation assumption in Eq. 31.36. We also assume that if the rotation radius is larger than some limit, the scrubbing torque will be zero.

Simulation results are shown in Figs. 31.9–31.11 with both the Stewart platform control frequency and the base robot control frequency chosen as 1,000 Hz. The position errors and velocity errors satisfy the usual spacecraft rendezvous and docking motion requirements [1], and we validate our control methodology when the frequency of the Stewart platform becomes as high as 10 Hz. We point out that when the Stewart platform motion is slow, we can achieve the satisfactory tracking with lower control frequency, while when the Stewart platform frequency increases, the tracking performance gets worse. A more advanced robust control method needs to

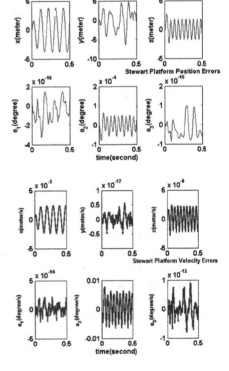

Fig. 31.9 Stewart platform position errors

Fig. 31.10 Stewart platform velocity errors

Fig. 31.11 Constraint
force/momentum

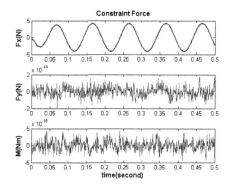

be designed in the near future to counteract the higher frequency constraint forces, which result from the faster relative motion and are treated as unknown disturbances in the control law design. In addition, since we already have equations of motion for the overall system, we may also design a control law based on the total system dynamic information. Although this control law automatically takes into account the constraint forces effect, it is very complicated and difficult to be solved. We believe our methodology to design robust control law for each subsystem in an uncoupled fashion is easier to implement, and also more robust with respect to system uncertainties, compared with designing a gigantic coupled control law based on the full system dynamic information.

31.4 Summary

We develop a high fidelity dynamic model for a mobile robotic system. Complete dynamic equations of motion are formulated and the dynamic interactions between the manipulator and the moving base are included. The rigorous approach we present in this chapter provides us with a tool to do high fidelity system analysis, control law verifications, and hardware in the loop emulation.

References

1. J. Davis, J. Doebbler, K. Daugherty, J. Junkins, and J. Valasek. Aerospace vehicle motion emulation using omni-directional mobile platform. South Carolina, Aug. 2007. AIAA Guidance, Navigation, and Control (GN&C) Conference.
2. X. Bai, J. Davis, J. Doebbler, J. Turner, and J. Junkins. Dynamics and Control of the Texas A&M Robotic Motion Emulation System. Technical report TAMU-AERO-2007-07-26-1, Texas A&M University, College Station, TX 77843-3141, July 2007.
3. X. Bai, J. D. Turner, and J. L. Junkins. Dynamic analysis and control of a stewart platform using a novel automatic differentiation method. Keystone, CO, Aug. 2006. AIAA/AAS Astrodynamics Specialist Conference.

4. X. Bai, J. D. Turner, and J. L. Junkins. Automatic differentiation based dynamic model for a mobile stewart platform. Greenwich, London, England, July 2006. 7th International Conference on Dynamics and Control of Systems and Structures in Space.

5. H. Yu, M. Spenko, and S. Dubowsky. Omni-directional mobility using active split offset castors. Journal of Mechanical Design, 126(5):822–829, Sep. 2004.

6. R. Fierro and F. L. Lewis. Control of a nonholomic mobile robot: Backstepping kinematics into dynamics. Journal of Robotic Systems, 14(3):149–163, Dec 1997.

7. Y. Kanayama, Y. Kimura, F. Miyazaki, and T. Noguchi. A stable tracking control method for an autonomous mobile robot. In Robotics and Automation, Proc. of the IEEE International Conference on, volume 1, pages 384–389, Cincinnati, OH, May 1990.

8. Y. Yamamoto and X. Yun. Coordinating locomotion and manipulation of a mobile manipulator. In Decision and Control, Proc. of the 31st IEEE Conference on, volume 3, pages 2643–2648, Tucson, AZ, Dec. 1992.

9. A.M. Bloch, M. Reyhanoglu, and N.H. McClamroch. Control and stabilization of nonholonomic dynamic systems. Automatic Control, IEEE Transactions on, 37(11):1746–1757, Nov. 1992.

10. I. Kolmanovsky and N.H. McClamroch. Developments in nonholonomic control problems. Control Systems Magazine, IEEE, 15(6):20–36, Dec. 1995.

11. J. C. Chiou and S. D. Wu. Constraint violation stabilization using inputoutput feedback linearization in multibody dynamic analysis. Journal of Guidance, Control, and Dynamics, 21(2):222–228, 1998.

12. Animatics SmartMotor 2315DT. http://www.animatics.com.

13. N. A. M. Hootsmans and S. Dubowsky. Large motion control of mobile manipulators including vehicle suspension characteristics. In Robotics and Automation, Proc. of the IEEE International Conference on, volume 3, pages 2336–2341, Sacramento, CA, April 1991.

14. J. H. Chung and S. A. Velinsky. Robust interaction control of a mobile manipulator dynamic model based coordination. Journal of Intelligent and Robotic Systems, 26(1):47–63, Sep. 1999.

15. H. Schaub and J. L. Junkins. Analytical Mechanics of Space Systems. AIAA, 2003.

Chapter 32
All Circuits Enumeration
in Macro-Econometric Models

André A. Keller

Abstract This contribution seeks to improve the knowledge of the cycle structure of macro-econometric models. This is essential for various purposes such as model building, economic analysis and simulations of economic policies. Given a standardization of the system which maps each equation to one endogenous variable, the causal structure of the resulting digraph is studied. The retained application is a recent academic model of 130 equations by Fair for the United States. The directed graph of this macro-econometric model consists of 53 strong connected components (one of them has a maximum of 77 vertices). The model is highly interdependent with 7,990 circuits, of which 8 include a maximum of 42 vertices.

Keywords Graph theory · Circuits Enumeration · Macro-Econometric Models · Cycle structure

32.1 Introduction

This contribution is an application of graph theory to the structural analysis of macroeconomic models. Applications are helpful to extend the set of theoretical aspects tackled in both fields of economic modeling and graph theory [1]. More precisely, this contribution seeks to improve the knowledge of the cycle structure of macro-econometric models. This is essential for various purposes such as model building, economic analysis and simulations of economic policies. Given a standardization of the system which maps each equation to one endogenous variable, the causal structure of the resulting digraph is studied. The retained application is a recent academic model of 130 equations by Fair [2] for the United States. The

A.A. Keller
Faculté des Sciences Economiques, Sociales et Juridiques, Université de Haute Alsace,
Mulhouse, France
e-mail: andre.keller@uha.fr

S.-I. Ao et al. (eds.), *Advances in Computational Algorithms and Data Analysis,*
Lecture Notes in Electrical Engineering 14,
465

directed graph of this macro-econometric model consists of 53 strong connected components (one of them has a maximum of 77 vertices). The model is highly inter-dependent with 7,990 circuits, of which 8 include a maximum of 42 vertices. More-over a base of 14 edge-disjoint circuits has been found. The computations have been carried out using the computer software $MATHEMATICA^{\circledR}$ 5.1 [3], its extension *DecisionAnalysis'Combinatorica* [4] and other source programs [4–8].

32.2 Elementary Circuit Theory

32.2.1 Preliminaries and Notations

A digraph of order n and size m is a pair $G = (V, E)$, where V is a finite set of vertices (with cardinality $|V(G)| = n$) and a set of ordered pairs from V (with cardinality $|E(G)| = m$). The oriented edges $e_k \in E, k = 1 \ldots, m$ ($k \in \mathbb{N}_m$) are arcs. A forward arc is such as $e_k = (x_{k-1}, x_k)$ and $e_k = (x_k, x_{k-1})$ denotes a backward arc. A chain c of G (or edge-sequence) from vertex v_0 to v_k is an alternating sequence of vertices and arcs $c = (v_0 e_1 v_1 e_2 \ldots v_{k-1} e_k v_k)$, such that $tail(e_i) = v_{i-1}$ and $head\ (e_i) = v_i$ for $i = 1, \ldots, k$ ($i \in \mathbb{N}_k$). The chain is elementary if each vertex v appears only once. A directed graph G is the triple (V, E, f) where f is a function to be defined. More precisely, a directed graph is defined when $f : E \mapsto V \times V.$[1] For $f(e) = (v_1, v_2)$, the edge e is incident out of the vertex v_1 and incident into the vertex v_2. The vertex v_1 is said adjacent to v_2. Two adjacent edges have a common vertex, such as $e_1 = (v_0, v_1)$ and $e_2 = (v_1, v_2)$ since they share v_1. Simple graphs, without self-loops and parallel edges, will be considered in this study.[2] The out-degree denotes the number of edges incident out of some vertex v, whereas the in-degree is the number of edges incident into some vertex v. The directed distance $d(u, v)$ is the length of the shortest path between the two vertices u and v.[3] The underlying undirected graph $G^0 = (V, E^0)$ of $G = (V, E)$ is obtained by ignoring the direction of the arcs. The connectivity of such an underlying graph refers to the concept of weakly connectivity. Deleting an edge e yields $G' = (V, E', f')$ where $E' = E - \{e\}$, f' being a restriction of f on E' [6]. Deleting a vertex v renders $G' = (V', E', f')$, where $V' = V - \{v\}$, $E' = \{e \in E | f(e) \neq (v, u)$ or (u, v), for $u \in V\}$ and $f' : E' \mapsto V' \times V'$ a restriction of f to E'. The $n \times n$-adjacency matrix A of G has entries

$$A_{ij} = \begin{cases} 1 & \text{if } (v_i, v_j) \in E, \\ 0 & \text{otherwise.} \end{cases}$$

---

[1] The function f for an undirected graph is defined by $f : E \mapsto \{\{u, v\} | u, v \in V\}$.

[2] Self-loops are edges e such that for one vertex v, $f(e) = (v, v)$. Parallel edges are such that $e_1 \neq e_2$ and $f(e_1) = f(e_2)$.

[3] The distance $d(u, v)$ satisfies the three properties (i) $d(u, v) = \infty$ if no path exists from u to v, (ii) $d(u, v) = 0 \Rightarrow u = v$, and (iii) the directed triangle inequality $d(u, v) + d(v, w) \leq d(u, w)$.

The adjacency list representation of an n-vertex graph consists of recording n lists, one list for each vertex containing all its adjacent vertices.

32.2.2 Circuit Definitions and Properties

A chain c is closed if its initial and terminal vertices coincide ($v_0 = v_n$). A cycle is a closed simple chain. An elementary circuit is an elementary path with identical extremal vertices, such as $(v_1, v_2), (v_2, v_3), \ldots, (v_{n-1}, v_n), (v_n, v_1)$. A circuit is isometric, if for any two vertices u and v, it contains a shortest path from u to v and a shortest path from v to u. A short circuit has only one of the two last properties. A circuit is relevant, if and only if it cannot be expressed as a linear combination of shorter circuits.

32.2.3 Enumeration Problems

There are two orientations for enumerating problems on sets of objects. Indeed, one can either estimate how many objects are in the set or look for an exhaustive finding of the present objects. Both orientations will be briefly considered here.

Counting problem

In a complete directed graph K_n the number of elementary circuits is given by

$$\sum_{i=1}^{n-1} \binom{n}{n-i+1} (n-i)!$$

This number belongs to the open interval $((n-1)!, n!)$. For the complete graph K_3, we find $\sum_{i=1}^{2} \binom{n}{4-i} (4-i)! = 5$. Moreover, we have 3 circuits of length 2 $\{(1,2,1),(2,3,2),(3,1,3)\}$ and 2 circuits of length 3 $\{(1,2,3,1),(1,3,2,1)\}$. The Fig. 32.1 shows the growth of the number of elementary circuits in the complete graphs from K_3 to K_{10}. These numbers grow faster than the exponential 2^n. Let us compute the number of circuits for each length l, $2 \leq l \leq n$ in a given complete graph $G = (V, E)$ with cardinalities $|V| = n$ and $|E| = m$.[4] The estimation of the number of circuits is given (see Appendix) by

$$\frac{n}{l} \frac{(n-1)!}{(n-l)!} \left(\frac{k}{n-1}\right)^l,$$

---

[4] For complete graphs K_n, the number of edges is given by the binomial $\binom{n}{2}$.

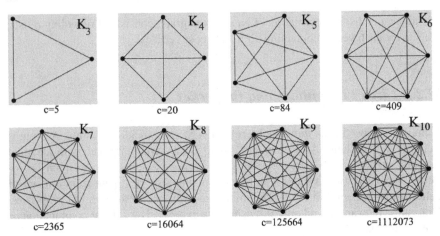

Fig. 32.1 Elementary circuits in complete graphs K_3 to K_{10}

Table 32.1 Number of circuits by length in complete graphs

Length	K_3	K_4	K_5	K_6	K_7	K_8	K_9	K_{10}
2	3	6	10	15	21	28	36	45
3	2	8	20	40	70	112	168	240
4	–	6	30	90	210	420	756	1,260
5	–	–	24	144	504	1,344	3,024	6,048
6	–	–	–	120	840	3,360	10,080	25,200
7	–	–	–	–	720	5,760	25,920	86,400
8	–	–	–	–	–	5,040	45,360	2,26,800
9	–	–	–	–	–	–	40,320	4,03,200
10	–	–	–	–	–	–	–	3,62,880
Total	5	20	84	409	2,365	16,064	1,25,664	11,12,073

where k denotes the integer part of m/n. For complete graphs K_3 to K_{10}, the number of elementary circuits by length is shown in Table 32.1. The Fig. 32.2 shows an almost-complete digraph $G = (6, 24)$. The discrepancies between the estimated and the true number of circuits have been computed and show close results. However, the errors become extremely high with larger non-complete graphs.

Enumerating problem

Following Prabhaker and Narsingh 1976 [6], every circuit enumeration algorithms to digraphs can be put into three classes of methods: search algorithms, algorithms using powers of the adjacency matrix, algorithms using line graphs (or edge-digraph). The search algorithms look for circuits in a set containing all the circuits.[5]

---

[5] The problem of counting cycles is treated by Harary and Palmer 1973 [9].

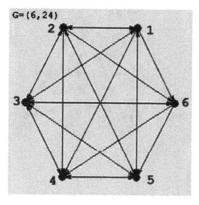

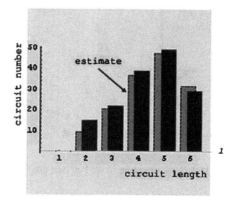

Fig. 32.2 Estimated number of the circuits of length l

In the integer powers $p \geq 1$ to $p = n$ of the adjacency matrix A, the nonzero elements A_{ij}^p are p-sequences from vertex v_i to v_j. The problem is to avoid non simple edge-sequences, where all vertices are not distinct, except for the extremities. The line graph $L(G)$ of a graph G has a vertex associated to each edge of G and an edge of $L(G)$ is drawn if and only if the two edges of G are adjacent.[6] The line graph of a p-circuit is also a p-circuit. There is a one-to-one correspondence between the circuits of G and $L(G)$.

Graph-search methods [4, 8]

The determination of some property of a graph often needs to explore all the edges and vertices in a particular way. Two traversals play an important role for algorithms: the depth-first search (DFS) and the breadth-first search (BFS). The DFS is used to solve path problems. The DFS procedure starts with a vertex and scan its neighbors, until it finds the first unexplored neighbor. The Fig. 32.3a shows an application to the Petersen graph. The DFS procedure starts from the vertex 6, visits recursively the adjacent vertices. The Fig. 32.3b is the DFS tree, with thick edges. Such spanning trees tend to have few but long branches. The thin non tree edges are the back edges. Each back edge (v, u) forms a cycle together with the path u-v in the DFS tree. The Fig. 32.3c shows a maximal list of two edge-disjoint cycles for this graph, $(4, 1, 3, 5, 2, 4)$ and $(10, 6, 7, 8, 9, 10)$.[7] The breadth-first search (BFS) explores all the adjacent vertices to the current vertex. It is the basis for finding a shortest path and the all-pairs shortest paths connecting each pair of vertices. The Fig. 32.4a shows the BFS of the Petersen graph. The edges are shaded according to

---

[6] The line graph of a complete graph of order n and size $m = n(n-1)/2$, contains m vertices and $\frac{1}{2}\sum_{i=1}^{n} d_i^2 - m$ edges, where d_i are the degrees.

[7] The DFS of a graph, whose representation is an adjacency matrix requires a running-time of $\mathcal{O}(n)$. It requires $\mathcal{O}(n+m)$ for an adjacency-list representation.

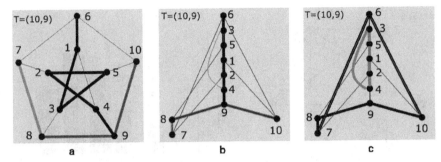

Fig. 32.3 DFS procedure to the Petersen graph

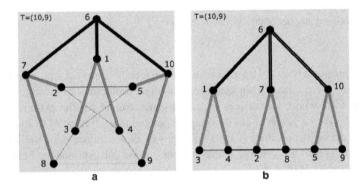

Fig. 32.4 BFS search with Petersen graph

their distance from the root vertex 6 of the Petersen graph. The thick edges show the shortest paths found by the algorithm. The BFS tree in Fig. 32.4b provides a representation of all the shortest paths from the root 6. Kocay and Kreher [5] show that the complexity of the BFS(.) is at most $\mathcal{O}(m)$.

32.2.4 Backtracking Procedure

Suppose that someone is facing with the search of a *True* option. The options *False* and *True* are organized as with the tree of Fig. 32.5. The procedure consists of starting at the root of the decision-tree, continuing until an option is found, and backtracking if necessary until the *True* option is found. In pseudo-code, the algorithm is described by the boolean function If *solve(n)* is *True*, vertex *n* is solvable. It is on a path from the root to some goal vertex and is a part of the solution. If *solve(n)* is *False*, then no path will include the vertex *n* to any goal vertex. A simple example illustrates this procedure. The visited nodes have been highlighted in Fig. 32.5.

The procedure may be

1: Start at root **1**. The decisions are nodes **2** and **7**. Choose node **2**.
2: At node **2**, the decisions are nodes **3** and **4**.

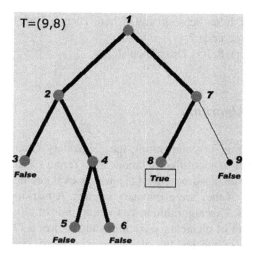

Fig. 32.5 Example of backtracking

Algorithm: BACKTRACKING PROCEDURE
input: an undirected graph $G = (V, E)$ and any arbitrary
vertex s. The tree consists of root s, leaves and a goal vertex
output: a path to a goal vertex
boolean solve(vertex s){
   if s is a leaf vertex{
      if the leaf is a goal vertex **then**{
         print s
         return *True*
      }
      else return *False*
   } **else**{
      for each vertex-child c of vertex s{
         if solve(c) succeeds **then**{
            print s
            return *True*
         }
      }
      return *False*
   }
}

3: Option **3** is *False*. Then go back to node **2**.
4: At node **2** choose node **4** (since node **3** was already tried).
5: At node **4**, try option **5**. It is *False*. Then go back to node **4**.
6: At node **4**, try option **6**. It is *False*. Then go back to node **4**.

7: Go back to node **2**. It has been already visited. Then go back to root **1**.

8: At the root **1**, choose node **7**.

9: At node **7**, try option **8**. It is *True*. Then stop.

32.2.5 Tarjan's Algorithm

The Tarjan's algorithm for enumerating the elementary circuits of a digraph is given by Tarjan [10] in algol-like notation. The Tarjan's algorithm is based on the Tiernan's backtracking procedure [11] which explores the elementary paths and checks to verify if they are elementary circuits. A marking procedure avoids unnecessary searches. This algorithm has a running-time of $\mathcal{O}(n \times m(c+1))$ where c denotes the number of elementary circuits and requires a $\mathcal{O}(n+m+S)$ space, where S is the sum of the lengths of all the elementary circuits.[8] A computer program[9] adopts these principles and rules. An adjacency-list representation is used to introduce the graph and the computer program renders a list of lexicographically ordered circuits. A main program is detecting the circuits, the subroutine SELECT is ordering each circuit from the smallest index, the subroutine CLASS is eliminating the repeated circuits by length and ordering the circuits lexicographically, the subroutine PRINT is editing the ordered circuits for each length.[10] The Fig. 32.6 shows a simplified diagram of the computer program.

32.2.6 Upper Bounds on Time and Space

Prabhaker and Narsingh 1976 [6] survey all known algorithms for enumerating all circuits in the 1970s. The Table 32.2 shows the running-time and storage performances of some algorithms using the backtrack method. The Johnson's algorithm is the fastest one, since the running time between two consecutive circuits never exceeds the size of the graph $(n+e)$. The elementary circuits are constructed from a root vertex u in the subgraph $G-u$ and vertices larger than u.[11] To avoid duplicate

---

[8] A worst-case time bound is given by Johnson 1975 [12]. The digraph $G = (3k+3, 6k+2)$ has exactly $3k$ elementary circuits with a maximum length of $2k+2$. The exploration from a given vertex takes $\mathcal{O}(k^3)$ time.

[9] Computer program BAOBAB (Fortran V,77) by A.A. Keller, Ph.D. Paris 1977. Translated (f2c) to C++.

[10] The Danut's program [13] has been adapted for this study. This computer program is for enumerating paths or circuits of a digraph, having a specified starting vertex for all possible lengths. Thus for a complete graph K_5, the computer program finds 320 circuits, whose 236 of them are circular permutations of existing circuits must be deleted. Hence, we have exactly 84 circuits in a K_5 complete graph. This computer program may be used for looking for a longest circuit.

[11] According to the Tiernan's principle [11].

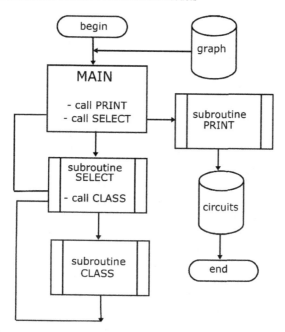

Fig. 32.6 Overview of a computer program for circuits

Table 32.2 Complexity of backtrack algorithms to circuits[a]

Author	Time bound	Space bound
Johnson 1975 [12]	$(n+e)c$	$n+e$
Tarjan 1973 [10]	$n \times e \times c$	$n+e$
Tiernan 1970 [11]	$n(const)^n$	$n+e$
Weinblatt 1972 [14]	$n(const)^n$	$n \times c$

[a]The graphs have n vertices, e edges and c circuits.

circuits, a vertex v is blocked when it is added to some elementary path beginning by u. This vertex stays blocked as long as every path from v to u intersects the current elementary path at a vertex other than u.

32.3 Structure of a Macro-Econometric Model

32.3.1 Graph Properties of the Model

The circular embedding is drawn with one of the longest circuit (the last of the eight longest circuits): big white points are those of the longest circuit, other black points denote the set of the remaining vertices of the giant strongly connected component

(*SCC*) and the remaining vertices of the graph.[12] The acyclic DAG clarifies the whole structure and classifies the variables according to their rank of influence from the top (exogenous character) to the bottom of the DAG. The resolution and the analysis of the model will take advantages from this information. Both graph embeddings are shown in Fig. 32.7. The original graph $G = (130, 396)$, with 130 vertices and 396 edges, is shown in Fig. 32.7a. The acyclic graph $H = (53, 170)$ in Fig. 32.7b has been obtained by contracting the vertices each *SCC* of G. This function runs in linear time by mapping the vertices of each *SCC* to one vertex of H. Let $G = (V, E)$ be an n-vertex graph and L a subset with k vertices to contract [4]. Contracting G gives a graph H with $n - k + 1$ vertices. Each vertex v in L is mapped to vertex $n - k + 1$ in H. Every other vertex v in G is mapped to vertex $v - i$ in H, if there are i vertices in L smaller than v. This mapping is constructed in $\mathcal{O}(n)$ time and creates the edges of H, in time proportional to the number of edges in G.

Several graph properties derive from the all-pairs shortest-path matrix. The ij-th entry is the length of a shortest path in $G = (V, E)$ or more frequently in a giant *SCC* of G between vertices i and j. Indeed the eccentricity of a vertex v is the length of the longest shortest path from v to other vertices. The radius is the smallest eccentricity of any vertex, while the center is the set of vertices whose eccentricity is the radius. On the contrary, the diameter is the maximum eccentricity of any vertex. These properties are given for the graph of the Fair model whose radius is 5 and diameter 11. The articulation vertices also play an important role. An articulation vertex is a vertex whose deletion will disconnect the graph. In a k-connected graph, there does not exist a set of $k - 1$ vertices whose deletion disconnects the graph. The properties of the Fair model are given by representing those particular vertices of the giant *SCC* in Fig. 32.8a.

The Fig. 32.8b also shows the maximum clique of the graph. Indeed, the largest clique induces a complete graph. The maximum clique has been highlighted and the economic meaning is given.

32.4 Circuits of the Digraph

The model is highly interdependent with 7,990 circuits, of which 8 include a maximum of 42 vertices. All the circuits are listed in a lexicographic order with increasing length. The Fig. 32.9 shows the number of circuits in the graph of the Fair model and their distribution according to the length of the circuits.[13]

---

[12] This embedding has been obtained applying permutation matrices, such as $P.A.P'$, where P is the permutation matrix (P' the matrix transposed) and A the adjacency matrix. The resulting digraph is isomorphic to the initial one.

[13] For this application with $G = (130, 396)$, the apparent computation time is about two hours on a personal computer which processor is AMD Athlon $3,400 + 2.2$ GHz.

G=(130,396)

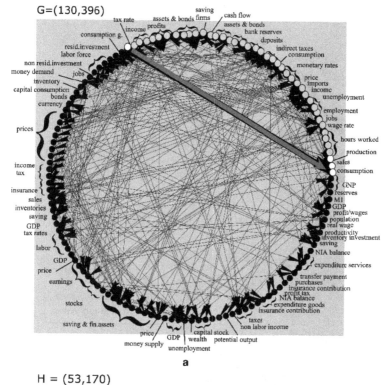

a

H = (53,170)

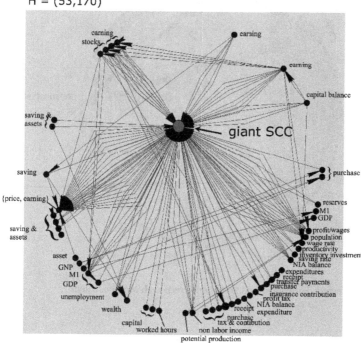

b

Fig. 32.7 a Digraph and **b** DAG of the Fair model

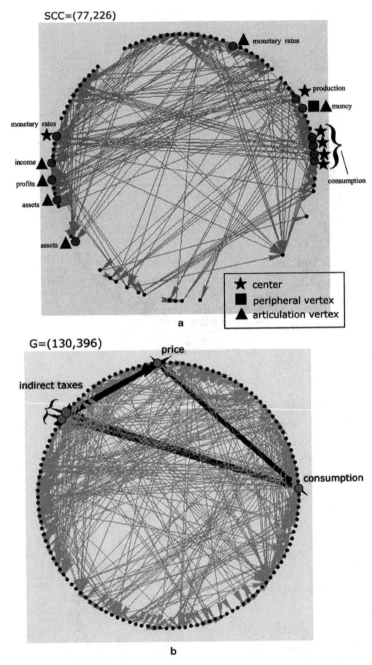

Fig. 32.8 **a** Centers, Peripherals and articulation vertices **b** Maximal clique

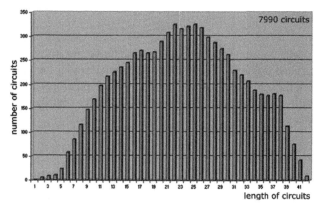

Fig. 32.9 Circuits distribution

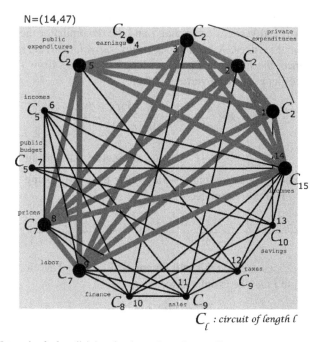

Fig. 32.10 Network of edge-disjoint circuits and maximum clique

32.5 Edge-Disjoint Circuits

The maximal list of edge-disjoint circuits for the Fair model consists of 14 circuits. The Fig. 32.10 shows the network of these circuits which are connected by an edge when the pair of circuits have at least one vertex in common. The maximum clique which is the maximum complete subgraph has been highlighted. It denotes a strong connection within those circuits.

32.6 Conclusion

This study has been concerned with the cycle structure of macro-econometric models, using associated digraphs. The highly interdependency of the model is proved when finding all the 7,990 elementary circuits. The graph properties underline the great vulnerability of the model. These results helpfully contribute to the model building process.

Appendix: Estimates of the *l*-Length Circuits Number

Let us consider an almost-complete digraph $G = (V, E)$ with cardinalities $|V| = n$ and $|E| = m$. The average number of arcs is the integer part of $k = m/n$. Starting from one vertex i, the total paths of length 1 is $\frac{n-1}{n-1}k$, where the numerator states the number of extremal vertices other than i and the denominator the total number of possible extremal vertices. At an another vertex $j, j \neq i$, there are $n - 2$ extremal vertices left. Hence, the number of paths of length 1 starting from j is $\frac{n-2}{n-1}k$. The number of paths of length 2 from vertex i is then

$$\frac{n-1}{n-1}k \times \frac{n-2}{n-1}k = \frac{(n-1)(n-2)}{(n-1)^2}k^2.$$

Hence, we have $\frac{(n-1)(n-2)\dots(n(l-1))}{(n-1)^{l-1}}k^{l-1}$ paths of length $l - 1$ starting from vertex i. The average number of arcs joining a vertex other than i to i being $\frac{k}{n-1}$, the total circuits of length l is firstly estimated to

$$\frac{k}{n-1}\frac{(n-1)(n-2)\dots(n(l-1))}{(n-1)^{l-1}}k^{l-1} = \frac{(n-1)!}{(n-l)!}(\frac{k}{n-1})^l.$$

However for n vertices each circuit will be counted l times. Therefore the number of circuits of length l may be estimated to

$$\frac{n}{l}\frac{(n-1)!}{(n-l)!}(\frac{k}{n-1})^l.$$

References

1. Keller, A.A. (2007), Graph theory and economic models : from small to large size applications, *Electr. Notes Discrete Math.*, 28:469–476.
2. Fair, R.C. (1994), *Testing Macroeconometric Models*, Cambridge, MA: Harvard University Press.
3. Wolfram, S. (2003), *The MATHEMATICA® Book*, 5th ed, Champaign Illinois, Wolfram Media Inc., http://www.wolfram.com.

4. Pemmaraju, S. and Skiena, S. (2003), *Combinatorics, and Graph Theory with Mathematica®*, Cambridge UK: Cambridge University Press, 2003.
5. Kocay, W. and Kreher, D.l. (2004), *Graphs, Algorithms, and Optimization*, New York: Chapman & Hall/CRC.
6. Prabhaker, M. and Narsingh, D. (1976), On algorithms for enumerating all circuits of a graph, *SIAM J. Comput.*, 5(1):90–99.
7. Schrijver, A. (2003), *Combinatorial Optimization : Polyedra and Efficiency*, Volume A (Paths, Flows, Matchings), Berlin: Springer.
8. Sedgewick, R. (2002), *Algorithms in C*, Part 5 (Graph Algorithms), New York: Addison-Wesley.
9. Harary, F. and Palmer, E.M. (1973), *Graphical Enumeration*, New York: Academic.
10. Tarjan, R.E. (1972), Depth-first search and linear graph algorithms, *SIAM J. Comput.*, 1: 146–160.
11. Tiernan, J.C. (1970), An efficient search algorithm to find the elementary circuits of a directed graph, *Comm. ACM*, 13:722–726.
12. Johnson, D.B. (1975), Finding all the elementary circuits of a directed graph, *SIAM J. Comput.*, 4(1):77–84.
13. Danŭt, M. (1993), On finding the elementary paths and circuits of a digraph, *Sci. Bull. P.U.B., Ser. D*, 55(3–4):29–33.
14. Weinblatt, H. (1972), A new search algorithm for finding the simple cycles of a finite directed graph, *J. Assoc. Comput. Math.*, 19:43–56.
15. Berge, C. (1985), *Graphs*, 2nd ed, New York: North-Holland.
16. Bermond, J.C. and Thomassen, C. (1981), Cycles in digraphs – a survey, *J. Graph Theor.*, 5:1–43.
17. Chartrand, G. and Lesniak, L. (2004), *Graphs & Digraphs*, 4th ed, New York: Chapman & Hall/CRC.
18. Ehrenfeucht, A., Fosdick, L.D. and Osterweil, L.J. (1973), An algorithm for finding the elementary circuits of a directed graph, University of Colorado at Boulder, Department of Computer Science, Report ♯CU-CS-024-73.
19. Erdös, P. and Pósa, L. (1965), On independent circuits contained in a graph, *Canad. J. Math*, 17:347–352.
20. Even, S. (1979), *Graph Algorithms*, Potomac, Maryland: Computer Science Press.
21. Gleiss, P.M., Leydold J. and Stadler, P.F. (2001), Circuit bases of strongly connected digraph, Department of Applied Statistics and Data Processing, Wirtschaftsuniversität Wien, *Preprint Series*, preprint 40, http://epub.wu-wien.ac.at.
22. Hopcroft J. and Tarjan, R.E. (1973), Algorithm 447: Efficient algorithms for graph manipulation, *Comm. ACM*, 16:372–378.
23. Horton, J.D. (1987), A polynomial-time algorithm to find the shortest cycle basis of a graph, *SIAM J. Comput.*, 16(2):358–366.
24. Tarjan, R.E. (1973), Enumeration of the elementary circuits of a directed graph, *SIAM J. Comput.*, 2(3):211–216.

Chapter 33
Noise and Vibration Modeling for Anti-Lock Brake Systems

Wei Zhan

Abstract A new methodology is proposed for noise and vibration analysis for Anti-Lock Brake Systems (ABS). First, a correlation between noise and vibration measurement data and simulation results need to be established. This relationship allows the engineers to focus on modeling and simulation instead of noise and vibration testing. A comprehensive ABS model is derived for noise and vibration study. The model can be set up to do different types of simulations for noise and vibration analysis. If some data is available from actual testing, then the test data can be easily imported into the model as an input to replace the corresponding part in the model. It is especially useful when the design needs to be modified, or trade-off between ABS performance and noise and vibration is necessary. The model can greatly reduce the time to market for ABS products. It also makes system level optimization possible.

Keywords Anti-Lock Brake Systems · Modeling and simulation · Noise and vibration · Vehicle systems

33.1 Introduction

Noise and vibration is one of the important measures [1] for many automotive systems such as Anti-Lock Brake Systems. It is estimated that the annual noise and vibration related cost to the automotive industry is over one billion dollars. In order to improve noise and vibration, one must be able to effectively evaluate noise and vibration for a given system. This is a challenging task since most of the noise and vibration evaluations are done in subjective manners [2]. Once a noise and vibration

W. Zhan
Department of Engineering Technology and Industrial Distribution, A&M University,
College Station, TX 77843-3367
e-mail: Zhan@entc.tamu.edu

S.-I. Ao et al. (eds.), *Advances in Computational Algorithms and Data Analysis,*
Lecture Notes in Electrical Engineering 14,

evaluation method is chosen, the evaluation could be conducted many times as design changes are made during the entire product development process. This is a very time consuming and inefficient way of designing products.

CAE is widely used for noise and vibration study. Many noise and vibration simulators are available for automotive systems. However, the brake noise and vibration modeling and simulation efforts are limited to foundation brake [3–9]. For ABS systems, noise and vibration is still being studied and evaluated mainly by testing. Usually the focus in the early development stage is on the ABS performance [10, 11]. The noise and vibration aspect is limited to simple theoretic analysis. After the desired performance is achieved, the engineers then try to improve noise and vibration based on feedback from customers and test engineers. The vast majority of the noise and vibration work for ABS systems is done in noise and vibration labs or vehicles. More often than not, it is found that there are conflicting demands from ABS performance and noise and vibration. Sometimes, the design process has to be repeated several times in order to have better overall result. The process is time consuming and costly. The result is neither reliable nor consistent.

In this paper we propose a new systematic approach for noise and vibration study for ABS systems. First, extensive analysis should be done using existing test data. A good understanding of the relationship between system components and noise and vibration needs to be established. Modeling and simulation should replace most of the actual testing and the subjective evaluation. Noise and vibration/performance optimization must be done at the system level, not at the subsystem level.

The advantage of this new approach is clear: The end result is optimized for the overall system; the project time is greatly reduced; the result is more consistent since the human factor is taken out of the evaluation process; the cost is a fraction of other methods; this also allows advanced feasibility study before any hardware/test track is available.

33.2 Modeling of ABS for Noise and Vibration Study

A typical hydraulic schematic for ABS system is depicted in Fig. 33.1 (Only one wheel is shown). When the driver applies force to the brake pedal, the master cylinder (MC) pressure increases. This causes the wheel pressure to increase. If the wheel pressure exceeds certain limit, the wheel will be locked up. A locked-up wheel would not allow for optimal lateral and longitudinal forces to achieve vehicle stability, steerability and short stopping distance. Once the ABS algorithm detects this condition, it sends a command to close the isolation valve. This prevents any brake fluid from flowing to the wheel. Next, the dump valve can be opened so that the brake fluid in the wheel can flow to the Low Pressure Accumulator (LPA). This completes a wheel pressure regulation cycle. In the next cycle, the isolation valve is pulsed open to allow small amount of fluid to flow to the wheel. The pressure

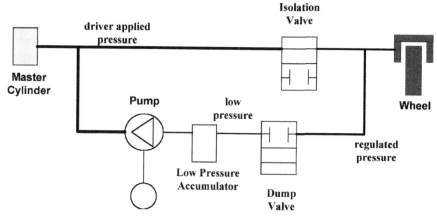

Fig. 33.1 A typical PMW control command

regulation cycles continue until the end of the ABS event. The brake fluid dumped to LPA needs to be moved back to MC so that there is enough brake fluid for the next pressure regulation cycle.

The noise is mostly caused by the "water hammering" effect when the valves are opened and closed and when the pump moves brake fluid from LPA to MC.

The vibration of the brake pedal is caused by the brake fluid movement due to the valve and pump activities. If there is more fluid flowing from MC to the wheel than what is being pumped from LPA to MC, then the driver will feel the brake pedal dropping. In the worst case, the brake pedal can drop all the way to the floor. When the pump delivers more fluid to MC than the amount of brake fluid flowing to the wheels, then the driver will feel the brake pedal rising. During an ABS event, the brake pedal moves up and down during the entire event. The brake pedal vibration or "pedal feel", a commonly terminology used by test engineers, is a measure for the quality for the brake pedal movement.

Let us first analyze the relationship between system components and noise and vibration. Noise for ABS is directly related to the pressure pulsation in MC. The MC pressure is determined by the brake pedal force applied by the driver and the pump and isolation valves activities. The pump is driven by a DC permanent magnetic motor, which is in turn controlled by the ABS algorithm. The valves are controlled by the ABS algorithm. The valve commands and the MC pressure determine the pressure at the wheels. The LPA can affect the wheel pressures in the sense that when it is full, the wheel pressure cannot be reduced [12]. The MC pressure and LPA volume determine the pump load, which directly affects the motor speed. The brake pedal vibration is determined by the MC pressure and the pump flow. The system level model is shown in Fig. 33.2.

Next we define each block in the system model. Without loss of generality, we assume the brake system has a front/rear split, i.e., the front wheels are in one MC circuit and the rear wheels are in the other MC circuit. The model can be easily modified for other types of brake systems.

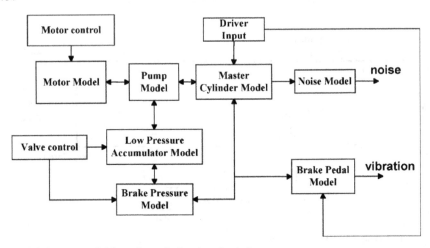

Fig. 33.2 System model for noise and vibration simulation

33.2.1 Noise and Brake Pedal

Initial tests are necessary to establish the relationship between noise and vibration and certain system parameters. This can usually be done using existing test data. We expect the fundamental structure of the relationship between the pressure, flow rate and noise and vibration to be the same for different vehicles models. The only difference between different vehicle models may be the coefficients in the model.

Therefore, once the fundamental structures of the noise and vibration models are established for one vehicle, we only need limited number of test traces to calibrate the model for noise and vibration analysis. This may seem as time consuming as the commonly used trial-and-error method. But the advantage of modeling and simulation for noise and vibration evaluation becomes apparent when some part of the system design needs to be modified. In this case, we only need to modify the model of the corresponding part and do not need to repeat all the noise and vibration tests.

One of the main contributors to noise and vibration is the pressure pulsation due to the pumping of the brake fluid from LPA to MC. We propose to analyze test data to find the correlation between the magnitude and frequency of the pressure pulsation and the noise level. The relationship between the brake pedal vibration and hydraulic fluid movement can be modeled using the test data and curve-fitting as follows

$$N = f_1(P_{MC}, F)$$

$$V = k_1 \max\{p_d(t)\} + k_2 \frac{1}{T} \int_0^T p_d(t)dt \qquad (33.1)$$

$$p_d(t) = b_1[x(t) - x(0)] + b_2\dot{x}(t)$$

where, P_{MC} is the MC pressure; T is the duration of the event; F is frequency of the P_{MC} oscillation; V is the measure of vibration of brake pedal; N is the noise measure; $x(t)$ is the brake pedal displacement; and k_1, k_2, b_1, b_2 are constants.

33.2.2 Motor and Pump

There are two positive displacement pumps in an ABS hydraulic unit. The pumps are driven by the motor through an eccentric shaft as shown in Fig. 33.3. The distance between the rotating center and the geometric center is $e(in)$. As the angle θ goes from $0°$ to $180°$, the right pump draws fluid from the low pressure pump inlet and the left pump pushes fluid out to the pump outlet. As the angle increase from $180°$ to $360°$, the left pump draws fluid while the right pump pushes out fluid. The hydraulic schematic for the positive displacement pumps used in ABS is illustrated in Fig. 33.4. The pump piston moves along the horizontal line in both directions. When the pump piston moves to the right, a vacuum is created inside the pump. In this case, the outlet check valve will be closed since the high pressure at the outlet pushes the ball toward the seat. The inlet check valve will be open since the low pressure at the inlet is higher than the vacuum inside the pump. The fluid then flows

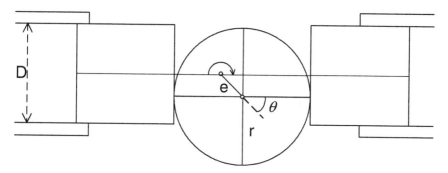

Fig. 33.3 Positive displacement pumps driven by the motor shaft

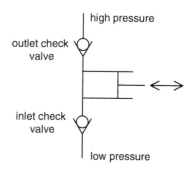

Fig. 33.4 Hydraulic schematics for a pump

from the low pressure area to the pump. The pump piston then starts to move in the opposite direction. The volume reduces, and the pressure inside the pump increases. This causes the inlet check valve to be closed. As the pump piston continues to move to the left, the pressure inside the pump will eventually be greater than the high pressure at the outlet. Thus, the outlet check valve will open and the fluid flows from the pump to the outlet. This completes a pump cycle.

The pump inlet pressure is usually very low. The resulting torque applied to the shaft is negligible during the pump intake stroke. The friction torque is also relatively small. We use a small constant value to model the friction torque. The main load to the motor comes from the pump that is pushing out fluid to the high pressure pump outlet.

When the pump is loaded, the force applied to the contact point between the pump piston and shaft is determined by

$$F = \tfrac{1}{4}\pi D^2 P_o \tag{33.2}$$

where F is the force (lb), D is the diameter (in), and P_o is the pump outlet pressure (psi). The torque from the left pump piston is thus determined by

$$T_f = 0.113 F e \sin(\theta), \text{ for } 0^\circ \le \theta < 180^\circ \tag{33.3}$$
$$T_f = 0, \text{ for } 180^\circ \le \theta < 360^\circ \tag{33.4}$$

where 0.113 is the conversion factor from lb-in to Nm. Eqs. (33.2)–(33.4) can be combined into a more compact form:

$$T_f = \frac{0.113}{8}\pi D^2 e \left(|\sin(\theta)| + \sin(\theta)\right) P_o \tag{33.5}$$

When the pump is not loaded, apparently, the torque from the front pump is zero. Therefore, Eq. (33.5) is modified as

$$T_f = P_o \varepsilon_f$$
$$\varepsilon_f = \frac{0.113}{8}\pi D^2 e \left(|\sin(\theta)| + \sin(\theta)\right) l_f \tag{33.6}$$

where l_f is determined by the front LPA volume: $l_f = 0$ if the front LPA is empty else $l_f = 1$. Similarly, the torque from the rear pump can be modeled as

$$T_r = P_o \varepsilon_r$$
$$\varepsilon_r = \frac{0.113}{8}\pi D^2 e \left(|\sin(\theta)| - \sin(\theta)\right) l_r \tag{33.7}$$

The total load to the motor is thus determined by

$$T_L = T_l + T_r + T_0 = (\varepsilon_f + \varepsilon_r)P_o + T_0 \tag{33.8}$$

where T_0 is the small friction torque that is assumed to be constant.

The motor connected to the power source can be modeled as

$$\frac{di_a(t)}{dt} = \frac{1}{L_a}e_a(t) - \frac{R_a}{L_a}i_a(t) - \frac{1}{L_a}e_b(t)$$
$$T_m(t) = K_i i_a(t)$$
$$e_b(t) = K_b \frac{d\theta(t)}{dt} \tag{33.9}$$
$$\frac{d^2\theta(t)}{dt^2} = \frac{1}{J}T_m(t) - \frac{1}{J}T_L(t)$$

where, K_i is the torque constant, in Nm/A; K_b is the back-emf constant, in V/(rad/second); $K_b = K_i$ (This can be easily derived based on the fact that energy goes in is equal to energy comes out); $i_a(t)$ is the current, in A; R_a is the coil resistance, in Ω; $e_b(t)$ is the back emf, in V; $T_L(t)$ is the load torque, in Nm; $T_m(t)$ is the motor torque, in Nm; $\theta(t)$ is the rotor displacement, in rad; L_a is the coil inductance, in H; $e_a(t)$ is the applied motor voltage, in V; and J is the rotor inertia, in kg-m^2. The first equation in Eq. (33.9) is derived using the Kirchhoff's Voltage Law: the sum of the voltage drops across the resistor, the inductor, and the back emf is equal to the applied voltage. The second equation in Eq. (33.9) is simply the linear approximation of the torque-current curve. The third equation in Eq. (33.9) is from the fact that the back emf voltage is proportional to the angular velocity of the rotor. The fourth equation in Eq. (33.9) is derived using Newton's Second Law.

When the motor is disconnected from the voltage source, the current quickly goes to 0. As a result, the only torque applied to the motor is the load torque. Therefore, the motor dynamics is governed by the Newton's Second Law:

$$\frac{d^2\theta(t)}{dt^2} = -\frac{1}{J}T_L(t) \tag{33.10}$$

When the motor is switched from on to off or from off to on, the final rotational position of the rotor at one stage becomes the initial condition for the next stage.

The pump flow rates are calculated as

$$F_{of}(t) = \varepsilon_f \frac{d\theta(t)}{dt}$$
$$F_{or}(t) = \varepsilon_r \frac{d\theta(t)}{dt} \tag{33.11}$$
$$F_o(t) = F_{of}(t) + F_{or}(t)$$

where $F_{of}(t)$ and $F_{or}(t)$ are the pump output flow rates for front and rear circuits; and $F_o(t)$ is the total flow rate from the pumps to MC.

A Simulink model (Fig. 33.5) for the pump motor subsystem is built using Eqs. (33.6)–(33.11). This is a switched nonlinear system since the motor dynamics during motor on and off periods are different as shown in Eqs. (33.9) and (33.10). It is worth noting that the nonlinearity of the system is caused by the motor load from the pump.

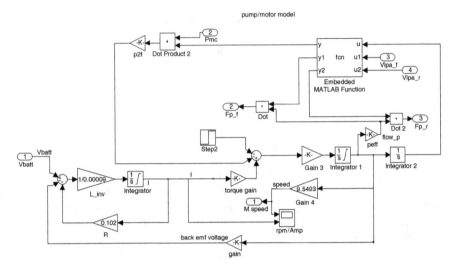

Fig. 33.5 Motor/pump model

The lower bound in the integrator "inte" is set to 0. Thus, the current transient does not create a negative torque. An interested reader can verify that the model shown in Fig. 33.5 represents the switched nonlinear control system defined by Eqs. (33.6)–(33.11) when one is only interested in the motor speed and pump flow rates.

33.2.3 Algorithm: Motor and Valve Control

The ABS algorithm calculates the commands for motor and valves in every algorithm loop. The command for the motor, which is the output of the motor speed control algorithm, is an on-off signal to connect or disconnect the motor from the power source. We can use the on-off command as an input or include the motor speed control algorithm in the model. The commands for valves can be simply on or off. The commands for valves are used by the LPA and Brake Pressure models to estimate pressure at the wheels and the volume in LPAs. These commands can be the inputs of the system model; it can also be the interface between this model and an external model such as an ABS model that includes the ABS algorithm.

33.2.4 Low Pressure Accumulator

The volumes in the LPAs are determined by the fluid coming from the wheels and the fluid going to the pumps. This can be modeled as

$$\frac{dV_{LPAf}(t)}{dt} = F_{bof}(t) - F_{of}(t)$$

$$\frac{dV_{LPAr}(t)}{dt} = F_{bor}(t) - F_{or}(t)$$

(33.12)

where $V_{LPAf}(t)$ and $V_{LPAr}(t)$ are the brake fluid volumes in the front and rear LPAs; $F_{of}(t)$ and $F_{or}(t)$ are defined in Eq. (33.11). $F_{bof}(t)$ and $F_{bor}(t)$ are the sum of the $F_{bo}(t)$ for the front wheels and rear wheels respectively, each defined in Eq. (33.16).

33.2.5 Brake Pressure

The brake pressure can be determined using the valve commands and the brake PV curve.

$$P_b(t) = PV(V_b(t)) \tag{33.13}$$

$$V_b(t) = V_{bi}(t) - V_{bo}(t) \tag{33.14}$$

$$\frac{dV_{bi}(t)}{dt} = C_1 \sqrt{P_{MC}(t) - P_b(t)} C_{appl}(t) \tag{33.15}$$

$$F_{bo}(t) = \frac{dV_{bo}(t)}{dt} = C_2 \sqrt{P_b(t)} C_{dump}(t) \tag{33.16}$$

where PV is the brake PV curve; C_1 and C_2 are the flow coefficients for the isolation/dump valves; $V_{bi}(t)$ is the volume flow into the brake and $V_{bo}(t)$ is the flow out of the brake; $C_{appl}(t)$ and $C_{dump}(t)$ are the commands for isolation and dump valves (they are digital signals equal to either 0 or 1). The Simulink model for one wheel is shown in Fig. 33.6.

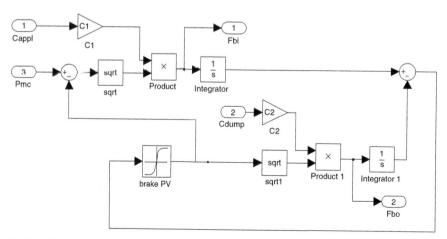

Fig. 33.6 Brake model

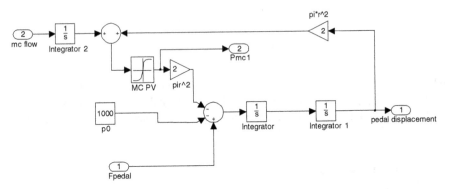

Fig. 33.7 MC model

33.2.6 Master Cylinder

MC is the main pressure source during an ABS event. We can assume the driver applies a constant force, ramp apply, spike apply, pedal pumping or any pre-determined profiles. The MC model determines the pressure head that the pump is working against.

$$m\frac{d^2x(t)}{dt^2} = F_{pedal}(t) - \pi r^2 P_{MC}(t)$$

$$P_{MC}(t) = PV_{MC}(V_{MC}(t)) \tag{33.17}$$

$$V_{MC}(t) = \int_0^t F_o(t)dt - \sum V_{bi} + \pi r^2 x(t)$$

where P_{MC} is the MC pressure, F_{pedal} is the brake pedal force applied by the driver; $V_{MC}(t)$ is the volume in MC; r is the MC piston bore radius; $x(t)$ is the MC piston displacement; and $F_o(t)$ is defined in Eq. (33.11). PV_{MC} is the PV curve of MC. The Simulink model for MC is shown in Fig. 33.7.

The system level model is shown in Fig. 33.8.

33.3 Simulation Scenarios

The model developed in the previous section can be used in many different ways for ABS noise and vibration analysis. For instance, during development testing, the LPA volume information is usually captured. In this case, the volume could be fed to the system model as an input. The brake pressure is usually captured also. In this case, the pressure calculation in Eq. (33.13) can be changed to an input.

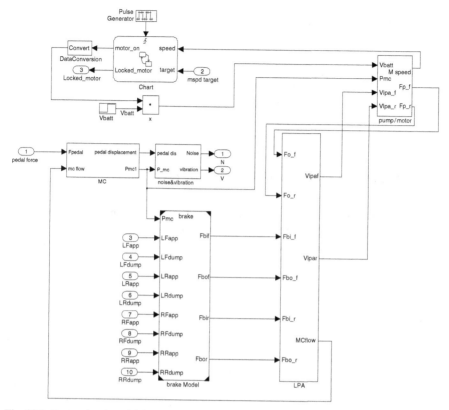

Fig. 33.8 System level model for noise and vibration study

33.3.1 Evaluating Impact of Motor Speed

Let's say two different traces of the motor speed are obtained from vehicle testing by a calibration engineer. Assuming other part of the system are the same, we can easily evaluate the noise and vibration impact of the motor speed difference. In this case, we do not need to model the motor/control. Simply replace the simulated motor speed by the actual test data in the Motor/Pump model (Fig. 33.5).

33.3.2 Evaluating Impact of Pump Size and Efficiency

If the pump diameter or efficiency is changed during the design process, we can use the system model to recalibrate the motor speed controller. The noise and vibration result can then be re-evaluated by simulating the system model.

33.3.3 Evaluating Impact of ABS Algorithm Calibration

We can use the system model to optimize the ABS algorithm control parameters. This includes the valve control and motor control among others. Typically, an ABS algorithm goes through several rounds of calibration. We can use the model to come up with noise and vibration results for different calibration sets. This would then allow us to do trade-off study between ABS performance and noise and vibration.

33.3.4 Evaluating Impact of Master Cylinder Design

Some design parameters of the MC can be varied to give a suggestion for the optimized values. During the product development process, it is not uncommon that the actual hardware is changed. The model can be used to give a quick estimation of what the impact would be.

33.4 Conclusions

It is time consuming to try different combinations of hardware/software if we have to do noise and vibration testing for each design. We propose a new approach for noise and vibration analysis using simulation. An ABS system is analyzed. A set of equations is derived for an ABS system. A simulation model is built based on these equations. The purpose of the simulation method is to provide a first cut estimation of noise and vibration results for ABS products. This method can be used in different ways, such as comparison of different controllers or trade-off study for ABS performance and noise and vibration. In addition to the simulation that can be done using the system model proposed in this paper, we can further improve the simulation capability by including more components in the model. For example, the hardware durability is a very important aspect of ABS product. Very few papers that address the performance issue together with durability and noise and vibration can be found [13]. For ABS systems, durability is mainly determined by the temperature at the semiconductor switches and the current through the motor coil. The relationship is not straightforward. We propose to use test data to create an empirical model that uses motor current as input and durability performance as the output. Eventually the whole vehicle model can be added using software such as Carsim or Matlab. Then we can do virtual testing on the desk top for overall system evaluation including performance, noise and vibration, and durability. This will be a very significant progress. We can also do hardware-in-the-loop simulation [14]. The advantage of these simulations is that we can design the test condition arbitrarily. We are still in the preliminary stage of developing this new method. A lot of work still needs to be done in order for this new approach to work. The effectiveness of this approach needs to be validated by a real system.

References

1. J. Fry, P. Jennings, R. Williams and G. Dunne, "Understanding how customers make their decision on product sound quality", Proceedings of the 33rd International Congress and Exposition on Noise Control Engineering, Prague, Czech Republic, paper 170.
2. N. Otto, S. Amman, C. Eaton and S. Lake, "Guideline for Jury Evaluations of Automotive Sounds", Proceedings of the 1999 Noise and vibration Conference, Traverse City, MI, USA, 1999-01-1802.
3. W. Nack, A.M. Joshi, "Friction Induced Vibration: Brake Moan", SAE paper 951095, 1995.
4. H. Misra, E. Johnson, L. Komzsik, "NVH Optimization on NEC Supercomputers Using MSC.Nastran", Proceedings of the 1st Worldwide MSC Aerospace Users' Conference, Long Beach, CA, June 7–10, 1999.
5. L.W. Dunne, "MSC.Nastran V68X – A Tool for NVH Response Optimization", Proceedings of the 3rd International Conference on High Performance Computing in the Automotive Industry, edited by M. Sheh, Eagan, Minn., USA, pp. 77–88, February 1997.
6. B. L. Boyle, D. G. Ebert, P. S. Gritt, J. Kubokawa, and J. Mack (eds.), "Brake Technology: ABS/TCS Systems, NVH, and Foundation Brakes", SAE, 2000.
7. A.M. Balvedi, S.N.Y. Gerges, and S. Tousi, "Identification of Brake Squeal Noise via Sound Intensity and Acoustical Measurement", Proceedings of INTER-NOISE 2002, edited by A. Selamet, R. Singh and G.C. Maling, Jr. (Dearborn, MI, August 19–21, 2002).
8. A. Papinniemi, J.C.S. Lai, J. Zhao, and L. Loader, "Brake Squeal: a literature review", Applied Acoustics, 63, 391–400 (2002).
9. K.B. Dunlap, M.A. Riehle, and R.E. Longhouse, "An Overview of Automotive Disc Brake Noise", SAE Paper 1999-01-0142 (1999).
10. S. Anwar, An anti-lock braking control system for a hybrid electromagnetic/electrohydraulic brake-by-wire system, Proceedings of the 2004 American Control Conference, Boston, Massachusetts, USA, 2004.
11. C.B. Patil, R.G. Longoria, J. Limroth, "Control prototyping for an anti-lock braking control system on a scaled vehicle", 42nd IEEE Conference on Decision and Controls Proceedings, Hawaii, USA, pp. 4962–4967, 2003.
12. H. Fennel, A. Kolbe, K. Honus, Method and Circuit Arrangement for Controlling the Flow Rate of a Hydraulic Pump. U.S. Patent No. 5,704,766, January 6, 1998.
13. D. Littlejohn, T. Fornari, G. Kuo, B. Fulmer, A. Mooradian, K. Shipp, J. Elliott, and K. Lee, "Performance, Robustness, and Durability of an Automatic Brake System for Vehicle Adaptive Cruise Control", 2004 SAE World Congress, 2004–01–0255.
14. M.W. Suh, J.H. Chung, C.S. Seok, and Y.J. Kim, "Hardware-in-the-Loop Simulation for ABS based on PC", International Journal of Vehicle Design, Vol. 24, No. 2, pp. 157–170, 2000.

Chapter 34

Investigation of Single Phase Approximation and Mixture Model on Flow Behaviour and Heat Transfer of a Ferrofluid using CFD Simulation

Mohammad Mousavi

Abstract In the present attempt, the effect of single and multiphase models on flow behaviour and thermal convection of a kerosene based ferrofluid in a cylindrical geometry is investigated. Constant temperatures on the top and bottom ends and adiabatic boundary conditions on sidewalls were applied. The domain heated from below and subjected to a uniform magnetic field parallel to temperature gradient. It was found at high solid volume fraction mixture method is more effective than single phase approach. To study the aggregation and particles diameters' effects on hydrodynamics of the system just mixture model can be applied. Using this model it was obtained that with higher magnetic particles' diameter heat transfer will decrease, and Rayleigh rolls will not be observed.

Keywords Single phase approximation · Mixture model · Heat transfer · Ferrofluid · CFD simulation

34.1 Introduction

Magnetic fluids are composed of magnetic nanoparticles and carrier fluid which is typically an oil or water base [1]. They are often treated as a homogenous colloidal suspension and modeled using standard single phase flow equations. In some conditions sedimentation and formation of particle aggregates may occur and two-phase modeling is more appropriate for system description. Convective heat transfer in magnetic fluids can be modeled using the two-phase or single phase approach. The first provides the possibility of understanding the functions of both the fluid phase

M. Mousavi
Department of Chemical Technology, Lappeenranta University of Technology, Lappeenranta, Finland
e-mail: Mohammad.Mousavi@lut.fi

S.-I. Ao et al. (eds.), *Advances in Computational Algorithms and Data Analysis*,
Lecture Notes in Electrical Engineering 14,
© Springer Science+Business Media B.V. 2009

and the solid particles in the heat transfer process. The second assumes that the fluid phase and particles are in thermal equilibrium. This approach is simpler and requires less computational time. Thus it has been used in several theoretical studies of convective heat transfer with nanofluids [2].

The mixture model is a simplified multiphase model that can be used to model multiphase flows where the phases move at different velocities, but assume local equilibrium over short spatial length scales. The coupling between the phases should be strong. It can also be used to model homogeneous multiphase flows with very strong coupling and the phases moving at the same velocity. The mixture method can model n phases (fluid or particulate) by solving the momentum, continuity, and energy equations for the mixture, the volume fraction equations for the secondary phases, and algebraic expressions for the relative velocities [3].

A particular stability problem is Rayleigh–Benard instability in a horizontal thin layer of fluid heated from below [4]. A detailed account of thermal convection in a horizontal thin layer of Newtonian fluid heated from below has been given by Chandrasekhar [5].

The investigation of instabilities of complex fluids such as ferrofluids under external fields is motivated by several important aspects. First, detection of the onset allows very precise measurements of relevant material parameters, which very often cannot be measured directly. Second, depending on the nature of the complex fluid, a host of different external fields, like electric and magnetic fields, temperature and concentration gradients, flow and mechanical fields can be used to drive the system out of equilibrium. Third, in complex fluids even first instabilities can show non-trivial features, thus allowing a rather detailed theoretical description of complicated instability types as well as a quantitative comparison with experiments. Advanced topics are concerned with the spatial structures above onset, their defects and dynamics [6].

Finlayson [7] was the first studied thermomagnetic convection instability in the presence of homogeneous vertical magnetic field. A thorough understanding of the relation between an applied magnetic field and the resulting heat transfer is necessary for the proper design and control of thermomagnetic devices [8]. The advantage of such devices is that no other source except temperature gradient is needed to achieve convection. Magnetic field can facilitate heat transfer in magnetic fluids.

Study of transport phenomena in ferrofluids involves use of computational fluid dynamics (CFD). Several recent publications have established the potential of CFD for describing the ferrofluids behavior like fluid motion and heat transfer with focus on mixture model [9–11]. The relationship between an imposed magnetic field, the resulting ferrofluid flow and the temperature distribution is not understood well enough, and the references regarding heat transfer with magnetic fluids is relatively sparse [12]. There are few simulation studies considering ferrofluids as single phase. The main objective of this work was to investigate the effects of single phase and mixture models on flow behavior and thermomagnetic convection in a simplified geometry using CFD tools.

34.2 Mathematical Formulation

In order to compare the effect of single phase approximation and mixture model on prediction of ferrofluids' behavior, both methods were studied. In the single phase approximation the ferrofluid is assumed incompressible, and the conservation equations of mass, momentum and energy are as follow:

$$\nabla \cdot \mathbf{u} = 0 \tag{34.1}$$

$$\rho_0 \left(\frac{\partial \mathbf{u}}{\partial t} + \mathbf{u}.\nabla \mathbf{u} \right) = -\nabla p + \rho\,(T)\,\mathbf{g} + \mu \nabla^2 \mathbf{u} + \mu_0\,(\mathbf{M}.\nabla)\,\mathbf{H} + \frac{\mu_0}{2} \nabla \times (\mathbf{M} \times \mathbf{H}) \tag{34.2}$$

$$\left[\rho c_{V,H} - \mu_0 \mathbf{H}. \left(\frac{\partial \mathbf{M}}{\partial T} \right)_{V,H} \right] \left(\frac{\partial T}{\partial t} + \mathbf{u}.\nabla T \right) + \mu_0 T \left(\frac{\partial \mathbf{M}}{\partial T} \right)_{V,H}.\frac{\partial \mathbf{H}}{\partial t} = k\nabla^2 T + \mu \Phi \tag{34.3}$$

where ρ, $\mathbf{u}$, t, p, T, $\mathbf{g}$, μ, μ_0, $\mathbf{M}$, $\mathbf{H}$, $c_{V,H}$, and k denote density, velocity vector, time, pressure, temperature, gravity, dynamic viscosity, magnetic permeability in vacuum, magnetization vector, magnetic field vector, heat capacity in constant volume, and thermal conductivity, respectively. Properties of the magnetic fluid except density are assumed constant. The last term in Eq. (34.2) represents dissipative which is often neglected for stationary fields [13]. Also the last term on the left hand side of Eq. (34.3) vanishes when stationary fields are applied and due to small velocities the viscous dissipation, $\mu \Phi$, may also be neglected [14]. By applying the Boussinesq approximation, $\rho(T) = \rho_0 [1 - \beta\,(T - T_0)]$, for density variation in the buoyancy term and $\mathbf{M}\,(T, \mathbf{H}) = \chi \mathbf{H}_0 - \beta_m \mathbf{M}_0\,(T - T_0)$ for magnetization, momentum and energy equations can be written as:

$$\rho_0 \left(\frac{\partial \mathbf{u}}{\partial t} + \mathbf{u}.\nabla \mathbf{u} \right) = -\nabla p + \rho_0 \beta\,(T - T_0)\,\mathbf{g} + \mu \nabla^2 \mathbf{u} + \mu_0 \beta_m \mathbf{M}_0\,(T - T_0)\,\nabla \mathbf{H} \tag{34.4}$$

$$\left(\frac{\partial \mathbf{T}}{\partial t} + \mathbf{u}.\nabla T \right) = \frac{k}{\rho c_{V,H}} \nabla^2 T \tag{34.5}$$

where β, β_m, and χ represent thermal expansion coefficient, pyromagnetic coefficient, and susceptibility, respectively. Subscript 0 represents initial conditions. The temperature dependence of β_m is:

$$\beta_m = -\frac{1}{\mathbf{M}} \left(\frac{\partial \mathbf{M}}{\partial T} \right) \tag{34.6}$$

Far from the Curie temperature and for strong fields the pyromagnetic coefficient of magnetic fluids is equal to thermal expansion coefficient [14].

Figure 34.1 illustrates three different methods to calculate susceptibility. Deviation between different theories for prediction of χ at weak field strength is the largest. In order to find susceptibility in single phase approximation and mixture

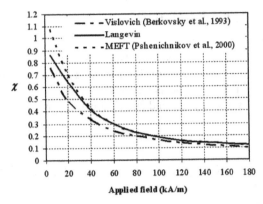

Fig. 34.1 Susceptibility of ferrofluids versus applied magnetic field calculated using three different theories

model, the modified variant of the effective field theory (MEFT) [15] and Langevin theory were used, subsequently. Vislovich [16] is an approximate method, and this theory was not uses in this study.

Mathematical formulation and the details of mixture model can be found in [1, 17].

34.3 Numerical Methods

Commercial software, Gambit 2.2 (Fluent Inc.), was used to create the geometry and generate the grid. To divide the geometry into discrete control volumes, more than 10^5 tetrahedral computational cells, 2×10^5 triangular elements, and 2.4×10^4 nodes were used. The grid is shown in Fig. 34.2.

A grid independency check has been conducted to ensure that the results are not grid dependent. To do this test, three different grids have been chosen. To do simulations in this part, it is assumed that the magnetic fluid treats as a two phase mixture of magnetic particles in a carrier phase. The mixture model uses a single fluid approach, and is an intermediate between the single phase approximation and full set of equations governing the dynamics of multiphase flow [17]. Details of grids and obtained numerical results at saturation magnetization $M_s = 48 \,\text{kA/m}$, magnetic particles' diameter $d = 5.5 \,\text{nm}$, and $\Delta T = \Delta T_{\text{critical}} = 25 \,\text{K}$ are shown in Table 34.1. In this study critical ΔT is the critical temperature difference for the natural convection due to buoyancy. To investigate heat transfer performance of the ferrofluid, the local Nusselt number, Nu, was calculated as numerical results. There was no significant variation in the Nusselt number resulted by the grid with 2.1×10^5 mesh volume and those obtained from the fine grid, so the second grid was selected for all calculations.

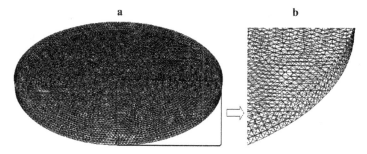

Fig. 34.2 a The grid used in simulations. **b** To obtain better visualization the highlighted part in **a** is magnified

Table 34.1 Effect of grid on Nusselt number after 200 s

Grid	#1	#2	#3
Number of nodes	1.9×10^3	2.4×10^4	5×10^4
Mesh volume	1.2×10^5	2.1×10^5	2.5×10^5
Nusselt number	4.1	1.238	1.234

The commercial code for computational fluid dynamics, Fluent, has been used for the simulations, and a user defined function was added to apply a uniform external magnetic field parallel to the temperature gradient. The finite volume method is adopted to solve three dimensional governing equations. The solver specifications for the discretization of the domain involve the presto for pressure and second-order upwind for momentum and energy in both models. In addition first-order upwind was used for volume fraction in mixture model. The underrelaxation factors, which are significant parameters affecting the convergence of the numerical scheme, were set to 0.3 for the pressure, 0.7 for the momentum, and 0.2 for the volume fraction. Using mentioned values for the under-relaxation factors a reasonable rate of convergence was achieved.

34.4 Results and Discussion

A kerosene-based magnetic fluid with magnetization 48 kA/m, particle magnetic moment 2.5×10^{-19} Am^2, vacuum permeability $4\pi \times 10^{-7}$ H/m and thermal expansion coefficient 0.00086 1/K was used in this study. Other properties are listed in Table 34.2. The magnetic fluid is in a cylinder with diameter and length 75 and 3.5 mm, respectively. Uniform external magnetic field was subjected parallel to the temperature gradient. Constant temperature boundary conditions were applied for both bottom and top of the cylinder (z direction) and sidewalls were insulated.

Both single phase approximation and mixture model show that heat transfer of the ferrofluid depends on time, and it will decrease with passing time. Strek and

Table 34.2 Properties of studied ferrofluid

Property	Single phase model	Continuous phase	Dispersed phase
Thermal conductivity	0.22 W/mK	0.149 W/mK	1 W/mK
Dynamic viscosity	0.008 kg/ms	0.0024 kg/ms	0.03 kg/ms
Heat capacity at constant pressure	3,259 J/kgK	2,090 J/kgK	4,000 J/kgK
Density	1,250 kg/m^3	$(1,248 - 1.56 \times T[K])$ kg/m^3	5,400 kg/m^3

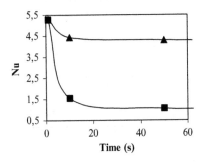

Fig. 34.3 Nusselt number versus time. Here $H = 60\,kA/m$, and $\Delta T_{critical} = 25\,K$. Triangles and squares represent $\Delta T / \Delta T_{critical} = 1.52$ and $\Delta T / \Delta T_{critical} = -1.52$, respectively. Positive values of $\Delta T / \Delta T_{critical}$ means that temperature of bottom of the cylinder is higher than the temperature of top of the cylinder

Fig. 34.4 Nusselt number versus gravitational Rayleigh number

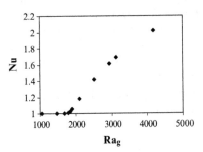

Jopek [12] also referred to this in their work. With using geometry and applying single phase method, as Fig. 34.3 represents after 10 s the system becomes stable and after that variation of Nusselt number with time is not significant. As can be expected the heat transfer is enhanced in the presence of natural convection. As Jafari et al. showed [1] with increase of aspect ratio, length to diameter of cylinder, instability in the system will increase.

The thermoconvective motion of a ferrofluid under an applied magnetic field was investigated for different conditions of gravitational and magnetic Rayleigh numbers. Ra_g and Ra_m are dimensionless parameters which provide a measure of the thermal efficiency of the investigated heating system. Figure 34.4 illustrates Nusselt number versus gravitational Rayleigh number in the absence of magnetic field and using single phase approximation. This figure corresponds to pure thermal convection and only temperature difference between top and bottom of the domain was varied. In this case warm flow due to buoyancy forces go up and cold flow come down. In addition magnetic particles are transferring toward decreasing temperatures. This was proved numerically by authors [17].

Ra_g can be defined as:

$$Ra_g = \frac{g.\beta.\Delta T.l^3}{\upsilon.\alpha} \tag{34.7}$$

where l, υ, and α represent length of the cylinder, kinematic viscosity, and thermal diffusivity, respectively. As Fig. 34.4 shows the critical value for Rayleigh number is near 1,708, so simulation results are in good agreement with theoretical value for critical Rayleigh number.

A series of simulations has been made for different magnetic Rayleigh numbers. This was done to study the influence of magnetic field on convection. Ra_m is a ratio of magnetic volume forces to viscous forces and used to characterize magnetic convection. Ra_m is defined as follow:

$$Ra_m = \frac{\mu_0.\beta_m.\mathbf{M}_{0.}\Delta T.\,l^3.\,\Delta\mathbf{H}}{\rho.\upsilon.\alpha}. \tag{34.8}$$

The presence of sufficiently strong magnetic field, changes the structure of the flow. In Fig. 34.5 the Nusselt number as a function of magnetic Rayleigh number has been presented. Comparing Figs. 34.4 and 34.5 shows that heat transfer in the case of thermomagnetic convection is more efficient than in the case of pure natural convection.

Using magnetic liquids the field gradients can be in the order of 10^4–10^7 A/m that will lead to magnetic Rayleigh numbers of the same order or even some orders of magnitude larger than the gravitational Rayleigh number. As Fig. 34.5 illustrates the critical magnetic Rayleigh number for the onset of convection is dramatically reduced as compared to the pure-fluid reference value or field free systems.

Similar simulations in the absence of natural convection and using single phase approximation have been done. Results are shown in Fig. 34.6. Comparison of obtained results shows that in the presence of natural convection heat transfer will increase. Authors represented that with applying mixture model behavior of the fluid

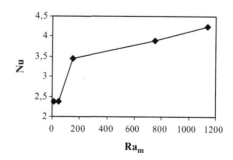

Fig. 34.5 Nusselt number versus magnetic Rayleigh number

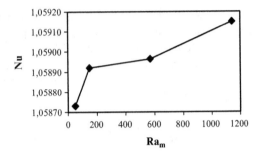

Fig. 34.6 Nusselt number versus magnetic Rayleigh number in the absence of natural convection and using single phase approximation

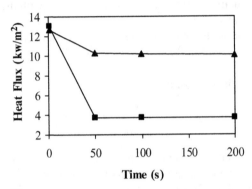

Fig. 34.7 Heat flux versus time. Triangles and squares are related to single phase approximation and mixture model, respectively

is the same as when single phase method is used [1]. In addition it was found with increase of temperature gradient the effect of natural convection will improve. Also the temperature gradient is more effective than the magnetic field on heat transfer of ferrofluids [17].

As Fig. 34.7 represents numerical results of single phase approximation and mixture method are compared to each other at $\mathbf{H} = 60\,\mathrm{kA/m}$, and temperature difference 38 K. For the mixture model it is assumed that solid volume fraction of magnetic particles is about 20%. Effect of small volume fraction of magnetic particles ($\leq 10\%$) has been investigated by authors [17], so here the amount of solid particles was increased in order to study its effect. As this figure shows in solid volume fraction greater than 10% mixture model is more suitable for prediction of ferrofluids' behavior because prediction of single phase approach is more than experimental results [18]. Comparing numerical results with what reported by authors [1] illustrate that at low solid volume fraction or very small diameter of magnetic particles, single phase approximation could be well adopted.

Convection patterns of the magnetic fluid at $\Delta T / \Delta T_{critical} = 1.52$, $\mathbf{H} = 120\,\mathrm{kA/m}$, and using single phase model at different times are shown in Fig. 34.8. Disordered convection rolls spontaneously appear and disappear. Similar behaviors have been

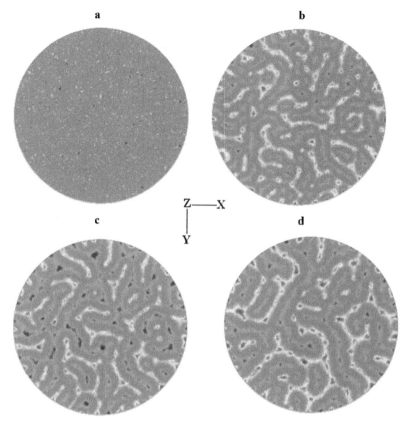

Fig. 34.8 Convection pattern in the ferrofluid at different times on a plane at $z = 0.00175$ m. **a** 1 s, **b** 50 s, **c** 100 s, and **d** 200 s

observed experimentally [18]. Number of Rayleigh rolls depends on different parameters such as temperature difference, magnetic field strength, its direction, and geometry (aspect ratio). Any variation of these quantities can induce a change in fluid behavior.

The temperature oscillations recorded corresponded to Fig. 34.8 are presented in Fig. 34.9. As it is visible, temperature signals contains high and low frequency oscillations. As Tynjälä [14] showed using wavelet analysis, temperature signals are more evident.

To study the effect of natural convection more simulations were performed in the conditions where temperature of top of the cylinder was higher than temperature of bottom of the geometry. Convection pattern and change of temperature on a plane at $z = 0.00175$ m are shown in Fig. 34.10 a and b. Comparison of Fig. 34.10 with Figs. 34.8 and 34.9 illustrate that in the presence of buoyancy forces because of formation of rolls and circulation of the fluid in the system, transport phenomena will improve. Similar treatment has been observed using mixture method [1].

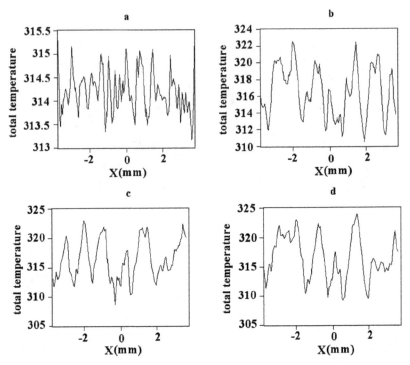

Fig. 34.9 Temperature oscillation of the ferrofluid at different times on a plane at z = 0.00175 m.
a 1 s, **b** 50 s, **c** 100 s, and **d** 200 s

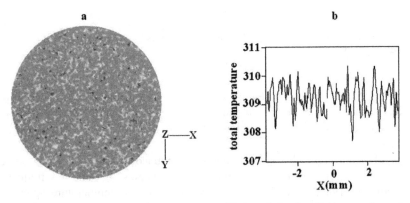

Fig. 34.10 Convection pattern and temperature oscillation of the ferrofluid on a plane at z =
0.00175 m after 200 s and in the absence of natural convection

According to the estimation of Rosensweig [19] particle size 10 nm for magnetic
fluids is on the threshold of agglomerating. It means that the investigation of effect
of particle's diameter on heat transfer of the ferrofluid is important. This effect is
studied here using mixture method. It is clear that in single phase method it is not

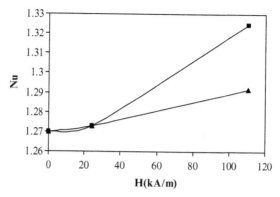

Fig. 34.11 Nusselt number versus magnetic field strength. Here $\Delta T/\Delta T_{critical} = 1.52$, and $\Delta T_{critical} = 25K$. Triangles and squares belong to d = 30 nm and d = 5.5 nm, respectively

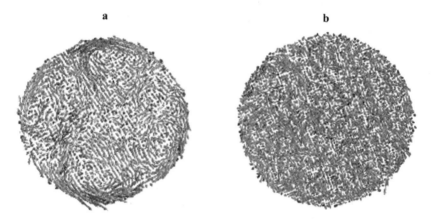

Fig. 34.12 Velocity vectors on a plane at z = 0.00175 m, and for $\Delta T_{critical} = 25K$ and $\Delta T/\Delta T_{critical} = 0.76$. **a** d = 5.5 nm, and **b** d = 30 nm

possible to study effect of magnetic particles' diameter because the whole system treats as one fluid. Numerical results related to effect of particles' diameter on heat transfer are shown in Fig. 34.11. According to this figure larger particles decrease the heat transfer probably because of the formation of aggregates in the system.

In order to obtain better understanding the effect of particle size on transfer phenomena flow pattern of the ferrofluid also was studied. As Fig. 34.12a illustrates in the presence of natural convection for smaller magnetic particles' diameter Rayleigh rolls can be observed. This rotations increase heat transfer, and their effect on thermal convection in ferrofluids is important in certain chemical engineering and biochemical situations [4]. For $d = 30$ nm such kind of vortices did not appear (Fig. 34.12b).

34.5 Conclusion

Hydrodynamics behaviour and thermomagnetic convection in a kerosene based ferrofluid have been studied using computational fluid dynamics. Both single and multiphase models were investigated, and obtained results show that both methods can
predict the behaviour of the system qualitatively. But using two-phase approach,
gives significant information on the behaviour of ferrofluids. For example effect
of particle's diameter on behaviour of the system can not be studied using single
phase approach. It was found that using both methods heat transfer of the ferrofluid
depends on time and it will decrease with passing time. In addition results show that
heat transfer in the presence of natural convection is more efficient. Obtained results
from the thermoconvective motion of a ferrofluid under an applied magnetic field
for different conditions of gravitational and magnetic Rayleigh numbers represent
that critical magnetic Rayleigh number for the onset of convection will dramatically
reduce compare to the pure-fluid reference value or field free systems.

References

1. A. Jafari, T. Tynjälä, S. M. Mousavi, and P. Sarkomaa, CFD simulation of heat transfer in
   ferrofluids, *Int. J. Heat Fluid Fl.* DOI: 10.1016/j.ijheatfluidflow.2008.01.007 (2008).
2. S. Mirmasoumi, and A. Behzadmehr, Numerical study of laminar mixed convection of a
   nanofluid in a horizontal tube using two-phase mixture model, *App. Therm. Eng.* **28**, 717–727
   (2008).
3. Fluent, *Fluent 6.2 users guide* (Fluent, Lebanon, 2007).
4. Sunil, P. Chand, P. K. Bharti, and A. Mahajan, Thermal convection in micropolar ferrofluid in
   the presence of rotation, *J. Magn. Magn. Mat.* **320**, 316–324 (2008).
5. S. Chandrasekhar, *Hydrodynamic and Hydromagnetic Stability* (Dover, New York, 1981).
6. H. Pleiner, *Instabilities in Complex Fluids* (March 5, 2007); http://www.mpip-mainz.mpg.de/
   ~pleiner/instab.html.
7. B. A. Finlayson, Convection instability of ferromagnetic fluids, *J. Fluid Mech.* **40**, 753–767
   (1970).
8. A. Mukhopadhyay, R. Ganguly, S. Sen, and I. K. Puri, A scaling analysis to characterize
   thermomagnetic convection, *Int. J. Heat Mass Tran.* **48**, 3485–3492 (2005).
9. S. M. Snyder, T. Cader, and B. A. Finlayson, Finite element model of magnetoconvection of a
   ferrofluid, *J. Magn. Magn. Mat.* **262**, 269–279 (2003).
10. T. Tynjälä, A. Hajiloo, W. Polashenski, and P. Zamankhan, Magnetodissipation in ferrofluids,
    J. Magn. Magn. Mat. **252**, 123–125 (2002).
11. C. Tangthieng, B. A. Finlayson, J. Maulbetsch, and T. Cader, Heat transfer enhancement in
    ferrofluids subjected to steady magnetic fields, *J. Magn. Magn. Mat.* **201**, 252–255 (1999).
12. T. Strek and H. Jopek, Computer simulation of heat transfer through a ferrofluid,*Phys. Stat.
    Sol. (b)*, DOI: 0.1002/pssb.200572720, 1–11 (2007).
13. H. W. Muller, and M. Liu, Structure of ferrofluid dynamics, *Phys. Rev. E* **64**, 061405–1–
    061405–7 (2001).
14. T. Tynjälä, *Theoretical and Numerical Study of Thermomagnetic Convection in Magnetic Flu-
    ids*, Ph.D. Thesis (Lappeenranta University of Technology Press, Finland, 2005).
15. A. F. Pshenichnikov and V. V. Mekhonoshin, Equilibrium magnetization and microstructure of
    the system of superparamagnetic interacting particles: numerical simulation, *J. Magn. Magn.
    Mat.* **213**, 357–369 (2000).

16. B. M. Berkovsky, V. F. Medvvedev, and M. S. Kravov, *Magnetic Fluids Engineering Applications* (Oxford Science Publication, Oxford, 1993).
17. A. Jafari, T. Tynjälä, S. M. Mousavi, and P. Sarkomaa, CFD simulation and evaluation of controllable parameters effect on thermomagnetic convection in ferrofluids using Taguchi technique, *Comput. Fluids*, DOI: 10.1016/j.compfluid.2007.12.003 (2008).
18. A. A. Bozhko, and G. F. Putin, Heat transfer and flow patterns in ferrofluid convection, *Magnetohydrodynamics* **39**, 147–169 (2003).
19. R. E. Rosensweig, *Ferrohydrodynamics* (Dover, New York, 1997).

Chapter 35
Two Level Parallel Grammatical Evolution

Pavel Ošmera

Abstract This paper describes a Two Level Parallel Grammatical Evolution (TLPGE) that can evolve complete programs using a variable length linear genome to govern the mapping of a Backus Naur Form grammar definition. To increase the efficiency of Grammatical Evolution (GE) the influence of backward processing was tested and a second level with differential evolution was added. The significance of backward coding (BC) and the comparison with standard coding of GEs is presented. The new method is based on parallel grammatical evolution (PGE) with a backward processing algorithm, which is further extended with a differential evolution algorithm. Thus a two-level optimization method was formed in attempt to take advantage of the benefits of both original methods and avoid their difficulties. Both methods used are discussed and the architecture of their combination is described. Also application is discussed and results on a real-word application are described.

Keywords Parallel Grammatical Evolution · Variable length linear genome · Two-level optimization method · Backward coding

35.1 Introduction

Optimization is an important aspect of many scientific and engineering problems. Recent optimization techniques model principles of natural evolution. Evolutionary algorithms apply selection and mutation operators to a population of states to guide the population to an optimal solution of the objective function.

Classic optimization methods often lead to unacceptably poor performance when applied to real-world circumstances. A more robust optimization technique is required. Applying the logical aspects of the evolutionary process to optimization

P. Ošmera

Institute of Automation and Computer Science, Brno University of Technology, Technicka 2, 616 69 Brno, Czech Republic,
e-mail: osmera@.fme.vutbr.cz

S.-I. Ao et al. (eds.), *Advances in Computational Algorithms and Data Analysis,*
Lecture Notes in Electrical Engineering 14,
© Springer Science+Business Media B.V. 2009

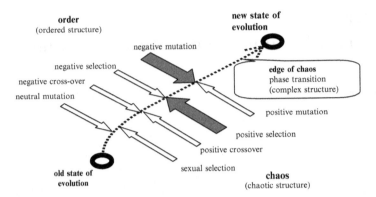

Fig. 35.1 Evolution on the edge of chaos

offers several distinct advantages. There exists a large body of knowledge about the process of natural evolution that can be used to guide simulations [1–4] (see Fig. 35.1). This process is well suited for solving problems with unusual constrains where heuristic solutions are not available or generally lead to unsatisfactory results. Often revolution has an interdisciplinary character. Its central discoveries often come from people straying outside the normal bounds of their specialties.

35.2 Overview of Evolutionary Algorithms

Evolutionary computation is generally considered as a consortium of genetic algorithms (GA), genetic programming (GP), evolutionary strategies (ES). There are new GPs: grammatical evolution, [5] grammatical swarm [6] and chemical genetic programming [7].

Differential evolution (DE) [8] is a rather unknown approach to numerical optimization, which is very simple to implement and requires little or no parameter tuning. After generating and evaluating an initial population the solutions are refined by a *DE* operator as follows. Choose for each individual genome *j* three individuals *k*, *l*, and *m* randomly from the population. Then calculate the difference of the chromosomes in *k* and *l*, scale it by multiplication with a parameter f and create an offspring by adding the results to the chromosome of *m*. The only additional twist in this process is that not the entire chromosome of offspring is created in this way, but that genes are partly inherited from individual *j*.

There are another Soft Computing methods [9]:

Multi-Agent Evolutionary Algorithms
Agent-based Multi-objective Optimization
Ant Colony Optimization, Team Optimization, Culture Algorithms
Fuzzy logic, Neural Networks, Fuzzy-rough Sets, Fuzzy-Neural Modeling
Hybrid Learning, Intelligent Control, Cooperative Co-evolutionary Algorithms
Parasitic Optimization, Bacterial Evolutionary Algorithms (BEA)

Artificial Immune Algorithms, Artificial Life Systems
Parallel Hierarchical Evolutionary Algorithms
Meta-Heuristics, Evolutionary Multi-objective Optimization
Evolvable control, Embryonic Hardware
Human-Computer Interaction, Molecular-Quantum Computing
Data Mining, Chaotic Systems, Scheduling, etc.

35.3 Parallel Grammatical Evolution

Grammatical Evolution (GE) [5] can be considered a form of grammar-based genetic programming (GP). In particular, Koza's genetic programming has enjoyed
considerable popularity and widespread use. Unlike a Koza-style approach, there is
no distinction made at this stage between what he describes as function (operator
in this case) and terminals (variables). Koza originally employed Lisp as his target
language. This distinction is more of an implementation detail than a design issue.
Grammatical evolution can be used to generate programs in any language, using
Backus Naur Form (BNF). BNF grammars consist of terminals, which are items
that can appear in the language, i.e. $+$, $-$, sin, log etc. and non-terminal, which can
be expanded into one or more terminals and non-terminals. A non-terminal symbol is any symbol that can be rewritten to another string, and conversely a terminal
symbol is one that cannot be rewritten. The major strength of GE with respect to
GP is its ability to generate multi-line functions in any language. Rather than representing the programs as parse tree, as in GP, a linear genome representing is used.
A genotype-phenotype mapping is employed such that each individual's variable
length byte strings, contains the information to select production rules from a BNF
grammar. The grammar allows the generation of programs, in an arbitrary language
that are guaranteed to be syntactically correct. The user can tailor the grammar to
produce solutions that are purely syntactically constrained, or they may incorporate domain knowledge by biasing the grammar to produce very specific form of
sentences.

GE system [5] codes a set of pseudo random numbers, which are used to decide
which choice to take when a non-terminal has one or more outcomes. Because GE
mapping technique employs a BNF definition, the system is language independent,
and, theoretically can generate arbitrarily complex functions. There is quite an unusual approach in GEs, as it is possible for certain genes to be used two or more
times if the wrapping operator is used. BNF is a notation that represents a language
in the form of production rules. It is possible to generate programs using the Grammatical Swarm Optimization (GSO) technique [6] with a performance similar to the
GE. Given the relative simplicity of GSO, the small population sizes involved, and
the complete absence of a crossover operator synonymous with program evolution
in GP or GE. Grammatical evolution was one of the first approaches to distinguish
between the genotype and phenotype. GE evolves a sequence of rule numbers that
are translated, using a predetermined grammar set into a phenotypic tree.

35.4 Parallel Grammatical Evolution with Backward Processing

Our approach uses a parallel structure of GE (PGE) [9–11]. A population is divided into several subpopulations that are arranged in the hierarchical structure. Every subpopulation has two separate parts: a male group and a female group. Every group uses quite a different type of selection. In the first group a classical type of GA selection is used. In the second group only different individuals can be included. It is a biologically inspired computing similar to a harem arrangement. This strategy increases an inner adaptation of PGE. The following text explains why we used this approach. Analogy would lead us one step further, namely, to the belief that the combination of GE with a sexual reproduction. On the principle of the sexual reproduction we can create a parallel GE with a hierarchical structure.

35.4.1 Backward Processing of the GE

The PGE is based on the grammatical evolution GE [5] where BNF grammars consist of terminals and non-terminals. Terminals are items, which can appear in the language. Non-terminals can be expanded into one or more terminals and non-terminals. Grammar is represented by the tuple $\{N, T, P, S\}$, where N is the set of non-terminals, T the set of terminals, P a set of production rules which map the elements of N to T, and S is a start symbol which is a member of N. For example, below is the BNF used for our problem:

$$N = \{expr, \ fnc\}$$
$$T = \{sin, \ cos, \ +, \ -, \ /, \ *, \ X, \ 1, \ 2, \ 3, \ 4, \ 5, \ 6, \ 7, \ 8, \ 9\}$$
$$S = <expr>$$

and P can be represented as four production rules:

1. $<expr> := <fnc><expr>$
   $\qquad\qquad <fnc><expr><expr>$
   $\qquad\qquad <fnc><num><expr>$
   $\qquad\qquad <var>$
2. $<fnc> := \quad sin$
   $\qquad\qquad\quad cos$
   $\qquad\qquad\quad +$
   $\qquad\qquad\quad *$
   $\qquad\qquad\quad -$
   $\qquad\qquad\quad U-$
3. $<var> := X$
4. $<num> := 0, 1, 2, 3, 4, 5, 6, 7, 8, 9$

The production rules and the number of choices associated with each are in Table 35.1. The symbol $U-$ denotes an unary minus operation.

Table 35.1 The number of available choices for each production rule

Rule no	Choices
1	4
2	6
3	1
4	10

There are notable differences when compared with the approach [12]. We don't use two elements *<pre_op>* and *<op>*, but only one element *<fnc>* for all functions with *n* arguments. There are not rules for parentheses; they are substituted by a tree representation of the function. The element *<num>* and the rule *<fnc> <num> <expr>* were added to cover generating numbers. The rule *<fnc> <num> <expr>* is derived from the rule *<fnc> <expr> <expr>*. Using this approach we can generate the expressions more easily. For example when one argument is a number, then $+(4,x)$ can be produced, which is equivalent to $(4 + x)$ in an infix notation. The same result can be received if one of *<expr>* in the rule *<fnc> <expr> <expr>* is substituted with *<var>* and then with a number, but it would need more genes.

There are not any rules with parentheses because all information is included in the tree representation of an individual. Parentheses are automatically added during the creation of the text output.

35.4.2 Reduction of the Search Space

If in the GE is not restricted anyhow, the search space can have infinite number of solutions. For example the function $cos(2x)$, can be expressed as $cos(x+x)$; $cos(x+x+1-1)$; $cos(x+x+x-x)$; $cos(x+x+0+0+0...)$ etc. It is desired to limit the number of elements in the expression and the number of repetitions of the same terminals and non-terminals.

In our system every application of an element from the set will decrease its number of possible uses (see Table 35.2). In the case, when this counter is 0, the element is removed from the set. It means that the number of production rules is decreased and the gene value will have quite a different influence on the result.

Consider two genes with values 20 and 81, without reducing the space the results would be:

$$20 \bmod 4 = 0 => A$$
$$81 \bmod 4 = 1 => B$$

When rule A is temporarily removed from the set:

$$20 \bmod 3 = 2 => D$$
$$81 \bmod 3 = 0 => B$$

Table 35.2 Number of available choices for each production rule

Rule	All choices	Available choices
A	2	0
B	5	3
C	6	3
D	3	3

Table 35.3 Sorted table of available choices for each production rule

Rule	All choices	Available choices
C	6	3
B	5	3
D	3	3
A	2	0

When a rule is removed from the set the result of modulo operation codes a different rule. Result 0 codes rule A before reduction and rule B after reduction. Therefore it is better to have a sorted list of rules (see Table 35.3).

Using the sorted list without reduction the results would be:

$$20 \bmod 4 = 0 => C$$
$$81 \bmod 4 = 1 => B$$

When rule A is removed from sorted list:

$$20 \bmod 3 = 2 => D$$
$$81 \bmod 3 = 0 => C$$

The same result of modulo operation 0 codes the same rule C. The probability that rule C is removed before rule A is low because number of all choices for rule C is higher. When number of available choices for rule C is zero then result of modulo operation has still different mapping from genotype to phenotype. The use of a list sorted by available choices simplifies analysis of genotype or phenotype.

35.4.3 Grammatical Evolution

The chromosome is represented by a set of integers filled with random values in the initial population. Gene values are used during chromosome translation to decide which terminal or nonterminal to pick from the set. When selecting a production rule there are four possibilities, we use *gene_value mod 4* to select a rule. However the list of variables has only one member (variable X) and *gene_value mod 1* always returns 0. A gene is always read; no matter if a decision is to be made, this approach

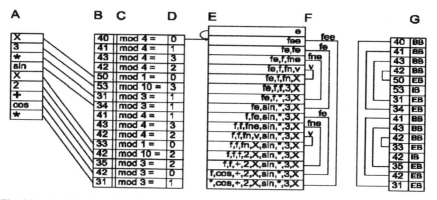

Fig. 35.2 Relations between genotype (column B) and phenotype (column A)

makes some genes in the chromosome somehow redundant. Values of such genes can be random, but genes must be present.

The figure (Fig. 35.2) shows the genotype-phenotype translation scheme. Body of the individual is shown as a linear structure, but in fact it is stored as a one-way tree (child objects have no links to parent objects). In the diagram we use abbreviated notations for nonterminal symbols: f – <fnc>, e – <expr>, n – <num>, v – <var> .

The column description in Fig. 35.2:

A. Objects of the individual body (resulting trigonometric function)
B. Genes used to translate the chromosome into the phenotype
C. Modulo operation, divisor is the number of possible choices determined by the gene context
D. Result of the modulo operation
E. State of the individual's body after processing a gene on the corresponding line
F. Blocks in the chromosome and corresponding production rules
G. Block marks added to the chromosome

Since operation modulo takes two operands, the resulting number is influenced by gene value and by gene context (Fig. 35.2C = Fig. 35.2 column C). Gene context is the number of choices, determined by the currently used list (rules, functions, variables). Therefore genes with same values might give different results of modulo operation depending on what object they code. On the other hand one terminal symbol can be coded by many different gene values as long as the result of modulo operation is the same $(31 \bmod 3) = (34 \bmod 3) = 1$. In the example (Fig. 35.2A) given the variables set has only one member X. Therefore, the modulo divider is always 1 and the result is always 0, a gene which codes a variable is redundant in that context (Fig. 35.2D). If the system runs out of genes during phenotype-genotype translation then the chromosome is wrapped and genes at the beginning are reused.

35.4.4 Processing the Grammar

The processing of the production rules is done backwards – from the end of the rule to the beginning (Fig. 35.2). E.g. production rule $<fnc>$ $<expr1>$ $<expr2>$ is processed as $<expr2>$ $<expr1>$ $<fnc>$. We use $<expr1>$ and $<expr2>$ at this point to denote which expression will be the first argument of $<fnc>$.

The main difference between $<fnc>$ and $<expr>$ nonterminals is in the number of real objects they produce in the individual's body. Nonterminal $<fnc>$ always generates one and only one terminal; on the contrary $<expr>$ generates an unknown number of nonterminal and terminal symbols. If the phenotype is represented as a tree structure then a product of the $<fnc>$ nonterminal is the parent object for handling all objects generated by $<expr>$ nonterminals contained in the same rule (Fig. 35.3). Therefore the rule $<fnc>$ $<expr1>$ $<expr2>$ can be represented as a tree (Fig. 35.4).

To select a production rule (selection of a tree structure) only one gene is needed. To process the selected rule a number of n genes is needed and finally to select a specific nonterminal symbol again one gene is needed. If the processing is done backwards the first processed terminals are leafs of the tree and the last processed terminal in a rule is the root of a subtree. The very last terminal is the root of the whole tree. Note that in a forward processing ($<fnc>$ $<expr1>$ $<expr2>$) the first processed gene codes the rule, the second gene codes the root of the subtree and the last are leafs.

When using the forward processing and coding of the rules described in the paper [5] it's not possible to easily recover the tree structure from genotype. It is caused with $<expr>$ nonterminals by using an unknown number of successive genes and

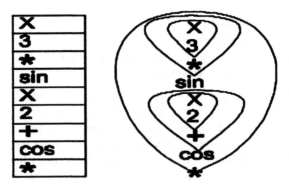

Fig. 35.3 Proposed backward notation of a function tree structure

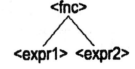

Fig. 35.4 Production rule
shown as a tree

the last processed terminal being just a leaf of the tree. The proposed backward processing is shown in Fig. 35.2E.

35.4.5 Phenotype to Genotype Projection

Using the proposed backward processing system the translation to a phenotype sub-tree has a certain scheme. It begins with a production rule (selecting the type of the subtree) and ends with the root of the subtree (in our case with a function) (Fig. 35.2F). In the genotype this means that one gene used to select a production rule is followed by n genes with different contexts which are followed by one gene used to translate <*fnc*>. Therefore a gene coding a production rule forms a pair with a gene coding terminal symbol for <*fnc*> (root of the rule). Those genes can be marked when processing the individual. This is an example of a simple marking system:

BB – Begin block (a gene coding a production rule)
IB – Inside block
EB – End block (a gene coding a root of a subtree)

The EB and BB marks are pair marks and in the chromosome they define a block (Fig. 35.2G). Such blocks can be nested but they don't overlap (the same way as parentheses). The IB mark is not a pair mark, but it is always contained in a block (IB marks are presently generated by <*num*> nonterminals). Given a BB gene a corresponding EB gene can be found using a simple LIFO method.

A block of chromosome enclosed in a BB-EB gene pair then codes a subtree of the phenotype. Such block is fully autonomous and can be exchanged with any other block or it can serve as completely new individual.

Only BB genes code the tree of individual's body, while EB and IB genes code the terminal symbols in the resulting phenotype. The BB genes code the structure of the individual, changing their values can cause change of the applied production rule. Therefore change (e.g. by mutation) in the value of a structural gene may trigger change of context of many genes or all following genes.

This simple marking system introduces a phenotype feedback to phenotype; however it doesn't affect the universality of the algorithm. It's not dependent on the used terminal or nonterminal symbols; it only requires the result to be a tree structure. Using this system it's possible to introduce a progressive crossover and mutation.

35.4.6 Crossover

When using grammatical evolution the resulting phenotype coded by one gene depends on the value of the gene and on its context. If a chromosome is crossed at random point, it is very probable that the context of the genes in second part will change. This way crossover causes destruction of the phenotype, because the newly added parts code different phenotype than in the original individual.

This behavior can be eliminated using a block marking system. Crossover is then performed as an exchange of blocks. The crossover is made always in an even number of genes, where the odd gene must be BB gene and even must be EB gene. Starting BB gene is presently chosen randomly; the first gene is excluded because it encapsulates (together with the last used gene) the whole individual.

The operation takes two parent chromosomes and the result is always two child chromosomes. It is also possible to combine the same individuals, while the resulting child chromosomes can be entirely different.

Given the parents:

1. $cos(x+2)+sin(x^*3)$
2. $cos(x+2)+sin(x^*3)$

The operation can produce children:

3. $cos(sin(x^*3)+2)+sin(x^*3)$
4. $cos(x+2)+x$

This crossover method works similar to direct combining of phenotype trees, however this method works purely on the chromosome. Therefore phenotype and genotype are still separated. The result is a chromosome, which will generate an individual with a structure combined from its parents. This way we receive the encoding of an individual without backward analysis of his phenotype. To perform a crossover the phenotype has to be evaluated (to mark the genes), but it is neither used nor know in the crossover operation (also it doesn't have to exist).

35.4.7 Mutation

Mutation can be divided into mutation of structural (BB) genes and mutation of other genes. Mutation of one structural gene can affect other genes by changing their context therefore structural mutation amount should be very low. On the other hand the amount of mutation of other genes can be set very high and it can speed up searching an approximate solution.

Given an individual:

$$sin(2+x)+cos(3^*x)$$

And using only mutation of non-structural genes, it is possible to get:

$$cos(5-x)^*sin(1^*x)$$

Therefore the structure doesn't change, but we can get a lot of new combinations of terminal symbols. The divided mutation allows using the benefits of high mutation while eliminating the risk of damaging the structure of an individual.

Fig. 35.5 The population model

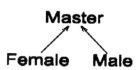

35.4.8 Population Model

The system uses three populations forming a simple tree structure (Fig. 35.5). There is a Master population and two slave populations, which simulate different genders. The links among the populations lead only one way – from bottom to top.

35.4.8.1 Female Population

When a new individual is to be inserted in a population a check is preformed whether it should be inserted. If a same or similar individual already exists in the population then the new individual is not inserted. In a female population every genotype and phenotype occurs only once. The population maintains a very high diversity; therefore the mutation operation is not applied to this population.

Removing the individuals is based on two criterions. The first criterion is the age of an individual – length of stay in the population. The second criterion is the fitness of an individual, using the second criterion a maximum population size is maintained. Parents are chosen using the tournament system selection.

35.4.8.2 Male Population

New individuals are not checked so duplicate phenotypes and genotypes can occur, also the mutation is enabled for this population. Mutation rate can be safely set very high (30%) provided that the structural mutation is set very low (less then 2%). For a couple of best individuals the mutations are non-destructive. If a protected individual is to be mutated a clone is created and added to the population. If the system stagnates in a local solution the mutation rate is raised using a linear function depending on the number of cycles for which the solution wasn't improved. Parents are chosen using a logarithmic function depending on the position of an individual in a population sorted by fitness.

35.4.8.3 Master Population

The master population is superior to the male and female populations. Periodically the subpopulations send over their best solutions. Moreover the master population performs another evolution on its own. Parents are selected using the tournament system. The master population uses the same system of mutations as the male population, but for removing individuals from the population only the fitness criterion is used. Therefore master population also serves as an archive of best solutions of the whole system.

35.4.9 Fitness Function

Around the searched function there is defined an equidistant area of a given size. Fitness of an individual's phenotype is computed as the number of points inside this area divided by the number of all checked points (a value in $< 0, 1 >$). This fitness function forms a strong selection pressure; therefore the system finds an approximate solution very quickly.

35.4.10 Results of Testing

Given sample of 100 points in the interval $[0, 2\pi]$ and using the block marking system described in 10.5, PGE has successfully found the searched function $sin(2*x)*cos(2+x)$ on the majority of runs. The graph (Fig. 35.6) shows maximum fitness in the system for ten runs and an average (bold). The system also found a function with fitness better then 0.8 in less then 40 generations using three populations with size of 100 individuals each. On the other hand the same system with phenotype to genotype projection disabled (Fig. 35.7). The majority of runs didn't find the searched function within 120 generations.

We have simplified the generation of numbers by adding a new production rule, thus allowing the generation of functions containing integer constants. The described parallel system together with phenotype to genotype projection improved

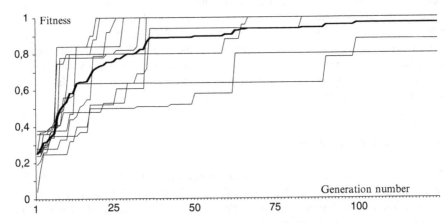

Fig. 35.6 Convergence of the PGE using backward processing (average in bold)

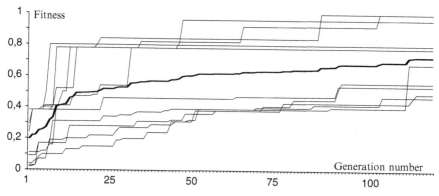

Fig. 35.7 Convergence of the PGE using forward processing (average in bold)

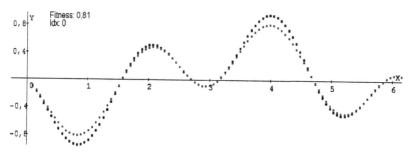

Fig. 35.8 A generated function with fitness 0.81 − $sin(sin(−9 + x))^*cos(−11 + 2^*x)$ (light) and searched function $sin(2^*x)^*cos(2 + x)$ (bold)

the speed of the system. The progressive crossover and mutation eliminates destroying partial results and allowed us to generate more complicated functions (e.g. $sin(2^*x)^*cos(2 + x)$) (Fig. 35.8).

We have described a parallel system, Parallel Grammatical Evolution (PGE) that can map an integer genotype onto a phenotype with the backward coding. PGE has proved successful for creating trigonometric identities.

Parallel GEs with the sexual reproduction can increase the efficiency and robustness of systems, and thus they can track better optimal parameters in a changing environment. From the experimental session it can be concluded that modified standard GEs with two sub-populations can design PGE much better than classical versions of GEs.

The PGE algorithm was tested with the group of six computers in the computer network (see Fig. 35.10). Five computers calculated in the structure of five subsystems MR1, MR2, MR3, MR4, and MR5 and one master MR. The male subpopulation M of MR in the higher level follows the convergence of the subsystem. In Fig. 35.9 is presented ten runs of the PGE-program. The shortest time of computation is only ten generation. All calculation were finished before 40 generation. This is better to compare with backward processing on one computer (see Fig. 35.6). The forward processing [5] on one computer was the slowest (see Fig. 35.7).

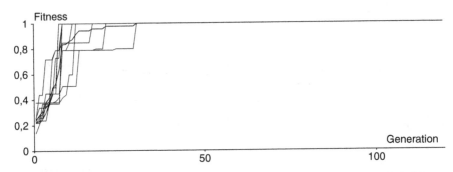

Fig. 35.9 Convergence of the PGE with six PC using backward processing (average in bold)

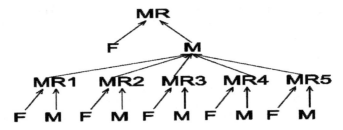

Fig. 35.10 The parallel structure of PGE with six computers

35.5 Logical Function XOR as a Test Function

Input values are two integer numbers a and b; a, b $< 0, 1 >$. Output number c is the value of logical function XOR. Training data is a set of triples (a, b, c):

$$P = \{(0,0,0); (0,1,1); (1,0,1); (1,1,0)\}.$$

Thus the training set represents the truth table of the XOR function. The function can be expressed using $\_$, $\hat{}$, $\neg$ functions:

$$a + b = (a \hat{} \neg b) \_ (\neg a \hat{} b) = (a \_ b) \hat{} (\neg a \_ \neg b) = (a \_ b) \hat{} \neg (a \hat{} b)$$

The grammar was simplified so that it does not contain conditional statement and numeric constants, on the other hand three new terminals were added to generate functions $\_$, $\hat{}$, $\neg$. Thus the grammar generates representations of the XOR functions using other logical functions.

1.
```
function xxor($a, $b) {
$result = "no_value";
$result = ($result)|
((((~(~(~(~(~($result))))))|(($a)&(($a)&(~($b)))))&($a))
|((~($a))&($b)));
```

return $result;
}
Number of generations: 32
2.
function xxor($a,$b) {
$result = "no_value";
$result = ($result)|((((~$b&($a&($a& ~$b)))&$a)|
(~$a&$b));
return $result;
Number of generations: 53

35.6 Two-Level Optimization

Although grammatical evolution is a very powerful method it does have its' weak spots. One of them is generation of real numbers. Since real numbers can be described using a context-free grammar it is possible for grammatical evolution to generate them [5]. The problem arises when generation of real numbers is combined with generation of functions. Generation of a string using grammatical evolution requires approximately twice as much genes as there are terminals in the string.

Therefore generation of an example function:
$y = 2 \cdot \ln(a + x \cdot b)/(x + c)$ would require approximately 18 genes. On the other hand generation of a real number with e.g. four digits would require about eight genes. If the above function will be generated including real constants (each having e.g. four digits) it would require about 36 genes. Another problem is when chromosome operation is to be designed, since the chromosome distance of numbers generated using grammatical evolution is not natural. For example under some circumstances number 0 can have the same distance to number 1 as to number 9. This has a negative impact on the complexity of the function, which is being raise by the number of used genes. Also there is some motivation to generate more complex function rather then more precise parameters, which is not practical.

A hard problem can be solved using a two-level optimization. The first level of the optimization is performed using grammatical evolution. The output can be a function containing several symbolic constants. Such function therefore cannot be evaluated and assigned a fitness value. In order to evaluate the generated function a secondary optimization has to be performed. For secondary optimization differential evolution is used. Input for the second level of optimization is the function with symbolic constants and the output is vector of the values of those constants.

A simplified flowchart diagram of the two-level optimization is shown on Fig. 35.11. Basically it consists of two nested population loops. The inner loop is using standard differential evolution [8]. The outer loop is a single population of parallel grammatical evolution [10]. The resulting Grammatical Differential Evolution (GDE) takes advantage of both the original methods [11].

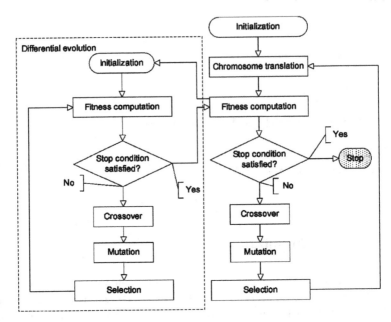

Fig. 35.11 Two-level optimization

35.7 Conclusion

We have simplified the generation of numbers by adding a new production rule, thus allowing the generation of functions containing integer constants. The described parallel system together with phenotype to genotype projection improved the speed of the system. The progressive crossover and mutation eliminates destroying partial results and allowed us to generate more complicated functions (e.g. $\sin(2^*x)^*\cos(2+x)$).

We have described the Parallel Grammatical Evolution (PGE) that can map an integer genotype onto a phenotype with the backward processing. PGE has proved successful for creating trigonometric functions.

Parallel GEs with hierarchical structure can increase the efficiency and robustness of systems, and thus they can track better optimal parameters in a changing environment. From the experimental session it can be concluded that modified standard GEs with only two sub-populations can create PGE much better than classical versions of GEs.

The parallel grammatical evolution can be used for the automatic generation of programs. We are far from supposing that all difficulties are removed but first results with PGEs are very promising.

The computer simulation helps identify the conditions under which the evolution of living world is running forever. Parallel genetic algorithms (GAs) can increase the efficiency and robustness of systems, and thus they can track better optimal parameters in a changing environment. It is not easy to say which individual modifications in parallel and hierarchical structure are the best. If we join them together by the parallel GA, then – in the higher level – it is not important which method will contribute more to the final solution.

The increased awareness from other scientific communities, such as biology and mathematics, promises new insights and new opportunities. There is much to accomplish and there are many open questions. Interest from diverse disciplines continues to increase and simulated evolution is becoming more generally accepted as a paradigm for optimization in practical engineering problems.

The parallel grammatical evolution can be used for the automatic generation of programs. This can help us to find information as a part of complexity. I am far from supposing that all difficulties are removed but first results with PGEs are very promising [5, 11].

Acknowledgements This work has been supported by Czech Grant Agency grant No: 102/06/1132 Soft Computing in Control.

References

1. Prigogine, I. and Stengers, I.: Order Out of Chaos, Flamingo, London, 1985.
2. Ošmera, P.: *Complex Adaptive Systems*, Proceedings of MENDEL'2001, Brno, Czech Republic (2001), pp. 137–143.
3. Ošmera, P.: *Complex Evolutionary Structures*, Proceedings of MENDEL'2002, Brno, Czech Republic (2002), pp. 109–116.
4. Ošmera, P.: *Evolvable Controllers Using Parallel Evolutionary Algorithms*, Proceedings of MENDEL'2003, Brno, Czech Republic (2003), pp. 126–132.
5. O'Neill, M. and Ryan, C.: Grammatical Evolution: Evolutionary Automatic Programming in an Arbitrary Language, Kluwer, Dordrecht, 2003.
6. O'Neill, M., Brabazon, A., and Adley, C.: *The Automatic Generation of Programs for Classification Problems with Grammatical Swarm*, Proceedings of CEC 2004, Portland, Oregon (2004), pp. 104–110.
7. Piaseczny, W., Suzuki, H., and Sawai, H.: *Chemical Genetic Programming – Evolution of Amino Acid Rewriting Rules Used for Genotype-Phenotype Translation*, Proceedings of CEC 2004, Portland, Oregon (2004), pp. 1639–1646.
8. Patarlini, S. and Krink, T.: *High Performance Clustering with Differential Evolution*, Proceedings of CEC 2004, Portland, Oregon (2004), pp. 2004–2005.
9. Li, Z., Halang, W. A., and Chen, G.: Integration of Fuzzy Logic and Chaos Theory; paragraph: Osmera, P.: Evolution of Complexity, Springer, 2006 (ISBN: 3-540-26899-5), pp. 527–578.
10. Ošmera, P., Popelka, O., and Panaek, T.: *Parallel Grammatical Evolution with Backward Processing*, Proceedings of MENDEL'2005, Brno, Czech Republic (2005), pp. 27–28.
11. Popelka, O.: *Two-Level Optimization Using Parallel Grammatical Evolution and Differential Evolution*, Proceedings of MENDEL'2007, Prague, Czech Republic (2007), pp. 88–92.
12. Hu, X., Shi, Y., and Eberhart, R.: *Recent Advances in Particle Swarm*, Proceedings of CEC.

Chapter 36
Genetic Algorithms for Scenario Generation in Stochastic Programming
Motivation and General Framework

Jan Roupec and Pavel Popela

Abstract Traditional deterministic max-min and min-min techniques are significantly limited by the size of scenario set. Therefore, this text introduces a general framework how to generate and modify suitable scenario sets by using genetic algorithms. As an example, the search of absolute lower and upper bounds by using GA is presented and further enhancements are discussed. The proposed technique is implemented in C++ and GAMS and then tested on real-data examples.

Keywords Genetic Algorithm · Scenario Generation · Stochastic Programming · Goal-oriented

36.1 Introduction

Stochastic programs have been developed as useful tools for modeling of various application problems. The developed algorithms usually require a solution of large-scale linear and nonlinear programs because the deterministic reformulations of the original stochastic programs are based on empirical or sampling discrete probability distributions, i.e. on so-called scenario sets. The scenario sets are often large, so the reformulated programs must be solved. Therefore, the suitable scenario set generation techniques are required. Hence, randomly selected reduced scenario sets are often employed. Related confidence intervals for the optimal objective function values have been derived and are often presented as tight enough. However, there is also

J. Roupec (✉)
Brno University of Technology, Faculty of Mechanical Engineering, Institute of Automation and Computer Science, Technická 2, 616 69 Brno, Czech Republic
e-mail: roupec@fme.vutbr.cz

P. Popela
Brno University of Technology, Faculty of Mechanical Engineering, Institute of Mathematics, Technická 2, 616 69 Brno, Czech Republic

S.-I. Ao et al. (eds.), *Advances in Computational Algorithms and Data Analysis*,
Lecture Notes in Electrical Engineering 14,

demand for goal-oriented scenario generation to learn more about the extreme cases. Traditional deterministic max-min and min-min techniques are significantly limited by the size of scenario set. Therefore, this text introduces a general framework how to generate and modify suitable scenario sets by using genetic algorithms. As an example, the search of absolute lower and upper bounds by using GA is presented and further enhancements are discussed. The proposed technique is implemented in C + + and GAMS and then tested on real-data examples.

36.2 Stochastic Programming

To introduce stochastic program, firstly, we denote a mathematical program as $? \in \arg\min\{f(\mathbf{x}) \mid \mathbf{x} \in C\}$. Then, we naturally obtain an underlying program $? \in \arg\min\{f(\mathbf{x}, \xi) \mid \mathbf{x} \in C(\xi)\}$, as we replaced several original constant parameters by random elements. There is ξ, a random vector defined on the probability space (Ξ, Σ, P), and $f : \mathrm{IR}^n \times \Xi \to \mathrm{IR}$, is a measurable function for each decision $\mathbf{x} \in \mathrm{IR}^n$ that must belong to the feasible set C. To be able to solve optimization problem correctly, the deterministic reformulation must be further specified. Usually, we cannot wait for observation ξ^s of ξ, and we must decide here-and-now (HN). In this case, we have to utilize a suitable HN-reformulation and we have chosen the most typical one:

$$? \in \arg\min_{\mathbf{x}} \left\{ E_\xi f(\mathbf{x}, \xi) \mid \mathbf{x} \in C(\xi) \text{ a.s.} \right\}, \tag{36.1}$$

where E denotes an expectation functional and abbreviation a.s. means almost surely. However, there are also different approaches to random parameter modeling, see, e.g. Kall and Wallace [1] for details.

It is difficult to solve the stochastic program described by Eq. (36.1) in the case when the random vector has the continuous probability distribution. Then, the approximation techniques based on discretization are used [1]. So, we focus on a case of finite support of ξ. Hence, for discrete random vector ξ, instead of solution difficulty related to multidimensional integration to compute E, we prefer to deal with the computational complexity caused by the so called scenario-based (SB) HN-reformulation. The scenario set generation techniques are discussed in recent papers [2, 3].

Particularly, E is computed explicitly as we may denote $p_s = P(\xi = \xi^s)$ and write the expectation as a sum. Therefore, the SB-reformulation is a large deterministic mathematical program:

$$? \in \arg\min_{\mathbf{x}} \left\{ \sum_{\xi^s \in \Xi} p_s f(\mathbf{x}, \xi^s) \;\middle|\; \mathbf{x} \in C = \bigcap_{\xi_s \in \Xi} C(\xi^s) \right\} \tag{36.2}$$

The main problem is of how to solve the given program (Eq. 36.2) in an efficient way. There are several special algorithms developed in the area of stochastic programming that are considered efficient and mostly based on various decomposition ideas [1]. Recently, in the case of integer decision variables, various heuristics are studied. The additional problem how to choose suitable realizations ξ^s appears when only incomplete information about the probability distribution is available. In this case, we have to deal with large discrete sets of scenarios, and so, we will show that heuristics may help again. Therefore, we review the principal concepts of the area of evolutionary and genetic algorithms in the next section.

36.3 Genetic Algorithms

The basic idea of a genetic algorithm (GA) is quite simple. GA works not only with one iterated solution in time but with the whole population of solutions in one algorithm iteration. The population contains many (ordinary several hundreds) individuals – bit strings representing solutions. Evolutionary algorithms deal with similar strings of more general form, e.g., containing integer numbers or characters. The mechanism of GA involves only elementary operations like strings copying, partially bit swapping or bit value changing. GA starts with a population of strings and thereafter generates successive populations using the following three basic operations: reproduction, crossover, and mutation. Reproduction is the process by which individual strings are copied according to an objective function value (fitness). Copying of strings according to their fitness value means that strings with a higher value have a higher probability of contributing one or more offspring to the next generation. This is an artificial version of the natural selection. Mutation is an occasional (with a small probability) random alteration of the string position value. It is needed since, in spite of reproduction and crossover effectively searching and recombining the existing representations, they occasionally become overzealous and lose some potentially useful genetic material. The mutation operator prevents such an irrecoverable loss. The recombination mechanism allows mixing of parental information while passing it to their descendants, and mutation introduces innovation into the population.

In spite of simple principles, the design of GA for successful practical using is surprisingly complicated. GA has many parameters that depend on the problem to be solved. In the first, it is the size of population. Larger populations usually decrease the number of iterations needed, but dramatically increase the computing time for each iteration. The factors increasing demands on the size of population are the complexity of the problem being solved and the length of the individuals. Every individual contains one or more chromosomes containing value of potential solution. Chromosomes consist of genes. The gene in our version of GA is a structure representing one bit of solution value. It is usually advantageous to use some redundancy in genes and so the physical length of our genes is greater than one bit. This type of redundancy was introduced by Ryan [4].

To prevent degeneration and the deadlock in local extreme the limited lifetime of each individual can be used. This limited lifetime is realized by the "death" operator, [5] which represents something like continual restart of GA. This operator enables decreasing of population size as well as increasing the speed of fitness improvement. It is necessary to store the best solution obtained separately – the corresponding individual need not to be always present in the population because of the limited lifetime.

Many GAs are implemented on a population consisting of haploid individuals (each individual contains just one chromosome). However, in nature, many living organisms have more than one chromosome and there are mechanisms used to determine dominant genes. Sexual recombination generates an endless variety of genotype combination that increases the evolutionary potential of the population. Because it increases the variation among the offspring produced by an individual, it improves the change that some of them will be successful in varying and often-unpredictable environments they will encounter. Using diploid or "multiploid" individuals can often decrease demands on the population size. However the use of multiploid GA with sexual reproduction brings some complications, the advantage of multiploidity can be often substitute by the "death" operator and redundant genes coding.

New individuals are created by operation called crossover. In the simplest case crossover means swapping of two parts of two chromosomes split in randomly selected point (so called one point crossover). In GA we use the uniform crossover on the bit level. The strategy of selection individuals for crossover is very important. It strongly determines the behavior of GA.

Genetic algorithms commonly use heuristic and stochastic approaches. From the theoretical viewpoint, the convergence of heuristic algorithms is not guaranteed for the most of application cases. That is why the definition of the stopping rule of the GA brings a new problem. It can be shown, [6] that while using a proper version of GA the typical number of iterations can be determined.

36.4 Sampling Techniques

We focus on cases introduced by Eqs. (36.1) and (36.2). We want to compute the optimal HN-solution $\mathbf{x}_{\min}^{HN}$ of Eqs. (36.1) or (36.2) and related objective function value $z_{\min}^{HN}$. One possible way is to approximate these programs and reduce a program size, i.e. number of included scenarios, by random sampling. We denote a random sample from $\boldsymbol{\xi}$ as

$$\boldsymbol{\xi}_{[.]} = \left(\boldsymbol{\xi}_{[1]}, \dots, \boldsymbol{\xi}_{[v]} \right)^{T}. \tag{36.3}$$

There are $\boldsymbol{\xi}_{[i]}, i = 1, \dots, v$ random variables identically distributed as $\boldsymbol{\xi}$ and they are stochastically independent. The realization of this random sample is usually denoted as

$$\boldsymbol{\xi}_{[.]}^{s} = \left(\boldsymbol{\xi}_{[1]}^{s}, \dots, \boldsymbol{\xi}_{[v]}^{s} \right)^{T}. \tag{36.4}$$

We often simplify our notation as follows:

$$\boldsymbol{\xi}_{[.]}^{s} = \left(\boldsymbol{\xi}^{1}, \dots, \boldsymbol{\xi}^{v} \right)^{T}. \tag{36.5}$$

For computational purposes, we may utilize sample realization (Eq. 36.5) and easily write expectation estimate of the objective function by using a sample mean realization. We assume the existence of the finite optimal solution of Eq. (36.1):

$$E_{\xi} \{f(\mathbf{x}, \xi)\} = \frac{1}{v} \sum_{s=1}^{v} f(\mathbf{x}, \xi^s). \tag{36.6}$$

By solution of sample program and we obtain:

$$z_{min}^v = \frac{1}{v} \sum_{s=1}^{v} f(\mathbf{x}_{min}^v; \xi^s) = \min_{\mathbf{x}} \left\{ \frac{1}{v} \sum_{s=1}^{v} f(\mathbf{x}, \xi^s) \mid \mathbf{x} \in C \right\}. \tag{36.7}$$

Randomly generated observations of ξ may then serve to compute the estimate of the objective function. Therefore, scenarios are selected by random procedure. Then, the scenarios ξ^s are used to build a scenario tree, and this reduced program is solved instead of the original one. However, its blindfold and exaggerated use can lead to misleading results. So, in addition, Monte Carlo techniques may be necessary to obtain an estimate of how good is such a simplification. Then, repeated computations inform us about the result stability and sensitivity.

Mak et al. [7] prove the following inequalities:

$$E_{\xi} \{\varsigma_{min}^v\} \leq E_{\xi} \{\varsigma_{min}^{v+1}\} \leq z_{min}^{HN} \leq z = E_{\xi} \{f(\mathbf{x}; \xi)\}, \tag{36.8}$$

where $\mathbf{x}$ is any feasible solution from C. They assume that a random sample from ξ denoted as

$$\xi_{[.]} = \left(\xi_{[1]}, \ldots, \xi_{[v_u]} \right)^T \tag{36.9}$$

is available and the ς_{min}^v denotes a random optimal objective function value depending on the random sample (Eq. 36.9). They also have v_l random samples, each having size v, therefore

$$\forall i = 14, \ldots, v_l : \xi_{[i.]} = \left(\xi_{[i1]}, \ldots, \xi_{[iv]} \right)^T. \tag{36.10}$$

They use inequalities (Eq. 36.8) and the central limit theorem to derive the following bounds:

$$P \left(\frac{1}{v_l} \sum_{i=1}^{v_l} \min_{\mathbf{x}(\xi_{[i.]})} \left\{ \frac{1}{v} \sum_{s=1}^{v} f\left(\mathbf{x}\left(\xi_{[i.]}\right); \xi_{[is]} \right) \mid \mathbf{x}\left(\xi_{[i.]}\right) \in C \text{ a.s.} \right\} - \frac{t_{1-\alpha/2} s_l(v_l)}{\sqrt{v_l}} \right.$$

$$\leq E_{\xi} \{\xi_{min}^v\} \leq z_{min}^{HN} \leq E_{\xi} \{f(\mathbf{x}; \xi)\} \leq$$

$$\left. \leq \frac{1}{v_u} \sum_{s=1}^{v} f\left(\mathbf{x}; \xi_{[s]} \right) + \frac{t_{1-\alpha/2} s_l(v_l)}{\sqrt{v_l}} \right) \approx 1 - \alpha$$

$$\tag{36.11}$$

The symbol $t_{1-\alpha/2}$ denotes the $1 - \alpha/2$ quantile of $N(0;1)$ distribution. Symbols $s_l(v_l)$ and $s_u(v_u)$ denote usual estimates of standard deviations $\sqrt{\mathrm{var}\,\varsigma_{\min}^v}$ and $\sqrt{\mathrm{var}\,f(\mathbf{x};\boldsymbol{\xi})}$. Hence, we may set α, then substitute observations $\boldsymbol{\xi}_{[.]}^s$ and $\boldsymbol{\xi}_{[i.]}^s$ in the (Eq. 36.11), and we obtain reliable bounds.

36.5 Extreme Scenario Sets

Let us consider a resulted sequence of optimum objective values for different sizes of samples. The aforementioned bounds in this case are also very promising and can be obtained when the computations for samples of the given size are repeated, e.g. 30 times. However, still one question remains. Are they so good because of small influence of randomness or only 'dangerous' scenarios are not participating in our samples?

So, in this case, we may try to realize the worst case analysis based on so called extreme scenario sets, generally defined as follows [8, 9]:

$$\min_{S \subset \Xi;\, |S|=n} \min \left\{ \frac{1}{S} \sum_{\boldsymbol{\xi}^s \in S} f(\mathbf{x};\boldsymbol{\xi}^s) \,\middle|\, \mathbf{x} \in \bigcap_s C(\boldsymbol{\xi}^s) \right\} \qquad (36.12)$$

$$\max_{S \subset \Xi;\, |S|=n} \min \left\{ \frac{1}{S} \sum_{\boldsymbol{\xi}^s \in S} f(\mathbf{x};\boldsymbol{\xi}^s) \,\middle|\, \mathbf{x} \in \bigcap_s C(\boldsymbol{\xi}^s) \right\} \qquad (36.13)$$

Because the objective function convexity with respect to $\boldsymbol{\xi}$ is not guaranteed in general case, this problem might be quite difficult to solve. Because of the problem size, it is also impossible to consider all scenarios as Rosa and Takriti [10] have done when tried to exclude certain scenarios computing the whole optimization program that sets their probabilities to zero. Therefore, we introduce GA as a suitable tool.

36.6 GA Framework

So, we assume that components of discrete random vector $\boldsymbol{\xi}$ are random variables with finite supports. Both cases of independent and dependent components have been studied. We suggest a usual algorithm for general case of finite marginal supports Ξ_i. Evolutionary and genetic algorithms were already used in stochastic programming by Berland [11] and a group of authors [12].

In our case, the changing set of scenarios S, see Eqs. (36.12) and (36.13), is considered as a GA population. The genetic algorithm framework, we have utilized can be defined as follows:

1. For each population, the objective function value is computed by running an external program in GAMS [13] that solves Eq. (36.5) for given population i.e. set of scenarios.

2. We obtain the sequence of approximate iterative solutions $\mathbf{x}_{min}$ of Eq. (36.12) and $\mathbf{x}_{max}$ of Eq. (36.13), respectively.
3. Therefore, for given scenario s and the optimal solution $\mathbf{x}_{min}$ (Eq. 36.12) and similarly for $\mathbf{x}_{max}$ in case of Eq. (36.13), we calculate fitness value for each scenario as a member of population:

$$\{f(\mathbf{x}_{min}; \boldsymbol{\xi}^s) \mid \mathbf{x}_{min} \in C(\boldsymbol{\xi}^s)\}. \tag{36.14}$$

4. Then the population is ordered by fitness values, and usual operations as reproduction, crossover, and mutation are realized and till the stopping rule is satisfied, we continue with step 1.

Detail on GA implementation are specified in the following way:

1. *Generation of the initial population*: At the beginning the whole population is generated randomly, the members are sorted by the fitness (in descendent order).
2. *Mutation:* The mutation is applied to each gene with the same probability. The mutation of the gene means the inversion of one randomly selected bit in the gene.
3. *Death*: Classical GA uses two main operations – crossover and mutation (the other operation should be migration). In GA described in this paper, we use the third operation – death. Every individual has the additional information – age. A simple counter that is incremented in each of GA iteration represents the age. If the age of any member reaches the preset lifetime limit LT, the member "dies" and is immediately replaced by a new randomly generated member. The age is not mutated nor crossed over. The age of new individuals (incl. individuals created by crossover) is set to zero. Lifetime limit in our application is set to 5.
4. Sorting is realized by the fitness function values.
5. *Crossover*: Uniform crossover is used for all genes (each bit of the offspring gene is selected separately from corresponding bits of genes of both parents).
6. When a stopping rule is not satisfied, go to step 2.

In crossover, we do not replace all members of the population. The crossover generates the number of individuals corresponding to the quarter of the population only. Created individuals are sorted into the corresponding places in the population according to their fitness in such a way that the size of the population remains the same. Newly created individuals of low fitness do not have to be involved in the population.

At the end of this section, we describe the basic loop of main algorithm implementation that searches extreme scenario sets:

During initialization, a main program controlling the whole computational process is started. It reads the optimization engine name (GAMS [13] in our example), main source filename, included filename with scenarios, filename for SB program, filename for GA input, number of scenarios, dimension, and probability distribution information. The main program calls the optimization engine, and it solves the SB-program. Input data for GA is generated; it includes fitness values related to scenarios. Then, the GA starts and generates a new set of scenarios. When a stopping rule is not satisfied, files are updated and optimization engine is called again.

36.7 The Use in Applications

To show how the aforementioned approach can be implemented, a scenario-based (SB), two-stage stochastic linear program, modeling a principal part of a melt control process in a suitable furnace (cupola, induced, or electric-arc) has been chosen as an example because it allows to use real world data and can be easily modified [14].

$$? \in \arg\min \left\{ \mathbf{c}^T \mathbf{x} + \sum_{s=1}^{S} p_s Q(\mathbf{x}; \boldsymbol{\xi}^s) \quad | \mathbf{x} \geq 0 \right\}$$

$$Q(\mathbf{x}; \boldsymbol{\xi}^s) = \min_{\mathbf{y}^s} \left\{ \mathbf{q}^T \mathbf{y}^s \mid \mathbf{T}_1^s \mathbf{A}_1 \mathbf{x} + \mathbf{A}_2 \mathbf{y}^s \geq \mathbf{l}_2, \right. \tag{36.15}$$

$$\left. \mathbf{T}_1^s \mathbf{A}_1 \mathbf{x} + \mathbf{A}_2 \mathbf{y}^s \geq \mathbf{u}_2, \, \mathbf{x}, \mathbf{y}^s \geq 0, \, s = 1, \dots, S \right\}$$

where $\mathbf{x}$ is a tonnage of n_1 charge materials and $\mathbf{y}^s$ represents a tonnage of n_2 alloying materials. Symbols $\mathbf{l}_2$ and $\mathbf{u}_2$ describe a minimum and maximum tonnage of m considered elements after alloying. Then, $\mathbf{A}_1$ is a matrix containing the known proportions a_{ij} of i–th element in j–th charge material and $\mathbf{A}_2$ is a matrix containing information about alloying materials. $\boldsymbol{\Xi} = \left\{ \boldsymbol{\xi}^1, \dots, \boldsymbol{\xi}^s \right\}$ is a finite support of distribution $\boldsymbol{\xi}$ that is composed of scenarios $\boldsymbol{\xi}^s$ (or shortly s), and $p_s = P(\boldsymbol{\xi} = \boldsymbol{\xi}_s) = 1/s$, $s = 1, \dots, S$. The random utilization of elements in the charge melt is therefore defined by $\mathbf{T}_1^s$. The alloying utilization is described by the unit matrix $\mathbf{I}$ that stays in front of $\mathbf{A}_2$. The overall expected cost of repeated similar melts is minimized, therefore, $\mathbf{c}$ and $\mathbf{q}$ are vectors representing costs per ton for input materials. The discussed melt control model development is described in [8, 9, 14–16]. The developed techniques are related to material engineering approaches in [17] and [18].

Utilization matrix $\mathbf{T}_1^s$ has a diagonal form. Firstly, we assumed only two random diagonal elements to simplify the results analysis. As the next step, we have assumed all diagonal elements random. The largest problem solved with real data dealt with 16 random elements of the diagonal of utilization matrix.

We have applied the aforementioned main algorithm together with GAMS source for Eq. (36.15) and for the discussed GA. As a result we obtained a sequence of randomly selected scenario sets together with sequences of searched pairs of extreme sets. The aforementioned confidence intervals are significantly tighter even in the case of increased confidence level, hence may be interpreted as too optimistic for the situation when the worst case analysis is the goal.

36.8 Results and Further Research

The main purpose of this text has been to motivate and discuss the use of genetic algorithms for scenario generation in stochastic programming. The algorithm proved the significant improvement in comparison with an older study [19].

Further research will be focused on postoptimality analysis realized with respect to the original set of scenarios. Mainly, the cases where the incomplete recourse [1] appears to be very important will be studied. We assume that the deeper understanding of the whole algorithm behavior will lead to the used GA improvement.

The used test problems are derived from the underlying linear programs; therefore, the proposed technique will be applied to nonlinear programming engineering applications, e.g., in reliable structural design.

Acknowledgements This work has been supported by the Czech Ministry of Education in the frame of MSM 0021630529 Intelligent Systems in Automation and by the GA CR projects: 103/05/0292 and 103/08/1658 and the CQR 1M06047 research center of the Czech Republic.

References

1. P. Kall and S. W. Wallace, *Stochastic Programming*. Wiley, Chichester, 1994.
2. J. Dupaová, G. Consigli, and S. W. Wallace, Scenarios for multistage stochastic programs, *Annals of Operations Research*, 100: 25–53, 2000.
3. K. Hoyland and S. W. Wallace, Generating scenario trees for multi stage problems. *Technical Report* 4/97, Department of Industrial Economics and Technology Management, Norwegian University of Science and Technology, Trondheim, 1996.
4. C. Ryan, Shades. Polygenic inheritance scheme, in *Proceedings of the 3rd International Conference on Soft Computing MENDEL '97*, Brno, Czech Republic, 1997, pp. 140–147.
5. J. Roupec, *Design of Genetic Algorithm for Optimization of Fuzzy Controllers Parameters* (in Czech), Ph.D. thesis, Brno University of Technology, Brno, Czech Republic, 2001.
6. J. Roupec, P. Popela, and P. Ošmera, The additional stopping rule for heuristic algorithms, in *Proceedings of the 3rd International Conference on Soft Computing Mendel '97*, Brno, Czech Republic, 1997, pp. 135–139.
7. W. K. Mak, D. P. Morton, and R. K. Wood, Monte Carlo bounding techniques for deterministic solution quality in stochastic programs, *Operations Research Letters*, 24: 47–56, 1999.
8. P. Popela, Advanced scenario tuning in stochastic programming, in *Proceedings of Czech-Slovak-Japan Workshop*, Bratislava, Slovakia, 1998, pp. 307–310.
9. P. Popela, *An Object-Oriented Approach to Multistage Stochastic Programming: Models and Algorithms*, Ph.D. thesis, Charles University, Prague, 1998.
10. C. H. Rosa and S. Takriti, A minimax formulation for stochastic programs using scenario aggregation, Technical report, SABRE Dec. Technol. and IBM T. J. Watson Center, July 17 1997.
11. N. J. Berland, *Stochastic Optimization and Parallel Processing*, Ph.D. thesis, University of Bergen, Bergen, 1993.
12. Y. Yoshitomi, T. Takeba, S. Tomita, and H. Ikenoue, Genetic algorithm approach for solving stochastic programming problems, in *Proceedings of International Conference on Stochastic Programming*, Vancouver, Canada, 1998.
13. A. Brooke, D. Kendrick, and A. Meeraus, *Release 2.25 GAMS A User's Guide*, 2nd edition, The Scientific Press Series, Boyd & Fraser, Boston, MA, 1992.
14. P. Popela, Application of stochastic programming in foundry, *Folia Fac. Sci. Nat. Univ. Masarykianae Brunensis, Mathematica*, 7: 117–139, 1998.
15. P. Popela, A multistage blending problem with an expert estimate of parameters (in Czech), in *Proceedings of 3rd Conference on Modern Mathematical Methods*, Ostrava, Czech Republic, 1994, pp. 101–105.
16. P. Popela and R. Setnička, Analysis of steel production (in Czech). Technical report, ŽĎAS, Žd'ár nad Sázavou, Czech Republic, 1995.

17. Z. Karpíšek, Heterogeneity characteristics of cast iron alloying elements, *Folia Fac. Sci. Nat. Univ. Masarykianae, Mathematica*, 7: 31–36, 1998.
18. F. Kavička, K. Stránský, J. Stětina, V. Dobrovská, J. Dobrovská, and B. Velička, Contribution to optimization of continuous casting of steel, *Acta Metallurgica Slovaca*, 5: 367–370, 1999.
19. P. Popela and J. Roupec, GA-based scenario set modification in two-stage melt control problems, in *Proceedings of the 5th International Conference MENDEL'99*, Brno, Czech Republic, 1999, pp. 112–117.

Chapter 37
New Approach of Recurrent Neural Network Weight Initialization

Roberto Marichal, J.D. Piñeiro, E.J. González, and J.M. Torres

Abstract This paper proposes a weight initialization strategy for a discrete-time recurrent neural network model. It is based on analyzing the recurrent network as a nonlinear system, and choosing its initial weights to put this system in the boundaries between different dynamics, i.e., its bifurcations. The relationship between the change in dynamics and training error evolution is studied. Two simple examples of the application of this strategy are shown: the identification of DC Induction motor and the detection of a physiological signal, a feature of a visual evoked potential brain signal.

Keywords DTRNN · Nonlinear system · Training · Bifurcation

37.1 Introduction

In this paper the particular model of network which is studied is the Discrete-Time Recurrent Neural Network (DTRNN) [1], also known as the Williams-Zipser architecture. Its state evolution equation is

$$x_i(k+1) = \sum_{n=1}^{N} w_{in} f(x_i(k)) + \sum_{m=1}^{M} w'_{im} u_m(k) + w''_i \qquad (37.1)$$

where

$x_i(k)$ is the ith neuron output.

$u_m(k)$ is the mth input of the network.

w_{in}, w'_{im} are the weight factors of the neuron outputs, network inputs and w''_i is a bias weight.

$f(.)$ is a continuous, bounded, monotonically increasing function such as the hyperbolic tangent.

R. Marichal (✉), J.D. Piñeiro, E.J. González, and J.M. Torres
System Engineering and Control and Computer Architecture Department, University of La Laguna, Avda. Francisco Sánchez S/N, Edf. Informatica, 38200, Tenerife, Canary Islands, Spain
e-mail: {rlmarpla, jpineiro, jmtorres, ejgonzal}@ull.es

S.-I. Ao et al. (eds.), *Advances in Computational Algorithms and Data Analysis*,
Lecture Notes in Electrical Engineering 14,
© Springer Science+Business Media B.V. 2009

From the point of view of dynamics theory it is interesting to study the invariant sets in state-space. The equilibrium or fixed points are the simplest of these invariants. These points are those which do not change in time evolution. Their character or stability is given by the local behavior of nearby trajectories. A fixed point can attract (sink), repel (source) or have both directions of attraction and repulsion (saddle) of close trajectories [2]. Following in complexity there are periodic trajectories, quasi-periodic trajectories or even chaotic sets, each with its own stability characterization. All these features are similar in a class of so-called "topologically equivalent" systems [3]. When a system parameter (such a weight, for example) is varied the system can reach a critical point in which it is no more topologically equivalent to its previous situation. This critical configuration is called a bifurcation [4], and the system will exhibit new qualitative behaviors. The study of how these multidimensional weight changes can be carried out will be another issue in the analysis. In this paper we focus in local bifurcations in which the critical situation is determined by only one condition, known as codimension 1 bifurcation, in a fixed point. There are several bifurcations of this kind (that can occur with the variation of only one parameter) in a discrete dynamical system: Saddle-Node, Flip and Neimark-Sacker bifurcations. The conditions that define the occurrence of these bifurcations are determined by the eigenvalues of the Jacobian matrix evaluated at the fixed point. For a Saddle-Node, one eigenvalue will be 1, for a Flip bifurcation, (-1) and finally, in a Neimark-Sacker bifurcation, two eigenvalues (complex conjugate) are on the unit circle.

With respect to recurrent neural networks as nonlinear systems studies, there are several results about their stability [5–7]. In the other hand, there are some works that analyze the chaotic behavior [8,9]. Finally, in some paper [10,11] is studied the Neimark-Sacker bifurcation and the quasi-periodic orbit apparition.

This paper is divided into six sections. In Section 37.2, the sample problems are explained: DC induction motor Identification and the visual potential evoked feature detection. In Section 37.3, the neural network dynamic resulting from standard training is analyzed. In Section 37.4 is studied the relationship between dynamics and training process error. In the last section, a new initialization weight strategy based on the bifurcation theory is proposed and experimental results in the sample problems are given.

37.2 Problems Description

37.2.1 DC Induction Motor Identification

This first problem is based on the identification of DC Induction Motor Identification [12]. In this problem, the output corresponds to axis motion, described by its angular velocity, and the input is the voltage applied to the rotor coil. The training input time series were of two kinds: sinusoidal and step. The desired output is a filtered version

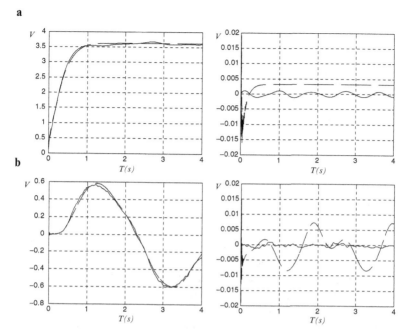

Fig. 37.1 Real and desired outputs with dead zones (right figures) and without them (left figures) for the input patterns: **a** step, and **b** sinusoidal. The discontinuous and continuous lines represent the neural network and desired output, respectively

of the measured real angular velocity response. In order to consider a representative range of input signals, sinusoidal frequencies varied from *0.25* to *1 Hz* and negative and positive steps were tried. Some of the real motor responses exhibited a dead-zone nonlinear response, i.e. within a range of the applied input voltage, the response is zero, the motor does not move. It was ensured that were present equal number of patterns with and without this phenomenon in the training set.

The best results were obtained with a high-dimension neural network (nine neurons). Figure 37.1 shows the results with both sinusoidal and step input signals. It can be appreciated that the neural network outputs corresponding to patterns without dead zones follow the desired outputs with good precision, and the pattern outputs with dead zones show a slightly inferior precision but still follow the desired response.

37.2.2 Potential Evoked Problem

The other test problem selected is the Evoked Potential feature detection. The Evoked Potential (EP) is defined as an electrical response of the Central Nervous System that holds a well-defined temporal relation with the stimulus that originated it. In this case the stimuli are of visual type. A flash light generates a series of stimuli

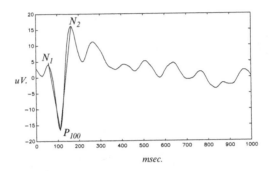

Fig. 37.2 Visual evoked potential

and the brain responses are acquired and averaged to get a waveform that in its first components frequently resembles a V letter. The points that demarcate this V are of specific diagnostic interest. The detection of a particular point (called P_{100} because often appears approximately *100 ms* after the stimulus), will be the problem to tackle. The detection must have into account the morphology of the signal (P_{100} context) and its time of occurrence because there are many other parts of the waveform with similar aspect [13]. Figure 37.2 shows a Visual Evoked Potential example.

Although a dynamic network of this kind can successfully represent complex temporal patterns, several drawbacks have been pointed out in the literature [1]. There are two main problems that appear when training this type of neural networks. Firstly, the error hyper surfaces tend to be rather complex and it is necessary to consider more sophisticated optimization algorithms. In this case, the variable metric method Broyden-Fletcher-Goldfard-Shanno (BFGS) [14] was used. Secondly, the problem of long-term dependencies arises frequently when the error function depends on the final state of the network, see for a description of this issue [15]. The particular trajectory in state space is irrelevant as far as the error function is concerned. Since the objective is to detect the P_{100} point, the desired output sequence is zero except at the location of this point. The error function will take into account every time instant. Incorporating the full trajectories to the error function alleviates the problem of long-term dependencies [16]. As shown in some studies [16] a good approach is to consider simultaneously feeding the network with shifted sections of the input sequence of length T (shown in Fig. 37.3). The shifted versions of the input sequence allow for the network to perceive at its inputs the relevant temporal context to take a decision just when the P_{100} arises.

37.3 Dynamic Analysis

37.3.1 DC Induction Motor Identification

In Fig. 37.4 it is shown the dynamics associated to the final values of the training weights. It is characterized by one saddle fixed point and two stable fixed points.

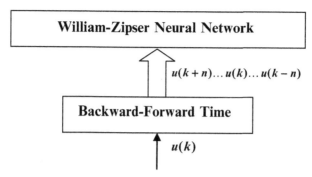

Fig. 37.3 Structure of the recurrent neural network model with shifted inputs. (PE problem)

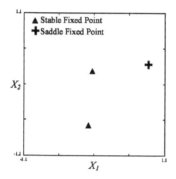

Fig. 37.4 Fixed points corresponding to the neural network dynamic of the DC motor problem

The stable fixed points presence can be explained by the learning of the two different kind of patterns commented earlier (with and without presence of dead zones). When the input is lower that the bias weight value, the state remains in one stable fixed point domain of attraction, reflecting the dead zone in the output. When it rises above this bias value, the state evolves towards to the domain of attraction of the other stable fixed point. In order to demonstrate this, a simple experiment was devised. It consists in considering different amplitude values of a step pattern input, letting the state evolve to a stationary value and recording the output final value. Figure 37.5 shows the step amplitude versus the final output value. This final value matches with one fixed point state coordinate.

37.3.2 Potential Evoked Problem

In this problem the dynamic is much more complex that in above case. In this problem appears a period two stable cycle and four saddle fixed points. In this case the

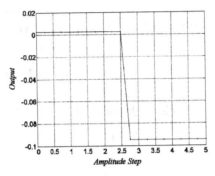

Fig. 37.5 Neural network output and amplitude step value

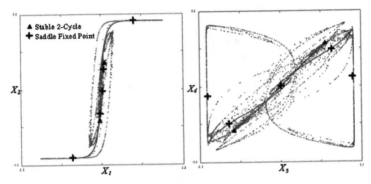

Fig. 37.6 Fixed points and saddle unstable manifolds (PE problem)

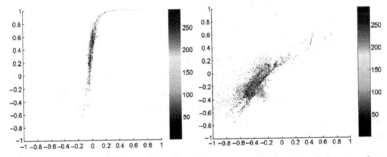

Fig. 37.7 State evolution with external input. X1 represents the output of the neural network. Iteration is coded by grey scale (PE problem)

unstable manifolds of the saddle fixed points (trajectories that diverge from saddle fixed points) are of dimension three and they are complex as can be appreciated in Fig. 37.6.

In other hand, Fig. 37.7 represents the state evolution with the input pattern present. It shows that the output corresponding to X_1 becomes saturated around the times in which appear the P_{100} pattern. The trajectories are complex, due to saddle fixed points and their unstable manifolds, as shown in Fig. 37.7.

37.4 Error vs. Dynamic

This section is devoted to analyzing the relationship between dynamics and error in the training process. Figure 37.9 shows the training error in the DC motor identification problem.

In this case two different behaviors appear. Figure 37.10 shows the dynamics corresponding to the biggest error point in Fig. 37.9. It is characterized by the presence of one stable fixed point, with a saturated state coordinate X_1, which is neural network output. This fact motivates the higher error value. The lower error point dynamics was analyzed in the previous section (see Fig. 37.4).

The neural network dynamics changes from only a stable fixed point to a situation in which it is a saddle fixed point, with two additional stable fixed points, produced by saddle-node bifurcation [4].

Similarly, in the Evoked Potential problem exists a relationship between the network dynamics and error evolution. Figure 37.8 shows that the error abruptly decreases. It is interesting to analyze the dynamics in the three error values represented in the figure.

Firstly, the greater error value dynamic is composed by three fixed points: stable, unstable and saddle, respectively, and three two-period orbits, two saddle and one stable (Fig. 37.11). Actually, the most important aspect with respect to dynamics is that stable fixed point and the stable two-period orbits are situated in the saturation. It means that states tend to saturation without external inputs.

In the other hand, in the middle error point, the corresponding dynamics is composed by a saddle fixed point and a quasiperiodic-orbit. The error value is relatively high because quasiperiodic-orbit is near the saturation values, as show in Fig. 37.12.

In the middle error dynamics appears a quasiperiodic orbit as a consequence of crossing a Neimark-Sacker bifurcation. In both problems, it is concluded that the error decreases by changing the dynamics and crossing a bifurcation of some kind.

Finally, the dynamics corresponding to lower error values was that described in the previous section.

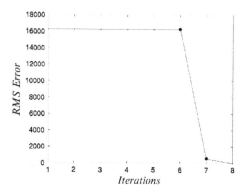

Fig. 37.8 Error evolution in the DC motor problem. The points correspond to the analyzed dynamics

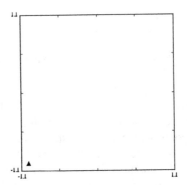

Fig. 37.9 Lower error point dynamic in DC Motor. The stable fixed point is represented by a triangle

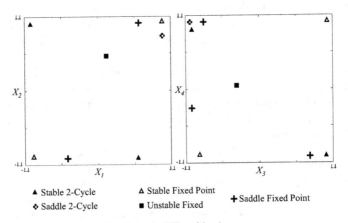

▲ Stable 2-Cycle   △ Stable Fixed Point   + Saddle Fixed Point
✧ Saddle 2-Cycle   ■ Unstable Fixed

Fig. 37.10 Greater value error point dynamic (PE problem)

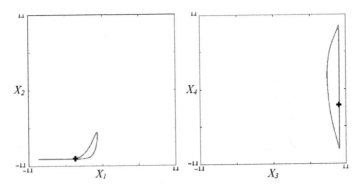

Fig. 37.11 Middle error point dynamic in PE problem. The saddle fixed point and quasiperiodic-orbit are represented by cross and points respectively

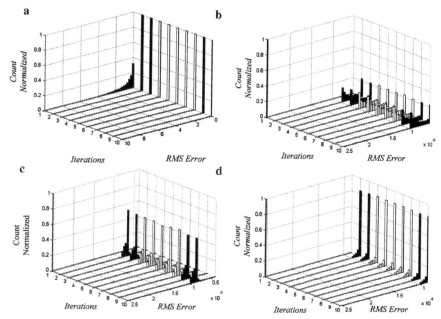

Fig. 37.12 Error normalizated distribution in the DC motor induced problem. **a** classical strategy, **b** fold bifurcation initialization, **c** flip bifurcation initialization and **d** Neimark-Sacker bifurcation initialization

37.5 Initialization Strategy

The classical weight initialization strategy is inspired by static neural networks; it is based on choosing small initial random weights [17]. In fact, this strategy avoids the neural network saturation and this situation is desirable because it does allow training with backpropagation algorithm in the static neural network or the state derivative along time in the dynamic neural network. However, with respect to the neural network dynamic this strategy is a bit restrictive because the origin is the only fixed point and is stable. In fact, the neural network is topologically equivalent to a linear stable system.

In general is necessary to extend the new strategy to a multidimensional dynamic neural network system. The bifurcation conditions are: an eigenvalue equal to *1* to reflect saddle-node bifurcation, an eigenvalue equal to *(−1)*, in the case of a flip bifurcation, and two conjugate complex eigenvalues in the unit circle in the Neimark-Sacker bifurcation.

In order to implement this new strategy, the initial weight matrix must be particularized. The algorithm of matrix initialization consists firstly in generating n random weights from a uniform distribution, where n corresponds to the number of neurons. Next, the first value is assigned to *1*, in case to begin on saddle-node bifurcation or *(−1)* in flip bifurcation. The third step consists in a permutation of this number and the inclusion of this vector as the diagonal in a diagonal matrix D. In the last step it is generated a $n \times n$ random rotation matrix T, so the definitive initial weight matrix is

$$W_{Output} = T^{-1} D T. \tag{37.2}$$

In order to guarantee that the neural network begins in the bifurcation it is necessary also that the bias and input weights are zero. The Neimark-Sacker bifurcation initialization is similar that above case, except that the matrix D include a rotational matrix characterized by two conjugate complex eigenvalues in the unit circle:

$$D = \begin{bmatrix} v_1 & 0 & \cdots & & \cdots & \cdots & 0 \\ 0 & \ddots & & & & & \vdots \\ \vdots & & \cos(\theta) & sen(\theta) & & & \vdots \\ \vdots & & -sen(\theta) & \cos(\theta) & & & \vdots \\ \vdots & & & & \ddots & & \vdots \\ 0 & \cdots & & \cdots & & 0 & v_n \end{bmatrix} \tag{37.3}$$

where θ is random value.

In the Fig. 37.13 the comparison results in the DC motor identification problem are sketched. The classical strategy, where the initial weight values come from

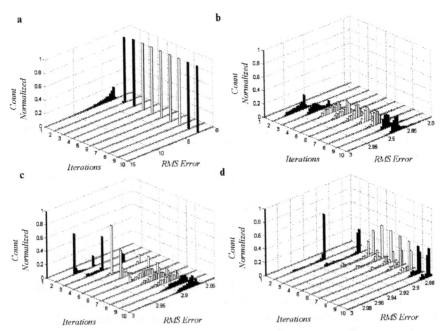

Fig. 37.13 Error normalizated distribution in the Potential Evoked problem. **a** classical strategy, **b** fold bifurcation initialization,**c** flip bifurcation initialization and **d** Neimark-Sacker bifurcation initialization

a uniform distribution in $(-0.01, 0.01)$, fold, flip and Neimark-Sacker bifurcation initialization results are displayed. It is shown that the best strategy corresponds to the fold bifurcation initialization. The second best is the flip bifurcation initialization that still improves the classical initialization. Inside each three-dimensional histograms appear many weights with similar error values. These are caused by the local minima in the error surfaces towards which the weights converge.

37.6 Conclusion

An algorithm for determining the initial recurrent network based on bifurcation theory was successfully developed. This new strategy was applied to two problems. The first problem, to allow easier visualization of the relation between dynamics and error and it was easy to explain the good results of the bifurcation strategy in this problem. The potential evoked problem was much more complex to train but it obtained similar results and it showed the benefits of inducing initial situations of bifurcation in training process.

References

1. C. M. Marcus and R. M. Westervelt, Dynamics of Iterated-Map Neural Networks, Physical Review A **40**, 501–504 (1989).
2. J, Cao, On Stability of Delayed Cellular Neural Networks, Physics Letters A **261**(5–6), 303–308 (1999).
3. P. Tiño, B. G. Horne, and C. L. Giles, Attractive Periodic Sets in Discrete-Time Recurrent Networks (with Emphasis on Fixed-Point Stability and Bifurcations in Two-Neuron Networks), Neural Computation **13**(6), 1379–1414 (2001).
4. R. Hush and B. G. Horne, Progress in Supervised Neural Networks, IEEE Signal Processing Magazine **10**(1), 8–39 (1993).
5. F. Pasemann, Complex Dynamics and the Structure of Small Neural Networks, Network: Computation in Neural System **13**(2), 195–216 (2002).
6. F. Pasemann, Synchronous and Asynchronous Chaos in Coupled Neuromodules, International Journal of Bifurcation and Chaos **9**(10), 1957–1968 (1999).
7. C. Robinson, Dynamical Systems: Stability, Symbolic Dynamics, and Chaos, Boca Raton, FL: CRC Press (1995).
8. J. Hale and H. Koçak, Dynamics and Bifurcations, New York: Springer (1991).
9. X. Wang, Discrete-Time Dynamics of Coupled Quasi-Periodic and Chaotic Neural Network Oscillators, International Joint Conference on Neural Networks, Baltimore, Maryland, USA (1992).
10. Y. A. Kuznetsov, Elements of Applied Bifurcation Theory, 3rd edition, New York: Springer (2004).
11. J. D. Piñeiro, R. L. Marichal, L. Moreno, J. Sigut, I. Estévez, R. Aguilar, J. L. Sánchez, and J. Merino, Evoked Potential Feature Detection with Recurrent Dynamic Neural Networks, International ICSC/IFAC Symposium on Neural Computation, Vienna (1998).
12. K. Ogata, Modern Control Engineering, 4th edition, Englewood Cliffs, NJ: Prentice Hall (2001).

13. D. Regan, Human Brain Electrophysiology: Evoked Potentials and Evoked Magnetic Fields in Science and Medicine, New York: Elsevier (1989).

14. G. Thimm and E. Fiesler, High-Order and Multilayer Perceptron Initialization, IEEE Transactions on Neural Networks **8**(2), 349–359 (1997).

15. R. L. Marichal, J. D. Piñeiro, L. Moreno, E. J. González, and J. Sigut, Bifurcation Analysis on Hopfield Discrete Neural Networks, WSEAS Transaction on System **5**, 119–125 (2006).

16. Y. Bengio, P. Simard, and P. Frasconi, Learning Long-Term Dependences with Gradient Descent Is Difficult, IEEE Transaction on Neural Networks **5**, 157–166 (1994).

17. R. Fletcher, Practical Methods of Optimization, 2nd edition, Chichester: Wiley (2000).

Chapter 38
GAHC: Hybrid Genetic Algorithm

Radomil Matousek

Abstract This paper introduces a novel improved evolutionary algorithm, which combines genetic algorithms and hill climbing. Genetic Algorithms (GA) belong to a class of well established optimization meta-heuristics and their behavior are studied and analyzed in great detail. Various modifications were proposed by different researchers, for example modifications to the mutation operator. These modifications usually change the overall behavior of the algorithm. This paper presents a binary GA with a modified mutation operator, which is based on the well-known Hill Climbing Algorithm (HCA). The resulting algorithm, referred to as GAHC, also uses an elite tournament selection operator. This selection operator preserves the best individual from the GA population during the selection process while maintaining the positive characteristics of the standard tournament selection. This paper discusses the GAHC algorithm and compares its performance with standard GA.

Keywords Genetic Algorithm · HC mutation · Hill climbing algorithm

38.1 Introduction

The term optimization refers to the class of problems in which one wants to minimize, or maximize, an objective function f by systematically selecting the independent parameters from a given set $\mathbf{X}$, which could consist of real or integer variables. The parameters are the variables $\mathbf{x}$ to be optimized for a given optimization task. The goal of an optimization algorithm is to find $\mathbf{x}_{opt}$ which fulfills Eq. (38.1)

$$\mathbf{x}_{opt} = \arg \underset{\mathbf{x} \in \mathbf{X}}{\mathrm{opt}} f(\mathbf{x}) \tag{38.1}$$

R. Matousek
Brno University of Technology, FME, Institute of Computer Science and Automation,
Technicka 2896/2, 61600 Brno, Czech Republic
e-mail: matousek@fme.vutbr.cz

S.-I. Ao et al. (eds.), *Advances in Computational Algorithms and Data Analysis,*
Lecture Notes in Electrical Engineering 14,
© Springer Science+Business Media B.V. 2009

where argopt returns the value (or values) of $\mathbf{x}$ in the set $\mathbf{X}$ that minimize (or maximize) the objective function f.

If the objective function is convex over the region of interest, any local extreme value of the function will also be a global extreme. For this class of problems, fast and robust numerical optimization techniques are readily available. However, if the objective function is nonlinear and/or multi-modal, deriving at the global solution can be difficult. In such a case, soft computing methods can offer a solution.

Examples of these methods are *Genetic Algorithms* (GA) and *Hill-climbing Algorithms* (HCA). A common feature of GA and HCA is the *binary representation* of the vector $\mathbf{x}$. In general [1], GA is a global search method, whereas HCA is a local optimization method [2]. By combining a local search technique, in this case HCA, with a global search method, i.e. GA, the resulting algorithm could be more powerful than the individual algorithms. In this work, HCA is incorporated into GA by using it as a mutation operator. An innovated version of tournament selection, denoted as an elite tournament, is also introduced in this research. Furthermore, it is shown that a careful design of the HCA can increase the optimization capabilities of the HCA.

38.2 Parameter Encoding

38.2.1 Binary Representation of the Task

The coding representation of a problem is based on binary vectors of length n.

$$\mathbf{a} = \{a_1, a_2, \ldots, a_n\} \in \{0, 1\}^n \tag{38.2}$$

In the context of genetic algorithms, these vectors are called individuals. A set of individuals is denoted as a population $\mathbf{P}$.

$$\mathbf{P} = (a_{i,j})_{\substack{i \in \{1,2,\ldots l\} \\ j \in \{1,2,\ldots n\}}} = \begin{pmatrix} a_{1,1} & a_{1,2} & \cdots & a_{1,n} \\ a_{2,1} & a_{2,2} & \cdots & a_{2,n} \\ \vdots & \vdots & \ddots & \vdots \\ a_{l,1} & a_{l,2} & \cdots & a_{l,n} \end{pmatrix} \tag{38.3}$$

The real or integer parameters $\mathbf{x}$ are derived from the binary vector (38.2), respective (38.3), by means of a specific transformation function Γ (Eq. (38.4)). Figure 38.1 demonstrates the parameter encoding scheme.

$$\mathbf{x} = \Gamma(\mathbf{a}) \tag{38.4}$$

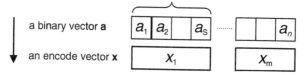

Sub-vector of vector **a** which encode a parameter x_1

a binary vector **a**

an encode vector **x**

Fig. 38.1 Parameters' encoding scheme

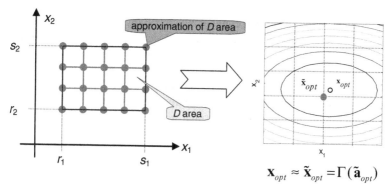

Fig. 38.2 The solution space sampling and its effect

38.2.2 Hamming Metric

A distance function ρ is a function which is used as a metric that defines a distance between elements of a set and fulfills the ground axioms of the metric (positive definiteness, symmetry, triangle inequality). A set with a metric is called a metric space, which is in this case a binary metric space. The *Hamming distance* ρ_H between two binary vectors of equal length is the number of positions for which the corresponding symbols are different. Let **a**, **b** are binary vectors of length n and a, b its elements, then the Hamming distance can be calculated as follows:

$$\rho_H(\mathbf{a}, \mathbf{b}) = \sum_{i=1}^{n} |a_i - b_i| \qquad (38.5)$$

38.2.3 Domain of Definition

The domain of an interval D of the encoded real or integer vectors **x** (38.4) is given by (38.6) and it represent the feasible area of the possible solution (38.1).

$$D = \prod_{i=1}^{m} [r_i, s_i] = [r_1, s_1] \times [r_2, s_2] \times \cdots \times [r_m, s_m] \qquad (38.6)$$

In case of binary representation, the D area can be approximated using an orthogonal grid (Fig. 38.2).

Each individual in the population $\mathbf{P}$ is evaluated using an objective function f on the domain definition D.

$$f : D \rightarrow \Re \qquad (38.7)$$

38.3 Implemented HCA

Hill climbing is an optimization technique which belongs to the class of local search methods [3, 4]. The algorithm starts with an initial solution to the problem, and subsequently makes small changes to the current solution, each time improving it slightly. After a number of iterations, the algorithm arrives at a point where it cannot improve the solution anymore, which is when it terminates. Ideally, at that point the solution found should be close to optimal solution, but it is not guaranteed that hill climbing will ever come close to this optimal solution.

38.3.1 Basic Concept

The most important step of HCA is to generate a close neighborhood of an initial (previous) solution. This solution can be denoted as a kernel of the HC transformation. The next solution, i.e. kernel, is chosen as the best solution from the set of neighborhood solutions (Fig. 38.3).

The Hamming metric is used for measuring the distances between the previous solution and the newly generated solution from the neighborhood. All solutions must belong to the discrete area D. The area $D \in \Re$ is given by the relation

$$\Gamma : \{0, 1\}^n \rightarrow D. \qquad (38.8)$$

Because $\mathbf{x} = \Gamma(\mathbf{a})$, one can find the optimal solution as follows:

$$\mathbf{a}_{opt} = \arg \min_{\mathbf{a} \in \{0,1\}^l} f(\Gamma(\mathbf{a})) \qquad (38.9)$$

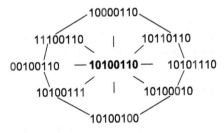

Fig. 38.3 An example of neighborhood generating

$$
\begin{aligned}
&\text{\% binary HCA (minimization)}\\
&\mathbf{a}_{opt} = \text{the first iteration (random generated vector)}\\
&\textbf{repeat}\\
&\qquad \mathbf{a}_{kernel} = \mathbf{a}_{opt}\\
&\qquad\quad \forall \mathbf{a}_{neighbor} = S(\mathbf{a}_{kernel})\\
&\qquad \mathbf{a}_{opt} = \underset{\mathbf{a}\in S\backslash(\mathbf{a}_{kernel})}{\arg\ \min}\ f\!\left(\Gamma(\mathbf{a})\right)\\
&\textbf{until}\ f\!\left(\Gamma(\mathbf{a}_{opt})\right)\gtreqless f\!\left(\Gamma(\mathbf{a}_{kernel})\right)
\end{aligned}
$$

Fig. 38.4 Pseudo-code of the Hill-climbing algorithm

For every valid kernel $\mathbf{a}_{kernel}$ it is possible to determine its neighborhood $\mathbf{a}_{neighbor} \in S(\mathbf{a}_{kernel})$. The choice of the transfer function S will determine the behavior of HCA. Figure 38.4 shows the pseudo-code for algorithm:

38.3.2 Transfer Function

The HCA used in this work is based on a given arbitrary but fixed set of transformations of binary vectors.

$$\mathbf{a} \in \{0,1\}^{n},\ n \in \mathbb{N} \quad (n\ldots\text{length of a binary string}) \tag{38.10}$$

Let H be a set of designed transformations and t be a given transformation.

$$H = \{t_{0}, t_{1}, \ldots, t_{n}\} \tag{38.11}$$

For the mapping of a binary vector $\mathbf{a}_{kernel}$ on the $\mathbf{a}_{neighbor} \in A_{neighbor}$ the a transformation $t \in H$ is used, as is shown in the following equations

$$A_{neighbour} = \{\mathbf{a}_{1}, \ldots, \mathbf{a}_{c}\},\ A_{neighbour} = \begin{pmatrix} \mathbf{a}_{1} \\ \vdots \\ \mathbf{a}_{c} \end{pmatrix},\ \text{proc} \in \mathbb{N}, \tag{38.12}$$

$$t : \mathbf{a}_{kernel} \rightarrow A_{neighbour},\ \text{i.e. } t : \{0,1\}^{n} \rightarrow (\{0,1\}^{n})^{c}, \tag{38.13}$$

where c is the cardinality of the set $A_{neighbor}$. The cardinality depends on the chosen transformation t and the length n of the binary vector $\mathbf{a}_{kernel}$.

$$c_{k}(t_{k}, n) = |A_{neighbour}| = \binom{n}{k},\ \text{for } k \in \{0, 1, \ldots, n\}, \tag{38.14}$$

where index k denotes the relation to the concrete element from the set H according (38.11).

For the realization of the set of transformations H, a set of matrices $\mathbf{M}$ is introduced. For every given transformation t_k there exists a corresponding matrix $\mathbf{M}$, which is of the order k and is denoted by $\mathbf{M}_k$. Please note that in this case the order has a different meaning from the standard square matrix order.

$$t_k, n \Leftrightarrow \mathbf{M}_k, \text{ for } k \in \{0, 1, \ldots, n\} \tag{38.15}$$

Definition: The matrix $\mathbf{M}$ of order k, referred to as $\mathbf{M}_k$, is a matrix where the rows represent all the points of the Hamming spaces $\mathbb{H}^n$ with distances k from the origin (i.e. the zero vectors of length n) in the sense of metrics measurement ρ_H. $\Box$.

Possible examples of $\mathbf{M}$ are outlined in the following equations (38.16). For the calculation of the $\mathbf{A}_{neighbour}$ matrix, it is necessary to introduce an operation for the so-called vector replication of a binary vector $\mathbf{a}_{kernel}$.

$$\mathbf{M}_0 = \begin{pmatrix} 0_{1,1} & 0_{1,2} & \cdots & 0_{1,n} \end{pmatrix}$$

$$\mathbf{M}_1 = \begin{pmatrix} 1_{1,1} & 0_{1,2} & \cdots & 0_{1,n} \\ 0_{2,1} & 1_{2,2} & & \\ \vdots & & \ddots & \\ 0_{c_1,1} & & & 1_{c_1,n} \end{pmatrix}$$

$$\mathbf{M}_2 = \begin{pmatrix} 1_{1,1} & 1_{1,2} & 0_{1,3} & \cdots & & 0_{1,n} \\ 1_{2,1} & 0_{2,2} & 1_{2,3} & & & 0_{2,n} \\ \vdots & & & & & \\ & & & & \ddots & \\ 0_{c_2,1} & & & & 1_{c_2,n-1} & 1_{c_2,n} \end{pmatrix}$$

$$\vdots$$

$$\mathbf{M}_{n-1} = \begin{pmatrix} 0_{1,1} & 1_{1,2} & \cdots & 1_{1,n} \\ 1_{2,1} & 0_{2,2} & & \\ \vdots & & \ddots & \\ 1_{c_{n-1},1} & & & 0_{c_{n-1},n} \end{pmatrix}$$

$$\mathbf{M}_n = \begin{pmatrix} 1_{1,1} & 1_{1,2} & \cdots & 1_{1,n} \end{pmatrix} \tag{38.16}$$

An operation matrix $\mathbf{A}_{kernel}$ is formed and contains identical copies of the binary vector $\mathbf{a}_{kernel}$ in rows. The number of rows in the matrix $\mathbf{A}_{kernel}$ is equal to the number of rows in the given matrix $\mathbf{M}$.

$$\mathbf{A}_{kernel} = \begin{pmatrix} \mathbf{a}_{1,kernel} \\ \vdots \\ \mathbf{a}_{c,kernel} \end{pmatrix} \tag{38.17}$$

Applying the principle of addition in modular arithmetic (mod 2) to (38.14), respectively (38.15), leads to the following transformation t (38.13).

$$t_k : \mathbf{A}_{neighbor} = \mathbf{A}_{kernel} \oplus \mathbf{M}_k, \ for \ t_k \in H \ ak = 0, 1, \ldots, n \tag{38.18}$$

The transformation t_k generates the complete set of vectors, which have the distance k from the origin in terms of the metrics ρ_H.

$$t_k \Rightarrow \rho_H(\mathbf{a}_{kernel}, \mathbf{a}_{neighbor}) = k, \ for \ \forall \mathbf{a}_{neighbor} \in A_{neighbor} \tag{38.19}$$

The generalization of Eq. (38.18) for a arbitrary but fixed choice of the elements from the set of $\mathbf{H}$ is evident. The set of chosen, and for HCA fixed, transformation is denoted as $\mathbf{H}_v$.

$$H_v \subseteq H \tag{38.20}$$

The set $\mathbf{H}_v$ definitely determines the transformation S, which is linked to transformations from this set.

$$H_v \Leftrightarrow S, \ thus \ S : \mathbf{A}_{NEIGHBOR} = \begin{pmatrix} \mathbf{A}_{kernel, k_1} \oplus \mathbf{M}_{k_1} \\ \vdots \\ \mathbf{A}_{kernel, k_v} \oplus \mathbf{M}_{k_v} \end{pmatrix} \tag{38.21}$$

Where $k_i \in I$ and I is the index set of the chosen elements from $\mathbf{H}$, which is defined by the choice $\mathbf{H}_v$. The cardinality $c_{NEIGHBOR}$ of the set $A_{NEIGHBOR}$, which is arrived at by means of the set of transformations H_v, is determined by the sum of all partial cardinalities in Eq. (38.14), i.e.

$$A_{NEIGHBOR} = \bigcup_{k_i} A_{neighbor, k_i} \tag{38.22}$$

$$c_{NEIGHBOR} = \sum_{k_i \in I} \binom{n}{k_i} \tag{38.23}$$

Remark 1: The equation for the sum of binomial coefficients is obtained for the limited case. This equation states that the whole space of $n + 1$ transformations by (38.11) represents a full search of the Hamming metrics space.

$$\sum_{i=0}^{n} \binom{n}{i} = 2^n \tag{38.24}$$

Remark 2: One consequence of the previous remark is rather theoretic; algorithms that carry out an exhaustive space searching are guaranteed to find the global solution in a given universe, but this is often not practical because of the combinatorial expansion of the search space.

$$\mathbf{a}_{global \ optimum} = \left[\arg \min_{a \in \{0,1\}^n} f(\Gamma(\mathbf{a})) \right], \ for \ |A_{NEIGHBOR}| \rightarrow 2^n \tag{38.25}$$

38.3.3 Implementation of the Proposed Hill Climbing Algorithm

For the practical implementation of the mutation operator in GAHC a variant of HCA denoted as HC12 (conjunction of the HC1 and HC2 algorithms) was used [4]. This HCA uses the set of transformations $H_v = \{t_0, t_1, t_2\}$.

Algorithm name	HC1	HC2	HC12
Set of transformations	t_0, t_1	t_0, t_2	t_0, t_1, t_2
M matrices	$\mathbf{M}_0, \mathbf{M}_1$	$\mathbf{M}_0, \mathbf{M}_2$	$\mathbf{M}_0, \mathbf{M}_1, \mathbf{M}_2$
Cardinality of neighborhood	$1 + n$	$1 + \binom{n}{2}$	$1 + n + \binom{n}{2}$
$\rho_H(a_{kernel}, \forall a_{neighbour})$	1	2	1 and 2

The designed variants of HCA prevent a decrease of the objective function value because the t_0 transformation is always contained. Examples of $\mathbf{A}_{neighbor}$ vector coverage of the solution space can bee seen in the figure below (Fig. 38.5).

Examples of stand-alone HCA search runs for the well-known Rosenbrock test function, also known as F2 or banana valley function, are shown in Fig. 38.6 (standard variant HC1) and in Fig. 38.7 (advanced variant HC12).

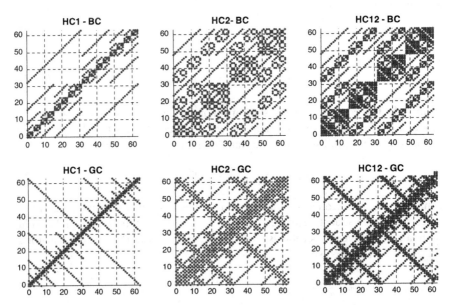

Fig. 38.5 Examples of coverage of a solution space for different transformations H_v and selected encoding methods, where BC denotes a direct binary encoding, whereas GC represents a Gray binary encoding

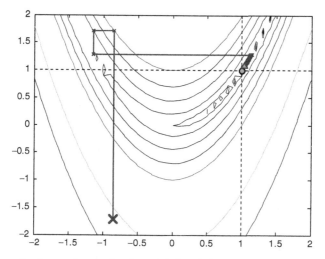

Fig. 38.6 Example of a search run using the stand-alone HCA variant HC1

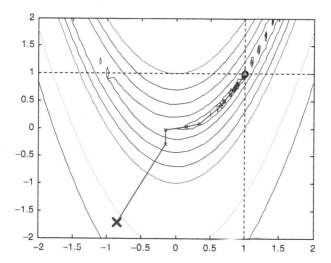

Fig. 38.7 Example of a search run using the stand-alone HCA variant HC12

In these examples, each parameter was encoded using 15-bit Gray code representation. The standard HC1 variant needed 1,527 iteration steps and found the solution $F2_{min}(x_1 = 1.0375, x_2 = 1.0764) = 0.0014$. The advanced HC12 variant needed, starting with the same initial solution, only 46 iteration steps and found a better solution $F2_{min}(x_1 = 0.9999, x_2 = 0.9998) = 0.0000$.

38.4 Elite Tournament Selection

Tournament selection is one of many methods for selecting offspring in genetic algorithms. It runs a "tournament" among a number of randomly chosen individuals from the population and selects the winner, i.e. the one with the best fitness, for crossover. This chapter introduces briefly an enhanced version of tournament selection, called *elite tournament* [5, 6]. This improved selection method overcomes the main problem of the standard tournament selection, which is that it does not guarantee the reproduction of the best solution contained in the current generation.

38.4.1 Background

The selection operator is intended to improve the average quality of the population $\mathbf{P}$ (38.3) by giving individuals of higher quality a higher probability to be copied into the next generation, i.e. selection operator $s: \mathbf{P_{BS}} \rightarrow \mathbf{P_{AS}}$. Here, $\mathbf{P_{BS}}$ refers to the population before selection and $\mathbf{P_{AS}}$ refers to the population after selection. In other words, the selection operator focuses the search on promising regions in search space.

The tournament selection operator is very popular because of its properties [1]. However, one shortcoming of this selection mechanism is that it does not guarantee the survival of the best individual through the selection process $s: \mathbf{P_{BS}} \rightarrow \mathbf{P_{AS}}$. In the case of elite tournament selection, this disadvantage of the classical tournament selection operator is eliminated, while preserving all the advantages of the classical tournament selection.

Tournament selection works as follows: Choose randomly t individuals from the population $\mathbf{P_{BS}}$ and copy the best individual from this tournament group into the population $\mathbf{P_{AS}}$. This step is repeated l times. The probability of selecting p_s of the best individual is

$$p_s(t,l) = 1 - \left(\frac{t-1}{l}\right)^{t \cdot l} \tag{38.26}$$

Elite tournament selection is the same as tournament selection; it copies l times the best individual from a group to population $\mathbf{P_{AS}}$, but the tournament groups are formed differently: the first individual in group i, where $i \in (1, 2, \ldots, l)$, is the ith individual from population $\mathbf{P_{BS}}$, the rest of group is selected randomly. The time complexity of this algorithm is the same as tournament selection, i.e. $\mathcal{O}(n)$. The probability of selection of the best individual is

$$p_s(t,l) = 1 \tag{38.27}$$

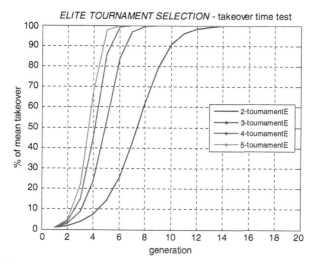

Fig. 38.8 Elite tournament selection (tournamentE) for different tournament sizes t = {2,3,5,5} using populations of 100 individuals. The graphs represent the proportion of the population consisting of the best individual as a function of generation

38.4.2 Evaluation and Analysis

For the analysis of the proposed algorithms the *takeover time* test was used. In the experiments a population size $n = 100$ was used. The following empirical results (Fig. 38.8) were obtained from 1,000 independent runs using the selection process only.

38.5 GAHC

Exploration versus exploitation is a well-known issue in Evolutionary Algorithms. An unbalanced search may waste many iteration steps, i.e. computational expensive fitness evaluations, or lead to premature convergence. A very successful strategy is to combining global and local search methods to improve global optimization capabilities. However, HCA, in the version HC12, is strictly speaking not a local search technique.

38.5.1 Background

The following primary requirements for GAHC design were identified:

- The global search capability of the GA must be preserved or improved. Premature convergence of the GA should not occur.

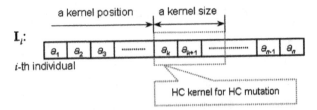

Fig. 38.9 The principle application of HCA as mutation operator for GAHC

- The local search feature of the GA should only require a few iteration steps, i.e. fitness evaluations, during the search for the optimum.
- The computation time for solving the optimization task should be less than for the stand-alone GA.

GAHC, an improve Genetic Algorithm, addresses this requirements by using an intelligent mutation technique, which can replace or supplement the standard GA mutation operator. Here, HCA is used as mutation operator:

1. Chose an individual $\mathbf{I}_i$
2. Determining the kernel position
3. Applying of the HCA on the kernel, using the entire individual for objective function evaluation

The individual, kernel position and kernel size can be chosen randomly or based on some heuristic, that could exploit some existing inside into the optimization problem at hand. The number of HCA kernels and their sizes are determined by the standard GA mutation operator. For example, if ten HCA kernels of 5 bit length would be used, this would result into a standard bit mutation of 50 bits. The HCA kernel is chosen to contain 10 bits for the GAHC in order to reduce computation time and to improving the global search behavior.

The size of the HCA kernel is depended on the bit length of an individual. The basic principle of the application of the HCA mutation operators can be seen from Fig. 38.9.

38.5.2 Evaluation and Analysis

In order to evaluate the new GAHC algorithm, Ratrigin's function F6 (Eq. (38.28)) and many another was chosen for the experiments.

$$F_6(\mathbf{x}) = 10n + \sum_{i=1}^{n} \left(x_i^2 - 10\cos(2\pi x_i) \right)$$

$$-5.12 \le x_i \le -5.12, \min F_6(\mathbf{x}) = F_6(0,\dots,0) = 0$$

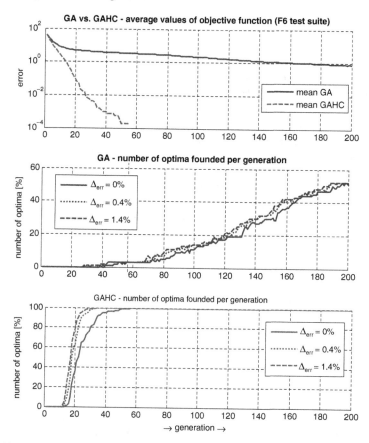

Fig. 38.10 A visualization of the GA and GAHC test runs. The second and third graphs represent the number of optima founded per generation for 1,000 runs

Rastrigin's function F6 is based on the power function with the addition of cosine modulation to introduce many numerous minima. Thus, the resulting test function is highly multimodal and the locations of the minima are equally distributed.

Results of the test runs for the GA and GAHC variant are presents in Fig. 38.10. The number of runs that were carried out in each test was 1,000. The population size was chosen to be 50 and each individual had a length of 50 bits. The GA parameters chosen were the followings:

Parameter	Value
GA test suite	F6, 5 optimized variables
GA selection	Elite tournament selection, 3
GA crossover	80% of individual in 3cat
GA mutation	Mutation probability 0.02
GAHC	10 HCA kernel of size 5 bits

38.6 Conclusion

This paper introduced GAHC, a novel improved evolutionary algorithm, which combines genetic algorithms and hill climbing.

The main principles of the Hill-Climbing algorithm were explained and a powerful variant, referred to as HC12, was introduced. These algorithms can be used as stand-alone optimization algorithms or they can be used as an effective and efficient mutation operator within a Genetic Algorithm.

Experiments, using the well-known test functions F6 for GAHC and F2 for HCA and HC12, have shown that if the Hill-Climbing method is used as a mutation operator, it improves the local search capabilities of Genetic Algorithms while preserving their global search capabilities.

Finally, was also demonstrated that a new tournament selection operator, referred to as elite tournament selection, further improved the performance of the new hybrid algorithm.

Acknowledgements This work was supported by the Czech Ministry of Education in the frame of MSM 0021630529 *Intelligent Systems in Automation.*

References

1. Goldberg, D. E., 1989, *Genetic Algorithms in Search, Optimization and Machine Learning.* Boston, MA: Addison-Wesley.
2. Mitchell, M. and Holland, J. H., 1994, When Will a Genetic Algorithm Outperform Hill Climbing? In J. D. Cowan, G. Tesauro, and J. Alspector (Eds.), *Advances in Neural Information Processing Systems 6.* San Mateo, CA: Morgan Kaufmann.
3. Halim, C., 2006, Developing Combined Genetic Algorithm – Hill-Climbing Optimization Method for Area Traffic Control, *Journal of Transportation Engineering*, Volume 132, Issue Number 8, pp. 663–671.
4. Matousek, R., 1995, *GA (GA with HC Mutation) – Implementation and Application (in Czech)*, Master thesis, Brno University of Technology, Brno, Czech Republic.
5. Bednar, J. and Matousek, R., *Elite Tournament Selection*, in the P. Osmera Proceedings of Mendel 2005 Soft Computing Conference, Brno, Czech Republic.
6. Matousek, R., 2004, *Selected Methods of Artificial Intelligence – Implementation and Application (in Czech)*, Ph.D. thesis, Brno, BUT, Czech Republic.

Chapter 39
Forecasting Inflation with the Influence of Globalization using Artificial Neural Network-based Thin and Thick Models

Tsui-Fang Hu, Iker Gondra Luja, Hung-Chi Su, and Chin-Chih Chang

Abstract We study globalization influences on forecasting inflation in an aggregate perspective using the Phillips curve for Hong Kong, Japan, Taiwan and the US by artificial neural network-based thin and thick models. Our empirical results support the hypothesis that globalization influences do generate a downward tendency in inflation through time in all cases with different levels. Moreover, the artificial neural network-based thin and thick models that have been developed upon the best linear model for each country have shown significant superiority over the naïve model in the most cases and over the best linear model in some cases. With the virtue of application flexibility, finding optimal values of all the parameters for so many artificial neural network models is very time-consuming and complex because of a large number of parameters considered. Thus we do not make any claim on the optimality of the artificial neural network models in this study. As a consequence, although our empirical results are moderately satisfactory, building artificial neural network models based upon the best linear model is a good compromise between practicality and optimality.

Keywords Artificial neural networks · Nonlinearity · Thick model

T.-F. Hu
P.O. Box 2166, State University, Arkansas 72467, U.S.A.

I.G. Luja
Department of Mathematics, Statistics & Computer Science, St. Francis Xavier University, Antigonish, NS B2G 2W5, Canada

H.-C. Su (✉)
Department of Computer Science, Arkansas State University, State University, Arkansas 72467
e-mail: suh@cs.astate.edu

C.-C. Chang
Department of Computer Science and Information Engineering, Chung Hua University, Hsinchu City 30012, Taiwan

S.-I. Ao et al. (eds.), *Advances in Computational Algorithms and Data Analysis*,
Lecture Notes in Electrical Engineering 14,
© Springer Science+Business Media B.V. 2009

39.1 Introduction

Inflation is widely taken as one of the main economic problems, because it involves in the allocation efficiency of the economy by means of changing the relative prices among commodities. A dramatic acceleration of inflation results in large costs to the public and business community. The mechanism of how the relative-price adjustments among commodities affect (or cost) the economy is very complicated and not yet well understood. Pursuing stable prices is the primary goal of a central bank while creating and maintaining an independent and efficient monetary system. However, the acceleration in the pace of international trade in good, service and financial assets relative to the growth rate of the domestic trade, broadly defined as globalization, has attracted a lot of attentions [1]. This is because globalization has broken some macro rules such as a decrease in the correlations between the labor cost and inflation and a decline in the elasticities between import prices and core inflation. During the past decade, the acceleration in the pace of the world trade growth is well above the acceleration in the pace of the world economic growth with a disparity from 2.99% in 1990 to 3.16% in 2006, especially in recent years, but the annual inflation in the advanced countries has been declining since 1990 and it is even well below the world economic growth rate. In other words, strong economic growth does not result in high inflation in the most advanced economies since 1990. Even for the developing countries, the annual inflation starts to decelerate since 1995. It is unavoidably conspicuous that China's investment- and export-led economic growth with a low labor cost has significantly contributed to the world economic growth in the recent years, which seems to continue. With such a strong engine (China-centric supply chain in Asia) in the supply side, it should possibly generate a downward tendency in inflation while pursuing the world economic growth. The global trade (exports and imports of goods and services) stood 38% of the total world GDP in 1990 and hit 61% in 2006 (see Fig. 39.1). Hence, it is hard to consider the domestic factors only while the central bank conducts the monetary policy when the domestic commodity and financial markets becomes increasingly open

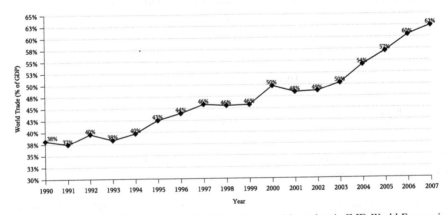

Fig. 39.1 World Trade (% of GDP), 1990–2007 (Elaborated from data in IMF, *World Economic Outlook Database* (Washington, DC, October 2007))

(and integrated to the world markets). As a result, globalization seems to become more and more important to this issue.

Inflation is a major issue for the central bank when deciding the monetary policy, but forecasting inflation is intuitively not an easy task for any economy, in which there are many mechanisms operating simultaneously, especially as the influence of globalization becomes stronger. In economics, the Phillips curve theory conveys that there is a tradeoff relationship between inflation and unemployment rate. Empirically, it seems that the Phillips curve is nonlinear [2]. An increasing body of research [3–7] using artificial neural networks (ANNs) provides evidence of the existence of nonlinear and complex relationships among (or within) economic time series. Consequently, an ANN can be used as an alternative approach to empirically analyze the Phillips curve theory and also can be contrasted with a typical linear approach. Atkeson and Ohanian [8] conclude that their linear models based on the non-accelerating inflation rate of unemployment (NAIRU) cannot provide more accurate forecasts than the naïve forecasts. Therefore, to a certain extent, the modern Phillips curve-based models are not as useful as expected. However, based on the study of McNelis and McAdam [6], the trimmed ANN-based thick model [9] outperforms the linear model for real-time and bootstrap forecasts in the Euro area as well as in some other countries considered in their study using the Phillips-Curve framework. Basically, they remove the 5% largest and smallest forecasts from several ANNs, and then use the average of the remaining forecasts as combined forecasts. It is practical to use a thick model on several neural networks in the first several forecasting periods, because there is no theoretical basis to select the best ANN among many ANNs that have different numbers of hidden layers and neurons in each hidden layer, or different network architectures, but have same input variables. Even after some periods of forecasting, the forecasting performances of different ANNs might be still too close to determine which one is the best. Therefore a thick model can be accordingly applied. However, the removals of the largest and smallest forecasts from several ANNs do not guarantee the best forecasting performance, insomuch as either the largest or the smallest forecast does not necessarily generate the largest forecasting error for each forecasting period. To a certain extent, the range for forecasts should be known a priori when using a trimming method. Moreover, an ANN-based thick (ANN thick) model might not always be a better choice over an ANN-based thin (single ANN; ANN thin) model, so we should consider the ANN thin model as well. Besides, the globalization factor, an increasingly important factor, is absent in their study.

The contribution of this study is two-fold. First, we consider globalization influences while forecasting inflation in an aggregate perspective using the Phillips curve by artificial neural network-based thin and thick models for Hong Kong, Japan, Taiwan and the US, which have data available for the globalization factor. Second, we compare the forecasting performances of ANN-based thin and thick models with the forecasting performances of a simple naïve and the best linear models. The chapter is organized as follows. Section 39.2 describes the theoretical underpinnings and methodology. In Section 39.3, we discuss and compare the empirical results. The chapter ends with some concluding remarks.

39.2 Methodology

39.2.1 Theoretical Underpinnings of Artificial Neural Networks

Artificial Neural Networks (ANNs) are one of the most frequently used techniques in the field of machine learning. The field of machine learning focuses on the study of algorithms that improve their performance at some task automatically through experience [10].

Suppose we are given training data as a set of n observations. Each observation is a pair $(\mathbf{x}_i, y_i)$ where $\mathbf{x}_i \in \mathfrak{R}^d$ and $y_i \in \mathfrak{R}$. Assume that the training data has been drawn independently from some unknown cumulative probability distribution $P(\mathbf{x}, y)$. The goal is to find a function $f : \mathfrak{R}^d \mapsto \mathfrak{R}$ that implements the optimal mapping. In order to make learning feasible, we have to specify a function space $\mathcal{F}$ from which the function is to be chosen. The functions in $\mathcal{F}$ are labelled by a set $\mathcal{P}$ of adjustable parameters. In particular, a set of n pairs of training data, denoted by $\{(\mathbf{x}_i, y_i)\}_1^n$, are given for the purpose of a binary classification task, where $\mathbf{x}_i \in \mathfrak{R}^d$ and $y_i \in \{1, -1\}$ is the class label. Assume that the data is linearly separable and let $\mathcal{F}$ be the set of linear decision boundaries of the form

$$f(\mathbf{x}) = \text{sign}(\mathbf{w} \cdot \mathbf{x} + b)$$

where $\mathbf{w} \in \mathfrak{R}^d$ and $b \in \mathfrak{R}$ are the adjustable parameters (i.e., $\mathcal{P} = \{\mathbf{w}, b\}$). Thus, choosing particular values for $\mathcal{P}$ results in a trained classifier (See Fig. 39.2).

For any trained classifier, the hyperplane corresponding to $\mathbf{w} \cdot \mathbf{x} + b = 0$ is the decision boundary. A linear decision boundary is a simple classifier that can be learned very efficiently. However, due to its small complexity it can correctly classify data that is linearly separable only.

In general, one way to measure the performance of a trained classifier $f \in \mathcal{F}$ is to look at the mean error computed from the training data. This is known as the empirical risk (or training error) and is defined as

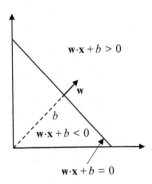

Fig. 39.2 A simple binary classifier

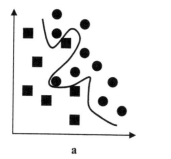 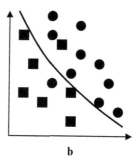

a b

Fig. 39.3 Generalization performance: **a** an overly complex classifier that results in zero error on the training data, but may not generalize well to unseen data; **b** a classifier that might represent the optimal tradeoff between error in the classification of training data and complexity of the classifier, thus capable of generalizing well on unseen data

$$R_{emp}(\mathcal{P}) = \frac{1}{n} \sum_{i=1}^{n} Q(\mathbf{x_i}, \mathcal{P})$$

where $Q(\mathbf{x_i}, \mathcal{P}) = 1$ if $f(\mathbf{x_i}, \mathcal{P}) \neq y_i$ and $Q(\mathbf{x_i}, \mathcal{P}) = 0$ if $f(\mathbf{x_i}, \mathcal{P}) = y_i$.

Minimizing the empirical risk is one of the most commonly used optimization procedures. However, even when there is no error on the training data, the classifier may not generate correct classifications on unseen data (See Fig. 39.3). This problem is known as overfitting. The ability of a classifier to correctly classify new data that is not in the training set is known as generalization. Having a classifier with good generalization is, of course, a much harder problem. The generalization performance of a particular trained classifier f can be measured by the expected risk (or just the risk) defined as

$$R(\mathcal{P}) = \int Q(\mathbf{x}, \mathcal{P}) \, dP(\mathbf{x}, y)$$

Choosing optimal values for $\mathcal{P}$ that minimize the expected risk is known as risk minimization. However, this is not a trivial problem because $P(\mathbf{x}, y)$ is usually unknown. There is a competition of terms. As the complexity of the classifier increases, the empirical risk tends to decrease. However, the generalization error usually increases with increasing complexity. Therefore, in order to control the expected risk, we have to control both the empirical risk and the complexity of the classifier. Note that these two tasks are in conflict with one another. For example, an ANN with a very simple structure may not be capable of correctly classifying most of the training data. That is, it may have high empirical error. On the other hand, an ANN with a very complex structure may correctly classify all the training data but may not generalize well on unseen data. In order to choose from among multiple classifiers, we can follow Ockham's razor: prefer the simplest classifier that is consistent with the training data. The principle of structural risk minimization is an attempt to identify the optimal balance between the quality of the approximation of the training data and the complexity of the approximating function (See Fig. 39.3).

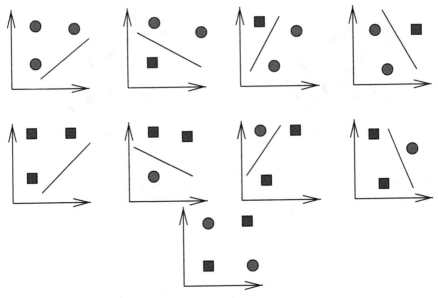

Fig. 39.4 The VC dimension of linear decision boundaries is 3 because they can shatter (any) 3 points in a 2-dimensional space but not (any) 4 points

The Vapnik Chervonenkis (VC) dimension [11] is a measure of the complexity of a set of classifiers $\mathcal{F}$. It is defined as the size of the largest subset of points that can be shattered (or arbitrarily labelled) by choosing classifiers from $\mathcal{F}$ with different values of $\mathcal{P}$ (See Fig. 39.4). Any given set of classifiers $\mathcal{F}$ has a fixed VC dimension. For example, an ANN with a fixed structure represents a set of classifiers (obtained by all possible values for the weights) with a fixed complexity (i.e., fixed VC dimension).

There is a number of bounds on the expected risk. Vapnik and Chervonenkis [11] proved that, given a set of n training examples and a set of classifiers $\mathcal{F}$, with probability $1 - \eta$ over the choice of training set, the expected risk of a trained classifier $f \in \mathcal{F}$ is bounded by

$$R(\mathcal{P}) \leq R_{emp}(\mathcal{P}) + \sqrt{\frac{h(1 + \ln \frac{2n}{h}) - \ln \frac{\eta}{4}}{n}}$$

where h is the VC dimension of $\mathcal{F}$ [11]. Therefore, in order to control the expected risk, by the principle of structural risk minimization, we have to control both the empirical risk (i.e., we have to minimize the error on the training data) and the VC dimension (i.e., we have to minimize the complexity of the classifier).

Thus, for such models with broad approximation abilities and few specific assumptions (e.g., ANNs), the distinction between memorization and generalization becomes important [12]. A validation set (disjoint from the training set) is commonly used to assess generalization performance. Cross-validation is a technique that reduces overfitting. For instance, in a κ-fold cross-validation, the collected data is divided into κ partitions. Each partition in turn is left out and the remaining $\kappa - 1$ partitions are used for training. The left out partition is then used to test generalization performance. The value reported is then the average of the κ tests.

Basically, ANNs are a class of non-linear, non-parametric models that can be trained to approximate general non-linear, multivariate functions. They are massively parallel systems comprised of many interconnected processing elements (known as nodes or neurons) based on neurobiological models of the brain. One of their main advantages in comparison to other models is that they have very few domain-specific assumptions and are highly adaptable (e.g., they learn from experience). Thus, they need no a priori assumption of a model and are capable of inferring complex non-linear input-output transformations. In addition to their typical use in (nonlinear) regression, they are commonly used in pattern recognition, where the ANNs assign a set of input features to one or more classes.

In the late 1950s, Rosenblatt [13] introduced the perceptron learning rule, the first iterative algorithm for training a simple ANN: the perceptron (see Fig. 39.5), which is a linear classifier. After initializing $\mathbf{w}$ and b randomly, each training point $\mathbf{x}_i$ is presented and the output value of y is compared against y_i. If y and y_i are different (i.e., $\mathbf{x}_i$ is misclassified) the values of $\mathbf{w}$ and b are adapted by moving them either towards or away from $\mathbf{x}_i$. Rosenblatt proved that, assuming the classes are linearly separable, the algorithm will always converge and find values for $\mathbf{w}$ and b that solve the classification problem. That is, the algorithm finds a hyperplane that divides the d-dimensional space into two decision regions.

A linear decision boundary (e.g., a single-layer perceptron) is a simple classifier that can be learned very efficiently. However, due to its small complexity it can correctly classify data that is linearly separable only. On the other hand, a more complex decision boundary can correctly classify general data that may not be linearly separable.

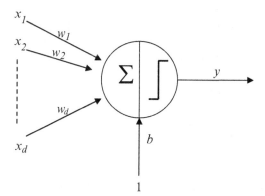

Fig. 39.5 One neuron single-layer d-input perceptron with threshold activation function. The output $y = \text{sign}(\mathbf{w} \cdot \mathbf{x} + b) = \text{sign}(\sum w_i x_i + b)$

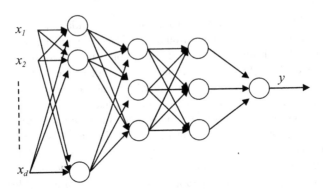

Fig. 39.6 Multilayer d-input perceptron with two hidden layers

The key idea responsible for the power, potential, and popularity of ANNs is the insertion of one or more layers of nonlinear hidden units (between the inputs and output) [12]. These kinds of nonlinearity allow for interactions between the inputs (e.g., products between input variables). As a result, the network can fit more complicated functions [12]. A multilayer perceptron consists of an input layer, at least one middle or hidden layer, and an output layer of computational neurons (see Fig. 39.6). The input signals are propagated in a forward direction on a layer-by-layer basis.

Learning in a multilayer network is similar to learning with a perceptron. That is, each training point x_i is presented to the network as input. The network computes the corresponding output y. If y and y_i are different, the network weights are adjusted to reduce this error. In a perceptron, there is only one weight for each of the d inputs and only one output. However, in a multilayer network, how can we update the weights to the hidden units that do not have a target value? The revolutionary (and somewhat obvious) idea that solved this problem is the chain rule of differentiation [12]. This idea of error backpropagation was first proposed in 1969 [14] but was ignored due to its computational requirements. The backpropagation algorithm was rediscovered in 1986 [15] at a time when computers were powerful enough to allow for its successful implementation. In the backpropagation algorithm, the error (i.e., the difference between the network output and the desired output) is calculated and then propagated backwards through the network from the output to the input layer. The weights are modified as the error is propagated. The sum of squared errors (over all training data) is the performance measure that the backpropagation algorithm attempts to minimize. When the value of this measure in an entire pass through all training data is sufficiently small, it is considered that the network has converged. The following section is going to present the ANN thin, ANN thick and inflation models.

39.2.2 ANN Thin, ANN Thick and Inflation Models

In contrast to traditional models that are "theory-rich" and "data-poor", ANNs are "data-rich" and "theory-poor" in a way that little or no a priori knowledge of the problem is present [12]. Thus, they need no a priori assumption of a model and are capable of inferring complex non-linear input-output transformations. In spite of its straightforward application, there is no theoretical basis to select the best ANN among several ANNs that have different parameter values such as the numbers of neurons in each hidden layer (or different network architectures) at the beginning of forecasting periods, although they may have same input variables. Moreover, the following situation could also possibly occur: ANNs have no significant difference in their out-of-sample forecasting performances after a period of time. Usually there is a method, called thick model, that could be used under these circumstances. A thick model combines several forecasts from different ANNs. In the combination of several forecasts, the weights for the forecasts can be equal or different. Alternatively, we can trim the outliers from the forecasts if we have prior information about the reasonable range of forecasts. Therefore, it is reasonable to use a thick model on several neural networks in the first several periods of forecasting or when it is still hard to decide which ANN is the best after some periods.

In our study, ANN thin and ANN thick models are both considered for each sample country. Additionally, for the ANN thick model, we implement two different kinds. One is ANN thick with equal weight (i.e., average value) model, and the other one is ANN thick with unequal weight model that gives greater weight to the relatively better ANNs and less weight to the relatively worse ANNs by nonlinearly optimizing (minimizing) the forecast error statistics from the first several out-of-sample test periods. In this study, we repeatedly estimate and update both the linear and several ANN models to obtain a one-step-ahead forecast (the real time forecast) while the sample period rolls forward a quarter at a time. For the purpose of avoiding obsolete information, we keep the sample size fixed by getting rid of the oldest data while the estimation is rolling over to the next period.

The Phillips curve framework with a globalization factor can be described as the following model for both the linear and ANN models:

$$\pi_{t+1}^4 - \pi_t^4 = f(\Delta\pi_t^4, ..., \Delta\pi_{t-i}^4, \Delta\pi_t, ..., \Delta\pi_{t-j},$$
$$\Delta u_t, ..., \Delta u_{t-m}, \Delta g_t, ..., \Delta g_{t-n})$$

where π_t^4 is the the 4-quarter change consumer price index (CPI) between quarters t and $t-4$ in an aggregate perspective at an annualized value as measured by $100[\ln(P_t) - \ln(P_{t-4})]$, π_t is quarterly inflation as measured by $400[\ln(P_t) - \ln(P_{t-1})]$, u_t is the quarterly unemployment rate in quarter t, g_t is the quarterly globalization factor (the ratio of exports plus imports to the overall GDP for a country) in quarter t, and i, m, and n are the lag lengths. The left side term is the difference between the 4-quarter change inflation in the current quarter and the next quarter. We have to avoid forecasting long horizons such as the difference between the inflation over the next 4 quarters and inflation over the previous 4 quarters, although some research does not avoid forecasting long horizons [6, 8]. The reason is that a long forecasting horizon does not cohere with the Phillips curve theory that only exists in the short or medium run. In other words, there is no permanent tradeoff between inflation and unemployment in the long run, because the long-run inflation is determined by the central bank, i.e., the relationship in the Phillips cure does not exist in the long run. In addition, we also try different forms of information from the past such as the 4-quarter change unemployment rate and 4-quarter change globalization factor to check if they also generate significant influences in each country. As a result, only the 4-quarter change globalization factor is more significant than the quarterly globalization factor in the case of the US.

In the inflation model, we focus on the influences running from the growth paces of the input variables to the growth pace of the output variable. Usually, the central bank is more concerned with 12-month change in CPI so that the comfort zone of the CPI changes can be observed easily. Due to the unavailability of monthly data for the globalization factor, we can only investigate four countries, which are Hong Kong, Japan, Taiwan and the United States with quarterly data. It will still mean the same for the central bank if we use the 4-quarter change in CPI. The whole sample period runs from the fourth quarter of 1982 to the third quarter of 2006, and the dataset is obtained from International Monetary Fund's International Finacial Statistics (IFS) CD-ROM. For each sample country, around three fourths of each dataset is used to estimate or train the forecasting models. Afterward, the remaining one fourth dataset (i.e., 20 periods as test periods starting from the first quarter of 2002 till the third quarter of 2006) is used to evaluate real time forecast performances among different forecasting approaches.

39.3 Empirical Results and Analysis

As mentioned previously, we repeatedly estimate and update both the linear and several ANN models to get a one-step-ahead forecast from these different forecasting models. Since the economy evolves through time, it is not appropriate to keep obsolete information in our forecasting models, so we get rid of the earliest period while the estimation is rolling over to the next period. Thus, the sample size is fixed. The derived benefit to do so is that we can examine the model's evolution (model stability) through time, insomuch as it is not quite appropriate to use the same model while repeatedly estimating the forecasting models for a growing economy. As a result, the explainable variables in each period might be different through time, and we can also obtain the time-varying coefficients for every input in the linear model and their relative importance through time. The linear model is a multivariate time series model. To ensure that the residuals from the linear models are white noise, we perform the Breusch-Godfrey Lagrange multiplier test, Ljung-Box Q-statistics and correlogram of squared residuals while building the linear model in each period. Selecting variables for the ANN models is based on the best linear model. The reason is that it is a very time-consuming and complex process to obtain all the optimal parameters (e.g., learning rate, number of neurons in each layer, number of training cycles, etc.) for so many candidate ANN models. Note that this is only a compromise between practicality and optimality. Nevertheless, our previous work does indicate great advantages when applying this method [7, 16].

Table 39.1 Forecast accuracy for inflation of sample countries

Model	Hong Kong		Japan		Taiwan		US	
	RMSE	MAE	RMSE	MAE	RMSE	MAE	RMSE	MAE
naïve	0.746	0.595	0.331	0.279	0.771	0.607	0.572	0.445
Linear-Thin	0.813	0.541	0.265	0.204	0.702	0.601	0.482	0.406
NN-Thin	0.730	0.620	0.360	0.284	0.745	0.570	0.464	0.397
NN-Thick-W_E	0.667	0.521	0.428	0.341	0.818	0.608	0.494	0.426
NN-Thick-W_o	0.655	0.516	0.390	0.327	0.791	0.592	0.486	0.421

An inverse relationship between unemployment rate and inflation (Phillips curve) exists in all cases, and the historical inflation significantly affects the current inflation as well in all cases. It is not surprising that the globalization factor is very significant in the linear model for each sample country, and it does generate the downward tendency in inflation through time with different levels. Table 39.1 presents all the summary forecast error statistics, root-mean-squared error (RMSE) and mean absolute error (MAE), for each sample country. As previously discussed, in the first few out-of-sample test periods, we might be able to decide which ANN performs best, but some ANNs still have similar performances after some test periods. If this situation happens, the thick models (ANN thick with equal weight and ANN thick with unequal weight models) can be accordingly applied. Except for the linear model, ANN thin and thick models, a simple naïve model that generates the benchmark forecast by assuming $E(\Delta \pi_{t+1}^4) = 0$ is also included. Basically, there are five different forecasting models, which are naïve, linear, ANN thin, ANN thick with equal weight (ANN-Thick-W_E), and ANN thick with unequal weight (ANN-Thick-W_O) models.

Some significant observations emerge from our empirical results in Table 39.1. First, there are two (Hong Kong and the US) out of four cases, in which ANN-based (i.e., thin or thick) models outperform the naïve and best linear models. This result is very encouraging because we do not make any claim on the optimality of the ANN models that are merely based upon these repeatedly-updated and well-specified linear models. As previously mentioned, although the ANNs have the same inputs as the best linear model, the variations on learning rates, number of neurons in each layer, and number of training cycles could generate a lot of different ANN models that might lead to different forecasting performances. Nevertheless, obtaining optimal values for so many ANN models is rather time-consuming and complex. In the case of Hong Kong (and all other cases indeed), we use the average of all forecasts from the different ANNs in the first ten out-of-sample test periods, because we could not determine which ANN is the best without any performance history. Afterward, we use the forecasting performances from the first ten test periods to obtain the weights in order to combine weighted forecasts for the next test period, and then we repeatedly update the weights for each of the remaining test periods. Actually, the weights vary just a little each time as the new information rolls in. This indicates that weights could become stable after some test periods in the case of Hong Kong if we have more data available. However, due to the constraint of data availability, we do not expect to get the optimal weights within some test periods, which could be possibly improved by using longer test periods with more experiences. Hong Kong is the only case in which the ANN thick with unequal weight (ANN-Thick-W_O) model outperforms all other models in this study with substantially lower RMSE (e.g., down 12% when compared with the naïve model) and MAE. In the case of the US, the ANN thin model outperforms all other models. Thirdly, in the cases of Taiwan, the ANN thin model performs as well as the best linear model, and it significantly outperforms the naïve model. Fourthly, in the case of Japan, the linear model has the best forecasting performance, and neither the ANN thin nor the ANN thick model can outperform the naïve model. Lastly, the ANN thick with unequal weight model in all cases do outperform the ANN thick with equal weight model despite the fact that they are not the best in some cases.

In short, our empirical results are not the most satisfactory but due to the application flexibility of ANNs, it is rather hard to obtain all the optimal parameters (e.g., learning rate, number of neurons in each layer, number of training cycles, etc.) for so many ANN candidate models because the whole process is very time-consuming and complex. The focus of this study was on a real time forecasting with many repeated estimations and sample countries, so selecting variables for the ANN models based upon the best linear model is only a compromise between practicality and optimality. However, our results are very encouraging.

39.4 Conclusions

In this paper, we study the globalization influences on forecasting inflation in an aggregate perspective using the Phillips curve for Hong Kong, Japan, Taiwan and the US by ANN-based thin and thick models. Our empirical results support the hypothesis that globalization influences do generate the downward tendency in inflation through time in all cases. Either the ANN thin or the ANN thick model that has been developed upon the best linear model for each country has shown significant superiority over the naïve model in most cases and over the best linear model in some cases in this study. With the virtue of application flexibility, finding optimal values for the large number of parameters involved in the ANN models is rather time-consuming and complex because of a large number of parameters considered. Thus we do not make any claim on the optimality of our ANN models. Although our empirical results are moderately satisfactory, building ANN models based upon the best linear model is a good compromise between practicality and optimality, which sheds strong light for future work such as cross-validations and sensitivity analysis for selecting variables to enhance the ANN's performances. In addition, whether or not obtaining the optimal parameter values (time-consuming and unpractical task) is essential to conspicuously enhance the ANN's performances to a very satisfactory level could be an issue to study as well.

References

1. International Monetary Fund (2006). World economic outlook, April 2006: *Globalization and Inflation*, Washington, DC.
2. Stock, J. H. and Watson, M. W. (1999). Forecasting inflation. *Journal of Monetary Economics*, 44(2):293–335.
3. Hinich, M. J. and Patterson, D. M. (1985). Evidence of nonlinearity in daily stock returns. *Journal of Business and Economic Statistics*, 3(1):69–77.
4. Sarantis, N. (2001). Nonlinearities, cyclical behaviour and predictability in stock markets: international evidence. *International Journal of Forecasting*, 17(3):459–482.
5. Ammermann, P. A. and Patterson, D. M. (2003). The cross-sectional and cross-temporal universality of nonlinear serial dependencies: evidence from world stock indices and the Taiwan stock exchange. *Pacific-Basin Finance Journal*, 11(2):175–195.
6. McNelis, P. and McAdam, P. (2004). *Forecasting Inflation with Thick Models and Neural Networks*. Working Paper Series 352. European Central Bank.
7. Hu, T., Su, H., and Gondra Luja, Iker (2007). Examining nonlinear interrelationships among foreign exchange markets in the pacific basin with artificial neural networks. In *PDPTA'07 – The 2007 International Conference on Parallel and Distributed Processing Techniques and Applications*, Las Vegas, USA, pp. 829–835.
8. Atkeson, A. and Ohanian, L. E. (2001). Are phillips curves useful for forecasting inflation? *Federal Reserve Bank of Minneapolis Quarterly Review*, 25(1):2–11.
9. Granger, C. W. J. and Jeon, Y. (2004). Thick modeling. *Economic Modeling*, 21(2):323–343.
10. Mitchell, T. (1997). *Machine Learning*. McGraw-Hill, New York.
11. Vapnik, V. N. and Chervonenkis, A. Ya. (1971). On the uniform convergence of relative frequencies of events to their probabilities. *Theory of Probability and its Applications*, 16(2):264–280.
12. Gershenfeld, N. A. and Weigend, A. S. (1993). *Time Series Prediction: Forecasting the Future and Understanding the Past*, pp. 1–70. Addison-Wesley, Reading, MA.
13. Rosenblatt, F. (1958). The perceptron: A probabilistic model for information storage and organization in the brain. *Psychological Review*, 65(6):386–408.
14. Bryson, A. E. and Ho, Y. C. (1969). *Applied Optimal Control*. Blaisdell, New York.
15. Rumelhart, D. E., Hinton, G. E., and Williams, R. J. (1986). Learning representations by back-propagating errors. *Nature*, 323:533–536.
16. Hu, T., Gondra Luja, I., and Su, H. (2007). Forecasting daily stock returns of east asian tiger countries with intermarket influences: A comparative study on artificial neural networks and conventional models. In *PDPTA'07 – The 2007 International Conference on Parallel and Distributed Processing Techniques and Applications*, pp. 836–842.

Chapter 40
Pan-Tilt Motion Estimation
Using Superposition-Type Spherical
Compound-Like Eye

Gwo-Long Lin and Chi-Cheng Cheng

Abstract The compound eyes of an insect can focus on prey accurately and quickly. From biological perspective, compound eyes are excellent at detecting motion. Based on the computer vision aspect, limited studies regarding this issue exist. Studies have verified that a trinocular visual system incorporating a third CCD camera into a conventional binocular is very helpful in resolving translational motion. Extended from this concept, this study presents a novel spherical compound-like eye of a superposition type for pan-tilt rotational motion. We conclude that as the number of ommatidium an insect has increased, capability for detecting prey increases, even for ambiguous patterns in each ommatidium. In this study, the compound eyes of insects are investigated using computer vision principles.

Keywords Biological imaging · Superposition-type spherical compound-like eye · Pan-tilt motion detection

40.1 Introduction

The configuration of insect compound eyes has always attracted the attention of researchers. Biology-based visual studies have recently flourished along with a boom in microlens technology. Development of image acquisition systems based on the compound eye framework has progressed more rapidly than ever before. The fabrication of micro-compound eyes has been described in literature with an orientation toward commercial applications. Well-known commercial applications include the Thin Observation Module by Bound Optics (TOMBO) compound eye developed by Tanita, [1] and the hand-held plenoptic camera developed by Ng [2]. The

G.-L. Lin and C.-C. Cheng
Department of Mechanical and Electro-Mechanical Engineering, National Sun Yat-Sen University
70, Lien Hai Rd., Kaohsiung City 80424, Taiwan, Republic of China
e-mail: gwolonglin@gmail.com; chengcc@mail.nsysu.edu.tw

S.-I. Ao et al. (eds.), *Advances in Computational Algorithms and Data Analysis,*
Lecture Notes in Electrical Engineering 14,
© Springer Science+Business Media B.V. 2009

TOMBO compound eye is a multiple-imaging system with a post-digital processing unit that has a compact hardware configuration with processing flexibility. The
hand-held plenoptic camera is similar to the plenoptic camera developed by Adelson
and Wang, [3] but with two fewer lenses, significantly shortening the optical path
and resulting in a portable camera. Furthermore, some related image-acquisition
systems contain one photoreceptor per view direction, [4, 5] a miniaturized imaging system, [6] an electronic compound eye, [7] curved gradient index lenses, [8]
an artificial ommatidia, [9] and a silicon-based digital retina [10]. All of these are
image-acquisition systems that have the framework of a compound eye.

The topic of how an insect compound eye accurately focuses on prey has not
been thoroughly investigated. Biologists believe that this ability is because of the
flicker effect; [11] that is, as an object moves across a visual field, ommatidia are
progressively turned on and off during which bees measure distance based on image
motion received by their eyes as they fly [11–16]. Because of the resulting "flicker
effect", insects respond far better to moving than stationary objects. Honeybees, for
instance, visit moving flowers more than they do still flowers. Thus, many studies
have exerted considerable effort in constructing images viewed from a compound
eye and reconstructing an environmental image from those image patterns. Most researches are limited to static images. However, this study focuses on the dynamic
vision of a compound eye. To achieve motion recovery for visual servo, ego-motion
estimation must be investigated first. Neumann [8] applied plenoptic video geometry to construct motion equations, and optimized an error function to acquire motion
parameters. Tisse [10] utilized off-the-shelf micro-optical elements to characterize
self-motion estimation. Nevertheless, neither study discussed this detection ability
for the dynamic vision of the compound eye, nor investigated the noise interference
problem. Lin and Cheng [17] recently presented a pan-tilt motion algorithm for
a single-row superposition-type spherical compound-like eye (SSCE) for recovery
of rotational motion utilizing pinhole perspective projection rather than a complex
mathematical interpretation. They indicated that the single-row SSCE generates image information, and markedly improves efficiency and accuracy when estimating
motion parameters. Based on this concept, recovery of rotational motion with an
SSCE will be examined to generate a complete SSCE framework, rather than limiting this investigation to a single-row SSCE.

40.2 The Compound-Like Eye in Computer Vision

According to mosaic theory, there are two compound eye types: apposition and superposition [18]. As the construction of these two eye types are clearly different,
the apposition eye acquires images from the ommatidium, and each ommatidium
is exploited to make up a complete ambiguous image. The superposition eye can
acquire a whole image by adjusting its ommatidia. Each ommatidium can itself capture an ambiguous image, and each image will differ based on different positions.
Strictly speaking, superposition here is neural superposition [19, 20]. These two eye

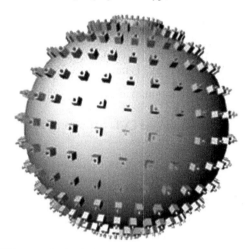

Fig. 40.1 Compound-like eye of spherical type

types are based on ecology, and can assist in determining how to produce very clear images from compound eyes. However, based on computer vision (CV), which configuration should be adopted? The compound-like eye in CV first proposed by Aloimonos [20] can be classified into two types: a planar compound-like eye; and, a spherical compound-like eye. Figure 40.1 depicts the spherical compound-like eye studied in this work.

The principal research task is to generate the configuration of spherical compound-like eye in CV. First, the configuration of a spherical compound-like eye must be defined. Suppose each ommatidium can look at an object. Based on this scenario, a number of CCD cameras treated as ommatidium are arranged on the sphere surface. The horizontal distance between adjacent ommatidia is fixed. To distinguish the apposition type compound-like eye, this specific arrangement has to be clearly defined as the superposition type. Each ommatidium will have a different and ambiguous view of an object in a noisy environment, not just a small view of an object. Besides, each ommatidium generates its image depending on its different location. The images generated by the SSCE will be vague and undistinguishable, and similar to blurred patterns viewed by insect ommatidia.

40.3 Pan-Tilt Ego-Rotational Motion for One CCD

For analytical purposes, a spherical compound-like eye can be modeled as a pan-tilt compound-like eye system. Assume total rotational angles for both pan and tilt are less than 180°. Based on this specific configuration, normal 3D rotational motion for one CCD can be transformed into 2D rotational motion. After acquiring the pan-tilt ego-rotational motion for one CCD camera, the superposition image of spherical compound-like eyes is discussed in the following section.

To simulate compound-like eyes for an ommatidium looking at an environment, assume an object is moving relative to the CCD platform, and the origin of the CCD platform is located at the optical center of the CCD camera and rotational center of the platform. When a single rigid object is moving, two features must be considered:

1. Using normal 3D rotational motion, pan-tilt rotational motion and image transformation can be established.
2. Based on the image observed by the CCD, the ego-rotational angle of the CCD camera can be resolved.

40.3.1 Pan-Tilt Rotation and Image Transformation

Given a world coordinate system, a rotation R applied to a 3D point $P = (X,Y,Z)^T$, which is accomplished through a displacement, $P \rightarrow P'$. A normal 3D rotation about an arbitrary axis through the origin of the coordinate system can be described by successive rotations $\psi, \theta,$ and ϕ about its Z, Y, and X axes, respectively. Then, transformation M for the arbitrary rigid motion in 3D space is given by M: $P' = RP = R(\psi)R(\theta)R(\phi)P$. Notably, the rotational operations do not commute.

For the composition of an SSCE, each CCD (ommatidium) will match with the cadence of pan-tilt motion and be placed on the sphere surface similar to Fig. 40.1. This formation approach can generate a compound-like eye pattern. Therefore, rotational motion of the SSCE is basically a simplification of a pure 3D rotational motion; thus, the rotation of the Z axis is set at 0, and only the motion behaviors in the X and Y axes are considered.

First, a 3D point P is moved to a new location by rotating the platform about the X axis by angle ϕ and about the Y axis by angle θ, given as $P \rightarrow P'$. Under perspective imaging, point P in 3D space is projected onto a location in the image plane $p = (x,y)^T$. Following the order of pan-tilt rotational motion $R(\theta)R(\phi)$, the rotational motion transformation of the 3D point can be viewed as moving an image point $p = (x,y)$ to a corresponding image point $p' = (x',y')$ based on a 2D image rotational mapping:

$$3D \text{ space point: } R(\theta)R(\phi): P(X,Y,Z) \rightarrow P'(X',Y',Z')$$
$$2D \text{ image point: } r(\theta)r(\phi): p = (x,y) \rightarrow p' = (x',y')$$

Therefore, the image point p of P moves to p', and is described by

$$\begin{bmatrix} x' \\ y' \end{bmatrix} = f \begin{bmatrix} x\cos\theta + y\sin\theta\sin\phi + f\sin\theta\cos\phi \\ y\cos\phi - f\sin\phi \end{bmatrix}$$
$$(-x\sin\theta + y\cos\theta\sin\phi + f\cos\theta\cos\phi)$$

where f is the CCD focal length. Notably, this image transformation does not require any information about the scene when the CCD rotates around its lens center.

40.3.2 Ego-Rotational Motion of Pan and Tilt

Two image locations, $p_0(x_0, y_0)$ and $p_1(x_1, y_1)$, are projections of a 3D point P at different times, t_0 and t_1. When the image point p_0 moves onto p_1 – assuming no translation occurs – the rotational angles ϕ and θ must be determined.

Image transformation is a forward procedure for image rotational mapping. Computing the amount of rotations from a pair of observations is obviously an inverse problem. To resolve this problem, an intermediate point $p_c(x_c, y_c)$ (Fig. 40.2) is assumed.

As the horizontal rotation and vertical rotation are applied to the CCD separately, Prazdny [21] proved that points in an image move along a hyperbolic path. Similar to the approach developed by Burger and Bhanu [22], the first step is $r(\phi)$, rotation around the X axis, moving the original point $p_0(x_0, y_0)$ to an intermediate point $p_c(x_c, y_c)$. The next step is $r(\theta)$, moving the intermediate point $p_c(x_c, y_c)$ to the final point $p(x_1, y_1)$ by camera rotation around the Y axis. Therefore, the angles of ego-rotational motion of pan and tilt can be obtained as

$$\phi = \tan^{-1}\frac{y_c}{f} - \tan^{-1}\frac{y_0}{f}, \theta = \tan^{-1}\frac{x_c}{f} - \tan^{-1}\frac{x_1}{f}$$

where the coordinates of the intersection point can derived as

$$x_c = fx_0 \left[\frac{\left(f^2 + x_1^2 + y_1^2\right)}{\left(f^2 + y_0^2\right)\left(f^2 + x_1^2\right) - (x_0 y_1)^2} \right]^{1/2}$$

$$y_c = fy_1 \left[\frac{\left(f^2 + x_0^2 + y_0^2\right)}{\left(f^2 + y_0^2\right)\left(f^2 + x_1^2\right) - (x_0 y_1)^2} \right]^{1/2}$$

Notably, this is an ego-rotational motion model of one CCD when two successive images at two different times are captured.

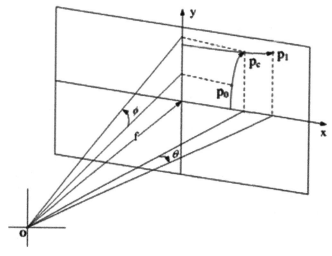

Fig. 40.2 The rotational path diagram of pan-tilt

40.4 Ego-Rotational Motion of Superposition Spherical Compound Like Eye

Based on images of an SSCE generated from CV, the rotational motion using an SSCE can be resolved using the following procedures:

1. Each camera in the SSCE whose image is ambiguous is the same as an ommatidium, and image fuzziness is independent. An ambiguous image can be achieved by adding random noises to an ideal image.
2. When the SSCE looks at an object, each ommatidium CCD perceives a different profile in the SSCE according to different locations of CCDs. In this manner, the compound-like eyes observe a entire ambiguous image, which is composed of many small, different, independent and ambiguous patterns.
3. When an object moves, the SSCE detects this rotation using two complete images, one before and one after the motion.
4. The process for generating the image for each CCD camera is the same as procedures 1–3.
5. Using those two vague images that include the rotational information, the corresponding intersection point for each camera can be estimated.
6. Any single CCD camera can generate a pair of pan and tilt angles with its intersection point. However, the number of cameras is positively correlated with the size of pan and tilt angles.
7. Taking the mean of all CCD camera ego-rotation angles, pan and tilt, unlike using standard least squares in ego-translation, [23] ego-motion angles of pan and tilt of the SSCE can be obtained easily.

In this manner, when the amount of ommatidium CCD cameras is increasing, the ego-motion angles of an SSCE will increase. That the SSCE can obtain accurate ego-rotation angles even in the case of large noise is explored.

40.5 Experimental Results

To verify the performance of noise immunity for the SSCE, a given synthesized cloud of 50 3D points (Fig. 40.3) is chosen as the test object. To simulate a realistic situation, noise is introduced into ideal data. Assume image components in the ideal motion field (x, y) are perturbed by additive zero-mean Gaussian noise. The two noise processes in the x and y image planes were independent, and each was spatially uncorrelated. Variances of noise in error analysis [24–26] were given proportional to magnitudes of velocity components. To reflect actual implementation in computing optical flows using image patterns at adjacent time instants, the noise on the positions of image pixels and their variances was assumed constant over the entire image plane. Therefore, the image points contaminated by noise before and after a movement are modeled as

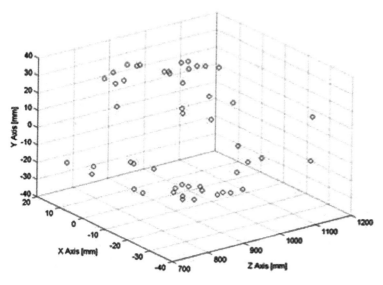

Fig. 40.3 A synthesized cloud of 50 3D points

$$\begin{bmatrix} x_1(i) + N_{x1}(i) \\ y_1(i) + N_{y1}(i) \end{bmatrix} \text{ and } \begin{bmatrix} x_2(i) + N_{x2}(i) \\ y_2(i) + N_{y2}(i) \end{bmatrix}$$

where i indexes the image point, $(x(i), y(i))$ locates the ideal image point for the i-th point, and $N(i)$ is a zero-mean Gaussian random noise at this position. The noise processes N_{x1}, N_{y1}, N_{x2} and N_{y2} are assumed to have the same statistical property, and are given by

$$E\{N_{x1}^2(i)\} = E\{N_{y1}^2(i)\} = E\{N_{x2}^2(i)\} = E\{N_{y2}^2(i)\} = \sigma^2$$

where σ is the standard deviation. In other words, all image points are contaminated by a random noise with the same variance.

For the sake of simplification for the subsequent validation and simulation, we define relative error in the pan-tilt rotational angles as

$$err = \frac{\sqrt{(\phi^* - \phi)^2 + (\theta^* - \theta)^2}}{\sqrt{\phi^2 + \theta^2}}$$

where (ϕ^*, θ^*) is the computed pan-tilt ego-rotational angle, and $(\phi, \theta) \neq 0$ is the actual pan-tilt rotational angle. To achieve statistically reasonable results, 300 trials were conducted by increasing the number of CCD cameras for each situation. For rotational motions, the relative errors of the SSCE are, ideally, equal to zero.

For the purpose of validation, the SSCE is divided into two types: a non-fixed total length and width SSCE; and, fixed total length and width SSCE. The former is rotated on the X and Y axes at the same time using a constant rotation angle interval, which extends to the left, right, upward, and downward with the same amount of CCD camera, whereas, the latter is rotated on the X and Y axes simultaneously using the varied rotation angle interval under the fixed total region of the SSCE with same number of CCD cameras.

40.5.1 Structure with Unconstrained Length and Width

The structural frame of the SSCE extends to four sides by a constant interval of $10°$ for each CCD camera. Based on this arrangement, the rotational model of the SSCE can be formed (Fig. 40.1). This outcome is the complete SSCE, which is different from the single-row SSCE. Basically, the complete SSCE is generated that can set any rotation angle and the number of CCD cameras; however, the total number of CCD cameras must be even. Consequently, the number of SSCEs becomes $(2n + 1) \times (2n + 1)$. However, for general applications, an odd number of CCD cameras can be utilized by shifting the center point of the SSCE.

This work selects only 1×1 to 9×9 SSCEs for the validation experiments and extend the variance of noise from 25–2,500. Under different numbers of CCD cameras in the complete SSCE, when noise variance changes from small to large, the performance of different CCD cameras in the complete SSCE using the proposed algorithm can be compared. The 9×9 SSCE under the rotational motion of pan $-9°$ and tilt $7°$ for the noise variance adding different noise levels at two time instants (red and black) (Fig. 40.4). The increasing noise variances make the image increasingly ambiguous, and the interference also increases. When noise variance is greater than 900, the noisy images (Figs. 40.4d–f) are all not easily distinguished from the original image. Even so, the proposed approach achieves very small relative errors for rotational motion (Table 40.1).

40.5.2 SSCE Structure with Constrained Length and Width

With fixed total length and width of the SSCE, this work validated the noise-resistance capability of a compound eye with an increased number of CCD cameras and a fixed interval between cameras; that is, the total length and width of the complete SSCE varies based on the number of CCD cameras. When the total length and width of the complete SSCE is fixed, then the interval between CCD cameras will be a variable and change based on the number of CCD cameras. When the number of CCD cameras increases in the SSCE sphere, angle interval decreases, meaning that the density of the SSCE increases simultaneously. Assume each extended angle is $80°$ in the vertical and horizontal directions and under the rotational motion of pan $7°$ and tilt $-5°$. Following the preceding manner of pan-tilt rotation, these different configurations are put into the extended angle area.

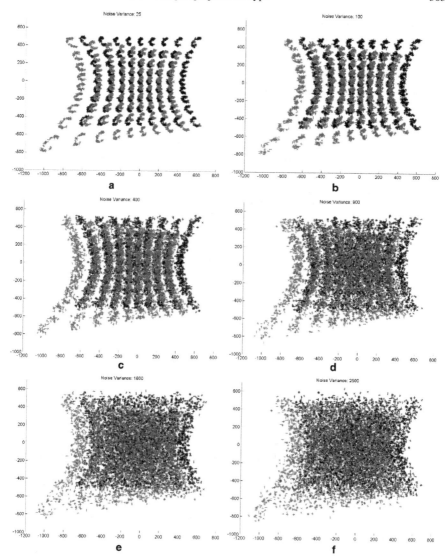

Fig. 40.4 The image points with different noise variances for the 9×9 complete SSCE at two time instants (red and black)

In this case, performance comparison was conducted by utilizing different numbers of cameras under different levels of additive noise, as described in Section 40.5.1. Table 40.2 presents the results of the complete SSCE with fixed total length and width for different cases, and lists the relative errors for the SSCE arrangements under the effect of different noise levels. The grace influence of immunity to noise was also demonstrated.

Table 40.1 Relative errors in percent of the ego-rotation for different SSCE configurations by non-fixed total length and width with various noise variances

Variances	1×1	2×2	3×3	4×4	5×5	6×6	7×7	8×8	9×9
25	1.19	0.58	0.38	0.28	0.23	0.18	0.15	0.13	0.11
100	2.39	1.13	0.77	0.57	0.45	0.36	0.31	0.25	0.22
400	4.68	2.38	1.52	1.08	0.91	0.72	0.63	0.50	0.45
900	7.14	3.58	2.24	1.67	1.35	1.08	0.92	0.82	0.69
1,600	9.45	4.56	3.09	2.47	1.89	1.43	1.23	1.02	0.88
2,500	11.84	5.93	3.86	2.83	2.26	1.88	1.55	1.32	1.11

Table 40.2 Relative errors in percent of the ego-rotation for different SSCE configurations by fixed total length and width with various noise variances

Variances	1×1	3	5×5	7×7	9×9	11×11	13×13	15×15
25	1.56	0.43	0.26	0.19	0.15	0.12	0.10	0.09
100	3.05	0.87	0.56	0.38	0.31	0.24	0.20	0.19
400	6.21	1.69	1.03	0.76	0.57	0.51	0.40	0.39
900	9.63	2.54	1.62	1.20	0.89	0.77	0.62	0.58
1,600	12.74	3.27	2.09	1.50	1.20	0.99	0.82	0.76
2,500	15.17	4.30	2.67	1.92	1.45	1.21	1.04	0.92

40.5.3 Discussion

Under these two different configurations—non-fixed and fixed total length and width of the complete SSCE – a number of conclusions are summarized

1. Regardless of whether total length and width of the SSCE are non-fixed or fixed, and rotation angles adopted, if the number of CCD cameras increases, noise resistance capability of the SSCE improves, even under the condition with high noise interference. For instance, with noise variance of 2,500 (Tables 40.1 and 40.2), relative error decreased from 11.84% to 1.11% and from 15.17% to 0.92%, at 9×9 in Table 40.1, 15×15 in Table 40.2, respectively depending on the arrangement of CCD cameras.
2. A dragonfly, which hunts during flight, has approximately 30,000 ommatidia in each eye, whereas butterflies and moths, which do not hunt during flight, have only 12,000–17,000 ommatidia [12]. The obvious difference between these insects is the amount of ommatidia. Based on experimental results (Table 40.1), when the number of ommatidium in an insect increases, response to moving objects improves. In other words, as number of ommatidium an insect have increased, its ability to detect prey increases.
3. Numerous insects have a forward- or upward-pointing regions of high acuity, related either to the capture of prey, or to the pursuit of females by flying males. Although both sexes have specialized predation behaviors, it is only the male that has the acute zone indicative of its role in sexual pursuit. Acute zones vary considerably. In describing acute zones it is useful to indicate how densely an

ommatidial array samples different regions in the surrounding environment [19]. Our experimental results exactly respond to this situation. When CCD camera density increases, the detection accuracy of compound-like eye improves (Table 40.2). Consequently, the compound-like eye is able to provide a very accurate detection capability in 3D ego-motion.

40.6 Conclusion

The compound eyes of flying insects are highly evolved organs. Although images received by their eyes are ambiguous, these insects are still capable of capturing prey so accurately and quickly. Inspired by these insects, pinhole image formation geometry has been applied to investigate the behavior of SSCEs when capturing moving objects. The concept underlying SSCE configuration and the ego-rotation model of SSCEs for pan-tilt motion are proposed.

Based on the number of ommatidia and acute zones in a compound eye, and through experiments in non-fixed and fixed total length and width SSCEs, this study determined that the total number and density of ommatidia are crucial factors to insect compound eyes. Those two influential factors very clearly correspond to experimental results obtained in this study. Notably, this work did not use any filters or optimization algorithm. Through these experiments, and based on the validation of the proposed algorithm, this work verified that insect compound eyes are powerful and excellent devices for detecting motion.

References

1. J. Tanida, T. Kumagai, K. Yamada, S. Miyatake, K. Ishida, T. Morimoto, N. Kondou, D. Miyazaki, and Y. Ichioka, "Thin observation module by bound optics (TOMBO): concept and experimental verification," *Appl. Optics*, vol. 40, no. 11, pp. 1806–1813, 2001.
2. R. Ng, Marc Levoy, M. Bredif, G. Duval, M. Horowitz, and P. Hanrahan, "Light field photography with a hand-held plenoptic camera," Stanford University Computer Science, Tech Report CSTR 2005–02, 2005.
3. E. H. Adelson and J. Y. A. Wang, "Single lens stereo with a plenoptic camera," *IEEE Trans. PAMI*, vol. 14, pp. 99–106, 1992.
4. T. Netter and N. Franceschini, "A robotic aircraft that follows terrain using a neuro-morphic eye," in *IEEE Proceedings of Conference on Intelligent Robots and Systems*, pp. 129–134, 2002.
5. K. Hoshino, F. Mura, H. Morii, K. Suematsu, and I. Shimoyama, "A small-sized panoramic scanning visual sensor inspired by the fly's compound eye," in *IEEE Proceedings of Conference on Robotics and Automation*, pp. 1641–1646, 1998.
6. R. Volkel, M. Eisner, and K. J. Weible, "Miniaturized imaging system," *J. Microelectronic Engineering*, Elsevier Science, Amsterdam, Netherlands, vol. 67–68, pp. 461–472, 2003.
7. R. Hornsey, P. Thomas, W. Wong, S. Pepic, K. Yip, and R. Krishnasamy, "Electronic compound eye image sensor: construction and calibration," in *Sensors and Camera Systems for Scientific, Industrial, and Digital Photography Applications V*, M. M Blouke, N. Sampat, R. Motta, eds., Proceedings of SPIE 5301, pp. 13–24, 2004.

8. J. Neumann, C. Fermuller, Y. Aloimonos, and V. Brajovic, "Compound eye sensor for 3D ego motion estimation," in *IEEE Proceedings of Conference on Intelligent Robots and Systems*, vol. 4, pp. 3712–3717, 2004.
9. J. Kim, K. H. Jeong, and L. P. Lee, "Artificial ommatidia by self-aligned microlenses and waveguides," *Opt. Express*, 30, pp. 5–7, 2005.
10. C. L. Tisse, "Low-cost miniature wide-angle imaging for self-motion estimation," *Opt. Express*, vol. 13, no. 16, pp. 6061–6072, 2005.
11. J. W. Kimball, "The compound eye," Kimball's Biology Pages, http://users.rcn.com/jkimball. ma.ultranet/BiologyPages/C/CompoundEye.html
12. M. Elwell and L. Wen, "The power of compound eyes," *Opt. & Photonics News*, pp. 58–59, 1991.
13. G. A. Horridge, "A theory of insect vision: velocity parallax," *Proceedings of the Royal Society of London B*, vol. 229, pp.13–27, 1986.
14. E. C. Sobel, "The locust's use of motion parallax to estimate distance," *J. Comp. Physiol. A*, vol. 167, pp. 579–588, 1990.
15. M. V. Srinivasan, S. W. Zhang, M. Lehrer, and T. S. Collett, "Honeybee navigation EN ROUTE to the goal: visual flight control and odometry," *J. Exp. Biol.*, vol. 199, pp. 237–244, 1996.
16. T. Collett, "Animal behaviour: Survey flights in honeybees," *Nature*, vol. 403, pp. 488–489, February 2000.
17. G. L. Lin and C. C. Cheng, "Single-row superposition-type spherical compound-like eye for pan-tilt motion recovery," in *2007 IEEE SSCI: 2007 IEEE Symposium on Computational Intelligence in Image and Signal Processing*, pp. 24–29, 2007.
18. W. S. Romoser, *The Science of Entomology*, Macmillan, New York, 1973.
19. M. F. Land and D.- E. Nilsson, *Animal Eyes*, Oxford University Press, New York, 2002.
20. Y. Aloimonos, New Camera Technology: Eyes from Eyes. Available: http://www.cfar.umd. edu/~larson/dialogue/newCameraTech.html
21. K. Prazdny, "Determining the instantaneous direction of motion from optical flow generated by a curvilinear moving observer," *Computer Graphics Image Processing*, vol. 17, pp. 238–248, 1981.
22. W. Burger and B. Bhanu, "Estimating 3-D egomotion from perspective image sequences," *IEEE Trans. PAMI*, vol. 12, pp. 1040–1058, 1990.
23. G. L. Lin and C. C. Cheng, "Single-Row superposition-type compound-like eye for motion recovery," in *IEEE International Conference on Systems, Man and Cybernetics*, pp. 1986–1991, 2006.
24. E. Simoncelli, E. Adelson, and D. Heeger, "Probability distributions of optical flow," in *Proceedings of the IEEE International Conference on Computer Vision and Pattern Recognition*, Mauii, Hawaii, 1991, pp. 310–315.
25. J. L. Barron, D. J. Fleet, and S. S. Beauchemin, "Performance of optical flow techniques," *International Journal of Computer Vision*, vol. 13, no. 1, pp. 43–77, September 1994.
26. N. Gupta and L. Kanal, "3-D motion estimation from motion field," *Artif. Intel.*, vol. 78, pp. 45–86, November 1995.